AIChE Symposium Series No. 325
Volume 97, 2001

FOUNDATIONS of MOLECULAR MODELING and SIMULATION

Proceedings of the First International Conference on
Molecular Modeling and Simulation
Keystone, Colorado, July 23-28, 2000

Editors

Peter T. Cummings
University of Tennessee

Phillip R. Westmoreland
University of Massachusetts

Production Editor, CACHE Publications
Brice Carnahan
University of Michigan

CACHE
American Institute of Chemical Engineers

2001

Library of Congress Cataloging-in-Publication Data

International Conference on Foundations of Molecular Modeling and Simulation (1st : 2000 : Keystone, Colo.)
 First International Conference on Foundations of Molecular Modeling and Simulation / Peter T. Cummings, Phillip R. Westmoreland, Brice Carnahan, volume editors.
  p. cm. -- (AIChE symposium series ; 325 = v. 97)
Includes bibliographical references and index.
 ISBN 0-8169-0839-7
 1. Molecular structure--Computer simulation--Congresses. 2. Molecules--Models--Computer simulation--Congresses. I. Title: Foundations of molecular modeling and simulation. II. Cummings, Peter T. III. Westmorland, Phillip R., 1951- IV. Carnahan, Brice. V. Title. VI. AIChE symposium series ; no. 325.
 QD480 .I55 2000
 541.2'2'0113—dc21

                                    2001022793

PREFACE

FOMMS 2000

Computational quantum chemistry and molecular simulation methods have now become key tools for meeting the demands and goals of the chemicals, fuels, biologics, and materials industries. Demands include increasing environmental regulation, the growing emphasis on specialty chemicals, the expanding diversity in customers. Goals include shortened times to market, reducing cost of development, increasingly precise control of final product properties, and the constant creation and design of new products to meet specific customer requirements. In the coming decades, we even can look forward toward the possibility of molecular design of products and materials, seamlessly integrated with the design of the processes to manufacture them. These developments have come about because of fundamental advances in theory, methods, and computing technology.

In recognition of these extraordinary developments and opportunities, a new conference series entitled "Foundations of Molecular Modeling and Simulation" (FOMMS) has been established by the non-profit educational foundation, CACHE Corporation, in collaboration with two programming groups – the Thermodynamics and Transport Properties group and the Computational Molecular Science and Engineering Forum – of the American Institute of Chemical Engineers. The first FOMMS conference was held at Keystone Resort, CO, on July 23-28, 2000, with the theme "Applications for Industry."

The target audience for the conference included practitioners (both from industry and academia), those interested in becoming practitioners (both in industry and academia), and industrial managers who wish to learn about the technology. One hundred and thirty five attendees participated in the conference - 29% from industry, 62% from academia and 9% from government laboratories. The majority of the attendees (64%) were from the U.S.; the countries with the next most participants were Japan (10%) and the United Kingdom (9%).

Content was balanced between molecular simulation and computational chemistry, along with a look to the future of chemical product design and the supporting computational environments. Twenty-two invited talks were presented. Two poster sessions, containing a total of 65 posters, provided opportunities for attendees to present their work. Large blocks of time, including receptions, were available for informal discussions, aiding interaction between conference participants. Surveys of and informal discussions with the participants indicated that the conference was a great success, both from technical and social viewpoints.

This proceedings volume contains manuscripts submitted by invited speakers and poster presenters and were reviewed by members of the organizing committee. The manuscripts were then copy-edited by Brice Carnahan at the University of Michigan.

The success of a conference such as FOMMS 2000 is the result of the work of many people, and we wish to record here our thanks to all involved. First and foremost, we thank our Co-Chair

Uli Suter   ETH-Zurich

and other members of the Organizing Committee:

Hank Cochran	Oak Ridge National Laboratory
Dave Dixon	Pacific Northwest National Laboratory
Jim Ely	Colorado School of Mines
Carol Hall	North Carolina State University
George Jackson	Imperial College, London
Sangtae Kim	Warner-Lambert
Tomoshige Nitta	Osaka University
John O'Connell	University of Virginia
Nigel Seaton	University of Edinburgh
Warren Seider	University of Pennsylvania

All of the members of the organizing committee reviewed manuscripts, participated in planning, and promoted the conference to the colleagues. Several members of the organizing committee deserve special mention: Dave Dixon and John O'Connell for their help in developing the program of invited speakers, Tomo Nitta for his enthusiasm in promoting the meeting in Japan (resulting in a very high number of Japanese attendees), and Jim Ely for local arrangements support.

Robin Craven did an excellent job of conference facilitation and we are indebted to her. The conference went smoothly and exceeded everyone's expectations in large part because of her efforts and organizational skill.

Peter Cummings wishes to express his thanks to his secretary, Teresa Finchum, and Clare McCabe, a research faculty member at the University of Tennessee. Teresa handled all of the early registration and responded to the questions of prospective attendees. Clare provided invaluable help in the preparation of the preprint volume distributed at the conference.

Next, we wish to express our thanks to our colleagues at CACHE Corporation for their support of molecular modeling within chemical engineering. In particular, Warren Seider has been an ardent supporter of our activities within CACHE, and the success of FOMMS 2000 is directly attributable to his foresight in inviting us to form a molecular modeling effort within CACHE in 1996. The quality of this publication is direct result of Brice Carnahan's dedication and expertise in editing, formatting, and assembly of the manuscripts. He was ably assisted by Decker Ringo, a Junior chemical engineering student at the University of Michigan.

Finally, we wish to thank all the participants in FOMMS 2000. Their enthusiasm and spirited participation in every aspect of FOMMS 2000 were the primary cause for the conference's success. As we go to press, preliminary plans are taking shape for FOMMS 2003. We expect an even more successful conference in 2003, as the word spreads about FOMMS 2000. We hope to see you all again in 2003!

Peter T. Cummings, Chair, FOMMS 2000
Phillip R. Westmoreland, Co-Chair, FOMMS 2000

TABLE OF CONTENTS

INVITED PAPERS

APPLICATIONS IN INDUSTRY

APPLYING THEORY

MODELING OF POLYMERS

BIOLOGICAL APPLICATIONS

FORCE FIELDS AND MOLECULAR SIMULATIONS

CONTRIBUTED PAPERS

MOLECULAR MODELING METHODS

PHASE EQUILIBRIA

TABLE OF CONTENTS

TRANSPORT PROCESSES

POLYMERS AND COMPLEX FLUIDS

AQUEOUS SYSTEMS

BIOLOGICAL SYSTEMS

MICROPOROUS MATERIALS

SURFACE STRUCTURE AND REACTIVITY IN A COMPLEX ENVIRONMENT

Anne M. Chaka
The Lubrizol Corporation
Wickliffe, OH 44092-2298

Xiao-Gang Wang
Delphi Corporation
Detroit, MI

Matthias Scheffler
Fritz-Haber-Institut der Max-Planck-Gesellschaft
D-14 195 Berlin, Germany

Abstract

Surfaces of industrial materials in a realistic environment can be exceedingly complex and therefore not easily analyzed using current experimental methods. Yet understanding these surfaces is crucial for predicting the performance of materials such as catalysts and thin films. Hence there is a need for theoretical models to enable understanding and interpretation of experimental data, and to determine which of many possibilities are the most probable surface structures encountered in a system in local thermal equilibrium with a given set of experimental conditions. In this paper we highlight work in two areas which illustrates how models based on density-functional theory calculations can be used to provide a framework of understanding to enable predictions regarding properties and structure of complex materials in a realistic environment. Part I reports the results of a systematic investigation of how the structure and bonding of iron mono- and disulfides in thin boundary layer films relate to the surface energy and tribological performance in gear applications. In Part II α-Al_2O_3 (0001) surfaces are allowed to exchange atoms with an atmosphere containing O_2, H_2, and H_2O over the range of physically realistic temperatures, pressures, and concentrations. Being in equilibrium with a realistic environment has a dramatic effect on the relative Gibbs free energies of the possible (1 x 1) surfaces, and is necessary to obtain theoretical predictions in agreement with experimental results.

Keywords

Iron sulfides, Antiwear, Tribology, Boundary layer, Density functional theory, *ab initio*, Corundum, α-Al_2O_3 (0001) surface, Gibbs free surface energy.

PART I. STRUCTURE AND PERFORMANCE OF THE IRON SULFIDE BOUNDARY LAYER

Introduction

For over 50 years, organosulfur compounds have been added to lubricating oils to protect ferrous metal surfaces via the formation of sacrificial layers of iron sulfide. This film formation is of critical importance in mechanical systems in which contact areas are too small and contact stresses too great for hydrodynamic lubrication to be effective. At the high pressures under these conditions the liquid film thickness becomes less than the asperity height, and the stress becomes increasingly borne by the asperities. Adding sulfurized olefins to lubricating oil results in the formation of an iron sulfide boundary layer with a shear strength lower than that of the metal surface (Prutton, 1946), and can thus prevent welding and metal-metal adhesion, and allow wear to proceed at a slow, controlled rate. Environmental and economic concerns have created a demand for new types of lubricant additives that are capable of re-engineering a metal surface *in situ* and reducing friction under a wide range of temperatures, pressures, speeds, and metallurgies. Achieving intelligent design of surface-active chemistry, however, requires an understanding of surface structure and tribological performance of the boundary layer films at the atomistic level. We need to understand how the current generation of additive chemistry performs in order to develop alternative chemistry that meets the environmental and economic demands of the future.

The atomic structure of the iron sulfide layer is extremely complicated, however, and it is not known how this structure relates to tribological performance. The iron sulfide boundary layer is approximately 50 Å thick, and consists of stoichiometric and nonstoichiometric regions of iron mono- and disulfides such as pyrite, troilite, pyrrhotite, marcasite, and FeS in a NiAs structure (Lubrizol, 1990). Nonstoichiometry can be due either to sulfur or iron vacancies. In addition, large amorphous regions are present as well. Under boundary layer lubrication conditions, oxide layers initially present on the metal surface are quickly worn away, and negligible oxygen is detected thereafter within the surface film.

Given the complexity of the iron sulfide boundary layer, what can we hope to learn from our calculations? The ultimate goal is to understand the relationship between surface structure and tribological performance. The macroscopic structure of the mechanical parts cannot be modeled atomistically, but the small microstructures which constitute the larger system can be isolated and the corresponding properties estimated to determine relative trends and points of weakness which contribute to performance of the overall system. What general features are common across the complex variety of crystalline and amorphous phases of iron sulfide? What are key differences? How do they relate to performance?

Our initial focus is on the crystalline regions, and we have chosen FeS in the NiAs structure and FeS_2 in the pyrite structure as representative of each of the two basic types of iron sulfide structures (mono- and di-) observed in the boundary layer film. Calculating the energy required to cleave the crystalline regions can provide insight as to whether the large crystalline domains cause problems in the boundary layer. Are the FeS_x crystalline regions 'harder' than bulk iron? Or can they cleave or shear easily and mitigate metal-metal adhesion?

Methodology

We have performed the first full potential geometry optimizations on these systems using the Full Potential-Linear Augmented Plane Wave (FP-LAPW) methodology

implemented in the Wien97 program (Blaha, 1999). The GGA functional proposed by Perdew, Burke, and Ernzerhof (PBE) (Perdew, 1996) was used throughout, as was a high planewave cutoff of 18 Rydbergs. All geometry optimizations and total energies reported were calculated using sufficient **k**-points to achieve energy convergence, corresponding to an 8x8x8 grid in the Brillouin zone. Core electrons were treated fully relativistic, and valence electrons were treated with scalar relativistic effects.

Mulliken populations (Mulliken, 1955) on the iron sulfides were obtained using density-functional theory (DFT) (Kohn, 1965) as implemented within the DMol 2.3.5 program (Delley, 1993). All population analyses were performed using Becke's (1988) gradient correction to the exchange and that of Lee, Yang, and Parr (1988) to the correlation (BLYP). All basis sets used are of double numeric (zeta) plus polarization (DNP) quality (Delley, 1993). Calculations on open-shelled systems were performed spin polarized

Structure and Bonding in Iron Sulfides

Although some interest in iron pyrite as a nontoxic semiconductor has resulted in several band structure calculations performed on the experimentally observed geometry (Birkholz, 1991; Zhao, 1993), there has been only one general study to date in which the structural parameters of transition metal sulfides have been investigated using density functional theory (Raybaud, 1997). The optimized geometries and lattice constants obtained by Raybaud, *et al*, using a planewave expansion for the basis set and ultrasoft Vanderbilt pseudopotentials (PW-PP) showed, in general, good agreement with experiment. The results for the iron sulfides, however, did not agree well with experiment. The errors in atomic volumes obtained for FeS in the NiAs structure and for iron pyrite were 17% and 7%, respectively. Some of the discrepancy for the antiferromagnetic NiAs structure may be due to magnetic effects, as only nonmagnetic calculations had been performed, but cannot explain the error for the nonmagnetic iron pyrite structure. Zeng and Holzwarth (1994) obtained agreement to within 1% of the lattice constant for the pyrite structure using GGA form of the exchange-correlation functional and norm-conserving Troullier and Martins pseudopotentials.

Our results are presented in Table 1. The monosulfide crystallizes in space group $P6_3/mmc$ in the hexagonal NiAs structure which consists of alternating layers of iron and sulfur atoms (Fig. 1). Whereas FeS in the NiAs structure is antiferromagnetic, we investigate the nonmagnetic, ferromagnetic, and antiferromagnetic configurations to determine how effective the FP-LAPW methodology is in predicting the correct electronic state. The antiferromagnetic configuration was indeed found to be the lowest in energy by 3.1 kcal/mol per FeS formula

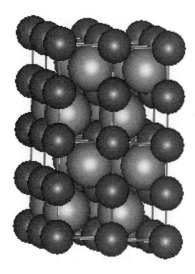

Figure 1. FeS in the NiAs structure showing the alternating layers of iron (dark grey) and sulfur (light grey) atoms.

Table 1. Lattice Parameters of FeS (NiAs Structure) and FeS₂ (Pyrite) with Antiferromagnetic (AFM), Ferromagnetic (FM), and Nonmagnetic (NM) electronic structures.

	a (Å)	c (Å)	c/a	V/atom (Å³)	V^{calc}/V^{expt}
FeS					
AFM					
Expt.[a]	3.445	5.763	1.67	14.81	1.0
LAPW[b]	3.445	5.715	1.66	14.70	0.99
FM					
LAPW[b]	3.444	5.715	1.66	14.70	0.99
NM					
LAPW[b]	3.255	5.239	1.61	12.02	0.81
PWPP[b]	3.328	5.124	1.54	12.29	0.83
FeS₂					
NM					
Expt.[d]	5.409			13.33	1.0
LAPW[b]	5.419			13.40	1.006
PWPP[b]	5.290			12.40	0.93

(a). Pinklea (1976).
(b). This work.
(c). Raybaud (1997).
(d). Coey (1979).

unit over the nonmagnetic structure, and 5.9 kcal/mol over the ferromagnetic.

The magnetic state also exhibits a dramatic effect on the lattice optimizations as well as the energetics, as one might expect. For the nonmagnetic state of FeS in the NiAs structure, we find an atomic volume that is 19% smaller than that observed experimentally, which is very similar to the 17% discrepancy obtained using the PW-PP methodology. The *c/a* ratio of 1.61 calculated by the FP-LAPW method was somewhat closer to the experimental ratio of 1.67, compared to 1.54 for the results of Raybaud, *et al.* (1997). The geometries for the ferromagnetic and antiferromagnetic structures both agree very closely with the experimental structure, differing by less than 1% from the atomic volume and *c/a* ratio. Hence, most of the error in the PW-PP calculations can be corrected by the inclusion of magnetic effects.

Iron pyrite is a nonmagnetic material, and hence different magnetic configurations were not investigated. The pyrite structure, which crystallizes in the cubic Pa3̄ space group shown in Fig. 2, consists of octahedrally coordinated iron atoms interconnected by bridges of sulfur dimers. The structural results obtained with the FP-LAPW methodology agree very well with experiment, as the optimized lattice constant is only 0.01 Angstroms larger than the observed structure. This error of less than 1% in the lattice constant and atomic volume obtained by the full-potential methodology represents a considerable improvement in accuracy over the ultrasoft pseudopotential calculations, and accuracy comparable to the Trouillier-Martins pseudopotentials used for these systems.

To obtain an understanding of the bonding in the iron mono and disulfides, the structural and electronic environments of both sulfur and iron are examined in

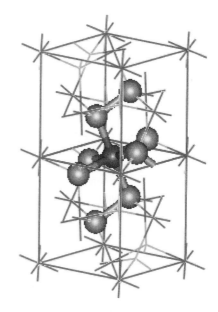

Figure 2. FeS₂ in the iron pyrite structure showing the iron octahedron bridged by sulfur dimers.

detail. The environment of sulfur in iron pyrite is consistent with a more covalent bonding environment than the monosulfide in the NiAs structure. The S-S distance of 2.141 Å in pyrite is essentially identical to the S-S single bond distance of 2.14 Å in H_2S_2 (Wimmewisser, 1968). Strong evidence for covalent bonding is provided by the σ-σ^* splitting in the sulfur 3s bonds in the band structure shown in Fig. 3. This is in stark contrast to the NiAs structure, in which there is negligible sulfur-sulfur interaction. The sulfur atoms are 3.445 Å apart, and the band structure indicates there is no splitting of the sulfur 3s bands due to interactions with neighboring sulfur atoms. Sulfur in iron pyrite has a tetrahedral structure, which is more consistent with covalent bonding than the less directional trigonal prism arrangement of six iron atoms around the sulfur in the NiAs structure.

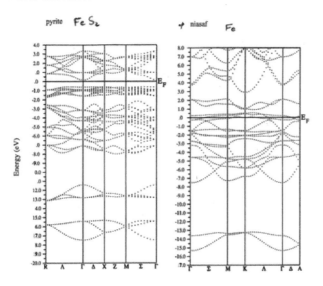

Figure 3.   Band structure of (a) FeS₂ Iron Pyrite, and (b) FeS in the NiAs structure.

Iron atoms in both the pyrite and the NiAs structure exists in an octahedral environment surrounded by six sulfur nearest neighbors. In the pyrite structure, the iron atoms are 3.825 Å apart and any interaction between them is negligible. The density of states (DOS) in Fig. 4 displays the classic t_g-e_g splitting which is characteristic of an isolated octahedral iron complex in solution. The structure is nonmagnetic and has a 1.0 eV bandgap, which is in relatively good agreement with the experimental result of 0.84 eV (Kou, 1978) and the calculated value of 0.9 by Folkerts *et al* (1987) using the augmented spherical wave methodology developed by Williams *et al.* (1979). There is evidence of strong metallic interactions in the monosulfide NiAs structure, however, as indicated by the antiferromagnetic coupling (μ = +/- 2.73) and absence of a bandgap. The high density of states at the Fermi level is shown in Fig. 4. The iron atoms within a layer are separated by 3.445 Å, but the interlayer spacing parallel to the *c*-axis is consistent with metallic bonding at 2.881 Å. Hence FeS in the NiAs structure behaves essentially like a one-dimensional metal.

The degree of change transfer in the iron sulfides can be estimated by Mulliken population analysis and used to gauge the relative ionic contribution to the cohesive energy. The charge transfer from the iron to the sulfur in FeS (0.771) is more than double that found in the iron pyrite structure. In iron pyrite, the charges on the iron and sulfur atoms are 0.302 and –0.151, respectively. Hence the monosulfide has much more ionic character than the disulfide, which is reflected in the less-directional trigonal prism arrangement of six iron atoms around the sulfur in the NiAs structure in contrast to the more-directional tetrahedral sulfur environment in iron pyrite.

Results

The results of the surface energy calculations are shown in Fig. 5. Energies for two iron cleavage planes are presented, the (100) and the lowest energy (110) surface. Because the iron pyrite has more covalent character than the NiAs structure, one would expect a higher cleavage energy. This is indeed the case for the pyrite (100) surface in which the S-S dimers are preserved.

The surface energies for the NiAs structure are an order of magnitude lower than for the pyrite. Cleaving along the (0001) plane perpendicular to the chain of iron atoms, and between the layers of iron and sulfur atoms, requires more energy than cleavage parallel to the Fe atom chains, reflecting the strength of the Fe-Fe metallic interactions.

Conclusion

Both the mono- and disulfide structures have a lower surface energy than iron, and hence will cleave much more readily, thus easily preventing metal-metal adhesion. In addition, the lower surface energy indicates

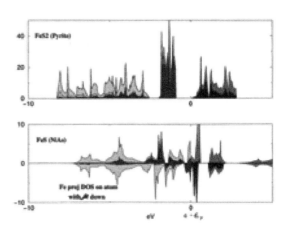

Figure 4.   Density of states for iron pyrite (top) and FeS in the NiAs structure. Total density of states are indicated in light grey, and the projected d-states are in black.

that the iron sulfides are not as hard as iron metal and the crystalline regions will not lead to abrasive wear. Monosulfides, with their combination of metallic and ionic bonding in a layered structure, cleave an order of magnitude more easily than the more covalent disulfides. FeS (NiAS) structure will preferentially cleave parallel to the short Fe-Fe metallic bonds

How well does this understanding for pyrite and the NiAs structure transfer to other iron sulfide structures? Marcasite also has the FeS₂ stoichiometry with S-S dimers bridging iron octahedra, but the cubic pyrite structure is tetragonally distorted as shown in Fig. 6

(Buerger, 1931). Because marcasite's key bonding characteristics are very similar to the pyrite structure, we would predict that the surface energy, and hence performance characteristics, to be very similar. The surface energy of 36.3 meV/Å^2 for the marcasite (100) plane, is comparable to the value of 25.9 meV/Å^2 for pyrite (100) is consistent with this prediction when the entire range of cleavage energies is considered.

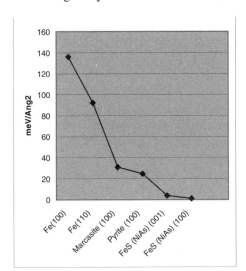

Figure 5.   Surface energies for iron and iron sulfides.

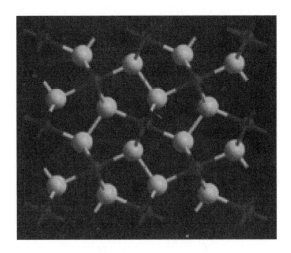

Figure 6.   Structure of FeS₂ in marcasite, which is a the tetragonally distorted pyrite structure.

PART II. α-AL₂O₃ (0001) SURFACE STRUCTURES AND STOICHIOMETRY IN EQUILIBRIUM WITH A REALISTIC ENVIRONMENT

Introduction

The nature of the corundum (α-Al₂O₃) surface has long been of considerable importance in a wide variety of technological applications. Yet, despite considerable experimental and theoretical efforts, the surface structure, and even the surface stoichiometry, are a matter of strong controversy. Recent experimental results (Renaud, 1998; Toofan, 1998; Ahn, 1997) indicate that there are several different surface terminations and structures possible, depending upon the thermal history and oxygen partial pressure employed. Experiments from various groups also report surface relaxations which are not consistent with each other in either magnitude or direction. Yet theoretical methods ranging from the empirical to the *ab initio* have for many years identified only one stable termination stoichiometry for the α-Al₂O₃ (0001) surface, a (1 x 1) aluminum-terminated structure (Mackrodt, 1987; Causa, 1989; Guo, 1992; Manassidis, 1993; Godin, 1994; Verdozzi, 1999). These discrepancies among experiments, and between theory and experiment are significant and need to be resolved.

What had not been previously considered in the theoretical treatment of the α-Al₂O₃ (0001) surface is the effect of the environment on the surface structure and stoichiometry. Under realistic conditions, a surface will exchange atoms with its surroundings. Hence the Gibbs free energy of the system must be determined with respect to its dependence on the chemical potentials of the components present in the material and the environment. For a metal oxide such as α-Al₂O₃ the O₂ partial pressure is obviously the most important factor in the analysis. In addition, the presence of stable hydrogen on the surface also needs to be addressed, as it can be incorporated into the bulk structure during growth, or may result from exposure of the surface to water vapor prior to placement in the UHV chamber. Recent dynamics calculations by Hass and coworkers (1998) have shown that if water is present on the corundum surface, it will dissociate and result in the formation of hydroxyl groups.

Corundum Structure

The corundum structure consists of five Al₂O₃ formula units arranged in a hexagonal unit cell, shown in

Fig. 7. Oxygen atoms are arranged in hexagonally closest-packed layers, with aluminum atoms occupying two-thirds of the octahedral holes (Wyckoff, 1982). The aluminum atoms between two layers of oxygen atoms can be thought of as one buckled aluminum layer or, for convenience in designating surface terminations here, as two aluminum layers. Three cleavage planes are possible along the (0001) direction, and are indicated by the arrows in Fig. 7. These surfaces are designated Al-O_3-Al-R for the aluminum-monolayer terminated surface, Al-Al-O_3-R for the aluminum double layer termination, and O_3-Al-Al-R for the oxygen termination, where R represents the appropriate corresponding bulk sequence of $-[O_3$-Al-Al$]_n$-. The α-Al_2O_3 (0001) surface has been found to maintain a (1 x 1) structure up to 1250K in vacuum (French, 1970). In addition to the three stoichiometric surface planes, we examine five additional surfaces that maintain the (1 x 1) periodicity. Three of these surfaces are nonstoichiometric oxygen deficient structures, which may be formed at higher temperatures as oxygen is evaporated from the surface. These are designated O_2-Al-Al-R and O_1-Al-Al-R with one and two oxygen vacancies, respectively, and one aluminum double layer with an additional aluminum atom, Al_3-O_3-Al-R. In addition we include the O_3-Al-Al-R with one (HO_3-Al-Al-R) and with three (H$_3O_3$-Al-Al-R) additional hydrogen atoms to evaluate the effects of possible contamination of the surface.

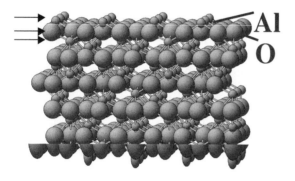

Figure 7. The corundum (α-Al_2O_3) structure, shown with aluminum atoms as small circles, and oxygen as large. The number of layers shown in the vertical direction corresponds to both the size of the bulk unit cell, and the aluminum monolayer terminated slab used in the calculation. Arrows indicate the stoichiometric slice places.

Methodology

To examine a system with many phases in equilibrium capable of exchanging particles, i.e. the bulk, surface, and gas phases, the free energy of species in all phases must be calculated. Details of the methodology for

calculating the Gibbs free energy of a surface can be found in (Wang, 2000). Key points are presented here.

The Gibbs free energy ζ of a system in equilibrium with the environment at temperature T and partial pressure p is:

$$\zeta = E^{\text{Total}} + \bullet G^{\text{Vib}} + \Sigma N_i \mu_i(T, p)$$

where E^{Total} is the self-consistent field (SCF) energy of the slab, ΔG^{Vib} is the vibrational contribution to the Gibbs free energy, N_i the number of the ith type of atoms in the slab, and $\mu_i(T, p)$ is the chemical potential of the ith type of atom at a given temperature and pressure. The sum is over all the types of the atoms in the system. The chemical potential at finite temperature and pressure, $\mu_i(T, p)$ can be expressed as $\mu_i(0K) + \Delta \mu_i(T, p)$, where 0K is taken as the reference state, and $\Delta \mu_i(T, p)$ is the change in free energy from that reference state to the system at a given temperature and pressure. $\Delta \mu_i(T, p)$ includes the changes in free energy due to the pV and TS terms for both solid and gas phase species at finite temperatures and pressures.

The SCF energy for the 0K values for surfaces and for bulk Al and Al_2O_3 is obtained using the generalized gradient approximation of Perdew, Burke, and Ernzerhof (1996) for the exchange-correlation potential and the Full-Potential Linear Augmented Planewave (FP-LAPW) method as implemented within the WIEN97 program. A uniform **k**-point mesh with ten points in the full Brillouin zone is used for each slab. For the total energies of the H_2, O_2, and H_2O, we use the experimental values for the energies of the atoms minus the zero-point vibrational energy (ZPE) to minimize error in comparison with the bulk and slab SCF calculations. The *differences* in free energies between the rest of the surfaces due to the contribution of the ZPE is negligible.

The JANAF Thermochemical Tables (Stull, 1971). are used to obtain $\Delta \mu_i(T, p)$ values for all crystalline and gas phase species.

Limits on the Range of the Chemical Potential

As each surface is considered here to be in chemical and thermal equilibrium with the bulk and the environment, the chemical potentials of each type of atom must be equal in all phases. This condition results in constraints on the range of accessible values of the chemical potential. Hence for the clean, non-hydrogen-containing surface:

$$2\mu_{Al} + 3\mu_O = E_{Al_2O_3 \text{ (bulk)}}$$

$$\mu_{Al} < E_{Al \text{(bulk)}}$$

$$\mu_O < \tfrac{1}{2} E_{O_2 \text{ (molecule)}}$$

where $E_{Al_2O_3 \text{ (bulk)}}$ is the total energy per bulk α-Al_2O_3 formula unit. The limits of the chemical potential for oxygen are determined by conditions of equilibrium with relevant oxygen-containing species in the system. The maximum oxygen chemical potential is that of a maximum concentration of O_2 molecules at 0K, which corresponds to O_2 condensing on the surface. This is the reference value of the oxygen chemical potential, which is the zero on the right hand side of the graph in Fig. 8. Attempts to move above this maximum limit would only result in formation of additional O_2 molecules at the surface. At thermal equilibrium we have

$$\mu_{Al_2O_3} = 2\,\mu_{Al} + 3\,\mu_O$$

Thus, μ_O is at a minimum when μ_{Al} is at a maximum, which corresponds to the chemical potential of an aluminum atom in bulk aluminum. Hence,

$$\mu_O = 1/3E_{Al_2O_3 \text{ (bulk)}} - 2/3E_{Al \text{ (bulk)}}$$

Below this range, an aluminum metal overlayer would form on the surface.

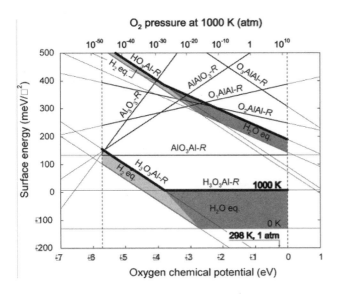

O₂ pressure at 1000 K (atm)

Figure 8. Surface energies of different α-Al_2O_3 (0001) surface terminations. The dashed vertical lines indicate the allowed range of the oxygen chemical potential, μ_O. The light and dark grey regions indicate the range where hydrogen on the surface is in equilibrium with H_2 and H_2O, respectively, from 0K up to 1000K and 1 atm pressure (heavy lines).

Results

The results for the α-Al_2O_3 (0001) surface Gibbs free energies with respect to the chemical potential of oxygen are presented in Fig. 8. In the absence of hydrogen, the stoichiometric aluminum-terminated surface is the most

stable across the range of the oxygen chemical potential. The oxygen-terminated surface is extremely high in energy due to the presence of a large surface dipole and a high degree of bond unsaturation, even at the maximum value of the oxygen chemical potential. In other words, in the absence of hydrogen, the metal-terminated surface will have a lower energy than the oxygen-terminated structure, even in the presence of an O_2 condensate at the surface. If hydrogen is added to the system, however, the oxygen dangling bonds can be saturated and the surface dipole can be drastically reduced. The net result is a surface energy which is much lower than the aluminum terminated surface.

For the saturated hydroxyl-terminated surface, if the concentration of O_2 is sufficiently high, i.e. μ_O is above −2.8 eV, then surface hydrogen will desorb as H_2O. Otherwise desorption will occur as H_2. At 1000K, this crossover value shifts to −3.8 eV.

Conclusions

We have calculated the Gibbs free energies of the possible (1 x 1) α-Al_2O_3 (0001) surfaces in equilibrium with an environment containing oxygen and hydrogen species. The experimental observation of an oxygen-terminated α-Al_2O_3 (0001) surface can be explained by the presence of hydrogen in the system. Real surfaces encountered in industrial applications may be much more heterogeneous and contain many defects and structural features of kinetic rather than thermodynamic origin. Hence having a model which can predict which structures and stoichiometries can be expected under equilibrium conditions in a multiphase environment throughout a range of physical temperatures and pressures provides a valuable framework from which to aid in the interpretation of experimental results on more complex systems.

References

Ahn, J., and J. W. Rabalais (1997). *Surf. Sci.,* **388**, 121.

Becke, A.D. (1988). *Phys. Rev.,* **A38**, 864.

Birkholz, M., S. Fiechter, A. Hartmann, and H. Tributsch (1991). *Phys. Rev.,* **B43**, 11 926.

Blaha, P., K. Schwarz, and J. Luitz (1999). *Wien97, A Full Potential Augmented Plane Wave Package for Calculating Crystal Properties.* Karlheinz Schwarz, Techn. Universitaet Wien, Austria. ISBN 3-9501031-0-4.

Buerger, M.J. (1931). *Am. Mineral.,* **16**, 361.

Causa, M., R. Dovesi, C., Pisani, and C. Roetti (1989). *Surf. Sci.,* **215**, 259.

Coey, J.M.D., and H. Roux-Buisson (1979). *Mater. Res. Bull.,* **14**, 711.

Delley, B. (1993). *J. Chem. Phys.,* **92**, 508; *DMol Version 2.3.5.* Biosym/MSI, San Diego, CA 92121.

Folkerts, W., G.A. Sawatzky, C. Haas, R.S. de Groot, and F.U. Hillebrecht (1987). *J. Phys. C: Solid State Phys.,* **20**, 4135-4144.

French, T.J., and G.A. Somorjai (1970). *J. Phys. Chem.*, **74**, 12.

Godin, T.J., and J.P. LaFemina (1994). *Phys. Rev.,* **B49**, 7691.

Guo, J., D.E. Ellis, and D.J. Lam (1992). *Phys. Rev.,* **B45**, 13647.

Hass, K.C., W.F. Schneider, A. Curioni, and W. Andreoni (1998). *Science,* **282**, 265.

Kohn, W., and L.Sham (1965). *J. Phys. Rev.*, **140**, 1133.

Kou, W.W. and M.S. Seehra (1978). *Phys. Rev.,* **B18**, 7062.

Lee, C., W. Yang, and R.G. Parr (1988). *Phys. Rev.,* **B37**, 785.

Lubrizol (1990). Experiments performed in Lubrizol Corporation laboratories.

Mackrodt, W.C., R.J. Davey, and S.N. Black (1987). *J. Cryst. Growth,* **80**, 441.

Manassidis, I., A. De Vita, and M.J. Gillan (1993). *Surf. Sci. Lett.,* **285**, L517.

Mulliken, R.S. (1955). *J. Chem. Phys.*, **23**, 1833-1846.

Perdew, J.P., S. Burke, and M. Ernzerhof (1996). *Phys. Rev. Lett.*, **77**, 3865.

Pinklea, S., C., Leconte, and E. Amma (1976). *Acta Cryst.,* **B32**, 529.

Prutton, C.F., Turnbull, and G. Dlouhy (1946). Mechanism of Action of Organic Chlorine and Sulfur Compounds in Extreme-Pressure Lubrication, J. Inst. Pet., 32, 90.

Raybaud, P., G. Kresse, J. Hafner, and H. Toulhoat (1997). *J. Phys. Condens. Matter,* **9,** 11085-11106.

Renaud, G. (1998). *Surf. Sci. Reports,* **32**, 1.

Stull, D.R., and H. Prophet (1971). *JANAF Thermochemical Tables*, 2nd edition. U.S. NBS.

Toofan, J., and P. R. Watson (1998). *Surf. Sci.,* **401**, 162.

Verdozzi, C., D.R. Jennison, P.A. Schultz, and M.P. Sears (1999). *Phys. Rev. Lett.,* **82**, 799.

Wang, X.G., A.M. Chaka, and M. Scheffler (2000). *Phys. Rev. Lett.,* **84**, 3650.

Williams, A.R., J. Kuebler, and C.D. Gelatt (1979). *Phys. Rev.,* **B19**, 6094.

Wimmewisser, M., and J. Haase (1968). *Z. Naturforsch* **23a**, 56.

Wyckoff, R. W. G. (ed.) (1982). *Crystal Structures,* 2nd edition, Vols.1-3. Krieger, Malabar.

Zeng, Y., and N.A.W. Holzwarth (1994). *Phys. Rev.,* **B50**, 8214.

Zhao, G.L., J. Callaway, and M. Hayashibara (1993). *Phys. Rev.,* **B 48**, 15781.

MOLECULAR MODELING AND SIMULATION OF CROP-PROTECTION CHEMICALS

Daniel A. Kleier
DuPont Agricultural Products
Newark, DE 19714

Abstract

In order for a compound to be successful as a commercial crop protection chemical it must be intrinsically active at some site critical to the well being of a pest, systemic enough to reach that site, and stable enough to survive environmental conditions. Molecular modeling and simulation can be a significant aid in designing and discovering compounds that satisfy these requirements.

Intrinsic activity often involves action at a protein responsible for facilitating a key biochemical process. When the structure of a target protein is known, molecular modeling is useful in designing compounds that are complementary to the active site, and hence disruptive to the biochemical process catalyzed or mediated by the protein. Molecular modeling can also be used to design pesticides that are active at cell membranes. For example, when a cell membrane protects a crop pathogen against entry of fungitoxic metals, molecular modeling can be used to design compounds with the right physico-chemical properties to overcome the protective barrier.

Intrinsic activity without delivery to the site of action is of no practical value. Compounds capable of moving from a site of application to a remote site of action can be designed using simulations of molecular transport within the plant vascular system. Efficacy also requires that compounds be stable enough to survive the often-harsh conditions experienced at the site where they are applied. For foliar application these harsh conditions include the uv-visible radiation present at the surface of a sun-drenched leaf. Compounds that are better able to withstand these conditions can be designed with the aid of electronic structure methods.

When enough data is available, statistical modeling can be used to express the influence of multiple structural descriptors on efficacy without invoking mechanical models for the individual steps of binding, transport and degradation. Molecular modeling is often used to supply values for the structural descriptors that are in turn used as independent variables in the statistical models.
Recent trends in discovery research suggest an ever-increasing role for molecular modeling and simulation. Modern discovery paradigms will require that multi-dimensional characterizations of large numbers of compounds be performed *in silico*.

Keywords

Crop-protection chemicals, Computer-aided design, Computer-aided discovery, Structure-based design, Property-based design, Molecular modeling, Simulation of transport, *in silico* screening.

Introduction: Discovery or Design

In 1909 physician Paul Ehrlich (Ehrlich and Bertheim, 1909) filed a patent on "compound 606" that had been discovered by screening 1000 compounds in rabbits for activity against syphilis. Salvarsan quickly became the

9

treatment of choice for syphilis and the modern science of chemotherapy was thus created. As a result of spectacular advances in high throuput chemical synthesis and screening, Ehrlich's discovery paradigm has experienced a renaissance. In 1977 Cushman and Ondetti (1979) designed an inhibitor of angiotensin converting enzyme (ACE) based upon the structure of a related enzyme. Captopril quickly became a useful treatment for high blood pressure, and the technology of rational design was born. Today rational design plays an important role in the generation, evolution and optimization of biologically active compounds.

Both paradigms, discovery by large numbers and design from chemical knowledge, are facilitated if not dependent upon molecular modeling and simulation. Discovery by large numbers is facilitated by high throuput in silico assays that quickly characterize and categorize large numbers of compounds. Rational design is facilitated by entering a virtual world in which potential bioactive agents are modeled in interaction with the catalysts, cofactors and environmental agents that mediate or moderate their activity.

Molecular Modeling in Structure-Based Pesticide Design

The rational design paradigm capitalizes on knowledge of chemical and biophysical processes essential for the viability of a target pest or disease. This knowledge includes an understanding of the mechanisms of essential biochemical processes, as well as information about the structures of the enzymes that mediate these processes.

Rice blast is a fungal disease caused by the plant pathogen *Magnaporthe grisea*. Control of this pathogen can be realized by inhibiting an enzyme in the melanin biosynthesis pathway known as scytalone dehydratase (SCDH). A crystal structure for SCDH with a salicylamide inhibitor bound (Lundqvist, et al., 1994) is illustrated in Fig. 1. This structure was the starting point for a rational design program (Chen, 1998) that sought to evolve the structure of the illustrated inhibitor.

Computer visualization of the co-crystal revealed hydrogen-bonding interactions between the salicylamide inhibitor and the enzyme as illustrated in Fig. 2. In addition to interacting with a number of amino acid residues, the inhibitor forms hydrogen bonds with a pair of water molecules.

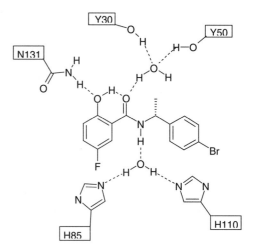

Figure 2. Hydrogen-bond network between inhibitor and active site residues.

Careful examination of the structure suggested that the water in the upper right might be displaced by a suitably evolved analog of the salicylamide. Work based upon the concept of isosterism had previously resulted in the design of benztriazine inhibitors (see Fig. 3, middle) that are ring-closed descendants of the parent salicylamides. Docking of sample benztriazines into the SCDH crystal structure suggested further modifications involving replacement of one of the benztriazine ring nitrogens (N:) with a carbon bonded to a cyano group (C-C#N:). This modification effectively moves the position of the replaced nitrogen by two bond lengths into the space occupied by one of the bound water molecules.

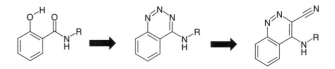

Figure 3. Molecular evolution of cyanocinnoline inhibitor from salicylamide.

Cyanocinnolines were synthesized in response to these virtual docking experiments, and binding affinity improvements of 3-30 fold could be attributed to this modification. Subsequent crystal structure determinations confirmed that cyanocinnolines displaced the active site water molecule as expected.

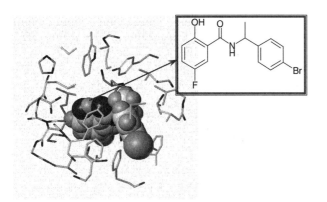

Figure 1. Active site of scyatalone dehydratase with bound salicylamide inhibitor.

Molecular Modeling in Property-Based Pesticide Design

A patent reporting broad-spectrum fungicidal activity for the 2-(pyrimidin-2-yl)-pyridine illustrated in Fig. 4 (Katoh, T., et al., 1989) was the starting point for a research project to design related compounds. Based upon structural resemblance to well known metal chelators such as bipyridyl, our project team hypothesized that metal chelation was involved in the mechanism of action of this and related compounds.

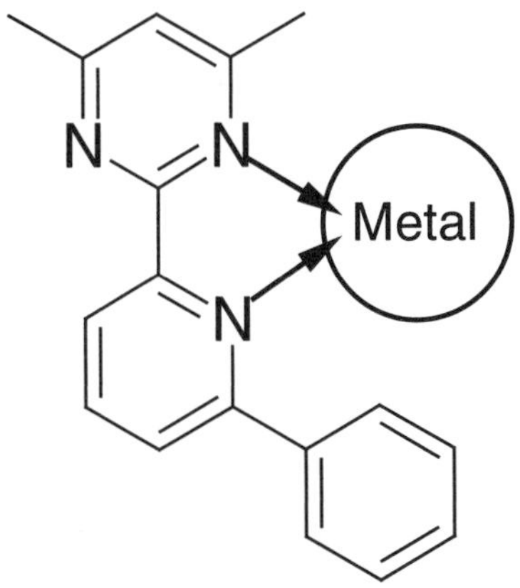

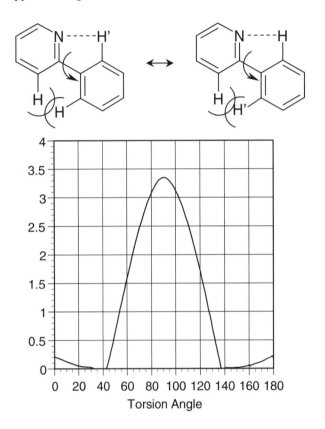

Figure 4. Possible chelation complex between metal ion and 2-(4,6-dimethyl-pyrimidin-2-yl)-6-phenyl-pyridine.

Molecular modeling added credibility to this hypothesis and then aided in the design of novel fungicides that had physico-chemical properties consistent with this mechanism of action. Later reports confirmed that metal chelation is involved in the mechanism of action of the pyrimidinyl pyridines. These compounds appear to facilitate the transport of copper ions across cellular membranes that normally protect fungi from the effects of naturally abundant levels of this metal (Baldwin, 1995).

In spite of the resemblance of the lead compound to bipyridyl, doubts remained about its ability to chelate metal ions. In particular, the phenyl group appears to sterically hinder approach of a metal ion to the nitrogen of the central ring. Furthermore, the electron withdrawing effect of the second nitrogen in the pyrimidine ring should reduce the ability of the first nitrogen of that ring to chelate metal ions. This latter concern is supported by the knowledge that pyrimidine is significantly less basic than pyridine. Electronic structure calculations were performed to address both of these doubts, and in the process novel structures were suggested for synthesis.

For these studies the three ring system of the lead compound was first divided into a pair of overlapping two ring systems: 2-phenyl pyridine and 2-(pyrimidin-2-yl)-pyridine. Conformational analyses were then performed

on both two ring systems at the Hartree-Fock level of approximation using the GradSCF program (Kormornicki, 1988) and a double zeta basis supplemented by polarization functions on non-hydrogen atoms. In order to account for correlation effects, second-order Moller-Plesset perturbation (MP2) calculations were employed on the stationary structures. The identity of transition state structures was established by the observation of a single imaginary frequency in the vibrational analysis.

The results for the conformational analysis of 2-phenyl pyridine as a function of inter-ring torsion angle appears in Fig. 5.

Figure 5. Potential Energy curve for ring twisting in 2-phenylpyridine.

A relatively flexible molecule is indicated by the calculations. The minimum energy conformations at θ = 40° and 140° allow approach of a metal ion to the pyridine nitrogen. In this orientation the phenyl ring may actually contribute to the stabilization of a metal complex by donating π-electrons to the metal. A 90°-twisted conformation can be realized at an energy cost of only 3.5 kcal/mole. Even though this is a modest energy penalty, molecular modifications that force the phenyl ring to be firmly twisted with respect to the pyridine should improve the free energy of metal binding. This pre-organization principle guided much of the synthesis effort for this project.

Electronic structure calculations were also used to investigate the metal chelating ability of 2-(pyrimidin-2-

yl)-pyridines relative to classic ligands such as bipyridyl and o-phenanthroline. This investigation was pursued in two steps. First, the energy required to pre-organize the ligand for metal chelation was determined, and then an estimate was made of the metal binding energy for this pre-organized ligand. Bidentate ligands characterized by an N-C-C-N substructure realize their most effective bite on a metal ion when they are organized in a coplanar conformation that finds both nitrogens are on the same side of the C-C bond. This so-called s-cis conformation optimally orients the lone pair electrons of both nitrogens toward the metal ion, and is often observed in metal complexes with ligands such as bipyridyl. It may come as a surprise, therefore, that the estimated energy cost for bipyridyl to realize the optimal s-cis conformation is a substantial 8 to 9 kcal/mole (Fig. 6, right).

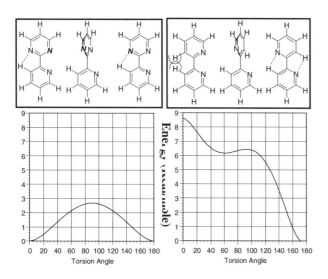

Figure 6. Potential energy curves for ring twisting in 2-(pyrimidin-2-yl)-pyridine (left) and bipyridyl (right).

By way of contrast, conformational analysis of 2-(pyrimidin-2-yl)-pyridine indicates that no such energy penalty need be paid for pre-organization. In fact, the pyrimidinyl-pyridine molecule has two energetically equivalent conformations (see Fig. 6, left) both of which possess N-C-C-N substructures in the optimal s-cis arrangement. Thus, from both conformational energy and entropy considerations, the pyrimidinyl-pyridine system appears to enjoy an advantage over bipyridine for metal chelation. But does this pre-organization advantage overcome the intrinsically smaller metal ion affinity expected for the pyrimidine of pyrimidinylpyridine compared with the pyridine of bipyridyl?

In order to rank order the intrinsic metal ion affinity of these ligands, we calculated the energy change for the pre-organized ligands to displace o-phenanthroline in a zinc complex (Table 1). In these calculations dithioglycolate was used as a spectator ligand to complete an approximately tetrahedral coordination environment around the zinc, and also gave the complexes a net

neutral charge. O-phenanthroline was used as a standard because of favorable pre-organization as well as the high basicity of its ligating centers.

Table 1. Reaction Energies for Phenanthroline Displacement from a Zinc Dithioglycolate Complex by Bidentate Ligands.

Reaction Energy (kcal/mole)

	Phenananthroline Displacement by Preorganized Ligand	Ligand Preorganization	Net
	-2.06	7.94	5.88
	1.56	0.00	1.56
	-1.31	0.00	-1.31

The calculations reported in Table 1 were performed using density functional theory (DFT) (Kohn and Sham, 1965) as implemented in the DGAUSS program. The non-local spin density (NLSD) method was used with non-local corrections evaluated self-consistently according to the Becke-Perdew formulation (Becke, 1988). A double-zeta split-valence basis set with the addition of polarization functions was used to expand the orbitals and an auxiliary basis represented the electron density. Geometry optimization was carried out with a loose gradient convergence threshold.

As judged by the phenanthroline displacement energies, pyridyl-pyridine (bipyridyl) is intrinsically a more powerful chelator than pyrimidinyl-pyridine by about 3.6 kcal/mole (Table 1, second column). This is the order expected based on the higher basicity of pyridine relative to pyrmidine. However, when ligand pre-organization energy is accounted for (Table 1, third column), the pyrimidinyl-pyridine system is predicted to be the superior chelator (Table 1, last column). As expected, addition of electron donating methyl groups to the pyrimidine ring further improves the predicted net displacement energy so that the dimethyl-pyrimidinyl-pyridine system is predicted to be a better chelator than phenanthroline itself. The methyl groups may actually play a dual role in enhancing the ionophoric characteristic of the fungicides. In addition to increasing intrinsic chelating ability of the ligand, they may shield the metal ion during transport across cell membranes. This may also be the role of the phenyl group attached at the 6-position of the pyridine.

In order to further explore this possibility, a molecular model of a 2:1 complex of zinc with a pair of 2-(4,6-dimethylpyrimidin-2-yl)-6-phenyl pyridine ligands was constructed, and optimized using the MOPAC program at the AM1 level of electronic structure theory (Dewar, et al., 1985). Examination of the modeled complex (Fig. 7) reveals a zinc ion coordinated by four nitrogen atoms in a roughly tetrahedral arrangement. In this complex the phenyl rings as well as the pyrimidine methyl groups present a largely lipophilic surface to the external environment which may aid in transporting the central ion across the lipid interior of a cell membrane.

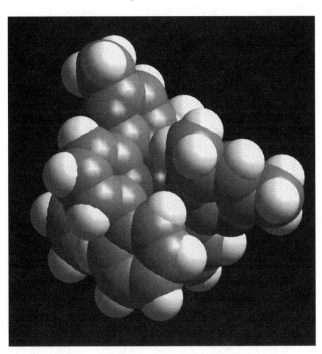

Figure 7. Structure of a 2:1 complex of zinc with a pair of 2-(4,6-dimethylpyrimidin-2-yl)-6-phenyl pyridine ligands.

Calculations like those described above prompted consideration of modifications to the fungicides that would improve chelating ability by locking the phenyl group into a twisted conformation relative to the pyridine ring, while maintaining or possibly even increasing the basicity of the ligating nitrogens. Molecular modeling at almost any level supports the notion that pre-organization of the 2-phenyl-pyridine can be realized by substituting a buttressing methyl group at a position adjacent to the phenyl group on the pyridine ring. This result is consistent with reports of high activity for the analogous methylated derivatives of the lead compound (Katoh, et al., 1989). Our research program considered alternative ways of enforcing the desired twist and settled upon incorporation of a bridge between the phenyl and pyridine rings (Daub and Piotrowski, 1998) as illustrated in Fig. 8.

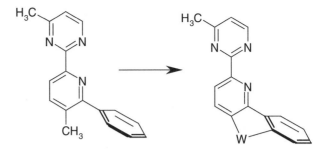

Figure 8. Conformationally constrained analogs of 2-(pyrimidin-2-yl)-pyridines.

Molecular modeling suggested that trimethylene and tetramethylene bridges would effect the desired twist while shorter bridges especially with double bonds would lock coplanar conformations that block metal chelation. The fungicidal activity of a series of bridged analogs (See Table 2) is consistent with this structure-activity hypothesis (Daub and Piotrowski, 1998).

Table 2. Effect of Bridge Type on Percent Disease Control at 100 gm/hectare.

Bridge, W	Wheat Eyespot	Wheat Leaf Blotch	Rice Blast
-(CH$_2$)$_3$-	94%	100%*	92%
-(CH$_2$)$_4$-	88	100*	67
-(CH$_2$)$_2$-	--	99	Inactive
-CH=CHCH$_2$-	Inactive	97	Inactive
-CH$_2$-	Inactive	90	Inactive
-CH=CH-	Inactive	Inactive	Inactive

* indicates that compound gave 100% control at application rate of 25 gms/hectare.

The trimethylene and tetramethylene bridged compounds are clearly the most active in this series. Compounds with bridges shorter than 3 atoms were significantly less active on all three species.

In summary, molecular modeling increased our confidence in the proposed mechanism of action, affirmed pre-organization and enhanced basicity as design principles, and aided in the design of a new class of highly active, conformationally constrained fungicides.

Molecular Simulations in Transport-Based Design

The distribution of a chemotherapeutic agent can be just as important in a treated crop as it is in a treated patient. And much as with drugs, the use of a systemic progenitor strategy can be a useful approach to pesticide design (Fukuto, 1985). In the plant vascular system, it is the phloem sap that plays the role of blood in carrying nutrients such as sucrose from mature leaves to other parts of a plant including roots, fruits and new leaves. Compounds that gain access to the phloem sap, and are retained there will be transported over long distances within the plant. Pro-pesticides that are phloem mobile are thus expected to move from leaves where they might have been applied to remote sites where their action can be expressed if the active ingredient can be released.

A simple dynamic model (Fig. 9) that can be used in the design of phloem systemic pro-pesticides is illustrated below (Grayson and Kleier, 1990). This model captures the essence of transport into and within the sieve tubes of a living plant.

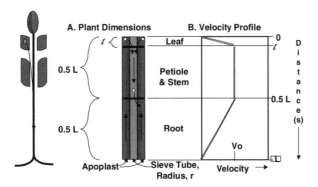

Figure 9. Simple model of the plant vascular system including velocity profile for movement of phloem sap.

It consists of two concentric cylinders separated by the semi-permeable sieve tube membrane (Figure 9A). Phloem sap flows down the sieve tube represented by the central cylinder with a velocity that varies with distance s from the leaf tip (Figure 9B). The transpiration stream moves through the surrounding cylinder with a higher volume flow and in a direction opposite to that of the phloem sap. Compound that has been applied to the apoplast of the leaf gains access to the phloem by permeating the sieve tube membrane. The rate of movement across the sieve tube membrane is expressed in terms of Fick's Law. Although there is net movement into the sieve tube in the leaf zone, the model provides for escape as compound moves downwards since the concentration in the surrounding apoplast below the leaf is assumed to vanish. The model also provides for the pH differential that typically characterizes the phloem sap of a plant relative to the surrounding apoplast. Since the pH of the phloem sap is generally higher than the surrounding tissue, acidic materials are trapped there as a result of a higher degree of dissociation, and the lower membrane permeance of the ionic species.

The plant parameters that characterize the model include the phloem sap velocity, plant length, leaf length, pH of the phloem and apoplast, as well as empirical parameters for the permeability of the sieve tube membranes. The compounds are characterized by physical properties including the partition coefficients of all ionic forms and the equilibrium dissociation constants

(pKa values) that connect these forms. The concentration of the xenobiotic is taken as the dependent variable and can be expressed in closed form as a function of time and distance s from the leaf tip. For most applications, results are determined in the steady-state limit, and total concentrations (acid plus conjugate base) are reported for a point deep in the root zone (s = 0.9L) relative to the concentration in the application zone.

These relative concentrations known as concentration factors (C_f) are plotted in Figure 10 as a function of partition coefficient (logKow) and pKa for a fixed set of plant parameters. Mono-functional acids have been assumed, and the plant parameters correspond to a "short" 15 cm plant in which the phloem sap is 2 pH units more basic than the surrounding apoplast. The partition coefficient plotted is that of the neutral (protonated) form of the acid. It was assumed that the corresponding negatively charged conjugate base has a logKow value 3.7 less than that of the neutral form.

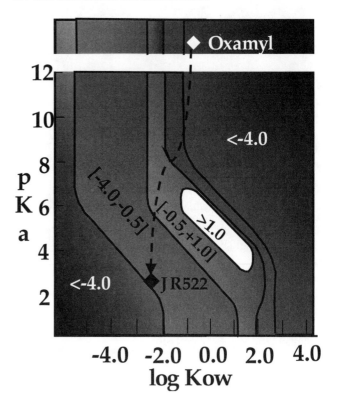

Figure 10. Contour plot of logCf as function of logKow and pKa illustrating position of oxamyl and JR522.

Validation studies and comparison with literature results suggest that compounds with predicted logCf values above −4 are likely to be phloem mobile. This corresponds to the filled contours that are labeled in black on the above plot.

The physical properties of the commercial nematicide, oxamyl, place it just outside this region where mobility is predicted mobility. Consistent with this placement, marginal phloem mobility has been reported for this compound (Alphey, et al., 1985). Guided by simulations like these, a pro-pesticide approach was taken with the goal of rendering oxamyl systemic in a wider variety of plants (Hsu, et al., 1995). The simulations indicated that improved phloem mobility could be expected in many plants for glucuronide derivatives such as JR522 (Figure 11) as a result of ion trapping associated with the lower pKa of the derivatives.

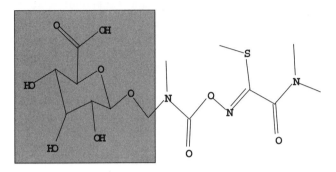

Figure 11. Hydroxymethylglucuronide derivative of oxamyl nematicide (JR522).

JR522 was synthesized in response to the simulations. When applied foliarly to tobacco plants that were engineered to express glucuronidase in a root specific fashion, JR522 showed better control of root-knot nematodes than did oxamyl. Simulations had thus played an essential role in the design of this pro-nematicide.

Molecular Modeling in Pesticide Optimization

Chemotherapeutic control of diseases in man differs from pest control in crops in a few important aspects. For example, herbicides applied to the surface of plant leaves experience a much different environment than drugs injected or taken orally. One particularly destructive element experienced by many compounds applied to a leaf surface is exposure to sunlight. Figure 12 illustrates the absorption spectrum of an experimental cinnolinium herbicide and the solar emission spectrum it experiences on an average sunny day in the field (Kleier, 1994). The overlap of the absorbance peak at about 360 nm with the solar emission spectrum is enough to cause significant losses due to photodegradaion under field conditions.

In order to address this problem, electronic structure calculations were used to simulate the uv-visible spectra of hypothetical cinnolinium compounds related to the lead compound shown in Figure 12. These calculations were performed using the semi-empirical CNDO/S method (Del Bene, 1968). Predicted shifts in the position of the problematical long-wavelength band are shown in Figure 13.

Initially, the project team intended to prepare analogs that would shift the long wavelength band out of harm's way by moving it below 300 nm (blue shift). However, our calculations indicated that this would require heavy

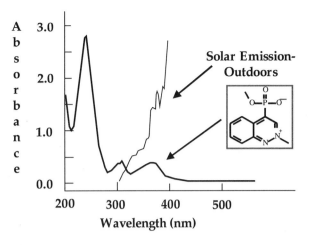

Figure 12. Absorbance spectrum (thick) of a 2-methyl-cinnolinium herbicide and solar emission spectrum outdoors (thin).

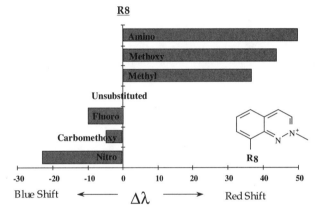

Figure 13. Position of long wavelength absorption band of 8-substituted 2-methyl-cinnoliniums.

substitution with strong electron withdrawing groups (e.g., nitro groups) which might introduce new problems of their own. However, the calculations were more encouraging for a red shift strategy who argued that the less energetic photons in this region of the spectrum would be less destructive to the sensitive herbicide. In particular, the calculations suggested that substitution of a methoxy group at the 8-position of the cinnolinium would shift the long wavelength band over 60 nm to the red. A number of 8-methoxy derivatives were prepared in response to these calculations, and for reasons that may never be completely understood, significantly greater stability was realized. The calculations had provided useful guidance in terms readily understood by project chemists.

Molecular Modeling as a Source of Descriptors for Statistical Modeling

An apparently exasperated diplomat once asserted that "There are three kinds of lies: lies, damn lies, and statistics" (B. Disraeli). This warning notwithstanding, empirical models built by statistical analysis of efficacy or related property data find many useful applications in lead optimization. Once established, they can be used to guide further synthesis. The structural descriptors used as independent variables in these models are often best provided by molecular modeling.

As an illustration, calculated electron affinities were used in the project to optimize the activity of the cinnolinium herbicides introduced above. Like the herbicide paraquat, these cations act by catalytically transferring electrons from the photosynthetic apparatus to molecular oxygen (G. Gardner, et al., 1992). Reduction potentials are thus important determinants of biological activity. As a rule of thumb, herbicides with this mechanism of action have reduction potentials in the range of –0.30 to –0.50 volts relative to the standard hydrogen electrode (Summers, 1980). In order to estimate reduction potentials for hypothetical cinnoliniums, an empirical relationship was first established between reduction potentials as measured by cyclic voltammetry and computed estimates of electron affinity (Kleier, 1993). One such relationship is given below:

$$E_{1/2} = 0.501*A - 2.164 \text{ v}$$
$$N=12; \ r^2=0.72$$

In this relationship $E_{1/2}$ is the predicted reduction potential and $\underline{A}$ is the electron affinity in volts as calculated using the delta SCF method with a minimum basis set. Once established, this relationship and others like it could be used to estimate reduction potentials for hypothetical cinnoliniums. By providing such estimates relative to the range usually associated with herbicidal activity, hypothetical compounds related to the lead were prioritized for chemical synthesis.

Path Forward

Recently the paradigm pendulum for pesticide discovery has swung away from rational design and towards discovery by random screening. This phase is characterized by the acquisition, synthesis and testing of compounds on an unprecedented scale. Large numbers of compounds are available either as grab bags of miscellaneous chemistry, or as libraries of systematically related chemistry. With just a few loci of variability, a single molecular scaffold can support a huge virtual library of synthetically accessible compounds.

It is this unprecedented scale that opens new opportunities for modeling and simulation to participate in the discovery process. No longer can a few human experts expect to effectively select from the wealth and richness of chemistry available to them without some sort of computer assistance. Increasingly explicit knowledge of the experts needs to be incorporated into a discovery machine. The engine that powers this machine will

include molecular modeling and simulation components for generation of the estimated properties needed to drive the compound selection and data analysis processes. Information technology will provide the supporting infrastructure required to handle the huge amount of data, both simulated and measured, that must be processed.

Because of the huge numbers involved molecular modeling faces a different set of challenges in this new phase. The emphasis is on timely characterizations of large numbers of compounds at a level appropriate to the application. Adequate characterization may range from simple topological or electrotopological characterizations (Kier and Hall, 1986, 1999; Pearlman, 1997) to more involved conformational models (Smellie, et al., 1995). Large-scale estimation of physico-chemical properties (partition coefficients, acid dissociation constants, solubility, reactivity, etc.) will enable sampling of libraries for compounds with improved chances for certain kinds of biological activity, as well as for acceptable uptake and transport.

Although structure-based pesticide design has realized only a fraction of its full potential (Reynolds, 1997), in silico screening of potential pesticides for goodness of fit to binding sites of target proteins is showing promise (Dixon and Blaney, 1998), and will grow in importance as more crystal structures for relevant targets become available. In favorable circumstances that can only become more common, advantage may be taken of sequence data for a target protein, and structural data for related proteins to build homology models (Egner, et al., 1995) that can then be used for high throughput in silico screening.

Rather than being used largely on an ad hoc basis, molecular modeling and simulation is moving more into the mainstream of the discovery workflow. Applications in the mainstream include compound selection from offered lists and design of virtual libraries, analysis of screening data for structure-activity relationships, design of analog programs to follow-up on active hits and leads, and optimization of biokinetic, environmental and safety properties.

Molecular modeling is likely to play an increasingly central role in selecting both diverse and focused subsets from compound collections. Molecular topological analysis (Pearlman, et al., 1997) is an appropriate level of characterization for the selection of diverse subsets, while estimates of physico-chemical properties will aid in limiting selections to regions of chemistry space consistent with acceptable biokinetic behavior (Kleier, 1998). Selections focused on specific targets may require more precise characterizations. Pharmacophore screening is one particularly promising class of focused selection. A pharmacophore is "the molecular framework that carries (phoros) the essential features responsible for a drug's (pharmacon) biological activity" (Ehrlich, 1909). Once quantified in terms of the spatial constraints between essential features, pharmacophores can be used as queries (Guner, 2000) to search 3D representations of large libraries for compounds carrying desired pharmacophoric features, or to deselect those with undesirable features.

Pharmacophore perception is a form of 3D structure-activity analysis that requires molecular modeling as input. Specific pharmacophores can be defined either by modeling the binding site of a target protein (Boehm, 1996), or, absent the structure of a target protein, by analyzing active ligands for energetically accessible conformations that arrange binding features in a similar fashion in 3D space (Sprague and Hoffmann, 1997). Although only a few examples of agrochemical discovery using pharmacophore modeling and 3D database searching have been reported (Bures, et al., 1991), the numerous success stories appearing in the pharmaceutical literature bode well for this paradigm. Increasingly the results of pesticide structure-activity studies are being expressed in terms of these chemically meaningful conceptualizations (Dinan, et al, 1999; Ren, et al., 2000; Liu, et al., 1999).

Follow-up to the large numbers of hits and leads expected from high throughput screening will benefit from computer aided approaches. Experimental designs for analog programs can often be formulated in terms of a testable structure-activity hypothesis. Such an hypothesis logically limits chemistry space to a manageable number of dimensions that can be explored in a systematic and informative manner. Depending upon the dimensions chosen for exploration, molecular modeling and simulation may play an important role in formulating these designs. Consider the example of a fungicide lead like that discussed above. From the structure of the lead an expert in the field might hypothesize that metal chelation is involved in the mechanism of action. Recognizing that these compounds must also transport to the site of action, a design might be based on varying metal chelation potential as well as lipophilicity in a statistically meaningful manner, subject to the constraint that the compounds in the design set not be too basic. Such a design would be aided by estimates of relative metal binding energies, partition coefficients and acid dissociation constants, all of which can be supplied by molecular modeling and simulation.

Opportunities also exist for molecular modeling to aid in addressing registration issues such as safety to the environment and non-target organisms. Methods for anticipating and remedying toxicity to non-target organisms could involve a large computational chemistry component. This is a grand challenge that requires not only an assessment of the safety of the parent compound, but also the anticipation of chemical and metabolic transformations likely to befall the substance under consideration. To date, models for predicting safety have generally been empirical (Enslein, 1988) or rule based in nature (Sanderson and Earnshaw, 1991). Increasingly, molecular modeling is providing the structural descriptors to be considered for inclusion as the independent variables in the empirical models (Klopman, 1994), while

molecular simulations are being used to anticipate molecular behavior such as susceptibility to metabolic transformations (DeGroot and Vermeulen, 1997).

This article has highlighted just a few of the application domains within the crop protection chemical business where molecular modeling and simulation can play a beneficial role in research and development. Similar opportunities exist in many other chemical businesses as well. Molecular modeling and simulation can provide property data that is difficult or impossible to measure, information relating chemical behavior to molecular structure, and valuable new knowledge gained by careful examination of results and exploratory in silico experimentation.

References

Alphey, T. J. W., R. H. Bromilow, and A. M. Abdalla (1985). Phloem translocation of foliar-applied oxamyl and its role in plant protection. *Nematologica,* **31**, 468-477.

Baldwin, B. C., A. J. Corran, and M. J. Robson (1995). The mode of action of pyridinyl-pyrimidine fungicides. *Pestic. Sci.,* **44**, 81-83.

Becke, A. D. (1988). A multicenter numerical integration scheme for polyatomic molecules, *J. Chem. Phys.,* **88**, 2547-2553.

Boehm, H.-J. (1996). Towards the automatic design of synthetically accessible protein ligands: peptides, amides and peptidomimetics. *J. Comp.-Aided Molec. Des.,* **10**, 265-272.

Bures, M. G., C. Black-Schaefer, and G. M. Gardner (1991). The discovery of novel auxin transport inhibitors by molecular modeling and three-dimensional pattern analysis. *J Comp.-Aided Molec. Design,* **5**, 323-334.

Chen, J. S. Xu, Z. Wawrzak, G. S. Basarab, and D. B. Jordan (1998). Structure based design of potent inhibitors of scytalone dehydratase: Displacement of a water molecule from the active site. *Biochem,* **37**, 17735-17744.

Cushman, D. W., and M. A. Ondetti (1979). History of the design of captopril and related inhibitors of angiotensin converting enzyme. *Hypertension,* **17**, 89-592.

Del Bene, J., and H.H. Jaffe (1968). The use of the CNDO method in spectroscopy. III. Monosubstituted benzenes and pyridines. *J. Chem. Phys,* **49**, 1221-1229.

De Groot, M. J., and N. P. E. Vermeulen (1997). Modeling the active sites of cytochrome P450's and glutathione-S-transferases, two of the most important biotransfromation enzymes. *Drug Metab Rev,* **29**,747-799.

Daub, J. P, and D. L. Piotrowski (1998) Pyridinylpyrimidine fungicides: synthesis, biological activity, and photostability of conformationally constrained derivative. In *Synthesis and Chemistry of Agrochemicals* V, ACS Symposium Series No. 686, 246-257.

Dewar, M. J. S., E. G. Zoebisch, E. M. Healy, and J. J. P. Stewart (1985). AM1: A new general purpose quantum mechanical model. *J. Am. Chem. Soc.,* **107,** 3902-3909.

Dinan, L., R. H. Hormann, and T. Fujimoto (1999). An extensive ecdysteroid CoMFA. *J. Comp.-Aided Molec Des*, 13, 185-207.

Dixon, S. and J. Blaney. Docking: Predicting the structure and binding affinity of ligand-receptor complexes. In Y. C. Martin and P. Willett (Eds.), *Designing Bioactive Molecules*. ACS, 175.

Egner, U., K. P. Gerbling, G. Hoyer, and G Kruger (1995). Design of inhibitors of photosystem II using a model of the D1 protein. *Pestic. Sci*, **47**, 145-158.

Ehrlich, P., and Bertheim (1911), Derivatives of oxyarylarsinic acids and process of making same. U.S. Patent 986,148.

Ehrlich, P. (1909). *Dtsch. Chem. Ges.*, **42**, 17-47.

Enslein, K. (1988) An overview of structure-activity relationships as an alternative to testing in animals for carcinogenicity, mutagenicity, dermal and eye irritation, and acute oral toxicity. *Tox. Ind. Hlth.*, **4**, 479-498.

Fukuto, T. R. (1985). A Trojan horse for pests. *ChemTech*, 362-367.

Gardner, G, J. J. Steffens, B. T. Grayson, and D. A. Kleier (1992). 2-Methylcinnolinium herbicides – effect of 2-methylcinnolinium-4-(o-methylphosphonate) on photosynthetic electron transport. *J. Ag. Food Chem.*, **40**,318-322.

Grayson, B. T., and D. A. Kleier (1990). Phloem mobility of xenobiotics. 4. Modelling of pesticide movement in plants. *Pestic. Sci.*, **30**, 67-79.

Guner, O. (2000) *Pharmacophore perception, development and use in drug discovery. IUL Biotechnology Series*, International University Line, La Jolla, CA.

Hsu, F. C., K. M Sun, D. A Kleier, and M. Fielding (1995). Phloem mobility of xenobiotics. 6. A phloem mobile pro-nematicide based on oxamyl exhibiting root specific activation in transgenic tobacco. *Pestic. Sci.*, **44**, 9-19.

Katoh, T., K. Maeda, M. Shiroshita, N. Yamashita, Y. Sanemitsu, and S. Inoue (1988). Preparation of pyridinylpyrimidine derivatives as plant fungicides. *European Patent Publication EP-A-259139*, 63.

Kier, L. B. and L. H. Hall (1986). *Molecular Connectivity in Structure-Activity Analysis*. John-Wiley and Sons, New York.

Kier, L. B. and L. H. Hall (1999). *Molecular Structure Description: The Electrotopological State*. Academic Press.

Kleier, D. A. (1993). Quantitative Structure-Activity Relationships: Adding Value to Herbicide Bioassays. In JC Streibig & P Kudsk (Eds.), *Herbicide Bioassays*. CRC Press, Boca Raton, FL, 97-110.

Kleier, D. A. (1994). Environmental effects on the photodegradation of pesticides. In H. G. Hewitt, J. Casely, L. Copping, and B. T. Grayson, (Eds.), *Comparing Glasshouse & Field Pesticide Performance, BCPC Monograph 59*. BCPC Publications, Farnham, Surrey, UK, 97-110.

Kleier, D. A. , B. T. Grayson, and F. C. Hsu (1998). The phloem mobility of pesticides. *Pesticide Outlook*, **9**, 26-30.

Klopman, G., M. Dimayuge, and J. Talafous (1994). META. 1. Program for evaluation of metabolic transformations of chemicals. *J. Chem. Info. Comp. Sci.*, **34**,1320-1325.

Kohn, W., and L. S. Sham (1965). Self-consistent equations including exchange and correlation effects. *Phys. Rev. A*, **140**, 1133-1138.

Komornicki, A. (1988). *QCPE Bulletin*, 8-9.

Liu, J., Z. Li, H. Yan, L. Wang, and J. Chen (1999). The design and synthesis of ALS inhibitors from pharmacophore models. *Bioorg. Med. Chem. Lett.*,**9**, 1927-1932.

Lundqvist, T., J. Rice, C. N. Hodge, G.S. Basarab, J. Pierce, and Y. Lindqvist (1994). Crystal structure of scytalone dehydratase - a disease determinant of the rice pathogen, *magnaporthe grisea*. *Structure(London)*, **2**, 937-944.

Pearlman, R. S., and K. M. Smith, R. (1998). Novel software tools for chemical diversity. *Perspect. Drug Discovery Des. 3D-QSAR in Drug Design: Ligand/Protein Interactions and Molecular Similarity*. Kluwer Academic, Dordrecht, Netherlands, **9/10/11**, 339-353.

Ren, T-R., H-W Yang, X. Gao, X-L Yang, J-J. Zhou, and F-H. Cheng (2000). Design, synthesis and structure-activity relationships of novel ALS inhibitors. *Pest. Management Sci*, **56**, 218-226.

C. H. Reynolds (1997). Molecular modeling in agrochemicals. *Chem & Industry*, Aug.4, 592-595.

Sprague, P. W., and R. Hoffmann (1997). Catalyst pharmacophore models and their utility as queries for searching 3D databases. In H. Van de Waterbeemd, B. Testa, and G. Folkers (Eds.), *Computer Assisted Lead Finding and Optimization*. VHCA, Basel, 230.

Sanderson, D. M. and C. G. Earnshaw (1991). Computer prediction of possible toxic action from chemical structure: the DEREK system. *Human Exp. Toxicol.* **10**, 261-273.

Smellie, A., S. D. Kahn, and S. Teig (1995). An analysis of conformational coverage. 1. Validation and estimation of coverage. *J. Chem. Inf. Comp. Sci.*, **35**, 285-294.

Summers, L. A. (1980). *The Bipyridinium Herbicides*, Academic Press.

CHEMICAL AND MATERIALS SIMULATION AT FORD MOTOR COMPANY

W. F. Schneider, K. C. Hass, M. L. Greenfield, C. Wolverton, A. Bogicevic,
D. J. Mann, and E. B. Stechel
Chemistry and Physics Departments
Ford Research Laboratory, MD 3083/SRL
Dearborn MI 48121-2053

Abstract

Chemical and materials simulation is finding growing application within Ford Motor Company. Over the last several years, significant contributions have been made in numerous areas, including atmospheric chemistry, photochemistry, catalysis, and surface chemistry. While these primarily chemical applications continue to be pursued, increasing emphasis is being placed on addressing problems in materials science. Recent efforts along these lines include simulation of friction at lubricated surfaces and prediction of alloy microstructure and phase diagrams. As the power of simulation continues to grow, it will assume an increasingly prominent position within industrial research.

Keywords

Chemical and materials simulation, Automotive industry, Atmospheric chemistry, Photochemistry, Catalysis, Surface chemistry, Tribology, Alloy microstructure.

Introduction

Ford Motor Company is one of the largest industrial manufacturing companies in the world. Ford has a long history of innovation in automotive technology, which has been fueled by a large and active in-house research effort. The responsibility for conducting the company's scientific research rests with the Ford Research Laboratory (FRL), which includes facilities in Dearborn, Michigan and Aachen, Germany. FRL is a part of the global product development organization and performs research related to all aspects of the Ford automotive business.

The Ford of the early twentieth century was renowned for its vertically integrated manufacturing system, which took in raw materials—from iron ore to soybeans to natural latex—and transformed them into all the materials necessary to produce an automobile. Today Ford relies on its supplier network to provide most of its manufacturing materials, but the company maintains vigorous research programs in many areas of chemistry,

chemical engineering, and materials science, including catalysts, polymers and composites, paints, and lightweight metal alloys. These activities serve essentially three purposes: first, to develop new materials, processes, and technologies that will meet evolving consumer and market demands and distinguish the company's products in the marketplace; second, to understand current materials, processes, and technologies well enough to optimize their dependability and robustness and reduce manufacturing costs and cycle times; and third, to interface with suppliers to ensure that their current and future products will meet the company's needs. Environmental drivers, including increasingly stringent emissions and fuel economy requirements, motivate much of this research. As a result, a significant research effort in FRL is also directed at understanding and quantifying the environmental impact of vehicles, especially on regional atmospheres and the global climate.

Molecular modeling research was conducted within the predecessor organizations of FRL as long ago as the early 1970's. The current effort began in the early 1990's, when opportunities appeared to apply molecular modeling to atmospheric chemistry—specifically the atmospheric degradation of alternative refrigerants chosen to replace chlorofluorocarbons—and to catalysis for emissions control—specifically NOx reduction catalysis. Early successes and recognition of the tremendous potential impact of simulation led to the gradual growth in program size, from an initial staffing of two researchers to the current level of six. As the program has grown, so has the range of problems explored, so that now "molecular modeling" within Ford is seen broadly to encompass all types of first principles, atomistic, and mesoscale simulations, and goes by the broader rubric of "chemical and materials simulation" (CAMS).

Because of the great complexity of the chemical and materials challenges faced by Ford, CAMS has thus far succeeded primarily in generating fundamental scientific insight and understanding as opposed to direct solution of technological problems. In what follows, four areas of significant past effort and impact will be reviewed, a number of current areas of emphasis briefly described, and the prospects for future impact at Ford considered.

Research Efforts

Atmospheric Chemistry

FRL has historically had a major effort in experimental atmospheric chemistry—for instance, much of the early understanding of tropospheric pollutant chemistry came out of work conducted at FRL (Weinstock 1969; Weinstock & Niki 1972). Gas-phase chemistry provides perhaps the most direct opportunity for coupling chemical simulation and experiment, and the two have grown to work closely together at FRL.

Hydrofluorocarbon (HFC) atmospheric chemistry is an excellent case in point. In the early 1990's, Ford faced the challenging prospect of phasing out chlorofluorocarbons (CFCs) used in automotive air conditioners because of the adverse effect of chlorine on stratospheric ozone. HFCs have a number of properties that made them appealing replacements: they have thermodynamic properties similar to the CFCs, they are chlorine-free, and they have much shorter atmospheric lifetimes than the CFCs. However, little was known about their atmospheric degradation mechanisms and products, and concerns arose about their potential impact on stratospheric ozone (Biggs et al. 1993). Research efforts at FRL and elsewhere focused on these questions for HFC-134a (CF_3CFH_2), the principal HFC candidate for automotive applications. This research included significant contributions by molecular modeling.

One important area of uncertainty was the thermochemistry of HFC-134a degradation products, including unstable or highly reactive species like CF_3OH, CF_3O radical, and $C(O)F_2$ (Fig. 1). These quantities are difficult to measure experimentally, and the available data were based on a small number of experiments, empirical group additivity analyses, and low-level molecular modeling. Ab initio calculations were used to provide a consistent set of heats of formation of CF_3O_x radical and CF_3O_xH species (Schneider & Wallington 1993) and to reconcile inconsistencies between these and the heat of formation of $C(O)F_2$ (Schneider & Wallington 1994). These results helped to prove the thermodynamic instability of the CF_3O/CF_3OH couple in the atmosphere and to exclude CF_3O radical from participation in catalytic ozone depletion cycles (Wallington & Schneider 1994). Molecular modeling was also used to rule out photochemical decomposition of CF_3OH in the stratosphere (Schneider et al. 1995), to investigate the kinetics of its homo- and heterogeneous decomposition (Schneider, Wallington, & Huie 1996), and to explain the observed branching ratios in the decomposition of CF_3CFHO radical (Schneider et al. 1998a). These results contributed to a consistent picture of HFC-134a atmospheric chemistry that demonstrated its environmental acceptability.

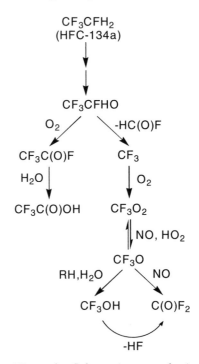

Figure 1. Schematic atmospheric decomposition pathways of HFC-134a.

The success of this effort can be traced to the close interaction between molecular simulation and experiment and the willingness of researchers to learn from each other. The uncertainties surrounding HFC-134a atmospheric chemistry have been largely addressed, but researchers using simulation and experiment continue to collaborate closely on other problems in atmospheric chemistry.

UVA Photostabilizers

Automotive paint systems have the dual tasks of protecting underlying surfaces from environmental assault and of providing an appealing and durable finish that will attract and satisfy customers. A major mechanism of paint system failure is photochemical degradation. To protect underlying layers and interfaces from such degradation, ultraviolet absorbers (UVAs) are added to the top-most layer of the paint system (the "clearcoat") to screen out damaging ultraviolet light. Unfortunately, the UVAs are themselves susceptible to long-term photochemical oxidation (Gerlock et al. 1995), with often-disastrous results for the paint system. Further, the mechanism by which UVAs dissipate ultraviolet energy has not been well understood, hindering efforts to generate more active and more durable photostabilizers. Experimentalists at Ford maintain an active dialog with UVA and paint suppliers to address these problems.

One of the most common classes of UVAs is that of 2-(2′-hydroxyphenyl)benzotriazoles, such as Tinuvin-P:

$$\text{enol} \xrightarrow{\; h\nu \;} \text{keto}$$

In its ground state, Tin-P exists in the enol form, with an intramolecular hydrogen bond preventing rotations about the single bond connecting the two ring systems. Absorption of an ultraviolet photon was believed to result in transfer of the hydroxyl proton to the triazole ring, forming a keto isomer, but how this excited state isomer dissipated energy to return to the ground state was unknown. In collaboration with researchers at the University of Delaware, ab initio (Estevez et al. 1997) and density functional theory (DFT) (Hass et al. 1996) calculations were used to probe the ground and excited state potential energy surfaces of Tin-P. The structure of the ground (S_0) enol state was confirmed. However, relatively simple configuration interaction-singles (CIS) calculations indicate that the lowest singlet excited state (S_1) does not relax to the simple planar keto structure previously supposed, but rather undergoes a complicated folding about the C-N bond in conjunction with $sp^2 \rightarrow sp^3$ rehybridization of the central triazole nitrogen. The CIS calculations suggested that the folded structure led to a conical intersection of the S_0 and S_1 states, explaining the facile internal conversion exhibited by Tin-P. The existence of this conical intersection was confirmed through application of more sophisticated computational tools to simplified models of Tin-P. Rehybridization and folding were thus found to be key to the remarkable photostabilizing ability of this molecule.

This relatively small computational contribution provided Ford experimentalists a much improved understanding of UVA function and persuaded the UVA supplier to accelerate and redirect its own research on improved UVAs, including boosting its internal computational effort. A subsequent collaboration between FRL and the supplier developed to demonstrate that a variety of computational approaches yielded the same ground state for another class of photostabilizer—a confirmation that refuted earlier claims (DeBellis & Hass 1999).

NO$_x$ SCR Catalysis

Probably the greatest challenge facing automobile manufacturers today are the simultaneous demands of increased fuel economy and decreased exhaust emissions. Today's remarkably low emissions levels have been achieved through the use of highly efficient "three-way" catalysts (so-called because of their ability to simultaneously catalyze removal of CO, NO$_x$, and hydrocarbon from engine exhaust) coupled with electronically controlled internal combustion engines. The mechanism of NO$_x$ reduction to N$_2$ on these supported noble metal catalysts is not well understood, and early computational efforts at Ford provided insight into this process (Tsai & Hass 1995). To achieve higher fuel economy, engine manufacturers are now considering lean-burn gasoline and diesel engines, which offer the promise of 10 – 15% reductions in fuel consumption. Unfortunately, three-way catalysts are ineffective in removing NO$_x$ from the O$_2$-rich exhausts of these engines.

One promising strategy for addressing this "lean NO$_x$" problem is selective catalytic reduction (SCR). Cu-exchanged ZSM-5 zeolites have amongst the highest known activities for NO$_x$ SCR, but have insufficient durability for most automotive applications. In collaboration with Arizona State University, DFT simulations have been used to probe NO$_x$ chemistry in these materials, with the expectation that improved understanding will lead to the development of more active and more durable catalysts. An area of fundamental uncertainty is the coordination environment and redox chemistry of the atomically dispersed, exchanged Cu cations. Cluster DFT calculations have been used explore the zeolite coordination chemistry of Cu$^+$ and Cu^{2+} ions (Hass & Schneider 1999; Schneider et al. 1996) and of the intermediate oxidation states CuO$^+$ and CuO$_2^+$ (Schneider et al. 1998b). A single, hydrogen-terminated AlO$_4$ tetrahedral site was found to provide a reasonable representation of the local coordination environment of CuO$_x^+$ (x = 0, 1, 2), and using this model, a molecularly detailed mechanism for conversion of NO to N$_2$ on a single Cu active site could be obtained (Fig. 2) (Schneider et al. 1997; Schneider, Hass, Ramprasad, & Adams 1998b). An important feature of this mechanism is that the spectroscopically identified adsorbates widely speculated to be reactive intermediates, including Cu-bound mono- and dinitrosyls, are found to be merely spectators (Ramprasad et al. 1997). Rather, the key precursor to N–N bond formation is a metastable isonitrosyl adsorbate, which combines with low activation barrier with a second NO to form N_2O and, by subsequent O atom transfer, N_2 (Schneider, Hass, Ramprasad, & Adams 1997). More recent work has explored the role of Cu ion pairs bridged by oxygen in NO chemistry (Goodman et al. 1998; Goodman et al. 1999), and has provided convincing evidence that at realistic Si:Al ratios and metal ion loadings, such pairs will exist and will likely participate in NO_x chemistry (Goodman et al. 2000).

Efforts to understand and improve NO_x SCR catalysts continue at FRL and elsewhere. While simulation has not led directly to improved catalysts, it has provided detailed molecular insights and understanding that allow the search to proceed from a much firmer chemical basis. These models also provide the groundwork for extensions to more complicated and potentially more effective catalytic systems, such as amine-promoted NO_x SCR or plasma-assisted NO_x SCR.

It would have been difficult or impossible to sustain this largely fundamental research project without significant contributions from academic collaborators. Seed money from Ford provided Arizona State collaborators support to initiate a research program in NO_x catalysis; the group was later able to obtain external funding to continue the project. Two graduate students from this group undertook various aspects of the NO_x problem as thesis projects. These students spent summers as interns at FRL, performing research and gaining exposure to the industrial work environment, and the rest of the year at their university. Such leveraging and personal interactions can greatly enhance the effectiveness of industrial/academic collaborations.

Alumina Surface Chemistry

Aluminas are of considerable technological significance, for instance as supports and active components in catalysts and as interfaces in adhesively bonded aluminum. Despite this importance, little is known about the surface chemistry of the various alumina phases. One of the most important questions to consider is the surface constituency in the presence of environmental contaminants, such as water or organics. As a first step towards addressing these questions, DFT calculations and Car-Parrinello molecular dynamics (CPMD) simulations were used to study the chemistry of water on a α-alumina surface (Hass et al. 1998; Hass et al. 2000). This work was performed as a collaboration between researchers at FRL and IBM-Zurich, the former contributing expertise in oxide surface chemistry and the latter in the CPMD methodology and parallel computing. Calculations explored the reactivity of the α-alumina surface at low and high (single monolayer) coverages of water. Individual water molecules were found to be highly reactive on this surface, readily dissociating to form surface hydroxyls. This reactivity persists at the higher coverage, and is in fact enhanced through cooperative interactions between multiple water molecules. These results allowed a rationalization of the experimentally observed coverage-dependent adsorption energy of water on alumina, and led to the conclusion that the most stable surface in the presence of water is fully hydroxylated. Finally, a novel structure for the fully hydroxylated surface was identified.

These first principles calculations were complemented by classical molecular dynamics (MD) simulations performed collaboratively with Wayne State University (Bolton et al. 1999; de Sainte Claire et al. 1997). The MD simulations probed the adsorption of alkane films on an α-alumina surface and perturbations of this adsorption by water. Alkanes were found to wet and order on the H-free alumina surface, with the degree of ordering and the surface dynamics sensitive to the empirical potentials and to the assumed surface corrugation (Bolton, Bosio, Hase, Schneider, & Hass 1999). Water droplets were found to very strongly physisorb and readily displace alkanes from the alumina surface (de Sainte Claire, Hass, Schneider, & Hase 1997). In agreement with the DFT calculations, water was found to easily contaminate alumina surfaces. These results form a basis for understanding the competition between adhesives and adventitious adsorbates in adhesive bonding at oxide surfaces.

Emerging Areas of Opportunity

The above-described projects involve primarily chemical applications of molecular simulation. An area of increasing opportunity is application of atomistic and mesoscale simulation to materials problems. Recognition of these opportunities has driven the growth of the simulation efforts at FRL. Several new areas of potential impact have been identified and are being actively pursued.

One of these is tribology—the study of friction, lubrication, and wear. An automobile contains many types of sliding interfaces. Some interfaces are lubricated to minimize friction, as in a valve train, while others are designed to provide some controlled amount of friction and slippage, as in a clutch plate or a brake rotor. Simulation can be used to probe the microscopic factors (e.g., interface type, lubricant size, shape, energetics, etc.) that affect sliding friction, with the goal of guiding the design of interfaces and lubricants with improved frictional characteristics. A first application in this area has been to "friction modifier" additives found in automatic transmission fluid. This additive adsorbs on clutch surfaces and reduces both static and low-speed dynamic friction, leading to a qualitative change in the low-speed dynamic friction coefficient: increases in friction with velocity are observed, in contrast to the decreases found in base oils. Simulation results to date (Greenfield & Ohtani 1999) confirmed that sharp increases in normal forces between fixed, lubricated surfaces at small separations result from additive chains being adsorbed as a film perpendicular to the surface. Such increases had been observed in surface forces apparatus (SFA) experiments on a similar system (Ruths et al. 1999). In ongoing work, the velocity dependence of the friction coefficient is being studied through both simulations (MD) and experiment (SFA).

Another promising area is the prediction of microstructure for the optimization of lightweight Al alloys. In order to optimize alloy design and processing conditions, to achieve Al-alloy castings with the necessary mechanical properties, researchers at FRL are developing the "Virtual Aluminum Castings" methodology: a suite of predictive computational tools that span length scales from atomistic to macroscopic to describe alloy microstructure, precipitation, solidification, and ultimately mechanical properties. First-principles atomistic computations have recently been added to the set of methodologies in this project. Until recently, microstructural issues in alloys, such as precipitation, were outside the realm of first-principles atomistic calculations. However, the development of a new method based on a cluster expansion of the energy enables the prediction of coherent phase equilibria and precipitate shapes in Al-alloys with system sizes up to 250,000 atoms (Wolverton 1999). These types of calculations can then be combined with larger-length-scale microstructural models, and, in conjunction with experimental efforts, can be used to suggest heat treatments that optimize thermal stability and hardness of industrial alloys.

A third area of opportunity is the study of oxygen transport in oxides. Metal oxides are used as oxygen storage components in automotive catalyst formulations and as oxygen transport components in high temperature gas sensors and solid oxide fuel cells (SOFCs). Current interest in fuel cells in particular is very high, and the SOFC is a promising complement to the popular but expensive polymer-membrane-based fuel cells. The key to the SOFC is the electrically insulating oxide that conducts oxygen between the anode and cathode. Doped zirconias are the prototypical oxygen transport materials: doping the Zr(IV) oxide with lower oxidation state materials, like Y(III) oxide, introduces vacancies into the oxide lattice that make oxygen conduction possible. The dependence of oxygen conduction on doping and the actual mechanism of conduction, however, are not well understood. To address these questions, DFT calculations and kinetic Monte Carlo simulations (Bogicevic et. al, 1998) are being developed to model the microstructure of the doped materials and its effects on oxygen transport.

A fourth topic currently being investigated via simulations relates to the structure and phase stability of γ-alumina and other transition aluminas formed upon dehydration of γ-AlOOH, or boehmite. Despite their importance for a variety of applications, far less is understood about these transition phases of alumina than is known about the stable phase, α-alumina, where most theoretical work to date has been directed. DFT calculations have been used to assess the role of hydrogen (and hence, water) on the energetics of the transformation sequence from boehmite to the transition phases. Additionally, calculations are being pursued to elucidate the effects of hydrogen on cation and anion mobilities in transition aluminas. It is anticipated that such calculations will provide valuable insight into the thermochemistry, structure, and porosity of everyday catalyst support materials, and perhaps even help understand the mechanisms of corrosion and pitting of aluminum.

Finally, on the boundary between chemical and materials simulation is recent work on NO_x trapping catalysts for emissions control. This alternative to NO_x SCR operates by adsorbing NO_x on a metal oxide during lean operation and releasing and reducing the NO_x to N_2 during periodic rich engine events. The traps are typically composed of alkaline earth oxide adsorbents and noble metal catalyst particles dispersed onto an alumina support. Unfortunately, these formulations are even more effective at scavenging SO_x than NO_x, so that even low levels of fuel sulfur quickly poison them. DFT calculations are being used to study the mechanism of NO_x and SO_x adsorption and oxidation on the alkaline earths, with the goal of designing formulations with high activity for NO_x adsorption and higher sulfur tolerance.

Future Prospects at Ford

Clearly, many opportunities exist to apply chemical and materials simulation at Ford or any automotive company. In the first several years of the *CAMS* effort, emphasis was placed on developing a scientifically sound and externally visible program. These efforts have been highly successful, contributing in numerous ways to key FRL projects and receiving considerable recognition both inside and outside the company. With the ever-increasing competitiveness of the global automotive industry come ever-greater expectations for the contributions of simulation: to impact projects quickly and meaningfully, to provide the company a competitive edge, and simultaneously to maintain a high level of scientific excellence. To satisfy these expectations, simulation will become more tightly integrated into, and involved earlier on in, experimental research programs. The *CAMS* effort will move beyond a supporting role to be seen as a full partner with experiment. It will become increasingly driven by problems rather than methodologies: practitioners will need to be familiar with and have readily available a wide range of simulation tools and resources. Much of this flexibility will continue to be sought through partnerships and leveraging with external academic, industrial, and government researchers. Finally, the *CAMS* teams will continue to be responsive to and educate internal customers and management to ensure that FRL realizes the benefits from the increasing power of simulation.

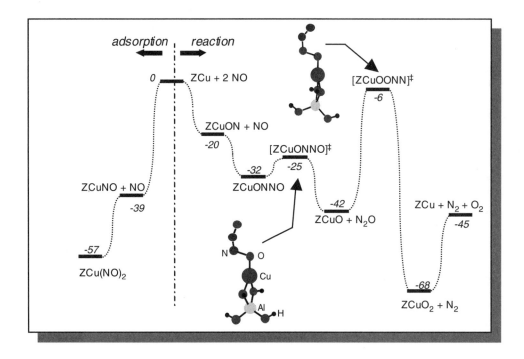

Figure 2. Calculated pathways for NO adsorption and NO reaction to N_2 and O_2. ZCu represents a zeolite-bound Cu cation, with the zeolite here modeled using a single, H-terminated AlO_4 tetrahedral-site. Relative energies, in kcal mol^{-1}, are from gradient-corrected DFT calculations.

References

Biggs, P., Canosa-Mas, C. E., Shallcross, D. E., Wayne, R. P., Kelly, C., and H. W. Sidebottom (1993). *Proceedings of the Step-Halocide/AFEAS Workshop*. Dublin, Ireland.

Bogicevic, A., Liu, S., Jacobsen, J., Lundqvist, B., and H. Metiu (1998). Island migration caused by the motion of the atoms at the border: Size and temperature dependence of the diffusion coefficient. *Phys. Rev. B*, 57, R9459.

Bolton, K., Bosio, S. B. M., Hase, W. L., Schneider, W. F., and K. C. Hass (1999). Comparison of explicit and united atom models for alkane chains physisorbed on α-Al_2O_3 (0001). *J. Phys. Chem. B*, **103**, 3885-3895.

de Sainte Claire, P., Hass, K. C., Schneider, W. F., and W. L. Hase (1997). Simulations of hydrocarbon adsorption and subsequent water penetration on an aluminum oxide surface. *J. Chem. Phys.*, **106**, 7331-7342.

DeBellis, A. D. and K. C. Hass (1999). Conformational study of a N-acylated hindered amine light stabilizer. *J. Phys. Chem. A*, **103**, 7665-7671.

Estevez, C. M., Bach, R. D., Hass, K. C., and W. F. Schneider (1997). Novel structural modifications associated with the highly efficient internal conversion of 2(2'-hydroxyphenyl)benzotriazole ultraviolet stabilizers. *J. Am. Chem. Soc.*, **119**, 5445-5446.

Gerlock, J. L., Tang, W., Dearth, M. A., and T. J. Korniski (1995). Reaction of benzotriazole ultraviolet light absorbers with free radicals. *Polym. Degrad. Stab.*, **48**, 121-130.

Goodman, B. R., Hass, K. C., Schneider, W. F., and J. B. Adams (1999). Cluster model studies of oxygen-bridged Cu pairs in Cu-ZSM-5 catalysts. *J. Phys. Chem. B*, **103**, 10452-10460.

Goodman, B. R., Hass, K. C., Schneider, W. F., and J. B. Adams (2000). Statistical analysis of Al distributions and metal ion pairing probabilities in zeolites. *Catal. Lett.*, submitted for publication.

Goodman, B. R., Schneider, W. F., Hass, K. C., and J. B. Adams (1998). Theoretical analysis of oxygen-bridged Cu pairs in Cu-exchanged zeolites. *Catal. Lett.*, **56**, 183-188.

Greenfield, M. L. and H. Ohtani (1999). Molecular dynamics simulation study of model friction modifier additives confined between two surfaces. *Tribo. Lett.*, **7**, 137-145.

Hass, K. C. and W. F. Schneider (1999). Density functional studies of adsorbates in Cu-exchanged zeolites: model comparison and SO_x binding. *Phys. Chem. Chem. Phys.*, **1**, 639-648.

Hass, K. C., Schneider, W. F., Curioni, A., and W. Andreoni (1998). The chemistry of water on alumina surfaces: reaction dynamics from first principles. *Science*, **282**, 265-268.

Hass, K. C., Schneider, W. F., Curioni, A., and W. Andreoni (2000). First-principles molecular dynamics simulations of H_2O on α-Al_2O_3(0001). *J. Phys. Chem. B*, **104**, in press.

Hass, K. C., Schneider, W. F., Estevez, C. M., and R. D. Bach (1996). Density functional theory description of excited-state intramolecular proton transfer. *Chem. Phys. Lett.*, **263**, 414-422.

Ramprasad, R., Hass, K. C., Schneider, W. F., and J. B. Adams (1997). Cu-dinitrosyl species in zeolites: a density functional molecular cluster study. *J. Phys. Chem. B*, **101**, 6903-6913.

Ruths, M., Ohtani, H., Greenfield, M. L., and S. Granick (1999). Exploring the 'friction modifier' phenomenon:

nanorheology of alkane chains with polar terminus dissolved in n-alkane solvent. *Tribo. Lett.*, **6**, 207-214.

Schneider, W. F., Hass, K. C., Ramprasad, R., and J. B. Adams (1996). Cluster models of Cu binding and CO and NO adsorption in Cu-exchanged zeolites. *J. Phys. Chem.*, **100**, 6032-6046.

Schneider, W. F., Hass, K. C., Ramprasad, R., and J. B. Adams (1997). First-principles analysis of elementary steps in the catalytic decomposition of NO by Cu-exchanged zeolites. *J. Phys. Chem. B*, **101**, 4353-4357.

Schneider, W. F., Hass, K. C., Ramprasad, R., and J. B. Adams (1998b). Density functional theory study of transformations of nitrogen oxides catalyzed by Cu-exchanged zeolites. *J. Phys. Chem. B*, **102**, 3692-3705.

Schneider, W. F. and T. J. Wallington (1993). Ab initio investigation of the heats of formation of several trifluoromethyl compounds. *J. Phys. Chem.*, **97**, 12783-12788.

Schneider, W. F. and T. J. Wallington (1994). Thermochemistry of COF_2 and related compounds, *J. Phys. Chem.*, 98, 7448-7451.

Schneider, W. F., Wallington, T. J., Barker, J. R., and E. A. Stahlberg (1998a). CF_3CFHO radical: decompostion vs. reaction with O_2. *Ber. Bunsensges. Phys. Chem.*, **102**, 1850-1856.

Schneider, W. F., Wallington, T. J., and R. E. Huie (1996). Energetics and mechanism of decomposition of CF_3OH. *J. Phys. Chem.*, **100**, 6097-6103.

Schneider, W. F., Wallington, T. J., Minschwaner, K., and E. A. Stahlberg (1995). Atmospheric chemistry of CF_3OH: Is photolysis important? *Env. Sci. Tech.*, **28**, 247-250.

Tsai, M.-H. and K. C. Hass (1995). First-principles studies of NO chemisorption on rhodium, palladium, and platinum surfaces. *Phys. Rev. B*, **51**, 14616.

Wallington, T. J. and W. F. Schneider (1994). The stratospheric fate of CF_3OH. *Env. Sci. Tech.*, **28**, 320A-326A.

Weinstock, B. (1969). Carbon monoxide: residence times in the atmosphere. *Science*, **166**, 224-225.

Weinstock, B. and H. Niki (1972). Carbon monoxide balance in nature. *Science*, **176**, 290-292.

Wolverton, C. (1999). First-principles prediction of equilibrium precipitate shapes in Al-Cu alloys. *Phil. Mag. Lett.*, **79**, 683.

MOLECULAR SIMULATION: SOME RECENT APPLICATIONS TO PHASE AND CHEMICAL EQUILIBRIA

Keith E. Gubbins
Chemical Engineering Dept.
North Carolina State University
Raleigh, NC 27695-7905

Abstract

Recent advances in direct and indirect methods of molecular simulation for studying fluid phase and chemical equilibria are reviewed. For bulk fluids and mixtures, the emphasis is on equilibria for fluids of nonspherical molecules, including ionic fluids, aqueous mixtures, hydrocarbons and chain molecules. The application of these methods to confined systems is also discussed, with emphasis on phase separation and reaction equilibria in porous media.

Keywords

Molecular simulation, Phase equilibria, Chemical equlibria, Adsorption, Porous materials.

Introduction

The use of Molecular Dynamics (MD) or Monte Carlo (MC) simulation methods to calculate the so-called "mechanical" properties, such as internal energy or pressure, is straightforward. The situation is considerably more difficult for the "statistical" properties - free energy, chemical potential or entropy. The difficulty is that conventional methods sample parts of phase space where the Boltzmann factor $\exp(-U/kT)$ is large (here U is configurational energy), whereas for the statistical properties other regions of phase space make major contributions [1]. Various ways around this problem have been devised, including special sampling techniques and integration over a range of thermodynamic states. Reviews are given in refs. [1-6] and references therein. Recently, methods for simulating fluid phase equilibria directly have been developed, and are fast and convenient where they can be applied [5].

An important application of such simulations is to the prediction of phase and chemical equilibria in cases where experimental measurements are difficult or impossible; examples include bulk systems at extreme temperatures or pressures, or fluids confined within porous media.

Simulations of this sort also find other applications, e.g.: (a) testing statistical mechanical theories, where identical models for the molecules (and any surfaces present) are used in both simulation and theory, so that the comparison tests only the statistical mechanical approximations in the theory; (b) comparisons with experiment, which give information about the suitability of the assumed intermolecular potential.

The main limitation of the simulations is usually the reliability of the intermolecular potentials used. Since computers are still too slow to calculate reliable *ab initio* potentials for any but the simplest molecules, most workers use semi-empirical potentials, developed by using a combination of knowledge from theory (quantum mechanics, electrostatics) and experimental data. Such models [7] include universal force fields, transferable isotropic site-site models (e.g. the OPLS model, or optimized potentials for liquid simulations), and more sophisticated models [8] involving anisotropic site-site potentials and distributed multipoles.

In the remainder of this paper a brief survey is given of some developments in this area over the last ten years.

Both direct and indirect (via calculation of the chemical potential or free energy) methods for calculating phase equilibria are considered, with an emphasis on applications to fluids of complex molecules and on confined systems.

Bulk Fluids

Direct Method for Phase Equilibria: Gibbs Ensemble Monte Carlo

The Gibbs Ensemble Monte Carlo (GEMC) method, first proposed by Panagiotopoulos in 1987 [9], involves setting up boxes representing the two coexisting phases (I and II) that are in equilibrium with each other; however, these phases are not in physical contact. The two boxes representing the phases have volumes V_I and V_{II} and contain N_I and N_{II} molecules, respectively. The system of two boxes is at a uniform temperature T, and the usual periodic boundaries are used with each box to minimize surface effects. The simulation requires three kinds of trial moves [9], which are designed to achieve (a) internal thermal equilibrium in each box through the usual Monte Carlo molecular moves, (b) mechanical equilibrium through equality of pressures between the two phases, by changing the volumes of the two phases (keeping the total volume of the two phases constant), and (c) chemical equilibrium via equality of chemical potential between phases, through molecule exchange between the two boxes. The derivation of the GEMC method is given in detail by Smit et al. [10], and the implementation of the method, programming considerations and applications have been recently reviewed [5,6]. The merit of the method is its directness and resulting speed of computation; in particular it is not necessary to calculate the chemical potentials, and the molecules do not have to diffuse across a physical interface in order for the system to reach equilibrium. The method is particularly attractive for mixture phase equilibria, where the indirect methods become tedious. The main limitation is the difficulty in making the molecule transfer step at high densities; this problem becomes more pronounced if the molecules are highly nonspherical in their interactions, and is acute for polymers. This limitation is shared by many of the other methods for calculating phase equilibria described below. The difficulty can be largely overcome by various biased sampling methods for liquids, but so far the method has not been successfully applied to solids or liquid crystals.

Applications of the method to pure fluids and mixtures up to 1995 have been reviewed [5,6]. Studies of vapor-liquid equilibria in pure fluids of more complex molecules have included the restricted primitive model of a 1:1 electrolyte (charged hard spheres of equal diameter and unit charge) [11], and chain molecules [12,13]. The usual sampling methods fail for chain molecules, and it is necessary to bias the sampling in a way that 'looks' for available space (see section below).

For mixtures, the method has been applied to high pressure fluid phase equilibria in hydrogen-helium mixtures [14]. Further calculations for this system have been made to 2500 K and 700 kbar [15]. Agreement with the existing experimental data is good. These studies provide a further example of the use of these methods to extrapolate existing data into regions that are difficult to reach experimentally. An application to a more complex mixture has been made for water with methanol [16]. The SPC (simple point charge) model was used for water, and OPLS (optimized potential for liquid state) for methanol with potential parameters taken from the literature and used without adjustment. The compositions of the two phases agree with experimental data within a few percent.

Direct Method for Chemical Equilibria: Reactive Monte Carlo

The Reactive Monte Carlo (RMC) method can be used to determine the equilibrium state of a reactive mixture. This method was developed by Johnson et al. [17] in the early 1990's. Essentially the same method was independently proposed by Smith and Triska [18], and termed the Reaction Ensemble (RE) method. An attractive feature of this method is that it is not necessary to calculate chemical potentials; the RMC approach is designed to minimize the Gibbs energy at constant pressure, and so determines the true equilibrium condition, irrespective of any rate limitations. The constant pressure version of the RMC method involves the following trial moves [17]:

(a) a change in the position or orientation of a molecule chosen at random;
(b) a random change in the volume;
(c) a forward reaction step, which involves choosing at random reactant molecules and changing them to product molecules;
(d) a reverse reaction step

The forward and reverse reaction moves are accepted with an appropriate probability function. The probability functions do not involve the chemical potentials of the components. These steps do require the calculation of molecular partition functions for the species involved. An important feature of this method is that it can be applied to reactions in which the number of moles changes, in contrast to other methods for simulating reaction equilibria. This method has been successfully applied to equilibria for the dimerization reaction $2NO=(NO)_2$ [17], and to the ammonia synthesis reaction [19]. Recently, Lísal et al. [20] have applied this method to the study of reaction and phase equilibria in the isobutene/methanol/methyl-tert-butyl ether (MTBE) system.

The Reactive Monte Carlo method and its applications has been the subject of a recent review by Johnson [21].

Indirect Methods

An alternative to the Gibbs ensemble MC and Raective MC methods is to calculate the free energy or chemical potentials for a range of state conditions. The phase transition and chemical equilibrium conditions can then be determined by the usual methods of Gibbs thermodynamics. Such methods are called indirect ones; they involve more computational effort than the direct methods, in general. However, they are useful because: (a) one often wants to know the values of the chemical potentials, e.g. in studying surface phase transitions or conformational changes (the direct methods do not yield the chemical potential unless special steps are taken in coding), and (b) the direct methods (and some of the indirect methods) may fail for high densities because of the molecule insertion/exchange step. This is particularly the case for liquid crystals and solids.

These indirect methods include the test particle method, Grand Canonical Monte Carlo (GCMC), modified sampling methods, and thermodynamic integration over states. They have been reviewed elsewhere [2-6]. Both the test particle and GCMC methods involve attempts to insert a molecule into the fluid, and so suffer from the same difficulty as the direct methods at high densities. In the modified sampling methods an attempt is made to overcome this problem by modifying the MC sampling procedure, so that the probability of successful insertion attempts is greatly increased. The aim of these methods is to try to find the 'holes' in the fluid and put the molecule there.

Of particular importance is the use of histogram reweighting methods. This method, due to Ferrenberg and Swendsen [22], can be applied in a variety of ensembles, but is most often used in the Grand Canonical Monte Carlo (GCMC) and Semi-Grand Canonical (SGMC) methods. In GCMC the chemical potential, volume and temperature (μ, V and T) are the independent variables that are fixed in the simulation. In this ensemble, the number of molecules, N, and the energy of the system, U, fluctuate. The essence of Ferrenberg and Swendsen's method is to construct a histogram of the distribution of N and U during the course of the simulation, and thus determine the distribution function $f(N,U)$ at the state condition (μ,V,T). From this distribution function it is possible to calculate the thermodynamic average of any property $X(N,U)$, including the grand partition function itself, and hence the grand free energy. Moreover, once $f(N,U)$ is known at a given state condition (μ,V,T), it is possible to calculate any property $X(N,U)$, including free energy, at all other state conditions (μ',V,T') [22,23]. Thus, in principle, one long GCMC run should be sufficient to obtain the complete phase diagram for the system. In practice it is usually necessary to carry out several runs at different (μ,V,T) values in order to obtain $f(N,U)$ with sufficient accuracy over the range of N and U needed. Weighting or biasing is frequently used in order to improve the statistics

in regions of phase space that are important in the desired averages. Histogram reweighting has been used to determine the vapor-liquid coexistence curve for polarizable Stockmayer fluids [23], polarizable water models [24,25], and carbon dioxide-water mixtures [24]. It can also be applied in other MC ensembles, including the Gibbs [24] and semi-grand ensemble [26].

The Gibbs-Duhem integration method developed by Kofke [27] is a valuable complement to the Gibbs ensemble method and has been shown to be useful for solid phases, including solid phases in binary mixtures of simple fluids [28]. More recently, pseudo-ensemble methods have been developed as hybrid simulation techniques in which the evolution of a conventional ensemble (such as GCMC, GEMC or NPT) is modified to satisfy preset conditions [29]. Pseudo-ensembles constitute an iterative scheme to search for the roots of the phase co-existence equations. The central theme involves an extrapolation of the ensemble parameters from state points near phase co-existence to determine the co-existing state conditions. The extrapolations are exact when performed in the framework of the histogram re-weighting method. An important advantage of the pseudo-ensemble method is that it can be used to perform flash, bubble point, and dew point calculations, unlike the GEMC methods that performs only flash calculations. Variations of the pseudo-ensemble formalism in conjunction with configurational bias methods [4] have been effectively used to study phase behavior of polymeric systems, including gels and networks; a review on this subject by Escebedo and de Pablo can be found in Ref. [30].

For the most difficult systems, e.g. liquid crystals and solids, thermodynamic integration may be necessary. This involves making a series of simulations for a range of thermodynamic temperatures, densities or intermolecular potentials, and then integrating over thermodynamic states using standard thermodynamic or statistical mechanical equations to obtain the chemical potential [4,6]. An alternative procedure is to calculate the Landau free energy as a function of some order parameter, Φ [31,32]. The probability $P(\Phi)$ of observing the system with an order parameter value between Φ and $\Phi+d\Phi$ is determined in the simulation, and assumes a bi-modal distribution when two phases are in coexistence. The Landau free energy is obtained from this probability distribution, and subsequent integration of this over the order parameter gives the grand free energy. The success of this method hinges on an appropriate choice of order parameter; the order parameter is a density variable or rotational invariant that takes on distinctly different values in different phases of the system. It has been used to study the fluid-solid phase transition for simple fluids and metals [33,34].

To obtain a complete phase diagram it is often necessary to employ several simulation techniques for different parts of the phase diagram. For the Gay-Berne model of a liquid crystal [35], for example, the vapor-liquid region was determined by the Gibbs ensemble MC method; isotropic liquid-nematic transitions were determined by thermodynamic integration; and the remaining transitions were found approximately from order parameters (orientational correlation parameter P_2, tilt angle ϕ and heat capacity C_v).

The extension of some of these methods to chain molecules in the last few years has sparked particular interest. Two approaches have been put forward, the *chain increment method* and the *configurational bias MC method*. The chain increment method, proposed by Kumar et al. [36], is based on a test particle equation that gives an exact expression for the incremental chemical potential for adding a monomer unit to a chain molecule. This increment is found to become essentially constant for chains longer than about 10-20 mers. If this limiting value is determined, and also the chemical potential for a relatively short chain, it is possible to calculate the chemical potential for long chains. The method has been applied [37] to obtain the phase diagram for a polymer melt of chains of length up to 100 mers. The chain increment method has the advantage that it can be applied to arbitrarily long chains.

The configurational bias MC method [38-40] involves the insertion of a short chain into the fluid, followed by the addition of other segments to the end of the chain until a chain of the desired length has been grown. The chain configuration is chosen by a suitable weighting process, using another modified test particle equation. The method can be used in Gibbs ensemble simulations to obtain phase equilibria directly. It has been used to study alkanes [39], and also alkanes dissolved in polyethylene and in supercritical solvents [38].

Confined Fluids and Solids

A fluid confined within a porous material can exhibit a variety of phase transitions. Such transitions are of two general types, those that also occur in the bulk phase but are modified in the pore by finite size effects and the strong fluid-wall forces (e.g. melting, condensation, liquid-liquid equilibria, etc.), and new transitions that arise solely from the fluid-wall forces (e.g. wetting, layering transitions, contact layer phases, hexatic phases). The methods used to study phase transitions in bulk fluids are, with little modification, suitable for investigation of phase transitions in pores. However, long-lived metastable states are often a more serious problem in confined systems. These lead to hysteresis effects in many cases, the state condition at which the transition occurs being different depending on the direction from which the transition is approached. The determination of the true thermodynamic transition point will often require the calculation of the free energy or chemical potential. Phase transitions that have been studied rather extensively by simulation are capillary condensation (condensation from vapor to liquid inside the pore; this usually occurs at a pressure much below the normal vapor pressure), wetting

(the point at which the liquid just wets the surface, i.e. the contact angle becomes zero), and layering transitions (transitions between low coverage and a full monolayer, between a full monolayer and two full layers, and so on). The melting transition, liquid-liquid separation for mixtures, and solubility of dilute components in a solvent have received less attention from the simulation community, but are the subject of current investigations.

Care is needed in defining exactly what is meant by a 'phase transition' in a pore. In cylindrical pores, for example, the molecular correlation length can grow to infinity along the axis of the cylinder, but is restricted to the pore diameter in the other two dimensions. Such a fluid does not have a true critical point in the usual sense. Nevertheless, in such a system there are often two states of distinctly different density and structure having the same free energy or chemical potential, and sharp changes occur between these two states.

We first briefly consider the most commonly used methods, followed by a brief discussion of the main types of transition. More extensive reviews of phase separation in porous materials are available [41,42].

Methods

Phase equilibria in pores, and also the equilibrium between the porous medium and the bulk phase, are governed by equality of temperature and the chemical potential of each component between the various phases. Pressure equality is not a requirement since pressure in confined systems is a tensor.

The GCMC method has been widely used to determine adsorption isotherms, heats of adsorption, and phase transitions in pores. The method (like the GEMC method) runs into difficulties with the molecule insertion step at high densities. This difficulty, which is more pronounced for significantly nonspherical molecules, can be overcome by using a biased sampling method [2-6], which attempts to insert the molecules into 'holes' in the fluid, thus improving the chance of successful insertion; the effects of such biasing on the statistical sampling is removed at a later stage. A further difficulty with GCMC (and some other methods) is that hysteresis is found in the calculation of adsorption isotherms where phase transitions occur. Thermodynamic integration can be used to determine the true transition point [43,40]. This method has been used by Peterson et al. [43,44] to determine capillary condensation lines.

Molecular dynamics (MD) simulations can be used to observe the interfaces in the pore between coexisting phases, and so have this advantage over the GCMC method. One procedure that has proved successful [43,45] involves first equilibrating the fluid in the pore at a high temperature, above the critical point in the case of gas-liquid or liquid-liquid transitions (or above the range in which the transition of interest occurs in other cases). The simulation is then stopped, the velocities of all the molecules scaled back by a constant factor (chosen to correspond to a suitably lower temperature), and then restarted. The new molecular velocities correspond to a temperature inside the spinodal region for the two phase region of interest. The fluid in the pore will then separate into the two phases. In general, mechanical equilibrium is quickly reached, whereas chemical equilibrium takes at least one order of magnitude longer, because of the slow diffusion in the dense phases. This quench MD method has the advantage of providing dynamic information about diffusion in the pore and the kinetics associated with the phase separation. Moreover it provides a clear picture of the interface itself, something that cannot be easily studied in the laboratory.

Other methods have proved valuable for special cases. For example, the semi-grand ensemble MC method is useful for the study of phase equilibria in dense fluid mixtures [26]; in this method there is no particle insertion step, but rather a species exchange, which is tolerated more easily at high densities. Thermodynamic integration methods for studying fluid-solid transitions usually fail for confined systems, because adsorbed layers near the wall of the pore freeze at a different point than the adsorbate in the pore interior. Order parameter methods are usually preferable for such studies [46].

Molecular Models of Porous Materials

Some porous materials are crystalline (e.g. aluminosilicates, aluminophosphates), so that a molecular model of the material can be immediately constructed on the computer from x-ray or neutron diffraction data. However, many others are disordered, and it is not possible to characterize the material fully from experimental measurements; examples are activated carbons, carbon and silica aerogels, silica xerogels, and metal oxides. In such cases one is faced with the problem of constructing a molecular model of the material before addressing the properties of an adsorbate phase confined within the pore structure. There are two general approaches to this problem. The first, which we term *mimetic simulation*, involves the development of a simulation strategy that mimics the synthetic process used to fabricate the material in the laboratory. The second, termed *reconstruction methods*, seeks to build a molecular model whose structure matches what experimental structure data is available.

Mimetic simulation methods have the advantages that they provide insight into the synthesis mechanism, and may thus suggest better ways to synthesize the material. Also they are less liable to bias on the part of the researcher than are reconstruction methods. An example of a mimetic simulation protocol is that used recently to prepare controlled pore glasses [47,48]. Such glasses are prepared by heating a mixture of oxides (SiO_2, B_2O_3, Na_2O) to a high temperature, about 1200°C, at which the mixture is a homogeneous mixture, and then quenching into the spinodal liquid-liquid region at about 700°C. Provided that the initial composition is not too far

Porous Silica   50%, 4.95 nm

Figure 1.   Model for controlled pore glass constructed using mimetic simulation [47,48].

from the critical mixing value, the mixture starts to separate into two phases, one of which (the B_2O_3-rich phase) becomes a connected phase of roughly cylindrical geometry. If left at this temperature the diameter of the cylindrical regions grows. When the desired diameter is reached, the mixture is quenched to room temperature to form a glass, the B_2O_3-rich phase is removed with an acid, and the resulting silica-rich structure is annealed. The porosity can be controlled over an appreciable range by controlling the initial composition. It turns out that the spinodal decomposition can be easily mimicked using quench MD, and the pore morphology and topology is insensitive to details of the intermolecular potentials [47]. Simple Lennard-Jones potentials, with the like pair 1-1 and 2-2 interactions being identical while the 1-2 interaction well depth is substantially weaker, are sufficient to produce pore structures that are remarkably similar to those produced experimentally. A binary mixture is equilibrated at a high, supercritical temperature using isothermal MD. It is then quenched into the spinodal region and phase separation commences. When the diameter of the cylindrical phase reaches the desired value the system is quenched to a very low temperature, the cylindrical phase is removed, and the system is annealed. The intermolecular potential parameters of the solid atoms are then adjusted to values suitable for silica. Glasses have been prepared by this method with diameters up to 7 nm, and with a range of porosities [47,48]. GCMC simulations of the adsorption isotherms for nitrogen and xenon in these glasses closely match those found experimentally [49]. An example of a glass prepared in this way is shown in Figure 1.

The main difficulty with mimetic simulation is that new simulation protocols must be developed for each new material. In some cases the synthesis process is so poorly understood or so complex that mimetic simulation is not practical at the present time. An example is the preparation of active carbons. Synthesis is by heating organic materials, such as wood, pitch, coal, coconut shell, in an inert atmosphere to a high temperature to carbonize it, followed by heating in an oxidizing atmosphere (e.g. H_2O, CO_2) to remove some carbon and form a high surface area material ('activation'). At the present time reconstruction methods must be used. At the simplest level this can involve construction of a simplified molecular model from visual observation of transmission electron micrograph (TEM) images of the material. This is the origin of the slit-pore model for activated carbons; this model consists of a collection of infinitely long slit-shaped pores of various pore widths.

Activated Carbon   50%, 1.0 nm

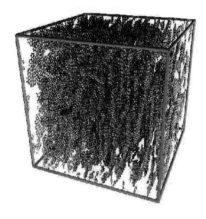

Figure 2. Model for activated carbon microbeads constructed from Reverse Monte Carlo simulation [52].

The model is simple to use, but omits effects due to edges of the carbon microcrystals, pore connectivity, and defects in the carbon microcrystals. A more realistic reconstruction method is based on the Reverse Monte Carlo (RMC) method [50,51]. In this method atoms, or microcrystals, are placed randomly in a box; the particles are then moved and reoriented in a Monte Carlo fashion. After each move the radial distribution function (or the structure factor) is calculated and compared with the experimental function. This is repeated after each MC move, the moves being accepted or rejected depending on whether the fit is improved or not. After a sufficient number of MC moves the simulation converges and a structure is obtained that matches, within some degree of error, the experimental result. The RMC method can accommodate a variety of refinements, for example trial changes in the density, size and shape of the microcrystals, introduction of defects, etc. Thomson and Gubbins [52] have used this method to build a structure for a microporous activated carbon. The initial structure was made up of graphene microcrystals having a range of sizes. In addition to moves and reorientations of the microcrystals, the microcrystals were allowed to change in size during the simulation. One resulting structure is

shown in Figure 2. The RMC method can also be used to match other experimental structure data on the material, for example TEM data.

Capillary Condensation and Layering Transitions

Since about 1986 there have been numerous simulation studies of capillary condensation in pores of simple geometry, e.g. slits and cylinders, and several studies of layering transitions. More recently, there have also been studies of capillary condensation in more complex and realistic pore structures [e.g. 47.48,53]. These are too numerous to review here, but a cross-section of examples of studies of this type can be found in the proceedings of the International Conferences on the Fundamentals of Adsorption, held every few years [54]. Most of these simulations have employed the GCMC method, but several use either Gibbs Ensemble MC or Quench MD.

Studies of the gas-liquid coexistence behavior (temperature vs. density) in narrow pores show that the coexistence curve is considerably narrower than for bulk fluids, and the critical point is lowered. The narrowing of the coexistence region occurs because the 'gas' phase in the pore usually consists of several layers of adsorbed fluid molecules on the pore walls, with a gas-like phase in the interior of the pore; since the adsorbed layers have a liquid-like density, the overall density in the pore is relatively high. The liquid side of the coexistence curve has a density similar (usually somewhat higher) to that of the bulk liquid. The lowering of the critical temperature in the pore is expected since the average coordination number, Z, of fluid molecules in the pore will be lower than that of the bulk liquid (because an appreciable fraction of molecules are near the wall). Mean field theory predicts that the critical temperature is proportional to Z.

Liquid-Liquid Equilibria

Simulation studies of liquid-liquid separation in pores have been reported for a simple Lennard-Jones mixture in which the unlike pair interaction is weak [26,55-57]. This is the simplest model for such a phase separation. Recent simulations have been made using semi-grand ensemble MC with histogram reweighting to obtain the liquid-liquid coexistence curves [26]. The critical mixing temperature is lowered as for the vapor-liquid case, and the coexistence curve (temperature vs. composition) is shifted towards the side of the component that more strongly wets the pore walls. These qualitative features agree with experiment. In addition to these equilibrium studies, the kinetics of phase separation has been studied using quench MD simulations [57], and the power laws governing the growth of the phases has been determined for simple geometries.

Melting/Freezing Transitions

Freezing and melting transitions in pores are of importance in frost heaving, in the distribution of pollutants in soils, in the manufacture of nanomaterials, and in several other manufacturing processes. Simulation studies, using GCMC and MD, have been made to study the melting and freezing of Lennard-Jones methane in both slit [46,58,59] and cylindrical [60] pores. Large hysteresis loops are found on heating or cooling, and it is necessary to calculate the free energies in order to determine the true melting point. Since adsorbed layers near the wall melt at different temperatures to adsorbed material in the inner parts of the pore, the usual thermodynamic integration methods cannot be used. Instead, order parameter methods have proved successful [46]. The freezing transition is found to be first order. For weakly adsorbing walls, such as silicas and oxides, the melting temperature in the pore is lower than that in the bulk. For carbons, however, the melting temperature is generally higher than the bulk value, and this effect is particularly marked for slit pores, since the molecules can more readily form the necessary lattice. Freezing in pores has been reviewed recently by Gelb et al. [42].

The stable phase in the pore, and also the phase transition temperatures, are best determined by the Landau free energy method [46]. An example is shown in Figure 3, where the Landau free energy is shown as a function of the order parameter Φ at several temperatures for LJ methane in slit carbon pores. The freezing temperature in the pore is higher than the bulk freezing temperature of 98 K.

Chemical Equilibria in Pores

The effect of confinement on reaction equilibria is complex. In general, for gas adsorption the density of the fluid is much higher in the pore than in the equilibrium bulk fluid, so that for reactions in which there is a change in the number of moles during reaction the increase in density will strongly affect the yield. For example, if the number of moles decreases during reaction, this effect would increase the yield. However, the outcome is also strongly affected by selective adsorption of the components, and this can counteract or enhance the density effect. Turner et al. [19] have recently used the Reactive Monte Carlo method to study the reactions $2NO=(NO)_2$ and $N_2+3H_2=2NH_3$ in a simple model (slit pores) of an activated carbon. In the case of the NO dimerization reaction, the two species have a similar affinity for the carbon wall, and the increased density in the pore is the dominant effect, leading to a large increase in the yield of the dimer. The influence of temperature on the yield is also greatly increased in the pore. For the ammonia synthesis reaction the dominant effect is selective adsorption. Although the increased density in the pore favors increased yield, it is overwhelmed by the strong selectivity of the carbon surface for nitrogen.

Hydrogen is only very weakly adsorbed, so that the yield decreases as a result of confinement.

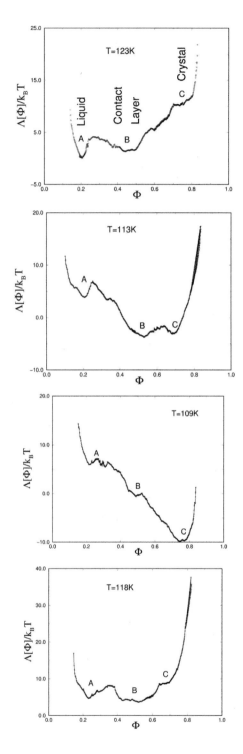

Figure 3. Landau free energy surfaces of LJ methane in a slit shaped graphite pore at four different temperatures showing three different phases: Phase A-liquid, Phase B- contact layer phase (contact layer frozen), Phase C-crystalline.

Conclusions

For fluid phase transitions the Gibbs ensemble MC and histogram reweighting methods are particularly useful. For transitions involving solid phases the order parameter method for obtaining free energies is likely to emerge as an important technique. Current research is likely to focus on applying and extending these methods to difficult systems such as ionic fluids, associating and reacting liquids, liquid crystals, polymers and surfactants. Much remains to be done to investigate chemical equilibria. In the case of confined systems, applications to more realistic and complex pore geometries should be emphasized. A persistent problem in such work remains the determination of sufficiently accurate intermolecular potentials for prediction of phase equilibria in real systems.

Acknowledgments

It is a pleasure to thank Ravi Radhakrishnan for helpful discussions. I am grateful to the Division of Chemical Sciences, U.S. Department of Energy (grant no. DE-FG02-98ER14847) for support of this work.

References

1. Allen, M. P. and D. J. Tildesley (1987). *Computer Simulation of Liquids*. Clarendon Press, Oxford, 49.

2. Mezei, M. and D. L. Beveridge (1987). *Ann. New York Acad. Sc.*, **482**, 1.

3. Gubbins, K. E. (1989). *Mol. Simulation*, **2**, 223.

4. Frenkel, D. and B. Smit (1996). *Understanding Molecular Simulation*. Academic Press, San Diego.

5. Panagiotopoulos, A. Z. (1995). In *Observation, Prediction and Simulation of Phase Transitions in Complex Fluids*, ed. M. Baus et al., Kluwer Academic, 463.

6. Gubbins, K. E. (1994). in *Models for Thermodynamic and Phase Equilibrium Calculations*, ed. S.I. Sandler. Dekker, New York, 507-600.

7. Gubbins, K. E. and M. Quirke (1996). in *Molecular Simulation and Industrial Applications*, ed. K.E. Gubbins and N. Quirke. Gordon and Breach, Amsterdam, 1-70.

8. Price, S. L. (1991). in *Computer Simulation in Materials Science*, eds. M. Meyer and V. Pontikis. Kluwer Academic, Dordrecht.

9. Panagiotopoulos, A. Z. (1987). *Mol. Phys.*, **61**, 813. A more detailed derivation is given in A. Z. Panagiotopoulos, N. Quirke, M. Stapleton, and D. J. Tildesley (1988); *Mol. Phys.*, **63**, 527.

10. Smit, B., de Smedt, Ph., and D. Frenkel (1989). *Mol. Phys.*, **68**, 931.

11. Panagiotopoulos, A. Z. (1992). *Fluid Phase Equilibria*, **76**, 97; Orkoulas, G. and A. Z. Panagiotopoulos (1994). *J. Chem. Phys.*, **101**, 1452; Calliol, J. M. (1994). *J. Chem. Phys.*, **100**, 2161.

12. Mooij, G. C. A. M., Frenkel, D., and B. Smit (1992). *J. Phys. Cond. Matter*, **4**, L255; Siepmann, J. I., Karaborni, S., and Smit, B. (1993). *Nature*, **365**, 330.

13. Laso, M., de Pablo, J. J., and U. W. Suter (1992). *J. Chem. Phys.*, **97**, 2817.

14. De Kuijper, A., Smit, B., Schouten, J. P., and J. P. J. Michels (1990). *Europhysics Lett.*, **13**, 679.

15. Schouten, J. A., de Kuijper, A., and J. P. J. Michels (1991). *Phys. Rev. B: Cond. Matter*, **44**, 6630.

16. Strauch, H. J. and P. T. Cummings (1993). *Fluid Phase Equilibria*, **83**, 213.

17. Johnson, J. K., Panagiotopoulos, A. Z., and K. E. Gubbins (1994). *Mol. Phys.* **81**, 717-733.

18. Smith, W. R. and B. Triska (1994). *J. Chem. Phys.* **100** 3019-3027.

19. Turner, C. H., Gubbins, K. E., and J. K. Johnson (2000). *J. Chem. Phys.*, submitted.

20. Lísal, M., Smith, W. R., and I. Nezbeda (2000). *AIChE Journal*, **46**, 866-875.

21. Johnson, J. K. (1999). *Adv. Chem. Phys.*, **105**, 461-481.

22. Ferrenberg, A. M. and R. H. Swendsen (1988). *Phys. Rev. Lett.*, **61**, 2635.

23. Kiyohara, K., Gubbins, K. E., and A. Z. Panagiotopoulos (1997). *J. Chem. Phys.*, **106**, 3338-3347.

24. Errington, J. R., Kiyohara, K., Gubbins, K. E., and A. Z. Panagiotopoulos (1998). *Internat. J. Thermophys.*, in press.

25. Kiyohara, K., Gubbins, K. E., and A. Z. Panagiotopoulos (1998). *Molec. Phys.*, **94**, 803-808.

26. Gelb, L. D. and K. E. Gubbins (1997). *Physica*, **244**, 112-123.

27. Kofke, D. (1993). *J. Chem. Phys.*, **98**, 4149-4162.

28. Hitchcock, M. R. and C. K. Hall (1999). *J. Chem. Phys.*, **110**, 11433-11444.

29. Escobedo, F. A. (1998). *J. Chem. Phys.*, **108**, 8761-8772.

30. Escobedo, F. A. and J. J. de Pablo (1999). *Phys. Rep.*, **318**, 85-112.

31. Landau, L. D. and E. M. Lifshitz (1980). *Statistical Physics*, 3rd edn. Pergamon Press, London.

32. Chaikin, P. M. and T. C. Lubensky (1995). *Principles of Condensed Matter Physics*. Cambridge University Press, Cambridge.

33. Van Duijneveldt, J. S. and D. Frenkel (1993). *J. Chem. Phys.*, **96**, 4655.

34. Lynden-Bell, R. M., van Duijneveldt, J. S., and D. Frenkel (1993). *Mol. Phys.*, **80**, 801.

35. De Miguel, E., Rull, L. F., Chalam, M. K., and K. E. Gubbins (1991). *Mol. Phys.*, **74**, 405.

36. Kumar, S. K., Szleifer, I., and A. Z. Panagiotopoulos (1991). *Phys. Rev. Lett.*, **66**, 2935; Kumar, S. K. (1992). *J. Chem. Phys.*, **96**, 1490; Szleifer, I. and A. Z. Panagiotopoulos (1992). *J. Chem. Phys.*, **97**, 6666.

37. Panagiotopoulos, A.Z. and I. Szleifer (1992). *Polymer Preprints*, **33**, 547; Sheng, Y.-J., Panagiotopoulos, A.Z., Kumar, S., and I. Szleifer (1994). *Macromolecules*, **27**, 400.

38. De Pablo, J. J., Laso, M., and U. W. Suter (1992). *J. Chem. Phys.*, **96**, 2395, 6157; de Pablo, J. J., Laso, M., Suter, U. W., and H. D. Cochran (1993). *Fluid Phase Equil.*, **83**, 323.

39. Siepmann, J. I. and D. Frenkel (1992). *Mol. Phys.*, **75**, 59; Frenkel, D. and B. Smit (1992). *Mol. Phys.*, **75**, 983.

39. Frenkel, D., Mooij, G. C. A. M. and B. Smit (1992). *J. Phys. Cond. Matter*, **4**, 3053.

40. Gubbins, K. E., Sliwinska-Bartkowiak, M., and S.-H. Suh (1996). *Molec. Simulation*, **17**, 333.

41. Gelb, L. D., Gubbins, K. E., Radhakrishnan, R., and M. Sliwinska-Bartkowiak (1999). *Rep. Progr. Phys.*, **62**, 1573.

42. Peterson, B. K., Gubbins, K. E., Heffelfinger, G. S., Marini Bettolo Marconi, U., and F. van Swol (1988). *J. Chem. Phys.*, **88**, 6487.

43. Peterson, B. K. and K. E. Gubbins (1987). *Mol. Phys.*, **62**, 215.

44. Heffelfinger, G. S., van Swol, F., and K. E. Gubbins (1987). *Mol. Phys.*, **61**, 1381.

45. Radhakrishnan, R. and K. E. Gubbins (1999). *Mol. Phys.*, **96**, 1249.

46. Gelb, L. D. and K. E. Gubbins (1998). *Langmuir*, **14**, 2097-2111.

47. Gelb, L. D. and K. E. Gubbins (1999). *Langmuir*, **15**, 305.

48. Gelb, L. D. and K. E. Gubbins (2000). paper in preparation.

49. McGreevy, R. L. and L. Putszai (1988). *Mol. Simulation*, **1**, 359.

50. Da Silva, F. L. B., Svensson, B., Akesson, T., and B. Jonsson (1998). *J. Chem. Phys.*, **109**, 2624.

51. Thomson, K. T. and K. E. Gubbins (2000). *Langmuir*, July.

52. Page, K. S. and P. A. Monson (1996). *Phys. Rev. E*, **54**, 6557.

53. See *Fundamentals of Adsorption* 6, F. Meunier, ed. Elsevier, Paris (1998).

54. Gózdz, W. T., Gubbins, K. E., and A. Z. Panagiotopoulos (1995). *Mol. Phys.*, **84**, 825.

55. Kierlik, E., Fan, Y., Monson, P. A., and M. L. Rosinberg (1995). *J. Chem. Phys.*, **102**, 3712.

56. Gelb, L. D. and K. E. Gubbins (1997). *Phys. Rev. E*, **56**, 3185-3196.

57. M. Miyahara and K. E. Gubbins (1997). *J. Chem. Phys.*, **106**, 2865-2880.

58. H. Dominguez, Allen, M. P., and R. Evans (1998). *Mol. Phys.*, **96**, 209.

59. M. W. Maddox and K. E. Gubbins (1997). *J. Chem. Phys.*, **107**, 9659-9667.

BRIDGING LENGTH SCALES IN SIMULATIONS OF VAPOR PHASE DEPOSITION PROCESSES

Klavs F. Jensen, Uwe Hansen, Seth T. Rodgers, and Rajesh Venkataramani
Department of Chemical Engineering
Massachusetts Institute of Technology
Cambridge, MA 02139

Abstract

The need to understand how vapor phase deposition processing conditions influence the performance of electronic, optical, and mechanical devices drives the coupling of molecular level simulations with traditional macroscopic transport phenomena descriptions. The development of predictive, efficient models that bridge across multiple length and time scales raises new challenges in terms of simulation strategies, numerical algorithms, and experimental validation. These issues are exemplified through studies linking quantum chemistry, molecular dynamics (MD), kinetic Monte Carlo (kMC), and macroscopic finite element simulations. Application examples are drawn from chemical and physical vapor deposition of aluminum as well organometallic chemical vapor deposition of gallium arsenide. Specifically, experimental observations and quantum chemistry predictions of elementary surface reactions are incorporated into MD and MC simulations to provide new understanding of microstructure evolution. These computations are subsequently integrated into self-consistent feature and reactor scale models. Comparisons with experimental data are given at each length scale for the complete system.

Keywords

Multiscale, Physical vapor deposition (PVD), Chemical vapor deposition (CVD), Finite element methods, Quantum chemistry, Molecular dynamics, Kinetic Monte Carlo.

Introduction

Thin film growth by physical and chemical vapor deposition (PVD and CVD) is a critical unit operation in fabricating mechanical, optical, and electronic devices. For example, layers of silicon, insulators, and metals are deposited by CVD and PVD and patterned by lithography and etching to form microelectronic circuits (Sze, 1983). Multiple layers of compound semiconductors (e.g., AlGaAs) with carefully controlled band structure and doping levels are synthesized by organometallic CVD or molecular beam epitaxy (MBE) to yield lasers and detectors (Stringfellow, 1986). Similarly, deposition, patterning, and etching steps are used to realize microstructure composites, known as micro-electromechanical systems (MEMS) for pressure sensors, accelerometers, and microfluidic applications (Wise, 1998).

Vapor phase deposition processes involve multiple length scales and different physical models may apply at these different scales (see Fig. 1). The processes are used to define active regions on the micron (mechanical and electronic devices) to nanometer scale (optoelectronic devices) on substrates, which are held in meter-sized deposition chambers. Moreover, the underlying physical and chemical processes occur at multiple time scales. Individual diffusion processes and chemical reactions controlling the film growth and defect formation occur on typical atomic motion time scales of 10^{-13}s. In contrast, the growth of the active layer takes $\sim 10^2$s and the total processing time for a multilayer structure could be $\sim 10^4$s.

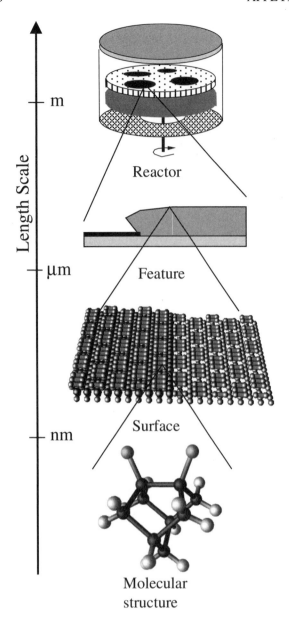

Figure 1. Schematic of different length scales in vapor phase deposition processes.

Decreasing sizes of active regions, combined with increasing functionality, integration, and materials choices, drive the need for process models based on detailed physical understanding. Prediction of performance of deposited structure and the ultimate device requires understanding of how process conditions (e.g., flow, feed concentration, and temperature) influence thin film synthesis on the atomic level (e.g., defect concentration and interface abruptness). However, most modeling techniques have focused on particular length and time scales; for example, computational fluid dynamics (CFD) models of the gas transport in the deposition chamber (Jensen, 1994, Kleijn, 1995), evolution of film morphology on the micron scale (Cale et al., 2000), and Monte Carlo simulations of atomic level surface diffusion and incorporation processes (Adams et

al., 2000, Gilmer et al., 2000). These models provide useful insights into the underlying physical phenomena, but primarily at the length and time scales for which they apply. Therefore, multiscale linking approaches are needed to combine cost-effective, physically accurate models of growth phenomena on different length and times. Such models have been developed for linking reactor and feature scale simulations (Merchant et al., 2000), but have generally not included chemistry and physical transport at the molecular level. Here we illustrate linking procedures spanning from atomic simulations, through surface evolution to reactor scale computational fluid dynamics.

Linking Macro to Micro Scale in PVD

The issue of coupling macroscopic system simulations with surface evolution on the micron scale is present in both PVD and CVD. For PVD systems, reactor scale simulations can be performed to determine the intensity and angular distribution of flux arriving at a substrate from a sputtered target (Kersch et al., 1995). These results subsequently become input to film morphology simulations. This sequential approach is generally applicable to PVD since very few atoms return to the reactor chamber after impacting the substrate surface, but it cannot be used in CVD where material is constantly exchanged between substrate and deposition chamber.

Using molecular dynamics (MD) data in growth modeling is clearly desirable to simplified continuum models based on empirical 'sticking coefficients'. In principle, the reaction rates obtained from MD or ab-initio calculations can be employed in a Monte Carlo (MC) model where the motion of each individual atom is traced during the course of the simulation (Gilmer et al., 2000). This approach provides the most detailed understanding of the surface growth processes, but it is be too computationally intensive to yield predictions of film topographies at length scales of microns in a reasonable time frame. For example, in ionized PVD of metals (Al and Cu), growth proceeds under simultaneous rare-gas bombardment and the number of deposited atoms and re-sputtered atoms are of similar magnitude. In such a case the overall growth rate is very low, but MC simulations would require tracking of all atom trajectories.

A new and computationally highly efficient scheme to model the growth of thin films is presented in the following (Hansen et al., 2000). The approach incorporates MD data (*i.e.*, angular and energy dependent surface reactions rates) within the level-set method (Sethian, 1999). Furthermore, effects of reemission are included so that the model provides a speed advantage compared to MC methods, by retaining accuracy of an atomic approach and resolving problems with low growth rates. The flux from the source of each species, and its angular and energy distribution are specified as inputs to our model. An equipment scale model could determine, in principle, these fluxes. After leaving the source, the impinging particles can experience three possible surface reactions, namely adsorption, sputtering and rejection. When the highly energized particle hits the surface it loses part of its kinetic energy and, depending on its impact angle and the surface orientation, it is adsorbed, rejected or sputters away other surface atoms. If the particle is rejected, it can subsequently hit another piece of the surface and be adsorbed or again rejected, and so forth. In such a scenario the total flux impinging at a certain point of the surface does not only stem from the source, there will be an additional contribution to the total flux due to re-emissions from the surface.

In order to capture these two different flux contributions, the growth rate is computed with an iterative approach:

1. Molecular dynamics calculations are performed to analyze surface interactions for each important energetic species in the system.
2. The flux arriving directly from the source is calculated for all points along the surface. The molecular dynamics results are employed to determine the angular distribution and amount of reemitted flux, (rejections and sputtering).
3. For each point at the surface the reemitted fluxes from the other surface points are taken into account. In addition, the distribution of the flux that is again reemitted is calculated.
4. Procedures 2 and 3 are repeated as long as the reemitted flux is less than a small fraction of the initial flux, such that further iterations will not change the velocity of the growth front.

The linking approach makes it possible to utilize MD data (e.g., reaction and sputtering probabilities as illustrated in Fig. 2) in topology simulations of Al deposition (Fig. 3). The same approach could be extended to Cu deposition by using the fundamental MD data developed by (Kress et al., 1999).

Linking Reactor to Feature Scale for CVD

In CVD, gas-phase reactions and transport processes determine the concentration of precursor species available for film formation and impurity incorporation. Gas-phase and surface reactions of even common CVD precursors are often poorly characterized since chemical kinetic experiments are expensive, in general, and difficult to perform. Moreover, reaction intermediates typically have short life times and low concentrations, which makes them difficult to detect by standard analytical methods. The lack of experimental data makes it desirable to predict reactions from first principles quantum chemistry computations combined with transition state theory. These techniques have been

effective in studying chemical reactivity and molecular properties of gas-phase reactions relevant to CVD (Allendorf et al., 1999, Lengyel et al., 2000, Melius et al., 2000, Simka et al., 1997, Willis, 1999, Willis et al., 2000).

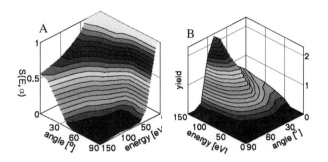

Figure 2. A. Calculated sticking probabilities for hyper-thermal Al atoms impinging on an Al(111) surface as a function of kinetic energy and incident off-normal angle. B. Calculated sputter yield for hyper-thermal Al atoms impinging on an Al(111) surface as a function of kinetic energy and incident off-normal angle.

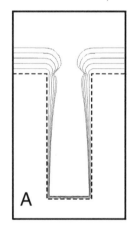
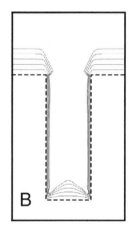

Figure 3. Simulated topographies for an aspect ratio 3 trench, width 0.2m. In panel (A) an Al film under PVD conditions, panel (B) shows results for ionized PVD with 80% ionization fraction.

CVD processes are typically implemented at conditions where flow and transport are described by macroscopic conservation equations for momentum, mass, and energy (Jensen, 1994). The resulting coupled nonlinear partial differential equations can then be solved with numerical techniques such as the finite volume or finite element method (FEM). However, the assumptions forming the basis for continuum models break down as the characteristic length scale becomes less than the gas phase mean free path (Cale et al., 2000, Rodgers et al., 1998).

The coupling between macroscopic transport to the growth interface and microscopic surface growth processes implies that a consistent set of boundary conditions between macroscopic and microscopic models must be derived. However, multiscale resolution is difficult since in many CVD operations, deposition occurs over integrated circuit (IC) device layouts that are heterogeneous patchworks on length scales ranging from centimeters to microns. For example, areas where device features are dense may border areas with no features, or masked regions may abut active deposition areas. Each region may present a somewhat different reactive surface area to film precursors, giving rise to concentration gradients at the edges of pattern fields. The resulting pattern-dependent non-uniformity is usually referred to as microloading. In these situations, the macroscopic and microscopic problems are linked and their solutions are mutually dependent. Capturing this interaction with models is difficult because microscale heterogeneity affects the macroscopic solution through the boundary conditions imposed at reactive surfaces, and the microscale problem is strongly dependent on macroscopic system conditions.

Size dependent surface heterogeneity, such as wafer, die, and feature scale effects in the fabrication of microelectronic circuits, could be represented, in principle, by an asymptotic analysis approach in which the model for each length scale is incorporated into increasing coarse grid descriptions (Merchant et al., 2000). However, such an approach is only applicable when the same general continuum equations apply over the whole range of length scales, which is rarely the situation. Moreover, it may be difficult to include microscale transport resistance, such as shadowing, that restricts access of film precursors to the growth interface.

In the case that transport within each feature is in the free molecular regime, the feature scale problem can be effectively integrated with the macroscopic reactor scale computation since there is no direct communication between features. All interactions occur through the macroscopic reactor environment. The effect of increased substrate surface area and microscale transport resistance inside features can then be explored by formulating a flux balance over a single feature (see Fig. 4).

The inbound flux is known from a macroscopic trial solution. The outbound flux is the microscopic system's response represented by the sum of all flux leaving the surface at different points, multiplied by the chance of its arrival at a specific point in the macroscopic domain. Thus, given molecular interactions at the substrate surface, fluxes at any control surface that encloses the substrate may be calculated employing transmission probabilities. The flux to this 'control surface' now becomes the boundary condition on the macroscopic calculation. While they can be evaluated employing Monte Carlo methods (Rodgers et al., 1998), a deterministic approach based on Markov chains has at least a thousand-fold advantage in speed over optimized Monte Carlo methods (Rodgers, 2000). The reaction-transport model can be coupled with a level set model (Sethian, 1999) for profile advancement allowing feature scale morphology to be tracked in time, as done for the above PVD case.

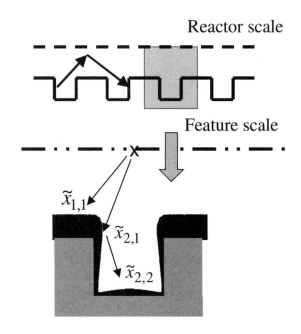

Figure 4. Schematic illustration of the coupling of feature and reactor scale simulations when free molecular flow governs transport in the features.

Once flux maps have been computed at the feature scale, contributions from each feature type present can be superimposed. Identical features at different locations are treated at no cost by simply 'shifting' the map. Thus, feature scale results may be combined to rapidly determine the total flux to a large layout. Iteration between the feature scale and reactor scale models is used to ensure a consistent set of boundary conditions on both the macroscopic and microscopic level. Fig. 5 shows the results of such a multiscale simulation for Al CVD from dimethyl aluminum hydride (DMAH). The depletion of precursor, DMAH, is largest over the region with small features, i.e., highest surface area (Curves 1 and 2). The small features close first, at which point the DMHA consumption becomes the same as for a planar substrate (Curve 3). Ultimately, the behavior is that of a nonpatterned substrate (Curve 4), for which most DMAH is consumed at the wafer edge because of higher mass transfer in that portion of the reactor. The kinetic parameters for this case are derived from combined quantum chemistry and surface spectroscopy investigations (Willis, 1999).

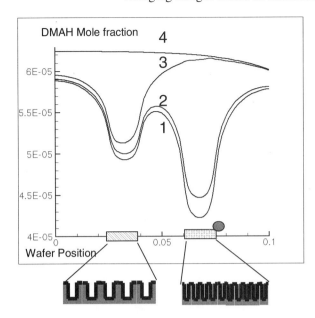

Figure 5. Example of combined linked reactor and feature scale calculations for Al deposition from dimethyl aluminum hydride (DMAH) (Rodgers, 2000). The numbers, 1-4, refer to increasing times during the deposition process.

Studies by Kisker et al. (Fuoss et al., 1992, Fuoss et al., 1995, Kisker et al., 1992, Kisker et al., 1995b) serve as a basis for the present example since these investigations are unique in providing both macroscopic and microscopic experimental data for an OMCVD process. Grazing incidence x-ray scattering (GIXS) is used in situ to monitor growth rate, transition between island-growth and step-flow growth, and the diffuse scattering from the surface. During step-flow growth (see Fig. 6a), the majority of the adatoms incorporate at a terrace edge. Island-growth (Fig. 6b) occurs when islands nucleate and coalesce on terraces. Step-flow mode is preferred since the films are smoother and have fewer defects than those grown under island-mode conditions.

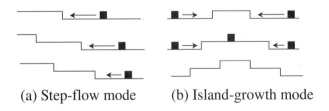

(a) Step-flow mode   (b) Island-growth mode

Figure 6. Schematic of (a) step-flow growth and (b) island (layer-by-layer) growth modes.

The growth mode is determined through a complex function of temperature, flux of adatoms to the surface, and initial miscut of the crystal. The diffusion length of adatoms increases with temperature. The flux of species to the surface determines the total surface concentration of adatoms, which impacts the probability of island nucleation. The initial miscut of the crystal determines the average initial terrace length. If the diffusion length of an adatom is larger than the terrace length and there are fewer adatoms than needed for nucleation, adatoms will incorporate at kink sites on terrace edges and step-flow growth will occur. If adatoms cannot reach terrace edges before nucleating or incorporating on islands in the middle of a terrace, then the island-growth mode results. Moreover, one typically aims for as low temperature deposition as possible to ensure materials stability and avoid impurity diffusion in the deposited layers. Thus, it is important to be able to predict the transition between island and step-flow modes as a function of macroscopic operating conditions. This is a typical multiscale simulation problem. Using each macro- and micro-scale models separately would not produce predictions of both growth rate and morphology. The reactor scale model cannot predict the morphology of the film, and the surface model needs as an input, the flux of gas phase species from the reactor to the surface.

FEM methods are used to solve the set of partial differential equations underlying the reacting flow problem in the reactor (Fotiadis et al., 1990). Momentum and energy balances are solved first to obtain flow and temperature fields in the reactor. These are employed subsequently with a chemical mechanism to predict species concentrations throughout the reactor. This decomposition of the reacting flow problem is a valid approach for OMCVD systems, where the concentration of the reacting species is usually dilute compared to the concentration of the carrier gas. The gas-phase chemical mechanism for the precursors (tertiarybutyl arsine [TBAs] and triethylgallium [TEG]) was simplified from published reaction mechanisms (Ingle et al., 1996, Mountziaris et al., 1993).

A kinetic Monte Carlo (kMC) approach (Kang et al., 1994) is used to simulate the evolution of the surface morphology during growth. A modified solid-on-solid (SOS) model represents the surface based upon the zincblende crystal structure of GaAs. The methodology is similar to that used for simulating molecular beam epitaxy (MBE) of GaAs (Clarke et al., 1987, 1988, Heyn et al., 1997) and OMCVD growth (Kasu et al., 1997). Use of the kMC method enables simulation of time and lengths scales not feasible with molecular dynamics approaches, but the approach is limited by the underlying set of rules and parameters. Lattice based kMC calculations provide a natural pathway for incorporating QC and MD results into process simulations. In principle, these techniques can provide energy barriers and paths for individual adsorption, diffusion, reaction, and desorption steps. However, the multitude of possible reaction steps involved in OMCVD means that such a comprehensive approach is not computationally feasible. Therefore, the transition probabilities are based on published experimental observations and density functional theory computations.

Two different surface models are considered (Venkataramani, 2000). **Model I** includes adsorption and diffusion of Ga species on the surface but ignores As dynamics. Arsenic is assumed to incorporate immediately when an As site becomes available. This assumption is typical for OMCVD grown GaAs, since As is 10-40 times in excess of the Ga precursors during growth. **Model II** explicitly keeps track of As dimers on the surface and involves As surface dynamics. The As_2 dimers on the surface block diffusion and access to growing islands. Including nearest neighbor barriers for As_2 desorption causes a modified c(4×4) reconstruction to appear on the surface (see Fig. 7) with sufficient As flux to the surface, as seen experimentally. This reconstruction on the surface allows for roughly 50% of the surface to always be available for Ga precursors to adsorb, and causes a zero dependence on the As flux, consistent with experiments.

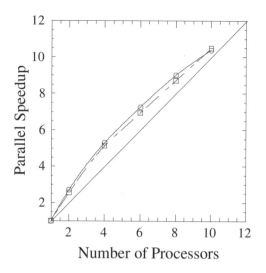

Figure 8. Parallel speedup (Time on N processors /Time on 1 processor).

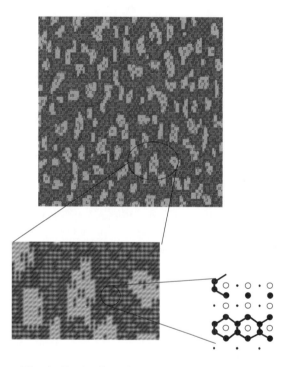

*Figure 7. Surface during growth as predicted with kMC (**Model II**). Note, the reduction of dimer density around the islands on the surface, as well as the modified c(4×4) reconstruction on the surface.*

A parallel kMC code was developed in order to simulate micron-sized surfaces (Venkataramani, 2000). Serial implementation of kMC scales poorly with surface size because of the many possible surface processes. The algorithm was made parallel by using ideas developed from the field of Parallel Discrete Event Simulations (Fujimoto, 1990). The parallel code achieved super-linear speedup, as shown in Fig. 8 when a large enough surface was simulated on each processor.

Reactor scale FEM and the surface kMC models are linked through the flux of species to the surface. The MC model uses the flux given by the FEM as input, and returns the computed flux. The FEM uses the flux calculated by the MC model as a boundary condition at the substrate. The solutions of both problems are iterated until a consistent flux to the surface is determined. The kMC model is embedded in the Newton iteration of the reactor scale FEM model, which leads to a strong linking between the models through the surface flux boundary conditions, and aids in convergence of the reactor scale model.

The developed multiscale simulation scheme predicts observed As concentrations in the gas phase above the substrate (Fig. 9), as well as reported growth rates (Fig. 10). Predictions of these macroscopic quantities are insensitive to the choice of surface model, i.e., not dependent upon As surface dynamics, as would be expected from experimental observations. The large excess of As to Ga precursors means that the experimental growth rate does not vary with As gas phase concentration. However, as discussed in the following sections, surface **Model I** is not able to predict the observed diffuse x-ray scattering, which is a direct refection of molecular surface processes.

The reported in situ x-ray scattering data by Kisker et al. (Fuoss et al., 1992, Fuoss et al., 1995, Kisker et al., 1992, Kisker et al., 1995a) provide a unique opportunity for evaluating the model. The investigators report scattering at the crystal truncation rod (CTR) and diffuse scattering. The first measurement is sensitive to the top bilayers of the growing crystal, and the scattering intensity oscillates with island-growth (layer-by-layer growth)•the intensity decreases with the nucleation of each new layer of islands and increases as the islands coalesce into a new layer. The oscillations disappear as the growth mode transitions to step-flow mode for which there is a characteristic surface length, the terrace length.

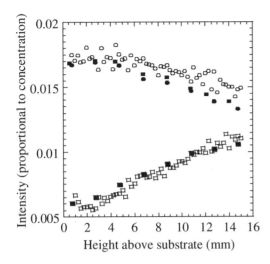

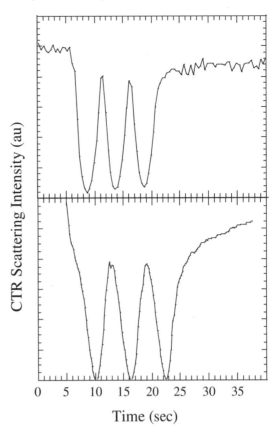

Figure 9. Comparisons of experimental and simulated As as a function of distance above the substrate. (○) Experimental at 50°C, (□) Experimental at 510°C, (Fuoss et al., 1995, Kisker et al., 1995a), (●) (■) Simulations.

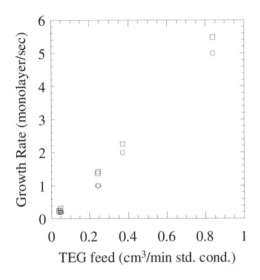

Figure 10. Comparison of experimental (o) and simulated (♣) growth rates.

Figure 11. Comparison of features of experimental (top) and simulated (bottom) CTR scattering oscillations.

Thus, the CTR oscillations provide a direct measurement of growth rates in island-growth, and a means for detecting the transition to step-flow mode. The diffuse scattering reflects the correlation length between islands, and by recording the diffuse scattering in two perpendicular directions, island-shaped anisotropy can be estimated. Since the kMC is based on the GaAs lattice, the CTR and diffuse scattering can be computed directly from the simulated lattice positions (Robinson et al., 1992). Figures 11 and 12 compare predicted CTR and diffuse scattering to the corresponding experimental diffuse scattering data.

The predicted CTR and diffuse scattering in Figs. 11 and 12, respectively, mirrors the features observed experimentally•oscillations in the CTR that correspond to layer-by-layer growth and oscillations in the diffuse scattering that are out of phase with the CTR. The initial signal for the CTR is different in the experimental and simulated surface, as the experimental surface has an As flux to the surface at all times. This creates some disorder on the surface that lowers the initial signal. The simulated surface begins with a clean surface (as shown by the high signal) and starts both As and Ga fluxes at the same time. During growth, the scattering shows the same features.

A direct comparison between model predictions and experiments for a range of conditions shows that surface **Model I** cannot predict the transition from island-growth to step-flow mode unless an unphysical high activation barrier for Ga diffusion (~2.3 - 2.7 eV) is employed. Similar observations have been reached in experimental OMCVD studies (Kisker et al., 1995a), while only from 1.3-1.58 eV for MBE surfaces (Shitara et al., 1992). However, these aggregate numbers are 'lumped' in that they are a function of many variables (miscut, temperature, growth rate). A better comparison is one between island sizes (correlation length) on the surface as a function of temperature and growth rate.

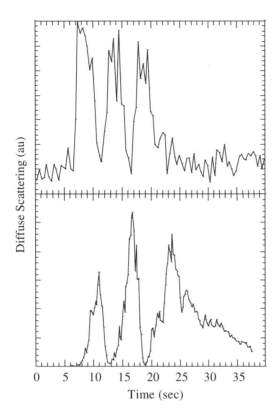

Figure 12. Comparison of features of experimental (top) and simulated (bottom) diffuse scattering.

Even though a range of parameter values is used, none of the predictions of **Model I** fit the diffuse scattering data, which suggests that additional physical phenomena are needed to describe the surface evolution•specifically the role of As dynamics as incorporated into **Model II**. Predictions with this model are able to represent the variation island correlation length with growth temperature and growth rate (Venkataramani, 2000). In this model, the temperature dependence is a convolution of the barriers for ethyl desorption, Ga diffusion, and As_2 desorption. If the transitions were independent, the effective barrier would be the sum of the individual barriers. The effective barrier is slightly less than the sum, indicating correlation between the processes. This can be interpreted in the following manner. In the early stages of growth all three barriers must be overcome in order for islands to form. Later stages of growth proceed with some percentage of the Ga precursors adsorbing near islands, bypassing the diffusion barrier and the As_2 desorption barrier. Additional surfaces are likely to be needed to form a complete understanding of OMCVD of GaAs. Nevertheless, this example clearly demonstrates the need for multiscale simulations to provide consistent models of OMCVD growth data. Moreover, the simulations provide the opportunity to build different surface models and test them against experimental data, analogous to the iterative process used to develop mechanisms for gas-phase reactions in the continuum regime.

Conclusions

The preceding examples demonstrate that coupling of molecular level simulations with traditional macroscopic transport phenomena descriptions allows the development of models for vapor deposition processes that bridge across multiple length and time scales. These models provide physical insight that could not have been obtained by individual models at particular length and time scales. For example, molecular level predictions of surface evolution in GaAs require both kMC simulations with molecular processes incorporation and macroscopic reactor calculations. The reactor simulations provide the flux of species to the surface, while molecular level surface kMC simulations define the boundary conditions for the reactor computations. The linking procedure currently requires considerable understanding of chemical and physical phenomena at all length and time scales in order to produce efficient, predictive models. The large amount of research into multiscale simulations by many groups will ultimately lead to scaling and linking procedures that should make the procedure applicable to a wide range of thin film processing. A parallel experimental effort is equally important. Without data on the same length scale as the predictions, it is rarely possible to validate the models. For example, with only growth rate data, **Model I** in the GaAs example was sufficient, but it could not predict diffuse scattering from nanometer scale surface topology. Thus, close collaboration between experimentalists, process engineers, and molecular scientists will be critical to the success of future multiscale simulations.

Acknowledgements

The work was supported by the National Science Foundation, Mitsubishi Chemical Corporation and the Semiconductor Research Corporation.

References

Adams, J. B., Z. Y. Wang, and Y. H. Li (2000). Modeling Cu thin film growth. *Thin Solid Films*, **365**, 201-210.

Allendorf, M. D., C. F. Melius, and C. W. Bauschlicher (1999). Heats of formation and bond energies in group III compounds. *Journal De Physique Iv*, **9**, 23-31.

Cale, T. S., T. P. Merchant, L. J. Borucki, and A. H. Labun (2000). Topography simulation for the virtual wafer fab. *Thin Solid Films*, **365**, 152-175.

Clarke, S. and D. D. Vvedensky (1987). Origin of reflection high energy electron diffraction intensity oscillations during molecular beam epitaxy: A computational modeling approach. *Phys. Rev. Lett.*, **58**, 2235-2238.

Clarke, S. and D. D. Vvedensky (1988). Growth kinetics and step density in reflection high-energy electron diffraction during molecular-beam epitaxy. *J. Appl. Phys.*, **63**, 2272.

Fotiadis, D. I., S. Kieda, and K. F. Jensen (1990). Transport phenomena in vertical reactors for metalorganic vapor phase epitaxy: I. Effects of heat transfer characteristics, reactor geometry, and operating conditions. *J. Crystal Growth*, **102**, 441.

Fujimoto, R. M. (1990). Parallel discrete event simulation. *Comm. ACM*, **33**, 30-53.

Fuoss, P. H., D. W. Kisker, F. J. Lamelas, G. B. Stephenson, P. Imperatori, and S. Brennan (1992). Time-resolved X-ray scattering studies of layer-by-layer epitaxial growth. *Phys. Rev. Lett.*, **69**, 2791-2794.

Fuoss, P. H., D. W. Kisker, G. B. Stephenson, and S. Brennan (1995). In-situ X-ray studies of organometallic vapor phase epitaxy growth. *Mat. Sci. and Eng.*, **B30**, 99-108.

Gilmer, G. H., H. C. Huang, T. D. de la Rubia, J. Dalla Torre, and F. Baumann (2000). Lattice Monte Carlo models of thin film deposition. *Thin Solid Films*, **365**, 189-200.

Hansen, U., S. Rodgers, and K. F. Jensen (2000). Modeling of metal thin film growth: Linking angstrom-scale molecular dynamics results to micron-scale film topographies. *Phys. Rev. B*, **62**, 2869-2878.

Heyn, C. and M. Harsdorff (1997). Simulation of GaAs growth and surface recovery with respect to gallium and arsenic surface kinetics. *Phys. Rev. B*, **55**, 7034.

Ingle, N. K., C. Theodropoulos, T. J. Mountziaris, R. M. Wexler, and F. T. J. Smith (1996). Reaction kinetics and transport phenomena underlying the low-pressure metalorganic chemical vapor deposition of GaAs. *J. Crystal Growth*, **167**, 543.

Jensen, K. F. (1994). Transport phenomena in epitaxy systems. In Hurle, D. (Ed.), *Handbook of Crystal Growth*, Elsevier, Amsterdam, 543-599.

Kang, H. C. and W. H. Weinberg (1994). Kinetic modeling of surface rate processes. *Surf. Sci.*, **299/300**, 755-768.

Kasu, M. and N. Kobayashi (1997). Surface kinetics of metalorganic vapor-phase epitaxy: Surface diffusion, nucleus formation, sticking at steps. *J. Cryst. Growth*, **174**, 513.

Kersch, A. and W. J. Morokoff (1995). *Transport Simulation in Microelectronics*. Birkhauser Verlag

Kisker, D. W., G. B. Stephenson, P. H. Fuoss, F. J. Lamelas, S. Brennan, and P. Imperatori (1992). Atomic scale characterization of organometallic vapor phase epitaxial growth using in-situ grazing incidence X-ray scattering. *J. Crystal Growth*, **124**, 1-9.

Kisker, D. W., G. B. Stephenson, P. H. Fuoss, and S. Brennan (1995a). Characterization of vapor phase growth using X-ray techniques. *J. Crystal Growth*, **146**, 104-111.

Kisker, D. W., G. B. Stephenson, I. Kamiya, P. H. Fuoss, D. E. Aspnes, L. Mantese, and S. Brennan (1995b). Investigation of the relationship between reflectance difference spectroscopy and surface structure using grazing incidence X-ray scattering. *Phys. Stat. Sol. (a)*, **152**, 9-21.

Kleijn, C. R. (1995). Chemical vapor deposition processes. In Meyyappan, M. (Ed.), *Computational Modeling in Semiconductor Processing*. Artech House, Norwood, MA, 97-229

Kress, J. D., D. E. Hanson, A. F. Voter, C. L. Liu, X. Y. Liu, and D. G. Coronell (1999). Molecular dynamics simulation of Cu and Ar ion sputtering of Cu (111) surfaces. *J. Vac. Sci. Tech.*, 2819-2825.

Lengyel, I. and K. F. Jensen (2000). A chemical mechanism for in situ boron doping during silicon chemical vapor deposition. *Thin Solid Films*, **365**, 231-241.

Melius, C. F. and M. D. Allendorf (2000). Bond additivity corrections for quantum chemistry methods. *J. Phys. Chem. A*, **104**, 2168-2177.

Merchant, T. P., M. K. Gobbert, T. S. Cale, and L. J. Borucki (2000). Multiple scale integrated modeling of deposition processes. *Thin Solid Films*, **365**, 368-375.

Mountziaris, T. J., S. Kalyanasundaran, and N. K. Ingle (1993). A reaction-transport model of GaAs growth by metalorganic chemical vapor deposition using trimethyl-gallium and tertiary-butyl-arsine. *J. Crystal Growth*, **131**, 283-299.

Robinson, I. K. and D. J. Tweet (1992). Surface X-ray Diffraction. *Rep. Prog. Phys.*, **55**, 599.

Rodgers, S. T. and K. F. Jensen (1998). Multiscale Modeling of CVD Processes. *J. Appl. Phys.*, **83**, 524-530.

Rodgers, S. T. (2000). *Multiscale Modeling of Chemical Vapor Deposition and Plasma Etching*. Massachusetts Institute of Technology

Sethian, J. A. (1999). *Level Set Methods and Fast Marching Methods*. Cambridge University Press

Shitara, T., D. D. Vvedensky, M. R. Wilby, J. Zhang, J. H. Neave, and B. A. Joyce (1992). Step-density variations and reflection high-energy-diffraction intensity oscillations during epitaxial growth on vicinal GaAs(001). *Phys. Rev. B*, **46**, 6815-6824.

Simka, H., B. G. Willis, I. Lengyel, and K. F. Jensen (1997). Computational chemistry predictions of reaction processes in organometallic vapor phase epitaxy. *Progress in Crystal Growth and Characterization of Materials*, **35**, 117-149.

Stringfellow, G. B. (1986). *Organometallic Vapor-Phase Epitaxy: Theory and Practice*. Academic Press.

Sze, S. M. (1983). *VLSI Technology*. McGraw Hill.

Venkataramani, R. (2000). *Multiscale Models of the Metalorganic Vapor Phase Epitaxy Process*. Massachusetts Institute of Technology.

Willis, B. G. (1999). *Complementary Computational Chemistry and Surface Science Experiments of Reaction Pathways in Aluminum Chemical Vapor Deposition*. Massachusetts Institute of Technology.

Willis, B. G. and K. F. Jensen (2000). Gas-phase reaction pathways of aluminum organometallic compounds with dimethylaluminum hydride and alane as model systems. *J. Phys. Chem. A*, **104**, 7881-7891.

Wise, K. D. (1998). Special issue on Integrated Sensors, Microactuators, and Microsystems (MEMS). *Proc. IEEE*, **86**, 1531-1533.

TOWARDS MULTISCALE SIMULATIONS OF FILLED AND NANOFILLED POLYMERS

Sharon C. Glotzer
Departments of Chemical Engineering and Materials Science and Engineering
University of Michigan
Ann Arbor, MI 48109

Francis W. Starr
Polymers Division and Center for Theoretical and Computational Materials Science
National Institute of Standards and Technology
Gaithersburg, MD 20899

Abstract

Simulation results of the effect of nanoscopic and micrometer-sized fillers on the structure, dynamics and mechanical properties of polymer melts and blends are presented. Molecular dynamics simulations show the tendency for polymer chains to be elongated and flattened near the filler surface. Additionally, the simulations show that the dynamics of the polymers can be dramatically altered by the choice of polymer-filler interactions. Time-dependent Ginzburg-Landau simulations of a phase-separating ultra-thin blend film show the influence on the mesoscale blend structure of immobilized filler particles when one component of the blend preferentially wets the filler. Preliminary finite element calculations predict the effect of mesoscale structure on ultra-thin film mechanical properties.

Keywords

FOMMS 2000, Fillers, Nanofillers, Polymers, Molecular modeling, Mesoscale modeling, Simulation.

Introduction

Major enhancements in mechanical, rheological, dielectric, optical, and other properties of polymer materials have long been achieved by adding fillers such as carbon black, talc, silica, and other inexpensive, inorganic materials (Wypych, 1999). Revolutionary advances in the design and fabrication of new materials that are lightweight and high strength, as well as, e.g., self-healing, self-reporting, and abrasion and thermally resistant, will come from advances in the fundamental understanding of nanocomposites, in which nanoscopic fillers ("nanofillers") are dispersed on nanometer scales within the polymer matrix (Roco, Williams, and Alivisatos, 2000). Nanofillers such as nanotubes, silica beads and cages, and clays, offer phenomenal advantages over more traditional fillers (such as carbon black, the ubiquitous filler used in tires) because greater property

improvement is achieved with far less material (see Fig. 1 for examples of filler and nanofiller geometries). Groundbreaking research at Toyota in the late 1980's and early 1990's led to a patented process allowing individual silicate sheets approximately 1 nm thick to be homogeneously dispersed in polymer (Giannelis, 1996). These new materials exhibited superior mechanical properties with substantially less inorganic content than conventional composites. For example, adding 1% by weight of ultra-fine, synthetic mica (30-nm diameter disks) to nylon gives super-tough nylon, while adding the same amount of traditional mica (micrometer-sized talc) gives only a slight improvement in toughness over the unfilled polymer.

The growing ability to design customized nanofillers of arbitrary shape and functionality provides an enormous variety of possible property modifications by introducing specific heterogeneity at the nanoscale (Roco, Williams, and Alivisatos, 2000; Schwab and Lichtenhan, 1998; Feher *et al.*, 1997). Indeed, future breakthroughs in the development of organic/inorganic hybrid nanocomposites will be possible by manipulating the inorganic phase on nanometer scales in order to achieve specific properties. Presently, little is known about controlling the dispersion, placement, and orientation of nanofillers within the polymer matrix, and thus the development of highly designed, nanostructured materials for specific applications is limited. Achieving such capability will require insight on many length scales, ranging from the interfacial interactions on molecular scales, to the ordering and assembly of inorganic phases on lengths scales from several tens of nanometers to tens of micrometers, to the manifestation of bulk material properties on macroscopic scales. Computer simulation will play an essential role in materials discovery and optimization, and in interpreting and guiding experiments to probe and manipulate these materials on molecular scales.

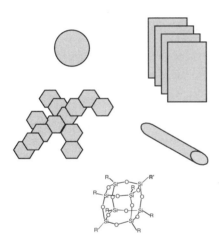

Figure 1. Examples of filler and nanofiller geometries. Clockwise from upper left: (a) spherical (e.g. colloidal silica); (b) sheets (e.g. clays); (c) rods, tubes, and fibers (e.g. carbon nanotubes); (d) cubes (e.g. cubic silsesquioxanes); and (e) fractal (e.g. fumed or precipitated silica, carbon black).

The multiple and highly disparate length and time scales inherent to polymer materials pose unique challenges to simulation. Filled polymers and nanocomposites pose a particular challenge, because they possess a hierarchy of length and time scales resulting not just from the polymer itself, but also from the structure of the filler or nanofiller. As a result, in modeling filled polymers, in principle one must capture phenomena on length scales that span typically 5-6 orders of magnitude, and time scales that can span a dozen orders of magnitude.

For example, consider carbon black (Fig. 2). At the smallest scale, carbon black is made of small, nearly graphitic, faceted particles that can range from 4 to several hundred nanometers across, depending on how they are produced (Wypych, 1999). During synthesis (while still molten), these so-called "primary particles" permanently fuse together into random structures, or "aggregates", which make up the smallest dispersible unit. These aggregates can in turn physically associate into larger and larger agglomerates, which are not permanent and can be broken up on shearing. If we consider filled polymer blends, the immiscibility of the blend provides even more structure to the material.

Here we describe three different simulation studies of filled and nanofilled polymers, each aimed at extracting fundamental information on different length and time scales. Each study involves a different simulation method appropriate for the particular phenomena of interest. In the first study, molecular dynamics (MD) simulations (Allen and Tildesley, 1987) are used to explore the effect of a single nanoscopic, model filler particle on the structure, dynamics, and glass transition temperature of a model polymer melt (Starr, Schrøder, and Glotzer, 2000a). In the second study, time-dependent Ginzburg-Landau (TDGL) methods (Glotzer, 1995) are used to study the effect of model nanometer to micron-sized fillers on the mesoscale structure of phase-separating polymer blends (Lee, Douglas, and Glotzer, 1999). In the third study, the finite element method (Reddy, 1993) is used in a preliminary study to probe the effect of fillers

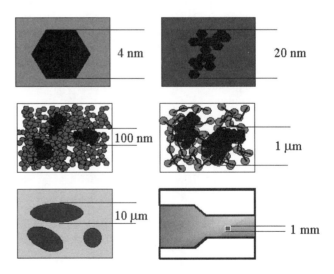

Figure 2. Example of the range of length scales on which phenomena must be modeled for carbon black filled polymers, ranging from the length scale of (from left to right) (a) the smallest primary particle; (b) an aggregate, the smallest dispersible unit; (c) and (d) agglomerates; (e) phase-separated domains; and (f) during flow (figure from Glotzer and Cummings, 2000).

on the mechanical properties of filled blend microstructures (Glotzer and Fuller, 2000). Although the studies are performed independently, we outline ways in which information gleaned from one simulation is used in another in order to bridge the length scales necessary to link molecular phenomena to macroscopic properties.

Molecular Simulations

Molecular dynamics simulations provide an ideal opportunity for direct insight into the effects of fillers on polymer structure and dynamics in the vicinity of the polymer-filler interface. As an example, we present results of an MD study of a polymer melt containing a single nanoscopic filler particle (Starr, Schrøder, Glotzer, 2000a). We can trivially adjust the interaction parameters between the filler and the surrounding polymer melt; the explicit control of the interactions helps to clarify which changes in the melt properties result from the type of interaction, and which properties change solely as a result of the steric hindrance introduced by the filler particle. This relatively simple single filler model provides an initial framework in which to interpret experiments on filled polymers (Lin et al., 1999; Cousin and Smith, 1994; Tsagaropoulos and Eisenberg, 1995; Kyu et al., 1996; Sombatsompop, 1999), and also possibly polymer thin films (Forrest and Jones, 2000; Jones et al., 2000l; Kraus et al., 2000; Theodorou, 1988; Kumar, Vacatello, and Yoon, 1989, 1990; Wang and Binder, 1991; Orts, et al., 1993; van Zanten, Wallace, and Wu, 1995, 1996; Forrest, Dalnoki-Veress, and Dutcher, 1997; Forrest, 1998; Anastasiadis, 2000), which report both increases and decreases of the glass transition temperature T_g, depending on the details of the system studied. In the case of filled polymers, future studies should consider the complicated geometrical effects that arise from the presence of multiple filler particles.

Our findings are based on extensive molecular dynamics simulations of a single nanoscopic filler particle surrounded by a dense polymer melt of 20-mer chains (Fig. 3). We use a "bead-spring" model for the polymers in which all monomers interact via a Lennard Jones (LJ) potential

$$U_{LJ} = 4\varepsilon\left[\left(\frac{\sigma}{r}\right)^{12} - \left(\frac{\sigma}{r}\right)^{6}\right]. \qquad (1)$$

The parameters ε and σ of the LJ potential are taken to be unity, such that all results are expressed in standard LJ reduced units[1]. We use the force-shifted form of the potential (cut-off at 2.5) to avoid any discontinuity in the potential or force (Allen and Tildesley, 1987). Nearest-neighbor monomers along the same chain are bound

together via a FENE anharmonic spring potential (Bird et al., 1987; Grest and Kremer, 1986; Rudisill and Cummings, 1991),

$$U_{FENE} = -\frac{kR_0{}^2}{2}\ln\left[1-\left(\frac{r}{R_0}\right)^2\right]. \qquad (2)$$

We use $k = 30$ and $R_0 = 1.5$. This simple polymer model has been studied in detail, and is known to be a good glass forming system, due to the incompatibility of the preferred FENE bond distance and the LJ potential minimum (Bennemann et al., 1999a, 1999b).

Our model filler has several general features typical of a primary carbon black particle, as well as some newer nanofillers (Wypych, 1999); it is highly faceted, but nearly spherical, and has facets with a size of about 10 nm. Specifically, the filler particle is icosahedral, with ideal force sites at the vertices, at 3 equidistant sites along each edge, and at 3 symmetric sites on the interior of each face of the icosahedron. This results in a facet size roughly equal to the end-to-end distance of the surrounding polymers. A particle is tethered to each of these sites by a FENE spring, which maintains a relatively rigid structure but allows for thermalization of the filler. We choose the filler sites to be more massive than the monomers and use a stiffer spring to minimize the oscillation of the tethered particles. Specifically, we take $m_f = 2.0$, $k_f = 45$, and $R_0^f = 1.0$. Filler sites interact with each other via a LJ potential with parameters $\varepsilon_{ff} = 2$ and $\sigma_{ff} = 1$. In order to avoid any spurious effects that might be induced by a large cavity inside the filler particle, replicas of the filler particle are embedded inside the filler, resembling an "onion" geometry. One replica has sites at the vertices and at one site along each edge, and the smallest replica has only sites at the vertices. This yields a total of 356 uniformly distributed force sites associated with the filler.

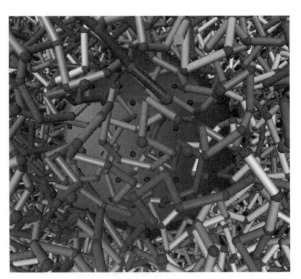

Figure 3. Model filler particle embedded in a model dense polymer melt.

---

[1] All values are reported in reduced LJ units. Standard units for T are recovered by multiplying T by ε_{mm}/k_B, where k_B is Boltzmann's constant. Time is in units of $(m_m\sigma_{mm}{}^2 / \varepsilon_{mm})^{1/2}$.

To determine which melt properties are a result of the steric constraints imposed by the filler, and which properties are affected by polymer-filler attraction, we consider two possible forms for the interaction between filler sites and monomers: (i) an excluded volume interaction only and (ii) excluded volume plus attractive interactions – a LJ interaction. The excluded volume interaction is modeled by dropping the attractive r^6 term in the LJ potential. The strength of the monomer-filler interaction is given using the Lorentz-Berthelot mixing rules $\varepsilon_{mf} = \sqrt{\varepsilon_{mm}\varepsilon_{ff}} = \sqrt{2}$ (Allen and Tildesley, 1987).

The simulation is performed in a cubic cell with periodic boundary conditions. The size of the cell is chosen such that the density ρ far from the filler particle varies by no more than 0.2% from $\rho = 1.0$, requiring a moderate set of preliminary simulations. This avoids any possible changes in properties resulting from a change in the bulk density. A box size of $L = 20.4$ satisfies this constraint at all T for the attractive system. In the non-attractive case, the characteristic first neighbor distance between filler and monomer depends on T (as will be discussed later), so at each T a different box size is required to achieve the correct density far from the filler, ranging from $L = 20.49$ at $T = 1.0$ to $L = 20.6$ at $T = 0.4$. Since we wish to know how the filler changes the melt properties relative to the pure system, we also simulate a pure system at density $\rho = 1.0$ at T ranging from 0.35-4.0.

profile of the monomers has a well-defined layer structure (Fig. 4). In the attractive case, we see a pronounced enhancement in the polymer density in the first layer, which we expect due to the relatively strong filler-monomer attraction; these density oscillations persist over a distance of roughly four monomers. The density profile depends weakly on temperature, becoming better defined as T decreases. It is somewhat surprising, in the case of excluded volume interactions, that there is an enhancement in the density in the first layer. However, notice that the location of the first layer is "pushed out" slightly, in comparison to the attractive filler case. The position of this peak increases with decreasing T, since the monomers have less kinetic energy, limiting the distance of closest approach.

The changes in the density profile must also be accompanied by some change in the local packing of the polymers. By focusing on the dependence of R_g on the distance from the filler surface (Fig. 5), we find a change in the overall polymer structure near the surface. R_g^2 increases by 50% on approaching the nanoparticle surface in both cases, and the perpendicular component $(R_g^{\perp})^2$ decreases by slightly more than a factor of 2. This indicates that the polymers become slightly elongated near the surface, and flatten significantly. The independence of the chain structure on the interaction suggests that the altered shape of the polymers is primarily due to geometric constraints of packing the chains close to the surface. Interactions seem to play a far more important role in changes of the *dynamics* of the system, as we see in the following.

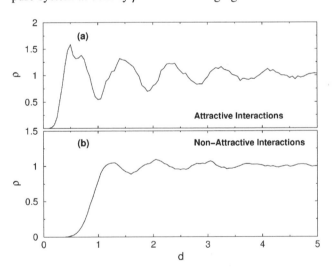

Figure 4. The local density of monomers as a function of the distance d from the filler surface. Formally, we define d as the difference between the radial position of a monomer and the radius of the inscribed sphere

$$r_{i\cos} = \frac{1}{12}\left(42 + 18\sqrt{5}\right)^{\frac{1}{2}} L \text{ of the icosahedral}$$

nanoparticle.

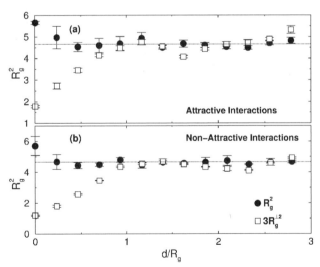

Figure 5. Radius of gyration, and its perpendicular component as a function of the distance of the center of mass of the polymer chain from the filler surface, as defined in Fig. 4.

We first consider changes in the polymer structure caused by the nanofiller. As in the case of polymers near a flat surface such as a wall (Kumar, Vacatello, and Yoon, 1989, 1990; Wang and Binder, 1991), the density

To quantify the effect of the nanofiller on dynamic properties, we calculate the intermediate scattering function

$$F(q,t) \equiv \frac{1}{NS(q)} \sum_{j,k=1}^{N} e^{-i\mathbf{q}\cdot(\mathbf{r}_j(t)-\mathbf{r}_k(0))}, \qquad (3)$$

which measures the decay of density fluctuations in the system, and is immediately accessible to neutron scattering experiments and other techniques (Hansen and McDonald, 1986). Here N is the total number of monomers. The relaxation time of $F(q,t)$, which we define by $F(q,\tau) = 0.2$ at the value of the wave-vector q_0 corresponding to the main peak in the static structure factor $S(q)$, provides a characteristic time which we can compare for systems at different T, or systems with different interactions. Relative to the pure melt, we find that τ increases in the filled system with attractive interactions, but decreases slightly for the filled system with only excluded volume interactions; in other words, the attractive system appears to slow the dynamics relative to the pure melt, while the excluded volume (non-attractive) system shows a slight enhancement of the dynamics.

From an experimental standpoint, the change in dynamics is most frequently indicated by a shift in the overall glass transition in the system. Based on our observations of τ, we would expect that the T_g of the attractive system would be somewhat larger than the pure melt, while the excluded volume system would exhibit a suppressed T_g. We check if these expectations are correct by estimating T_g using the Vogel-Fulcher-Tamman-Hesse form (Debenedetti, 1996)

$$\tau = \tau_0 e^{A/(T-T_0)}, \qquad (4)$$

where the parameter T_0 is known to be close to the experimentally measured T_g, and also exhibits the same changes found in T_g. For the pure system, we find $T_0 = 0.167$. Consistent with our expectations, we find that T_0 increases ($T_0 = 0.184$) for the attractive filler, and decreases ($T_0 = 0.156$) for the non-attractive filler. The fact that T_0 shifts in opposite directions for attractive versus purely excluded volume interactions demonstrates the importance of the surface interactions.

To elucidate how the filler influences the local dynamics of the monomers, we examine the relaxation of the self (incoherent) part $F_{self}(q,t)$ of $F(q,t)$ as a function of the monomer distance from the filler. We split $F_{self}(q,t)$ into the contribution from the separate layers previously observed in Fig. 6(a). In the attractive system, the relaxation of the layers closest to the filler is slowest, consistent with the system dynamics being slowed by the attraction to the filler. Conversely, for the non-attractive system (Fig. 6(b)), we find that the relaxation of inner layer monomers is significantly enhanced compared to the bulk, consistent with the observed enhancement of the system dynamics.

The altered dynamics persist for a distance slightly less than $2R_g$ from the surface. Our results demonstrate that interactions play a key role in controlling T_g and the

local dynamics of filled polymers. We expect the role of interactions to be largely the same when many filler particles are present in the melt, but there will be additional effects on dynamic properties due to the more complex geometrical constraints. These dynamic changes are likely also reflected in the thermodynamic properties, most notably in the configurational entropy (Gibbs and

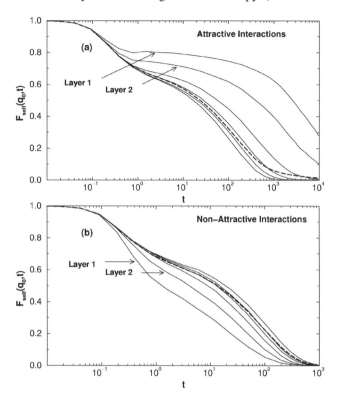

Figure 6. Relaxation of the self-part of the intermediate scattering function decomposed into the contribution of each layer of monomers around the nanofiller, as in Fig. 4. The surface monomers exhibit the slowest dynamics in the case of the attractive filler, while they exhibit the fastest dynamics in the case of the filler with only excluded volume interactions.

DiMarzio, 1958; Adam and Gibbs, 1965), long used to rationalize dynamic properties, and recently tested directly in simulations (Scala et al., 2000, Starr et al., 2000).

We finally compare with known results for ultra-thin polymer films, which suggests that the underlying physics in these systems is very similar, and is dominated by surface interactions[2]. In both systems, the polymers close

---

2 We note that the strongly heterogeneous nature of the dynamics of polymer melts, which has been established both by computer simulation (Bennemann, et al., 1999b, Glotzer, 2000) and experiment (Ediger, 2000) and which gives rise to a potentially large characteristic length scale near the glass transition, may play an additional role in T_g shifts in both filled

to the surface become elongated and flattened. This we expect from packing constraints near the surface. More significantly, it has been found that T_g may shift to either lower or higher T, depending on the substrate the film is deposited on. Recent simulations suggest that these shifts can be understood as a result of interactions with the surface (Varnik, Baschnagel, and Binder, 2000; Torres, Nealey, and de Pablo, 2000). This is precisely the interpretation presented here for the changes in T_g of the filled system. We further note that as the fraction of surface polymers to bulk polymers increases (e.g. by increasing the filler concentration), the changes in melt properties observed here are more dramatic (Starr, Schröder, and Glotzer, 2000b), in analogy with the increasing change in thin film properties with decreasing film thickness. Hence the pre-existing knowledge of thin-films may be useful for understanding and developing filled polymer materials.

Mesocale Simulations

The molecular dynamics results of the previous section can be used to design mesoscale simulations of filled polymer blends. Polymer blends, which are mixtures of two or more types of polymers, are generally immiscible, and will phase separate by a process called spinodal decomposition when the temperature, pressure, and blend composition are such that the blend is thermodynamically unstable. During this non-equilibrium process, domains rich in polymer A or polymer B (in the case of a binary blend) will form and coarsen in a self-similar way, eventually forming co-existing, macroscopic phases. In many instances, the phase-separating patterns can be trapped in the material by quenching the blend below T_g before the separation process is complete. When fillers are present in the blend, a preferential attraction of one of the polymers to the filler can break the symmetry of the spinodal decomposition process, producing novel mesoscale patterns and possibly even modifying domain growth laws.

A popular mesoscale method for simulating the structural evolution of phase-separation morphology in blends is the so-called time-dependent Ginzburg-Landau (TDGL) method[3]. This method is based on the Cahn-Hilliard (1958) and Cahn-Hilliard-Cook (CHC) (Cahn, 1961, 1962, 1963) models, and falls under the more general category of phase-field and reaction-diffusion models. In this approach, a free energy functional $F[\phi]$, which depends on a conserved local spatiotemporal concentration field $\phi(x,t)$ given by, e.g., the local fraction

of polymer A, is minimized to simulate a temperature quench from the miscible region of the phase diagram to the immiscible region where the blend is thermodynamically unstable. In this way, the resulting time-dependent structural evolution of the blend as it phase separates by spinodal decomposition can be investigated by solving the CHC/TDGL equation for the time dependence of the local blend concentration ϕ,

$$\frac{\partial \phi(\mathbf{x},t)}{\partial t} = \nabla \cdot M \nabla \frac{\partial F[\phi]}{\partial \phi(\mathbf{x},t)} + \xi(\mathbf{x},t). \qquad (5)$$

Here M is the mobility, which in general may depend on $\phi(x,t)$, and ξ is a thermal noise term. $F[\phi]$ consists of a bulk free energy term (e.g., the Flory-Huggins expression, or a Landau expression), and at minimum a square gradient term. A pedagogical discussion of this method and its application to blends and filled blends can be found in the review by Glotzer (1995) and in Lee, Douglas, and Glotzer (1999).

Numerous studies of blend phase separation under various conditions have been performed using this approach, including reactive blend phase separation (Glotzer, et al, 1995; Glotzer and Coniglio, 1994; Christensen, et al, 1996; Motoyama, 1996; Puri and Frisch, 1994; Fraaije, et al., 2000), phase separation of liquid crystal/polymer blends (Lapena, et al, 1999), phase separation under shear (Ginzburg, 1999), phase separation of block copolymers (Israels, 1995), and phase separation on patterned surfaces (Karim, et al, 1998).

The TDGL method has recently been applied to the study of phase separation in ultra-thin, filled blend films (Lee, Douglas and Glotzer, 1999; V. Ginzburg, et al., 1998, 1999, 2000). Ultra-thin blend films are thin enough (<< 100 nm) to suppress phase separation transverse to the solid substrate ("surface-directed spinodal decomposition") so that phase separation occurs quasi-two-dimensionally in the plane of the film. A preferential attraction for one of the blend components by the nanofiller can be modeled by adding a local surface interaction energy $F_s[\phi]$ to $F[\phi]$, which is expressed as an integral over the surface of the particle. A minimal model of phase separation near boundaries is given by Lee, Douglas, and Glotzer (1999), and references therein,

$$F_s[\phi] = \int_S d^{d-1}x \left[h\phi + \frac{1}{2} g\phi^2 + ... \right], \qquad (6)$$

subject to boundary conditions of zero flux and local equilibrium at the filler surface. Here, the coupling constant h in the leading term plays the role of a surface field that breaks the symmetry between the two phases and attracts one of the blend components to the filler surface. The value of h can be related to the interaction energy ε_{mf} in the previous section. The coupling constant g in the second term is neutral regarding the phases, and results from the modification of the interaction energy due to chain connectivity (Freed, 1996; Brazhnik, et al,

---

polymers and ultra-thin films; a study of this will be presented elsewhere.

[3] Another mesoscale method capable of modeling spinodal decomposition in polymers is dissipative particle dynamics; see Tildesley (2000), these proceedings, and references therein.

1994) and the missing neighbors near the surface when the equation is solved on a grid (Mills, 1971; Binder, 1983). The attractive interaction between filler and polymer A causes wetting of the filler by the polymer, analogous to the density enhancement shown in Fig. 4(a), which breaks the symmetry of the spinodal decomposition process, producing novel patterns.

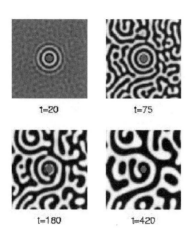

Figure 7. Simulation of the influence of a single isolated filler particle (central gray region in the figure) on polymer-blend phase separation in a critical composition blend, when the filler is immobile on the time scale of phase separation (from Lee, Douglas, and Glotzer, 1999). Shown is the concentration field, φ, at four different times following a temperature quench to the two-phase region. Here black denotes the polymer-A rich phase, which wets the filler, and white denotes the polymer-B rich phase.

Figure 8. At larger filler concentration, the target patterns produce interference patterns whose length scale depends on the spacing between particles relative to the spinodal wavelength. In this simulation, two different types of filler particles that each prefer a different blend component are present. (From Lee, et al, 1999.) Similar patterns have been observed in crosslinked blends (Tran-Cong, 1998).

Similar target patterns were observed via atomic force microscopy in a colloidal-silica-filled ultra-thin film blend of PS/PVME, in which the PS is preferentially attracted to the silica beads (Karim, et al., 1999). In the experiment, the filler was immobilized on the substrate supporting the film, and thus the conditions were very similar to those in the simulation. The patterns were trapped in the film by quenching below T_g before the spinodal decomposition process was completed.

The results shown here were obtained using a constant, composition- and spatially-independent mobility M, and represent a "minimal model" of the effect of fillers or nanofillers on phase separation morphology. To more accurately model the effect of filler interactions on the mesoscale structure of immiscible blends, we can use the results of the previous section in future studies to modify M so as to slow the polymer dynamics of the wetting phase, and speed up the dynamics of the non-wetting phase. This may have the effect of prolonging the lifetime of the target patterns, and/or increasing the distance over which they penetrate into the background spinodal pattern. It would also be interesting to incorporate T dependence in the mobility in order to capture certain aspects of glass transition phenomenology, and compressibility to accurately model the spatial dependence of the density variation of the wetting phase. In the present studies, the filler particles were considered immobile on the time scale of phase separation in order to compare with ultra-thin film experiments, where silica beads were essentially attached to the surface; introducing filler mobility into the model may modify the mesoscale structure (Ginzburg, 1998, 1999, 2000).

As will be shown in the next section, representative microstructures obtained from the TDGL/CHC approach can be used as input to finite element packages such as OOF[4] to predict elastic properties, thereby bridging several key length scales relevant to filled and nanofilled polymers.

Macroscale Mechanical Property Simulations

Fillers added to polymer blends can modify mechanical properties of the blend in several ways. First, the filler itself can impart additional strength and toughness through its own mechanical properties. Second, the filler can inhibit failure by blocking the propagation of cracks (White, et al. 2001). Third, the modification of polymer structure near the filler surface that we observed in our MD simulations can alter mechanical properties. And because mechanical properties depend on material microstructure, fillers can alter, e.g., the blend modulus through the modification of microstructure as demonstrated in the previous section.

---

[4] http://www.ctcms.nist.gov/oof

A new software tool developed at NIST, called OOF[4] (Carter, Fuller and Langer, 1998) facilitates the prediction of mechanical properties of materials by using real or simulated microstructures as input. OOF allows users to assign specific constitutive properties, such as modulus, etc., to the various parts of the microstructure, construct a finite element mesh to resolve the key features of the microstructure, and perform mechanical simulations using the finite element method. In this way, the effect of changes in microstructure on mechanical properties can be easily determined.

The current generation of OOF software assumes the behavior of the material is linear elastic (OOF was originally developed for ceramic materials, but is enjoying much broader application.) The next generation of the tool, OOF2, will allow incorporation of, e.g., nonlinear elasticity.

As one example of how OOF can be used for filled polymer blends, we have performed preliminary calculations of the elastic modulus of two critical

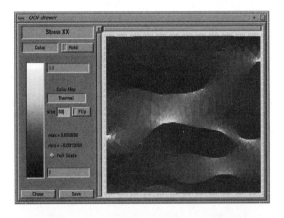

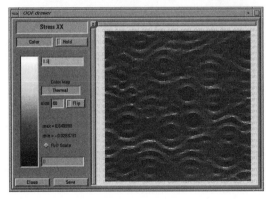

Figure 9. Example of OOF mechanical property calculations on two different blend microstructures resulting from phase separation in the absence of fillers (top) and presence of fillers (bottom). Color scheme: bright spots indicated regions of high local strain, and occur in the lower modulus phase (Glotzer, Fuller, and Han. 2000).

composition blend microstructures (Fig. 9). The first microstructure is obtained from a simulation of the TDGL/CHC equation in the absence of fillers, and the second is obtained from the same simulation but with fillers included, as in Fig. 8. Both simulations were stopped after the same number of time steps; in the absence of fillers, the blend microstructure coarsens to larger domain sizes. We assume both components are rubbery, and assume relative Young's moduli for the two phases of 10:1. We further assume the interfaces between the two phases to be sharp. We then apply a 10% longitudinal strain in the x-direction to both microstructures. If we neglect the additional modification of the mechanical properties due to the filler properties, and focus only on the effect of microstructure, we find that the bottom microstructure in Fig. 9 has an elastic modulus roughly 18% smaller than the modulus of the top microstructure in Fig. 9. The bright and dark features in the microstructures indicate regions of high and low local strain, respectively. As expected, the high strain regions correspond to the low modulus phase. Additional work is underway to extend this preliminary study, and will be reported elsewhere (Parekh, et al., in preparation).

Conclusions

We have described three different simulation studies of filled and nanofilled polymers, each aimed at extracting fundamental information on different length and time scales. Each study involved a different simulation method appropriate for the particular phenomena of interest. In the first study, MD simulations were used to explore the effect of a single nanoscopic, model filler particle on the structure, dynamics, and glass transition temperature of a model polymer melt. In the second study, TDGL/CHC methods were used to study the effect of model nanometer to micrometer-sized fillers on the mesoscale structure of phase-separating polymer blends. In the third study, finite element methods were used to probe the effect of fillers on the mechanical properties of blend microstructures. Although the studies were performed "off-line" from each other, we outlined ways in which information gleaned from one simulation can be used in another to bridge the length scales necessary to link molecular phenomena to macroscopic properties, depicted schematically in Fig. 10.

The link between mesoscale structure and macroscale mechanical properties was accomplished by using the output of the TDGL/CHC simulations as input for the OOF calculations. The link between the MD simulations and the TDGL/CHC simulations was less direct, and made only qualitatively through incorporation of filler-polymer interaction terms in the CHC free energy functional that were attractive or neutral to one phase or the other. A closer link can be made by incorporating more details from the MD simulation, including e.g. composition- position-, and/or temperature-dependent mobility, and compressibility. Not discussed here are links yet to be made between our "coarse-grained" MD simulations and more atomistically accurate simulations of polymer-filler interactions, using, e.g. united atom or explicit atom force fields for the filler and polymers derived from quantum chemistry calculations.

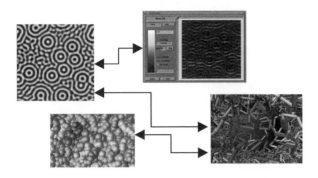

Figure 10.   One possible paradigm for the integration of length scales in the simulation of filled or nanofilled polymers. Counterclockwise from lower left: (a) Atomistic; (b) Coarse-grained MD; (c) Mesoscale continuum simulations; (d) finite element macroscale simulations.

Acknowledgements

We acknowledge useful discussions with E. Amis, P. Cummings, J. Douglas, E. Fuller, C. Han, A. Karim, J. Kieffer, B. Lee, A. Nakatani, N. Parekh, and T. Schrøder.

References

Adam, G. and J. H. Gibbs (1965). On the temperature dependence of cooperative relaxation properties in glass forming liquids. *J. Chem. Phys.*, **43**, 139-146.

Allen, M.P. and M. P. Tildesley (1987). *Computer simulation of liquids.* Oxford, Oxford.

Anastasiadis, S.H., Karatasos, K., Vlachos, G, Manias, E., and E. P. Giannelis (2000). Nanoscopic-confinement effects on local dynamics. *Phys. Rev. Lett.*, **84**, 915-918.

Bennemann, C., Paul, W., Baschnagel, J., and K. Binder (1999a). Investigating the influence of different thermodynamic paths on the structural relaxation in a glass-forming polymer melt. *J. Phys. Cond. Mat.*, **11**, 2179-2192.

Bennemann, C., Donati, C., Baschnagel, J. and S. C. Glotzer (1999b). Growing range of correlated motion in a polymer melt on cooling towards the glass transition. *Nature*, **399**, 246-249.

Binder, K. (1983). Critical behavior at surfaces. In Domb, C. and Lebowitz, J.L. (Eds.). *Phase transitions and critical phenomena*, **8**, 1-144.

Bird, R.B., Curtiss, C.F., Armstrong, R.C., and O. Hassager (1987). *Dynamics of Polymeric Liquids: Kinetic Theory, Vol. 2.* John Wiley and Sons, New York.

Brazhnik, P.K., Freed, K.F. and H. Tang (1994). Polymer melt near a solid wall. *J. Chem. Phys.*, **101**, 9143-9154.

Cahn, J.W. and J. E. Hilliard (1958). Free energy of a non-uniform system. I. Interfacial free energy. *J. Chem. Phys.*, **28**, 258-267.

Cahn, J.W. (1961). On spinodal decomposition. *Acta Metall.*, **9**, 795-801.

Cahn, J.W. (1962). On spinodal decomposition in cubic crystals. *Acta Metall.*, **10**, 179-183.

Cahn, J.W. (1963). Magnetic aging of spinodal alloys. *J. Appl. Phys.*, **34**, 3581-3586.

Carter, W.C., Langer, S.A., and E. Fuller (1998). *The OOF Manual: Version 1.0.* National Technical Information Service, Springfield.

Christensen, J., Elder, K. and H. C. Fogedby (1996). Phase segregation of a chemically reactive binary mixture. *Phys. Rev., E* **54**, 2212-2215.

Cousin, P. and P. Smith (1994). Dynamics mechanical properties of sulfonated polystyrene/alumina composites. *J. Poly. Sci. B: Poly. Phys.*, **32**, 459-468.

Debenedetti, P. G. (1996). *Metastable Liquids.* Princeton Univ. Press, Princeton.

Ediger, M. D. (2000). *Ann. Rev. Phys. Chem.* (in press).

Feher, F. J., Soulivong, D., Eklund, A. G. and K. D. Wyndham (1997). Cross-metathesis of alkenes with vinyl-substituted silsesquioxanes and spherosilicates: A new method for synthesizing highly-functionalizedSi/O frameworks. *Chem. Comm.*, **13**, 1185-1186.

Forrest, J. A., Dalnoki-Veress, K., and J. R. Dutcher (1997). Interface and chain confinement effects on the glass transition of thin polymer films. *Phys. Rev. E*, **56**, 5705-5716.

Forrest, J. A., Svanberg, C., Revesz, K., Rodahl, M., Torell, L. M., and B. Kasemo (1998). Relaxation dynamics in ultrathin polymer films. *Phys. Rev. E*, **58**, 1226-1229 .

Forrest, J.A. and R. A. L. Jones (2000). The glass transition and relaxation dynamics in thin polymer films. In Karim, A. and Kumar, S. (Eds.), *Polymer Surfaces, Interfaces, and Thin Films.* World Scientific, Singapore, 251-294.

Freed, K. F. (1996). Analytic theory of surface segregation in compressible polymer blends. *J. Chem. Phys.*, **105**, 10572-10582.

Giannelis, E. P. (1996). Polymer layered silicate nanocomposites. *Adv. Mater.*, **8**, 29.

Gibbs, J. H. and E. A. DiMarzio (1958). Nature of the glass transition and the glassy state. *J. Chem. Phys.*, **28**, 373-380.

Ginzburg, V. V., Peng, G., Qiu, F., A. C. and Balazs (1999). Kinetic model of phase separation in binary mixtures with hard mobile impurities. *Phy. Rev. E*, **60**, 4352-4359.

Glotzer, S. C. and A. Coniglio (1994). Self-consistent solution of phase separation with competing interactions. *Phys. Rev. E*, **50**, 4241-4244.

Glotzer, S. C. (1995*). Ann. Rev. Comp. Phys.*, **2**, 1.

Glotzer, S. C., DiMarzio, E. A. and M. Muthukumar (1995). Reaction-controlled morphology of phase separating mixtures. *Phys. Rev. Lett.*, **74**, 2034-2037.

Glotzer, S. C. (2000). Spatially heterogeneous dynamics in liquids: insights from simulation. *J. Non-Cryst. Solids*, **274**, 342-355.

Glotzer, S. C., Han, C., and E. Fuller (2000). (unpublished).

Hansen, J.P. and I. R. McDonald (1986). *Theory of Simple Liquids.* Academic Press, London.

Israels, R., Jasnow, D., Balazs, A. C., Guo, L., Krausch, G., Sokolov, J., and M. Rafailovich (1995). Compatibilizing A/B blends with AB diblock copolymers – effect of copolymer weight. *J. Chem. Phys.,* **102**, 8149-8157.

Jones, R. L., Kumar, S. K., Ho, D. L., Briber, R. M. and T. P. Russell (2000). Chain conformation in ultrathin polymer films. *Nature,* **400**, 146-149.

Karim, A., Douglas, J. F., Nisato, G., Liu, D-W., and E. J. Amis (1999). Transient target patterns in phase separating filled polymer blends. *Macromolecules,* **32**, 5917-5924.

Karim, A. Douglas, J. F., Lee, B. P. Glotzer, S. C., Rogers, J. A., Jackman, R. J., Amis, E. J., and G. M. Whitesides (1998). Phase separation of ultrathin polymer-blend films on patterned substrates. *Phys. Rev. E,* **49**, R6273-6276.

Kraus, J., Müller-Buschbaum, P., Kuhlmann, T., Schubert, D. W. and M. Stamm (2000). Confinement effects on the chain conformation in thin polymer films. *Europhys. Lett.,* **49**, 210-216.

Kumar, S. K., Vacatello, M., and D. Y. Yoon (1989). Off-lattice Monte-Carlo simulations of polymer melts confined between plates. *J. Chem. Phys.,* **89**, 5206-5215.

Kumar, S. K., Vacatello, M., and D. Y. Yoon (1990). Off-lattice Monte-Carlo simulations of polymer melts confined between plates 2. Effects of chain length and plate separation. *Macromolecules,* **23**, 2189-2197.

Lapena, A. M., Glotzer S. C., Langer S. A, and A. J. Liu (1999). Effect of ordering on spinodal decomposition of liquid-crystal/polymer mixtures. *Phys. Rev. E,* **60**, 29-32.

Lee, B. P., Douglas J. F.. and S. C. Glotzer (1999). Filler-induced composition waves in phase-separating polymer blends. *Phys. Rev. E.,* **60**, 5812-5822.

Lin, E. K., Kolb, R., Satija, S. K., and W. L. Wu (1999). Reduced polymer mobility near the polymer/solid interface as measured by neutron reflectivity. *Macromolecules,* **32**, 3753-3757.

Maurits, N. M., Sevink, G. J. A., Zvelindovsky, A. V., and J. G. E. M. Fraaije (1999) Pathway controlled morphology formation in polymer systems: Reactions, shear, and microphase separation. *Macromolecules,* **32**, 7674-7681.

Mills, D. L. (1971). *Phys. Rev. B,* **3**, 3887.

Motoyama, M. (1996). *J. Phys. Soc. Japan,* **65**, 1894.

Parekh, N., Fuller, E., Han, C. and S. Glotzer (2001). Mechanical properties of polymer blend microstructures using OOF. in preparation.

Peng, G. W., Qiu, F., Ginzburg, V. V., Jasnow, D., and A. C. Balazs (2000). Forming supramolecular networks from nanoscale rods in binary, phase-separating mixtures. *Science,* **288**, 1802-1804.

Puri, S. and H. L. Frisch (1994). Segregation dynamics of binary mixtures with simple chemical reactions. *J. Phys. A,* **27**, 6027-6038.

Qiu, F., Ginzburg, V. V., Paniconi, M., Peng, G. W., Jasnow, D., and A. C. Balazs (1999). Phase separation under shear of binary mixtures containing hard particles. *Langmuir,* **15**, 4952-4956.

Reddy, J. N. (1993). *An Introduction to the Finite Element Method.* Mc-Graw-Hill, New York.

Roco, M. C., Williams S., and P. Alivisatos (Eds.).(2000). *Nanotechnology Research Directions: IWGN Workshop Report Vision for Nanotechnology in the Next Decade.* Kluwer Academic Publishers, Dordrecht.

Rudisill, J. W. and P. T. Cummings (1991). The contribution of internal degrees of freedom to the non-Newtonian rheology of model polymer fluids. *Rheologica Acta,* **30**, 33-43 .

Scala, A., Starr, F. W., La Nave, E., Sciortino, F., and H. E. Stanley (2000). Configurational entropy and diffusivity of supercooled water. *Nature,* **406**, 166-169.

Schwab, J. J. and J. D. Lichtenhan (1998). Polyhedral oligomeric silsesquioxane (POSS)-based polymers. *Appl. Organomet. Chem.,* **12**, 707-713.

Sevink, G. J. A., Zvelindovsky, A. V., van Vlimmeren, B. A. C., Mauritis, N. M., and J. G. E. M. Fraaije (2000). Dynamics of surface directed mesophase formation in block copolymer melts. *J. Chem. Phys.,* **110**, 2250-2256.

Sombatsompop, N. (1999). Dynamic mechanical properties of SBR and EPDM vulcanisates filled with cryogenically pulverized flexible polyurethane foam particles. *J. Appl. Poly. Sci.,* **74**, 1129-1139.

Starr, F. W., Sastry, S., La Nave, E., Scala, A., Stanley, H. E., and F. Sciortino (2000). Thermodynamic and structural aspects of the potential energy surface of simulated water. *Phys. Rev. E,* **62** (in press).

Starr, F. W., Schrøder, T. B., and S. C. Glotzer (2000a). Effects of a nano-sized filler on the structure and dynamics of a simulated polymer melt and the relationship to ultrathin films.

Starr, F. W., Schrøder, T. B., and S. C. Glotzer (2000b). in preparation. cond-mat/0007486.

Sun, T., Balazs, A. C., and D. Jasnow (1997). Constrained free-energy functional of deformed polymer systems. *J. Chem. Phys.,* **107**, 7371-7382.

Tildesley, D. J. This volume.

Torres, J. A., Nealey, P. F., and J. J. de Pablo (2000). Molecular simulation of ultrathin polymeric films near the glass transistion. *Phys. Rev. Lett.,* **85**, 3221-3224.

Tran-Cong, Q. (1998). In Araki, T., Tran-Cong, Q., and Shibayama, M. (Eds.), *Structure and Properties of Multiphase Polymeric Materials.* Marcel Dekker, New York.

Tsagaropoulos, G. and A. Eisenberg (1995). Dynamic-mechanical study of the factors affecting the 2 glass-transition behavior of filled polymers -similarities and differences with random ionomers. *Macromolecules,* **28**, 6067-6077.

van Zanten, J. H., Wallace, W. E., and W. L. Wu (1996). Substrate interaction induced transitions in ultrathin polymer films. *Phys. Rev. E,* **53**, 2053-2059.

Varnik, F., Baschnagel, J. and K. Binder (2000). Molecular dynamics of supercooled polymer films. *J. Phys. IV,* **10**, 239-242.

Wallace, W. E., van Zanten, J. H., and W. L. Wu (1995). Influence of an impenetrable interface on a polymer glass-transition temperature. *Phys. Rev. E,* **52**, 3329-3332 .

Wang, J.-S. and K. Binder (1991). Enrichment of the chain ends in polymer melts at interfaces. *J. Phys. I France,* **1**, 1583-1590.

White, S. R, Sottos, N. R., Geubelle, P. H., Moore, J. S., Kessler, M. R., Sriram, S. R., Brown, E. N., and S. Viswanathan (2001). Autonomic healing of polymer composites. *Nature,* **409**, 794-797.

Wypych, G. (1999) *Handbook of Fillers.* ChemTec Publishing, Toronto.

MULTISCALE MODELING OF
LARGE BIOMOLECULES

Gary A. Huber
Department of Bioengineering
University of California, San Diego
La Jolla, CA 92093-0412

Abstract

This paper describes some of the recent developments in computer modeling of large biomolecules that take advantage of the multiple length and time scales associated with such systems. The developments described are particularly important for modeling interactions among two or more molecules. For computing electrostatic interactions, various methods based on multipole and multigrid methods show considerable promise. For computing the important internal motions of large molecules, we briefly describe a new method, the Hierarchical Collective Motions algorithm, which runs small simulations on overlapping segments of the large molecules and then pieces the resulting information together to obtain information on the important global motions. Combination of this method with efficient computation of electrostatics will be useful in modeling biomolecular interactions.

Keywords

Boundary elements, Multigrid, Finite elements, Normal modes, Biomolecular interaction.

Introduction

One of the main challenges in computational biology today is the successful modeling of associations of two or more large biomolecules. Such a goal is important for several reasons. Most aspects of cellular function depend upon molecular interactions, so computer simulations have the potential to increase basic understanding of cell biology. Also, biomolecular association is the basis of drug action in the body and of industrial bioseparation processes.

Biomolecular association simulations fall into three categories. First, many simulations attempt to find feasible configurations of a bimolecular complex; these have use for rapid screening of potential drugs (Morris et al., 1998). Second, some simulations attempt to compute binding constants at thermodynamic equilibrium (Ajay and Murcko, 1995; Kollman, 1993). Finally, there are simulations that attempt to compute the kinetics of binding reactions (Gabdoulline and Wade, 1997; Sept and Elcock and McCammon, 1999). To date, however, all computations of binding rates rely upon simplified models that use adjustable, ad hoc parameters.

The main difficulty with dynamic simulations is the huge number of variables that represents a realistic molecular system. Currently, most efforts focus on molecular dynamics (MD) simulations, in which the large molecule and a sufficiently large portion of its surrounding solvent are treated in atomic detail (Tara et al., 1999). Each atom is treated as a point mass and the atoms move according to Newton's law of motion. The forces on the atoms are computed from a force field, which has parameters that are general for all proteins, for example. Because of the algorithms used to update the atom's positions and velocities from Newton's equation of motion, the time interval between updates is limited to sizes on the order of a femtosecond. Time steps that are much larger lead to inaccurate results and numerical instabilities.

For example, a very important large biomolecule is actin, which is a polymer composed of monomers, each with 372 amino acids, which form double-helical filaments. Most protein molecules the size of an actin monomer undergo important biological motions that take place on timescales of nanoseconds or greater. Timescales of microseconds are associated with filament torsion (Rebello and Ludescher, 1998), and timescales on the order of seconds are associated with the flexing and diffusion of long actin filaments (Gittes et al., 1993). In general, the larger the molecule, the longer is the timescale over which important motions happen. So, in trying to simulate larger and larger molecules, two obstacles are encountered. First, one is obliged to run the simulation for a longer period to time, while using the tiny time intervals in position updates. Second, larger molecules mean that there is more to simulate, so the required computer time per update is larger. Recent nanosecond MD simulations of an important enzyme, acetylcholinesterase (604 amino acids), plus surrounding waters, took several thousand hours of processor time on a modern, massively parallel computer, the Cray T3E (Tara et al., 1999).

In order to simulate protein-protein encounters, drastic simplifications and shortcuts have been required. The molecules commonly are treated as rigid bodies, and the solvent is treated using continuum models instead of explicit molecules. A "reaction" is presumed to occur when the two molecules are aligned within certain tolerances defined by adjustable parameters. Although qualitative trends with differences in pH, ionic strength, and side-chain mutations have been reproduced, the simplicity of the model and the ad hoc parameters limit its predictive capabilities (Gabdoulline and Wade, 1997; Sept, Elcock and McCammon, 1999).

Fortunately, like most biological systems, protein molecules are organized in a hierarchical manner. The different levels have their own characteristic length and time scales, and often a system can be successfully described and modeled in terms of the upper levels alone. First of all, there is the familar succession from primary to quatenary structure. Also, there is evidence, both experimental and theoretical, that the potential energy landscapes of protein molecules have a hierarchical, "valley within a valley" structure (Frauenfelder, Sligar and Wolynes, 1991; Leopold, Montal and Onuchic, 1992). Finally, there is evidence from computer simulations that most of the internal motions of protein molecules can be described with only a few variables, which closely resemble "normal modes" from solid mechanics (Amadei and Linnsen, 1993). These features make it possible to devise algorithmic shortcuts to take advantage of the different levels of organization. These algorithms fall into two catagories: those that compute interactions among the molecules, and those that seek to simulate internal flexibility.

Solvent Models

One of the key strategies in devising simplified models is to treat the collection of atoms as a macroscopic continuum. Thus, macroscopic properties of bulk water, such as the dielectric constant, ionic strength, pH, surface tension, and viscosity play a role in these models (for a review, see Huber, 1998). Such treatment is possible, because the time and length scales of the fluctuations in the structure of the liquid water are much smaller than the time scales of protein-protein association. In protein-protein association simulations, the important forces mediated by the solvent are the electrostatic forces, hydrodynamic forces, and hydrophobic forces.

In calculating the forces due to the charges on the atoms, it can be useful to treat both the surrounding water and the protein molecule as continua with well-defined dielectric constants. (Sharp and Honig, 1990; Davis and McCammon, 1990; Honig and Nichols, 1995). The dielectric constant is a measure of how strongly the material is polarized and thus counteracts the applied electric field. Water has a relatively high value, since its dipolar molecules easily align themselves along an electric field. The dielectric constant of a protein molecule is not known from experiment, or even necessarily well-defined, although attempts at determination have been made from detailed molecular models (Simonsen et al., 1991). When treating the protein as a continuum, values from 2 to 4 have typically been used. The ionic strength can also be included in a continuum framework; positive ions tend to gather in regions of negative potential and vice versa, thereby mitigating the effect of the electric field. The usual assumption is that the ionic atmosphere around the protein rapidly approaches thermodynamic equilibrium in the electric field. The dielectric and electrolyte effects can be summarized in the Poisson-Boltzmann equation

$$\nabla \cdot (\varepsilon(\mathbf{x}) \nabla \phi(\mathbf{x})) =$$
$$\rho(\mathbf{x}) + \lambda(\mathbf{x}) \sum_i q_i c_i \exp(-q_i \phi(\mathbf{x})/kT)$$

where ϕ is the electric field, ε is the dielectric as a function of position, ρ is the charge density due to fixed charges (not due to the ion atmosphere or polarization of water), λ is a switching function that denotes where ions may exist, and q_i and c_i are the charge and bulk concentration of the ith ionic species. The charge density is usually treated as a collection of point charges on each atom (mathematically as collection of Dirac delta functions). The ionic switching function is 1 out in the water and 0 in the interior of the protein; the crossover point depends on how close the ions come to the surface of the protein. Likewise, the boundary between the water and protein values for ε depends on how the surface of the protein is defined. Very often, the Poisson-Boltzmann equation is linearized when the ionic strength is not very high.

Historically, most software packages that compute the electric field by solving Eq. 1 have used the finite-difference method (Nicholls and Honig, 1991; Madura et al.). This method is limited, because the solution domain must be large enough to include solvent far away from the protein, while the grid spacing must be small enough to resolve atomic details on the surface. The usual practice has been to compute and store the electric field around one of the molecules at the beginning of the simulation. As the simulation proceeds, forces and torques on the other molecule are computed using the product of its permanent charges and the electric field from the first molecule. This approximation is valid for a small molecule moving around a large molecule, but breaks down for equal-sized molecules, because the second molecule perturbs the electric field of the first. This problem is addressed in a method by Gabdoulline and Wade (1996), in which the electric field is computed for each body separately, a small number of effective charges within each molecule is computed to fit the electric field, and dielectric solvent effects are approximated by inducible dipoles in each molecule. Because the number of effective charges and inducible dipoles roughly correspond with the number of amino acids, the computational effort for an interaction calculation scales roughly as the square of the number of amino acids.

Other groups have used the boundary-element method (Vorobjev and Scheraga, 1997, Horvath et al., 1996), in which a linearized version of Eq. 1 is reformulated as an integral equation over the boundary of the molecules. The boundary is typically divided into small triangles, and the integral equation is approximated as a set of algebraic equations. Because the domain is reduced to two dimensions, the number of variables is smaller than with the finite difference method, but the triangles still must be small enough to resolve the details of the surface. Moreover, the set of equations is dense, causing the computational effort to scale as the cube of the number of elements, unless shortcuts are taken (as described below).

Adaptive Methods

The current methods of computing electrostatic interactions between two or more biomolecules are computationally intensive because there are two disparate length scales involved: the length scale of the surface details, typically on the order of an angstrom, and the length scale of the molecules' overall dimensions, which can be thousands of angstroms. Fortunately, adaptive, multiscale methods show great promise in addressing this problem.

One widely-used class of multiscale methods are multipole methods (Greengard and Rokhlin, 1987; Ding and Karasawa and Goddard, 1992). In many applications, one has a large number N of pairwise-interacting entities, such as charged atoms or boundary elements. If all pairs of interactions are computed directly, the computational effort scales as N^2. However, most physical interactions die off with increasing distance. As a result, it is often possible to compute one simple interaction term for a group of entities that are localized in space. As an example, suppose we have two groups of point charges. Each group is localized in space, and the two groups are far enough from each other, such that the distance between them is greater than the length scale of the volumes enclosing them. We want to compute the Coulombic forces of Group 2 on the charges of Group 1, and vice versa. For each group, we can compute a Coulombic multipole expansion, which is a simple representation of the electric potential far from the group; the charges in one group interact with the multipole expression from the other group. If there are N1 charges in Group 1 and N2 charges in Group2, then the computational effort scales as N1 + N2; the use of naive summation would scale as N1*N2.

For computations of pairwise interactions in general collections of objects in space, this idea is implemented in the Fast Multipole algorithm (Greengard and Rokhlin, 1987). The spatial domain is divided into small cubes, which are then grouped eight at at time into larger cubes. The larger cubes, in turn, are grouped eight at a time into even larger cubes, and so on, until there are 16 or less cubes at the top level. Then, multipole expansions are computed in each of the smallest cubes, and are used recursively to compute the multipole expansions of the larger cubes. Next, at the top level, the multipole expansion of each cube is transformed into a series expression (either Taylor series or spherical harmonics) for the field potential centered at all other top-level cubes that are not its immediate neighbors. Within each top-level cube, the expansion is translated to the centers of the eight children cubes. After that, the expansions of the next-top-level cubes are computed in a similar manner, and added to the expansions that came from the top-level cubes. This process is repeated recursively down to the bottom-level cubes, and the interactions of the objects in each bottom-level cube with those from non-adjacent cubes are computed from the total expansion in that cell. Finally, the interactions with other objects in adjacent cubes are computed directly. The required computer time scales linearly with the number of objects. This algorithm is a excellent example of taking advantage of the multiple length scales inherent in many problems. The most common application is computing Coulombic interactions among charged particles. In addition, the FM algorithm has been used to compute interactions between boundary elements in boundary element solutions (Purisima, 1998) to the Poisson equation (Eq. 1 without the ionic term). This allows the computational effort to scale as the square of the number of elements.

Another class of methods that takes advantage of multiple length scales inherent in biological problems are the multigrid methods (Briggs, 1987). They have frequently been used in conjunction with finite-difference methods, but recently they have shown great promise with finite-element methods (Baker, Holst, and Wang, 2000). In solving a linear system, the equation is reformulated in terms of the error and the residual, and the solution is found by subtracting successive estimates of the error from an initial guess. In order to compute the error, the problem is transformed to a coarser representation or discretization, the error is estimated using the coarser version of the problem (which takes less computer time), and transformed back to the original representation. For finite difference problems, the coarsening is done by eliminating alternate layers of grid points, and for finite element problems, it is done by lumping together adjoining tetrahedra. This procedure can be carried out recursively; the error on one level can be estimated by transforming the problem to yet a coarser level. Eventually, the problem is finally coarse enough that an exact solution is trivial, and then the error estimate is successively transformed back down to the finer levels. In general, several iterations are required for convergence, and the computational effort scales linearly with the number of variables.

Recently, a multigrid algorithm has been used in the framework of an adaptive finite-element method to solve the Poisson-Boltzmann equation (Holst, Baker, and Wang, 2000). The adaptive method starts out with a coarse representation of the problem, and successfully refines various portions of the problem domain by subdividing tetrahedra based on estimates of the solution error. This is especially useful for molecules, because small tetrahedra are used only near the surface, where material properties change over small length scales; the sizes of the tetrahedra increase exponentially as one moves farther from the molecule. Thus, this method takes full advantage of the differing length scales in the problem. It has an advantage over boundary-element methods in that it can solve the full, non-linear PB equation. We also mention, in passing, that a three-level multigrid formulation of the boundary element method has been used for solutions of the Poisson equation (Vorobjev and Scheraga, 1997).

The method of Gabdoulline and Wade (1997) discussed above demonstrates an important fact: that it is not necessary to compute a full PB equation solution for the system consisting of more than one molecule. The effects due to the perturbation of the electric field of one molecule by another can be treated using a fairly simple model. However, in order to compute the pairwise interactions among the charges and induceable dipoles, it may prove to be advantageous to use multipole methods. Before a simulation, one would compute the electric fields of the individual molecules, and then use such an algorithm during the simulation itself to compute the forces and torques on the molecules.

Multiple Scales of Internal Motions

As mentioned above, efforts have been made to take advantage of the multiple length and time scales of large molecules to compute their functionally important motions. Some of the first efforts were *normal modes analyses* (Go, Noguti, and Nishikawa, 1983). The potential energy as a function of the atom's positions is approximated as a quadratic function around one local minimum, and the resulting Newton's equation of motion is diagonalized. The main weakness of this approach for simulation is that it assumes that the state of the molecule stays near that one minimum, when in reality the molecule's state is moving among the basins of many local minima. Another problem is the computational effort required to diagonalize the resulting system of equations; the algorithm for finding eigenvalues and eigenvectors scales as N^3, where N is the number of variables. Several approaches for partitioning the molecule into smaller parts, separately diagonalizing the equations of each part, and reassembling the resulting information have been used to improve the scaling (Hao and Scheraga, 1994; Durand, Trinquier, and Sanjouand, 1994).

Another approach, the *essential dynamics method*, is to run a long MD trajectory, generate the correlation matrix of the atoms' positions in the trajectory, and diagonalize the correlation matrix to obtain the principal components of the motions (Amadei and Linnsen, 1993). This avoids the limitation of analyzing only one potential energy valley, but requires a long simulation.

The approach taken by Chun et al. (2000), called MBO(N)D, performs normal analyses on small parts of the molecule, and then treats the parts as flexible bodies connected by rods and springs. The detailed force field between atoms is still used in the simulation, but reformulating the equations of motion in terms of flexible bodies allows the user to use a much larger time step. Other attempts have been made to partition biomolecules into a small number of rigid bodies, based on known conformation changes and movements (Nichols et al., 1995; Wriggers and Schulten, 1997).

Other approaches directly simplify the problem by introducing simplified potential energy functions. Hinsen, Thomas, and Field (1999) use the α-carbons and assume *a priori* that the important motions are the Fourier modes of an elastic solid. Bahar, Atilgan, and Erman (1997) use an even more simple approach. Neighboring α-carbons are connected by Hookean springs, and the thermal factors in the X-ray crystal structure can be matched by adjusting the spring constant.

Unfortunately, none of the computer algorithms discussed above are capable of making predictions of phenomena that simultaneously involve several different levels, either because they cannot span the different levels without excessively long simulations, or some of the levels are simplified by using *ad hoc* parameters. In order to address this problem, we briefly describe a new computer algorithm below.

Hierarchical Collective Motions Algorithm

In order to bridge the length and time scales in large biomolecules, we have developed a computer algorithm, the Hierarchical Collective Motions (HCM) method. The molecule is divided into small overlapping segments, a simulation is run on each segment in isolation, and the information is pieced together in a hierarchical manner to find the motions of the whole molecule. Each segment is treated as a flexible body with internal degrees of freedom, represented by *internal variables*, as well as overall translational and rotational degrees of freedom, represented by *external variables*. Along the way, information about the details of the rapid, small-scale motions is decoupled from the large-scale motions. At present, the method has been formulated for the case of overdamped, diffusional motion that is characteristic of large biomolecules, and its feasibility has been demonstrated on a very simple model that qualitatively represents an actin filament.

The equation of motion for an overdamped system subject to thermal motion is usually written as the *Brownian dynamics equation* (Ermak and McCammon, 1978)

$$d\mathbf{x} \quad = \quad -\mathbf{D} \cdot \frac{\partial V}{\partial \mathbf{x}} dt + \sqrt{2dt}\, \mathbf{S} \cdot \mathbf{W}$$

where the vector $\mathbf{x}$ represents the internal and external variables, matrix $\mathbf{D}$ is the *diffusivity matrix*, V is the potential energy, dt is the size of the time increment, $\mathbf{W}$ is a vector of uncorrelated random Gaussian variables of zero mean and unit variance, and $\mathbf{S}$ is a matrix such that $\mathbf{D} = \mathbf{S} \cdot \mathbf{S}^T$. In general, the potential energy and diffusivity matrix are functions of $\mathbf{x}$. If the assumption is made that $\mathbf{D}$ varies only with the orientation of the segment and V can be represented as the product of univariate functions, then Eq. 2 can be written in terms of transformed internal variables, such that the new internal variables move independently of each other in the absence of net force and torque. Mathematically, each new internal variable obeys a 1-D version of Eq. 2. The new, uncoupled internal variables also carry the same physical units, so their average fluctuation amplitudes can be compared. In physical systems, one often sees that most of the new internal variables have small fluctuation amplitudes, while a few have relatively large amplitudes. Loosely speaking, the large-amplitude variables represent "interesting motions", since they often represent large-scale motions that involve the whole segment. Conversely, it is usually possible to simplify the computational treatment of the small-amplitude variables, which represent localized, "uninteresting" motions.

Because the above assumptions about V and $\mathbf{D}$ are not exactly true for most physical systems, it is necessary to project the actual dynamics of the segment onto the idealized system described above. This is done by running a simulation on the segment, using the same algorithm that would be used to do a brute-force simulation on the whole molecule. The trajectory information is then used to compute the parameters that best describe $\mathbf{D}$ and V. Because the time scales associated with a small segment are much less than that for the whole molecule, the simulations can be relatively short. The potential energy is calculated by estimating the probability density function of each uncoupled internal variable and assuming a Boltzmann distribution. The diffusion matrix is estimated by analyzing the changes in variables between adjacent configurations in the trajectory.

It is advantageous to represent a physical system as a set of uncoupled variables, not only for conceptual simplicity but also for purposes of simulation. The time-step size of most simulation algorithms, including Brownian dynamics, are limited by the requirement that the derivatives of V with respect to the state variables not change much between updates; this limitation is at the heart of the difficulties discussed above. With one-variable systems, however, it is possible to devise algorithms that can use arbitrarily large time steps.

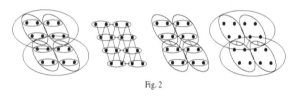

Fig. 2

Figure 1.   Illustration of HMC Method.

In Fig.1 is a description of a hypothetical 2-D molecule composed of 16 non-overlapping segments; this is used to illustrate the HCM method. The figure in the upper left illustrates how each pair of segments is joined together to form a new segment (represented by ellipses) at each level; the bottom 16 segments are represented as dots. These are the *primary segments*; they are used to describe the overall structure of the molecule. Not only are simulations run on the 16 bottom segments; they are also run on the 8, 4, and 2 segments at the higher levels. (The nature of the simulations on the upper-level segments is described below.) The whole molecule can also be considered a large segment. Segments that are next to each other but are not in the same primary segment on the next level still physically interact with each other. Thus, it is necessary to run simulations on these *secondary segments* as well; they are depicted on the different levels by lines joining their component segments. A similar scheme can be constructed in 3 dimensions.

The simulations on the bottom-level segments have been described; simulations on the second-level segments proceed in a similar manner, at the same level of detail as those for the bottom-level segments. However, the internal variables of the second-level segment are now comprised of the uncoupled internal variables of each component segment, plus 6 additional variables, the *intersegment variables*, that describe the relative positions and orientations of the two segments. The resulting trajectory is again projected onto the idealized system described above, and the uncoupled internal variables for the second-level segment are determined.

When the two component segments are each simulated in isolation, the uncoupled internal variables are, by definition, uncoupled. However, when the segments are simulated together in a second-level segment, the uncoupled internal variables of each segment are no longer necessarily uncoupled, because of physical interactions between the two segments. Fortunately, the small-amplitude motions discussed above are often localized in nature and are less likely to interact directly with motions on the other segment. So, it can be assumed that these motions continue to remain uncoupled. This is important, because it is not necessary to neglect them altogether, and they can still be independently stepped forward in time during a simulation. Deciding how many to decouple is done by successively decoupling variables starting from those with the smallest amplitudes, performing rapid simulations on the segment pair, and comparing the dynamics to a simulation with no decoupling. At this stage, the test simulations can be done very rapidly by taking large time steps along the uncoupled one-dimensional systems.

Next, simulations must be performed on segments made up of four bottom-level segments, or two second-level segments. These third-level simulations are performed differently then the previous ones. The four bottom-level segments are included in the simulation, and they are treated like flexible bodies. However, the decoupled variables in each of the four bodies are ignored. At this stage, simulations have already been run on all of the primary and secondary second-level segments within the third-level segment. From the resulting potential energy information, the forces and torques exerted by the lowest-level segments on each other, as well as the effective forces $\partial V/\partial x_i$ associated with their decoupled internal variables x_i are quickly computed at each step.

Once all forces and torques have been computed, the quartet of segments is stepped forward in time according to Eq. 2. The uncoupled internal and intersegment variables of the primary segments at the second level are computed, and these comprise the internal variables of the third-level segment (quartet).

This simulation is now much more efficient than the brute-force method. Even though the different motions are not yet uncoupled, larger time steps can be taken because the small amplitude motions have been decoupled and neglected for now. The internal variables are analyzed and decoupled as before. Running simulations on fourth- and higher-level segments follow the same procedure as the third-level segments. The

primary segments are used to construct the variables, and the secondary segments provide additional information on interactions among neighboring segments. Along the way, smaller-amplitude motions are decoupled, and when the top segment, comprising the whole molecule, is finally analyzed, all of the motions, large and small, have been decoupled into independent one-dimensional equations of motion.

In order to test this idea, a simple model that resembles an actin fiber was used. This model treated each monomer as a rigid body joined to each of its four immediate neighbors by three springs each, in order to favor the helical structure. The monomers were moved according to Brownian dynamics, and the translational and rotational diffusivities of each monomer followed Stokes Law, with no hydrodynamic interactions. A 16-mer was simulated and the largest bending mode was measured. Because of the stiffness of the springs, a billion time steps were required to resolve the slowest motions; the simulation took roughly 3 days on a Pentium III workstation.

The HCM method was performed on the same 16-mer; the bottom-level segments were adjacent pairs of monomers. A combined Monte-Carlo/Brownian dynamics algorithm was used to enhance sampling at each level. Because the structure of the oligimer repeats itself, only one monomer pair and one monomer quartet had to be simulated. The rms deviation and time autocorrelation function for the bending mode from the HCM method agreed closely with the results from the brute force Brownian dynamics simulation. The total processor time on the same workstation was roughly *15-20 minutes*. Further details are given in a forthcoming paper (Huber, 2000).

Although this was a simple model, not intended to closely resemble the real actin filament, it demonstrated the feasibility and efficiency of the HCM method. The model was non-linear and encompassed time scales that spanned several orders of magnitude. With substantial improvements to the actin model and extension of the algorithm to the Newtonian equations of molecular dynamics, the HCM algorithm has the potential to further probe the molecular basis of structure, dynamics, and function of actin and other large molecules.

Conclusion

Although proteins and other large molecules exhibit physical phenomena that span many diverse length and time scales, it is possible to design computer algorithms to take advantage of that fact. For modeling of biomoleculer interactions, the main challenges are computing the electrostatic forces between and the internal motions of the interacting molecules. In the near future, fast and accurate electrostatic interaction computations will be combined with flexible models of the molecules to create accurate models of biomolecular interactions. Ultimately, the ability to perform such simulations will lead to better models for describing drug action and bioseparations. Also, because these algorithms are not defeated by large disparities in length and time scales, there is hope that such methods could be used to simulate cell organelles and the cytoplasm interior.

References

Ajay and M. A. Murcko (1995). Computational methods to predict binding free energy in ligand-receptor complexes. *J. Med. Chem.*, **38**, 4953-4967.

Amadei, A. and A. Linssen (1993). Essential dynamics of proteins. *Proteins*, **17**, 412-425.

Bahar, I., Atilgan, A., and B. Erman (1997). Direct evaluation of thermal fluctuations in proteins using a single-parameter harmonic potential. *Fold. Des.*, **2**, 173-181.

Baker, N., Holst, M., and F. Wang (2000). Adaptive multilevel finite element solution of the Poisson-Boltzmann equation II: refinement at solvent accessible surface in biomolecular systems. *J. Comp. Chem.* (in press)

Briggs, B. L. (1987). *A Multigrid Tutorial*. SIAM, Philadelphia.

Chun, H., Padilla, C., Chin, D., Watanabe, M., Karlov, V., Alper, H., Soosaar, K., Blair, K., Becker, O., Caves, L., Nagle, R., Haney, and B. Farmer (2000). MBO(N)D: A multibody method for long-time molecular dynamics simulations. *J. Comp. Chem.*, **21**, 159-184.

Davis, M. E. and J. A. McCammon (1990). Electrostatics in biomolecular structure and dynamics. *Chem. Rev.*, **90**, 509-521.

Ding, H. Q., Karasawa, N., and W. A. Goddard (1992). Atomic level simulations on a million particles - the cell multipole method for coulomb and london nonbond interactions. *J. Chem. Phys.*, **97**, 4309-4315.

Durand, P., Trinquier, G., and Y. Sanejouand (1994). A new approach for determining low-frequency normal modes in macromolecules. *Biopolymers*, **34**, 759-771.

Frauenfelder, H., Sligar, S., and P. G. Wolynes (1991). The energy landscapes and motions of proteins. *Science*, **254**, 1598.

Gabdoulline, R. and R. Wade (1997). Simulation of the diffusional association of barnase and barstar. *Biophys. J.*, **72**, 1917-1929.

Gittes, F., Mickey, B., Nettleton, J., and J. Howard (1993). Flexural rigidity of microtubules and actin filaments measured from thermal fluctuations in shape. *J. Cell Biol.*, **120**, 923-34.

Go, N., Noguti, T., and T. Nishikawa (1983). Dynamics of a small globular protein in terms of low-frequency vibrational modes. *Proc. Natl. Acad. Sci. USA*, **80**, 3696-3700.

Greengard, L. and V. Rokhlin (1987). A fast algorithm for particle simulations. *J. Comp. Phys.*, **73**, 325-348.

Hao, H. M. and H. A. Scheraga (1994). Analyzing the normal mode dynamics of macromolecules by the component synthesis method: Residue clustering and multiple-component approach. *Biopolymers*, **34**, 321-335.

Hinsen, K., Thomas, A., and M. Field (1999). Analysis of domain motions in large proteins. *Proteins*, **34**, 369-382.

Honig, B. and A. Nicholls (1995). Classical electrostatics in biology and chemistry. *Science*, **268**, 1144-1149.

Horvath, D., van Belle, D., Lippens, G., and S. J. Wodak (1996). Development and parametrization of continuum solvent models. I. Models based on the boundary element method. *J. Chem. Phys.*, **104**, 6679-6695.

Huber, G. (1998). Future directions for combining molecular and continuum models in protein simulations. *Prog. in Biophys. & Mol. Biol.*, **69**, 483-496.

Huber, G. (2000). The hierarchical collective motions method for computing large-scale motions of biomolecules. *Biophys. J.*, submitted.

Kollman, P. (1993). Free energy calculations - applications to chemical and biochemical phenomena. *Chem. Rev.*, **93**, 2395-2417.

Leopold, P., Montal, M., and J. N. Onuchic (1992). Protein folding funnels: A kinetic approach to the sequence-structure relationship. *Proc. Natl. Acad. Sci. USA*, **89**, 8721.

Nichols, W., Rose, G., ten Eyck, L., and B. Zimm (1995). Rigid domains in proteins - an algorithmic approach to their identification. *Proteins*, **23**, 38-48.

Purisima, E. (1998). Fast summation boundary element method for calculating solvation free energies of macromolecules. *J. Comp. Chem.*, **19**, 1494-1504.

Rebello, C. A. and R. D. Ludescher (1998 Oct 13). Influence of tightly bound Mg2+ and Ca2+, nucleotides and phalloidin on the microsecond torsional flexibility of F-actin. *Biochem.*, **37**, 14529-38.

Sept, D., Elcock, A., and J. McCammon (1999). Computer simulations of actin polymerization can explain the barbed-pointed end asymmetry. *J. Mol. Biol.*, **294**, 1181-1189.

Sharp, K. A. and B. Honig (1990). Calculating total electrostatic energies with the nonlinear Poisson-Boltzmann equation. *J. Phys. Chem.*, **94**, 7684-7692.

Simonson, T., Perahia, D., and A. T. Brunger (1991). Microscopic theory of the dielectric properties of proteins. *Biophys. J.*, **59**, 670-690.

Tara, S., Helms, V., Straatsma, T., and J. McCammon (1999). Molecular dynamics of mouse acetylcholinesterase complexed with huperzine A. *Biopolymers*, **50**, 347-359.

Wriggers, W. and K. Schulten (1997). Protein domain movements: Detection of rigid domains and visualization of hinges in comparisons of atomic coordinates. *Proteins*, **29**, 1-14.

FORCE-FIELD DEVELOPMENT FOR
SIMULATIONS OF CONDENSED PHASES

A. Z. Panagiotopoulos[1]
IPST and Department of Chemical Engineering
University of Maryland
College Park, MD 20742-2431

Abstract

The focus of this review is on development of force fields for simulations of technologically important properties of fluids and materials. Of particular interest are models that can reproduce thermodynamic and transport properties over a broad range of temperatures and densities. Most existing force fields have been optimized to the configurational properties of isolated molecules and thermodynamic or structural properties of liquids near room temperature. Recently developed force fields are now available that also reproduce phase coexistence properties and critical parameters for selected systems, but they are not yet generally applicable to many systems of interest. Simulation methodologies for rapid determination of intermolecular potential parameters from experimental data are discussed. Two key unresolved questions remain, namely how to incorporate polarizability and other non-additive interactions, and the logistics of large-scale efforts to obtain parameters for broad classes of components.

Keywords

Force fields, Intermolecular potentials, Thermodynamic properties, Phase coexistence, Critical properties, Monte Carlo simulations, Molecular dynamics simulations.

Introduction

In 1965, Intel's co-founder Gordon Moore suggested that the computing power of a single chip doubles every eighteen months. We have now experienced several decades of growth in accordance with this statement. Some fundamental barriers to future growth may soon appear (Service, 1997), but today's desktop workstations have the processing power of supercomputers of just a few years ago. Molecular modelers have greatly benefited from this increasing computing power.

Molecular modeling and simulation algorithms have also improved over time. For example, configurational bias Monte Carlo methods (Siepmann and Frenkel, 1992; de Pablo *et al.*, 1992) enable sampling of conformations

and free energies of chain molecules with much higher computational efficiency than previous techniques.

Despite progress in computing hardware and simulation methodologies, prediction of properties of interest to industrial applications for condensed phases is not currently practiced on a routine basis. The lack of appropriate intermolecular potential functions is often quoted as *the* most important barrier for application of atomistic simulation methodologies to problems of industrial interest (Thompson, 1999, Table 4-2). A legitimate question is why this is still the case, given that many general force fields are now available that have been optimized to structural, bonding and thermodynamic properties of a large number chemical components. The

---

[1] Address after August 2000: Department of Chemical Engineering, Princeton University, Princeton NJ 08544-5263, USA

answer to this question is that many important properties are not currently considered during potential model development. In particular, most generally available force fields have not been tested for their ability to reproduce phase coexistence properties of pure fluids and mixtures over broad temperature ranges. Most existing force fields also have not been extensively tested for density or concentration conditions far removed from the pure liquid state at ambient temperature and pressure. Unfortunately, one cannot expect good performance of these force fields at conditions that have not been considered in their development. This, in turn, severely limits the reliability and predictive ability of molecular modeling methods.

The present review focuses on issues related to development of force fields for atomistic simulations of condensed phases, with particular emphasis on phase coexistence and other thermodynamic properties of fluids and amorphous materials. Issues related to force field development from *ab initio* calculations and hybrid methods are addressed by the contributions of M. Parrinello and R. Freisner in this volume.

The plan of this paper is as follows. Currently available general force fields for simulations of fluids are reviewed first. Computational methodologies for the routine determination of force fields accurate over broad temperature and density ranges are presented in the following section. Recently developed potentials for phase coexistence properties are described in detail. Possible additional force field features not included in current potentials and the issue of coordinated model development are discussed in the final section.

Currently Available General Force Fields

Many force fields for atomistic simulations have been developed over the past decades. No attempt is made in the present paper to provide an exhaustive review of all such efforts. Instead, the main approaches taken by previous researchers are briefly illustrated.

Most current force fields describe bonded interactions by stretching, bending and torsional potentials. The bonded interaction parameters are obtained by fitting to *ab initio* data, usually obtained at the Hartree-Fock level with large basis sets. Partial charges are used for describing electrostatic interactions. In recent years, the Ewald summation technique for handling the long-range character of the electrostatic interactions has become popular.

One approach for describing non-bonded interactions, pioneered in the early work by Jorgensen (1984), non-bonded potential parameters are obtained by fitting experimentally observed liquid densities and enthalpies and structural properties near room temperature. The latest versions of these models are "all-atom" force fields explicitly incorporating hydrogens.

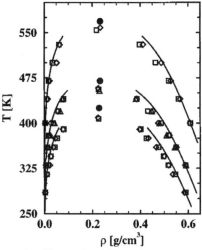

Figure 1. Vapor-liquid coexistence curves for variations of the OPLS all-atom force field for n-butane, n-pentane and n-octane. Points are simulation results and lines are experimental data. Reprinted with permission from Chen et al., 1998. © 1998 American Chemical Society.

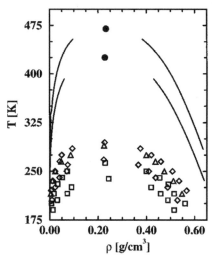

Figure 2. Vapor-liquid coexistence curves for variations of the MMFF force field for n-butane and n-pentane. Points are simulation results and lines are experimental data. Reprinted with permission from Chen et al., 1998. © 1998 American Chemical Society.

Examples include the OPLS (Jorgensen *et al.*, 1996), CHARMM (Mackerell *et al.*, 1995), AMBER (Cornell *et al.*, 1995) and COMPASS (Sun, 1998) force fields. The primary focus of these models is on biological and organic molecules in aqueous solution. As shown in Fig. 1 (from Chen *et al.*, 1998), for simple hydrocarbons, there is good, but not quantitative, agreement between calculated and experimental coexistence curves.

Another approach to obtaining force field parameters is to minimize the deviations between predicted and observed bond lengths, angles, vibrational frequencies, conformational energies, and minimum-energy conformations of clusters. A current example of this approach is the MMFF force field developed by Halgren (1996a-d). Unfortunately, even for simple hydrocarbon systems, this approach can result in completely unrealistic thermodynamic properties, as illustrated by the data in Fig. 2 (from Chen *et al.*, 1998).

The force fields described in this section share several common features. In particular, non-bonded dispersion interactions are described by Lennard-Jones 12-6 potentials. The only exception is the COMPASS force field, which uses a 9-6 potential. Only two-body interactions are included; three- and higher-order interactions are effectively incorporated in the values of the potential model parameters. The errors introduced by this approximation are most pronounced if calculations at densities and temperatures significantly different from the ones used in the parameterization. For example, the value of the dipole moment for the water molecule in these models is significantly higher than the gas-phase value. This is done in order to compensate for polarizability interactions that lead to a higher value of the dipole moment in liquid water, but lead to incorrect values for the second virial coefficient and energy at low densities.

Computational Methods

This section focuses on computational methodologies that have been found useful for rapid development of intermolecular potential models that can reproduce coexistence properties over a broad temperature range. A more extensive recent review of these methods is available (Panagiotopoulos, 2000a).

Gibbs Ensemble Monte Carlo

The Gibbs Ensemble Monte Carlo simulation methodology (Panagiotopoulos, 1987; Panagiotopoulos *et al.*, 1988; Smit *et al.*, 1989) enables direct simulations of phase equilibria in fluids. Gibbs ensemble simulations are performed in two separate microscopic regions, each within periodic boundary conditions. The thermodynamic requirements for phase coexistence are that each region should be in internal equilibrium, and that temperature, pressure and the chemical potentials of all components should be the same in the two regions. System temperature in Monte Carlo simulations is specified in advance. The remaining three conditions are satisfied by performing three types of Monte Carlo moves, displacements of particles within each region (to satisfy internal equilibrium), fluctuations in the volume of the two regions (to satisfy equality of pressures) and transfers of particles between regions (to satisfy equality of chemical potentials of all components). The method has been frequently used, in combination with

configurational bias sampling discussed in the following paragraph, to determine phase diagrams of a large number of fluids. Reviews of the Gibbs method and its applications are available (Panagiotopoulos 1995, 2000a).

Configurational Bias Sampling and Expanded Ensembles

The most common bottleneck in achieving convergence in methods that rely on particle transfers is the prohibitively low acceptance of transfer attempts. For dense fluid phases, especially for complex, orientation-dependent intermolecular potentials, configurations with "holes" in which an extra particle can be accommodated are highly improbable, and the converse step of removing a particle involves a large cost in energy.

Configurational-bias sampling techniques significantly improve sampling efficiency for Gibbs or grand canonical Monte Carlo simulations. The methods have been reviewed in detail by Frenkel (1995), Frenkel and Smit (1996), and Siepmann (1999). The methods trace their ancestry to biased sampling for lattice polymer configurations proposed by Rosenbluth and Rosenbluth (1955). Development of configurational-bias methods for canonical and grand canonical simulations and for continuous-space models took place in the early 1990's (Frenkel *et al.*, 1992; de Pablo *et al.*, 1992; Siepmann and Frenkel, 1992; Laso *et al.*, 1992).

Configurational-bias methods are based on segment-by-segment insertions or removals of a multisegment molecule. Several trial directions are attempted for every segment insertion, and a favorable growth direction is preferentially selected for the segment addition. This way, the acceptance probability of insertions is greatly enhanced.

For each segment growth or removal step, a correction factor (often called "Rosenbluth weight") is calculated. The product of the Rosenbluth weights of all steps is incorporated in the overall acceptance criterion for particle insertions and removals in order to correct for the bias introduced by the non-random growth along preferential directions.

Another approach for handling multisegment molecules is based on the concept of expanded ensembles (Lyubartsev *et al.*, 1992; Wilding and Müller, 1994; Vorontsov-Velyaminov *et al.*, 1996; Escobedo and de Pablo, 1996). Expanded ensembles for chain molecules construct a series of intermediate states for the molecule of interest, from a non-interacting (phantom) chain to the actual chain with all segments and interactions in place. These intermediate states can be semi-penetrable chains of the full length or shortened versions of the actual chain. Estimates of the free energy of the intermediate states are required to ensure roughly uniform sampling, as for thermodynamic and Hamiltonian scaling methods mentioned in the previous section. The advantage of expanded ensembles over configurational-bias methods is that arbitrarily complex long molecules can be sampled adequately, if sufficient computational effort is invested in constructing good approximations of the free energies of intermediate states.

Histogram Reweighting Methods

Early in the history of development of simulation methods it was realized that a single calculation can, in principle, be used to obtain information on the properties of a system for a range of state conditions (McDonald and Singer, 1976; Wood, 1968; Card and Valleau, 1970). However, the practical application of this concept was severely limited by the performance of computers available at the time. In more recent years, several groups have confirmed the usefulness of this concept, first in the context of simulations of spin systems and later for continuous-space fluids (Wilding, 1995; Panagiotopoulos *et al.*, 1998; Wilding, 1997; Potoff and Panagiotopoulos, 1998).

A histogram reweighting grand canonical Monte Carlo (GCMC) simulation for a one-component system is performed as follows (histogram reweighting for multicomponent systems is analogous to the one-component version). The simulation cell has a fixed volume V and is placed under periodic boundary conditions. The inverse temperature, $\beta=1/k_B T$ and the chemical potential, μ, are specified as input parameters to the simulation. Histogram reweighting requires collection of data for the probability $f(N,E)$ of occurrence of N particles in the simulation cell with total configurational energy in the vicinity E. This probability distribution function follows the relationship

$$f(N,E) = \frac{\Omega(N,V,E)\exp(-\beta E + \beta\mu N)}{\Xi(\mu,V,\beta)} \quad (1)$$

where $\Omega(N,V,E)$ is the microcanonical partition function (density of states) and $\Xi(\mu,V,\beta)$ is the grand partition function. Neither Ω nor Ξ are known at this stage, but Ξ is a constant for a run at given conditions. Since the left-hand side of Eq. 1 can be easily measured in a simulation, an estimate for Ω and its corresponding thermodynamic function, the entropy $S(N,V,E)=k_B \ln\Omega(N,V,E)$ can be obtained by a simple transformation of Eq. 1:

$$S(N,V,E)/k_B = \ln[f(N,E)] + \beta E - \beta\mu N + C \quad (2)$$

C is a run-specific constant. Eq. 2 is meaningful only over the range of densities and energies covered in a simulation. If two runs at different chemical potentials and temperatures have a region of overlap in the space of (N,E) sampled, then the entropy functions can be "merged'" by requiring that the functions are identical in the region of overlap.

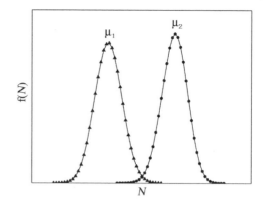

Figure 3. Schematic diagram of the probability $f(N)$ of occurrence of N particles for two GCMC runs of a pure component system at the same volume and temperature, but different chemical potentials, μ_1 and μ_2, respectively.

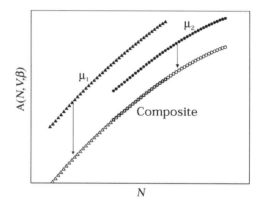

Figure 4. Illustration of the process of histogram transformation and combination. The abscissa is calculated as $\ln[f(N)] - \beta\mu N$ from the data of Figure 3.

To illustrate this concept, we make a one-dimensional projection of Eq. (1) to show histograms for two runs at different chemical potentials in Fig. 3. There is a range of N over which the two runs overlap. A transformation analogous to Eq. 2 gives the Helmholtz energy $A(N,V,\beta)$, within an additive constant, as shown in Fig. 4. The raw curves for μ_1 and μ_2 as well as a "composite" curve formed by shifting data for the two runs by the amount indicated by the arrows are indicated in Fig. 4. The combined curve provides information over the range of particle numbers, N, covered by the two runs.

Simulation data are subject to statistical uncertainties, which are particularly pronounced near the extremes of particle numbers and energies visited during a run. When data from multiple runs are combined as shown in Fig. 4, the question arises of how to determine the optimal amount by which to shift the raw data in order to obtain a global free energy function. Ferrenberg and Swendsen (1988) provided a solution to this problem that minimizes the differences between predicted and observed histograms. All thermodynamic quantities for the system over the range of densities and energies covered by the histograms can be obtained from the combined histograms.

In the absence of phase transitions or at temperatures near a critical point, the values of all observable quantities are independent of initial conditions, since free energy barriers for transitions between states are small or nonexistent. However, at lower temperatures, free-energy barriers for nucleation of new phases become increasingly larger. The states sampled at a given temperature and chemical potential depend on initial conditions, a phenomenon known as hysteresis. The histogram reweighting method can be applied to systems with large free-energy barriers for transitions between states, if care is taken to link all states of interest *via* reversible paths.

Thermodynamic and Hamiltonian Scaling

Two methods related to histogram reweighting are thermodynamic and Hamiltonian scaling Monte Carlo. Thermodynamic-scaling techniques proposed by Valleau (1991) are based on calculations in the constant pressure (*NPT*), rather than the grand canonical (μ*VT*) ensemble and provide information for the free energy over a range of volumes, rather a range of particle numbers. Thermodynamic scaling techniques can also be designed to cover a range of Hamiltonians (potential models) in the Gibbs (Kiyohara *et al.*, 1996) or Grand Canonical ensemble (Errington and Panagiotopoulos, 1998).

In their Hamiltonian-scaling form, the methods are particularly useful for optimizing parameters in intermolecular potential models to reproduce experimental data such as the coexisting densities and vapor pressures. Thermodynamic and Hamiltonian scaling methods require estimates for the free energy of the system as a function of conditions, so that the system can be forced to sample the range of states of interest with roughly uniform probability.

Gibbs-Duhem Integration

Most methods for determination of phase equilibria by simulation rely on particle insertions to equilibrate or determine the chemical potentials of the components. Methods that rely on insertions experience severe difficulties for dense or highly structured phases. If a point on the coexistence curve is known (e.g. from Gibbs ensemble simulations), the remarkable method of Kofke (1993a, 1993b) enables the calculation of a complete phase diagram from a series of constant-pressure simulations that do not involve any transfers of particles. The method enables calculations of solid-fluid coexistence (Agrawal and Kofke, 1995), for which other methods described in this paper are not applicable. The method and its applications have been recently reviewed (Kofke, 1999).

For one-component systems, the method is based on integration of the Clausius-Clapeyron equation over temperature,

$$\left(\frac{dP}{d\beta}\right)_{sat} = -\frac{\Delta H}{\beta \Delta V} \tag{3}$$

where "*sat*" indicates that the equation holds on the saturation line, and ΔH is the difference in enthalpy between the two coexisting phases. The right-hand side of Eq. 3 involves only "mechanical" quantities that can be simply determined in the course of a standard Monte Carlo or molecular dynamics simulation. From the known point on the coexistence curve, a change in temperature is chosen, and the saturation temperature at the new temperature is predicted from Eq. 3. Two independent simulations for the corresponding phases are performed at the new temperature, with gradual changes of the pressure as the simulations proceed to take into account the enthalpies and densities at the new temperature as they are being calculated.

Force Fields for Phase Coexistence Properties

In the recent years, availability of the methods described in the previous section has had an impact on force field model development by enabling determination of parameters that reproduce accurately phase coexistence properties over broad temperature ranges. The performance of models with respect to coexistence properties is of paramount importance for design of separation processes. However, there are many other reasons why such models are highly desirable. In particular, at low temperatures (near the triple point), the chemical potential of the liquid essentially determines the vapor pressure, so that models that do well for phase coexistence also describe accurately the free energy of the dense liquid. Near the critical point, phase coexistence involves phases with densities intermediate between those of a gas and a highly subcritical liquid. The requirements for good performance over a broad range of densities and temperatures are much more severe than the thermodynamic properties at a single temperature and density used in development of many currently available general force fields.

Hydrocarbons

Hydrocarbon molecules are ubiquitous in industrial processes and form the building blocks of biological systems. They are non-polar and consist of a small number of groups, thus making them the logical starting point for potential model development.

Three accurate united-atom potential sets for *n*-alkanes have appeared recently. The TRAPPE (Martin and Siepmann, 1998) and NERD models (Nath *et al.*, 1998) use the Lennard-Jones (12,6) potential to describe

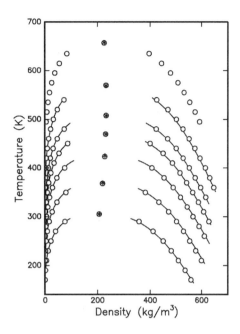

Figure 5. Phase diagrams of selected n-alkanes. The curves from bottom to top are for ethane, propane, butane, pentane, hexane, octane, and dodecane. Circles represent simulation results and lines are experimental data. Uncertainties are smaller than the size of the symbols. Reprinted by permission from Errington and Panagiotopoulos, 1999. ©1999, American Chemical Society.

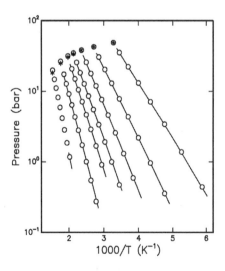

Figure 6. Vapor pressures of selected n-alkanes. The curves from right to left are for ethane, propane, butane, pentane, hexane, octane, and dodecane. Symbols are the same as for Fig. 5. Reprinted by permission from Errington and Panagiotopoulos, 1999. ©1999, American Chemical Society.

non-bonded interactions among methyl and methylene groups, while the model of Errington and Panagiotopoulos (1999) uses the exponential-6 functional form. All three reproduce the experimental phase diagrams and critical points. The exponential-6 model is slightly better with respect to representation of the vapor pressures. Figs. 5 and 6 illustrate the quality of representation of experimental data for the newer optimized models. De-viations from experimental data for the exponential-6 united atom model are comparable to those for a recently developed all-atom TRAPPE model (Chen and Siepmann, 1999).

United-atom potentials for branched alkanes have been developed by Cui *et al.*, (1997), Martin and Siepmann (1999) and for α-olephins by Spyriouni *et al.* (1999).

Polar Fluids

There have been several recent studies of the phase behavior of polar compounds such as *n*-alkanols (van Leeuwen, 1996), hydrogen sulfide (Kristof and Liszi, 1999) and carbon disulfide (Kristof *et al.*, 1996). However, no transferable force fields that can be used to obtain the phase behavior of polar fluids with reasonable accuracy are currently available. An example of the challenges in predictions of the behavior of polar fluids is provided by Visco and Kofke (1998), who studied two *ab initio* and one empirical potential model for hydrogen fluoride. The *ab initio* models did not model accurately the saturated liquid densities and none of the models predicted correctly the vapor pressure and heat of vaporization as functions of temperature.

Water

Because of the importance of water in biological and chemical systems, a large number of models have been proposed for atomistic simulations. Rigid fixed point charge models for water are often used in simulations of biological systems because of their simplicity and reasonable predictions of the structure and thermodynamics of liquid water at ambient conditions. Such models include the TIP4P (Jorgensen *et al.*, 1983), SPC (Berendsen *et al*, 1981) and SPC/E (Berendsen *et al.*, 1987) models. All represent water as a single Lennard-Jones (12,6) sphere within which are embedded positive and negative charges.

Table 1 compares the critical temperature T_c, density ρ_c, and vapor pressure, P_{VP}, at T=373 K of fixed point charge models for water. The "Exp-6" model (Errington and Panagiotopoulos, 1998) is also a fixed point charge model, but utilizes the exponential-6 functional form for the repulsion and dispersion interactions and was parameterized to the phase coexistence properties. The SPC/E model predicts a critical temperature near the experimental value, but underestimates the vapor pressure (and thus the chemical potential) by more than a factor of two. The SPC model has the correct vapor pressure, but severely underestimates the critical temperature. The TIP4P model has an even lower T_c and overestimates the vapor pressure. The Exp-6 model does very well for both T_c and P_{VP}, and has the closest ρ_c to the experimental value. However, the Exp-6 model does not do as well as the other models for structure of the liquid, in particular with respect to the oxygen-oxygen pair correlation function. It also has a dielectric constant lower than that the experimental value at room temperature.

Clearly, none of these simple fixed point charge models for water can adequately reproduce thermodynamic and structural properties over a broad temperature range. As also suggested earlier, the inclusion of only two-body interactions also results in a higher "effective" dipole moment relative to the isolated (gas phase) molecule.

Table 1. Comparison of Thermodynamic Properties of Fixed Point Charge Water Models. Simulation Data are from Errington and Panagiotopoulos (1998) for the SPC, SPC/E and Exp-6 Models and from Vlot et al., (1999) for the TIP4P Model.

Model	T_c (K)	ρ_c (kg/m^3)	P_{VP} (bar) at T=373 K
SPC	594 ± 1	271 ± 6	1.0 ± 0.2
SPC/E	639 ± 2	273 ± 9	0.37 ± 0.06
TIP4P	561 ± 3	290 ± 50	1.4 ± 0.1
Exp-6	646 ± 1	297 ± 5	0.95 ± 0.01
Experimental	647	322	1.01

Several polarizable models for water are available in the literature, but none seem to be better than the simple fixed point charged models with respect to the coexistence properties and critical parameters (Kiyohara et al., 1998). Some of the newer "fluctuating-charge" models developed from *ab initio* calculations (Stern *et al.*, 1999) have not yet been tested with respect to their phase coexistence behavior. Chen *et al.* (2000) performed an extensive search for optimized parameters for polarizable water models. Two optimized force fields were proposed. The SPC-pol-1 force field yields good saturated vapor and liquid densities, heats of vaporization and dielectric constants but does not represent well the liquid structure. The TIP4P-pol force field yields excellent liquid structures but does not perform well for the coexistence properties near the critical point. Chen *et al.* concluded that "we have to continue our quest for improved water force fields."

Mixtures

A key question for all force fields for atomistic simulations is their ability to predict properties of mixtures without use of additional adjustable parameters for the unlike-pair interactions. Extensive phase coexistence data are available for many mixtures, but

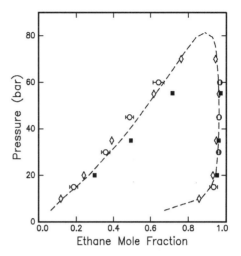

Figure 7. Phase diagram for a mixture of ethane and n-*heptane at* T=366 K. *Squares, diamonds and circles are used for the NERD, TraPPE and Exp-6 models, respectively. A dashed line is used for the experimental data. Reprinted by permission from Errington and Panagiotopoulos, 1999. ©1999, American Chemical Society.*

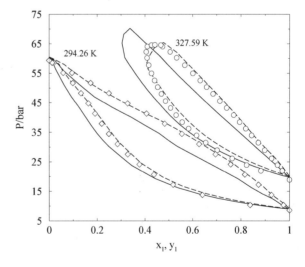

Figure 8. Propane/CO$_2$ pressure-composition diagram. Experimental data: T=327.59 K (open circles) T=294.26 K (open diamonds). GCMC simulations with the Lorentz-Berthelot (solid line) and Kong combining rules (dashed line). The average statistical uncertainties for the simulation data in pressure and composition are ±0.47 bar and ± 0.0050, respectively. Reprinted by permission from Potoff et al, 1999. ©1999 Taylor and Francis.

have not generally been used to validate proposed models. This is primarily because of the difficulties associated with simulations of phase coexistence for mixtures prior to the development of the computational methods discussed in the previous section.

For mixtures of non-polar components such as hydrocarbons or inert gases, there is considerable evidence that the newer force fields optimized to the phase equilibrium properties can be used for reliable predictions of mixture behavior. For example, Fig. 7 illustrates the predicted phase behavior of a mixture of ethane with n-heptane (Errington and Panagiotopoulos, 1999). Excellent agreement is obtained with experimental data without use of any mixture parameters.

An important unresolved question is that of combining rules for unlike-pair parameters of the repulsion and dispersion terms. The Lorentz-Berthelot combining rules (arithmetic mean for the diameters and geometric mean for potential well depths) have been used almost universally for unlike-pair parameters in predictive simulations. However, some recent results (Potoff et $al.$, 1999), suggest that different combining rules may be more appropriate for groups that differ significantly in size or polarity. Fig. 8 presents calculations for the pressure-composition diagram of propane with CO_2. The Kong (1973) combining rules, which result in less attractive unlike-pair interactions than the Lorentz-Berthelot rules, give better agreement with experimental data for this and similar mixtures. Potoff et $al.$ (1999) also find that for mixtures with large differences in polar character between the components, current models do not result in quantitative agreement with experiments.

Discussion and Future Directions

In this review, I have attempted to summarize the state-of-the art in force field development for simulations of condensed phases over large ranges of temperatures and densities. A number of computational methods that have been developed over the past few years enable the routine calculation of free energies and phase coexistence properties for model systems. These methods are now being used to obtain optimized intermolecular potentials for many substances. The resulting force fields have not yet attained the level of generality of currently available academic and commercial force fields, and they have not yet been incorporated in standard simulation codes. However, these developments are likely to occur in the near future.

Despite relatively rapid progress, there are many unresolved questions in the area. The first set of questions involve the functional form of the potential models. It is clear that for strongly interacting molecules, such as water, no simple two-body effective potential can quantitatively reproduce both structure and thermodynamics and additional interactions are

necessary. The most important of these is likely to be polarizability. Recent methodological developments (Martin et $al.$, 1998; Chen et $al.$, 2000) permit incorporation of polarizability in Monte Carlo calculations with a penalty of a factor of ten in CPU time relative to non-polarizable model calculations. This is still a rather large cost, so the quest for other methods to incorporate polarizability will likely continue.

An interesting new concept that may have some impact on potential model development is the use of finely discretized lattice models suggested by Panagiotopoulos and Kumar (1999) and Panagiotopoulos (2000b). The basic idea is that lattice models can be made to approximate their continuum analogs by increasing the ratio, ζ, of particle diameter to lattice spacing. The surprising result is that relatively low values of ζ are required for quantitative agreement of the calculated thermodynamic properties between lattice and off-lattice models. For Lennard-Jones and exponential-6 fluids, $\zeta=10$ results in agreement between lattice and continuum calculations for the phase envelope and critical parameters to within narrow statistical uncertainties. Even the structure of the fluids, which would have been expected to show some lattice artifacts, is identical between the two systems. The advantage of the finely discretized models is numerical speed. For ionic systems, the speed advantage is a factor of up to 100, while for the Lennard-Jones and exponential-6 interactions a factor between 10 and 20. A recent paper (Kumar, 2000) illustrates how this concept can be used to obtain quantitatively accurate lattice models for hydrocarbons.

Finally, a barrier to rapid force field development and testing with respect to free energies and phase coexistence predictions has been the lack of availability of portable, high quality codes appropriate for this purpose. Multiple groups (primarily academic) have been active in this area, but there has been relatively limited coordination and exchanges of data and codes.

Acknowledgements

Research in the author's group on which this manuscript is based was supported by the US Department of Energy, Office of Basic Energy Sciences, under grand DEFG02-98ER14014 and by the National Science Foundation under grant CTS-9509158. Profs. Sanat Kumar, Ilja Siepmann and Juan de Pablo provided preprints prior to publication.

References

Agrawal, R. and D. A. Kofke (1995). Solid-fluid coexistence for inverse-power potentials. *Phys. Rev. Lett.*, **74**, 122-125.

Berendsen, H. J. C., Postma, J. P. M., van Gusteren, W. F. and J. Hermans (1981). in *Intermolecular Forces*, Pullmann, B. (ed.); Reidel, Dodrecht, 331.

Card, D. N. and J. P. Valleau (1970). Monte Carlo study of the thermodynamics of electrolyte solutions. *J. Chem. Phys.*, **52**, 6232-40.

Chen, B., Martin, M. G., and I. J. Siepmann (1998). Thermodynamic properties of the Williams, OPLS-AA and MMFF94 all-atom force fields for normal alkanes. *J. Phys. Chem. B*, **102**, 2578-86.

Chen, B. and J. I. Siepmann (1999). Transferable potentials for phase equilibria. 3. Explicit-hydrogen description of normal alkanes. *J. Phys. Chem. B*, **103**, 5370-9.

Chen, B., Xing. J., and J. I. Siepmann (2000). Development of polarizable water force fields for phase equilibrium calculations. *J. Phys. Chem.*, **104**, 2391-40.

Cornell, W. D., Cieplak, P., Bayly, C. L., Gould, I. R., Merz, K. M. Jr, Ferguson, D. M., Spellmeyer, D. C., Fox, T., Caldwell, J. W., and P. A. Kollman (1995). A Second generation force field for the simulation of proteins, nucleic acids, and organic molecules. *J. Am. Chem. Soc.*, **117**, 5179-5197.

Cui, S. T., Cummings, P. T., and H. D. Cochran (1997). Configurational bias Gibbs ensemble Monte Carlo simulation of vapor-liquid equilibria of linear and short-branched alkanes. *Fluid Phase Equilibria*, **141**, 45-61.

De Pablo, J. J., Laso, M., and U. W. Suter (1992). Simulation of polyethylene above and below the melting point. *J. Chem. Phys.*, **96**, 2395-2403.

Errington, J. R. and A. Z. Panagiotopoulos (1998). Phase equilibria of the modified Buckingham exponential-6 potential from Hamiltonian scaling grand canonical Monte Carlo. *J. Chem. Phys.*, **109**, 1093-1100.

Errington, J. R. and A. Z. Panagiotopoulos (1999). A new intermolecular potential model for the n-alkane homologous series. *J. Phys. Chem. B*, **103**, 6314-22.

Escobedo, F. and J. J. de Pablo (1996). Expanded grand canonical and Gibbs ensemble Monte Carlo simulation of polymers. *J. Chem. Phys.*, **105**, 4391-4.

Ferrenberg, A.M. and R. H. Swendsen (1988). New Monte Carlo technique for studying phase transitions. *Phys. Rev. Lett.*, **61**, 2635-8.

Frenkel, D., Mooij, G. C. A. M., and B. Smit (1992). Novel scheme to study structural and thermal properties of continuously deformable molecules. *J. Phys.: Condens. Matter*, **4**, 3053-3076.

Frenkel, D. (1995) chapter in Baus, M., Rull, L. F., and J.-P. Ryckaert (editors) *Observation, Prediction and Simulation of Phase Transitions in Complex Fluids*. NATO ASI Ser. C, **460**, 357-419.

Frenkel, D. and B. Smit (1996). *Understanding Molecular Simulation*. Academic Press, London.

Jorgensen, W.L., Madura, J. D., and C. J. Swenson (1984). Optimized intermolecular potential functions for liquid hydrocarbons. *J. Am. Chem. Soc*, **106**, 6638-46.

Jorgensen, W. L, Maxwell, D. S., and J. Tirado-Rives (1996). Development and testing of the OPLS all-atom force field on conformational energetics and properties of organic liquids. *J. Am. Chem. Soc.*, **118**, 11225-36.

Halgren, T.A. (1996a). Merck molecular force field. I. Basis, form, scope, parametrization, and performance of MMFF94. *J. Comp. Chem.*, **17**, 491-519.

Halgren, T. A. (1996b). Merck molecular force field. II. MMFF94 van der Waals and electrostatic parameters for intermolecular interactions. *J. Comp. Chem.*, **17**, 520-552.

Halgren, T. A. (1996c). Merck molecular force field. III. Molecular geometries and vibrational frequencies for MMFF94. *J. Comp. Chem.*, **17**, 553-586.

Halgren, T. A. (1996d). Merck molecular force field. IV. Conformational energies and geometries for MMFF94. *J. Comp. Chem.*, **17**, 587-615.

Jorgensen, W.L., Chandrasekhar, J., and J. D. Madura (1983). Comparison of simple potential functions for simulating liquid water. *J. Chem. Phys.*, **79**, 926-35.

Jorgensen, W.L., Madura, J. D., and C. J. Swenson, (1984). Optimized Intermolecular Potential Functions for Liquid Hydrocarbons. *J. Am. Chem. Soc.*, **106**, 6638-46.

Kiyohara, K., Spyrouni, T., Gubbins K. E., and A. Z. Panagiotopoulos (1996). Thermodynamic-scaling Gibbs ensemble Monte Carlo: A new method for determination of phase coexistence properties of fluids. *Molec. Phys.*, **89**, 965-74.

Kiyohara, K., Gubbins, K. E., and A.Z. Panagiotopoulos, (1998). Phase coexistence properties of polarizable water models. *Molec. Phys.*, **94**, 803-8.

Kofke, D. A. (1993a). Gibbs-Duhem integration: A new method for direct evaluation of phase coexistence by molecular simulation. *Molec. Phys.*, **78**, 1331-6.

Kofke, D. A. (1993b). Direct evaluation of phase coexistence by molecular simulation via integration along the saturation curve. *J. Chem. Phys.*, **98**, 4149-4162.

Kofke, D. (1999). Semigrand canonical Monte Carlo simulations: Integration along coexistence lines. *Adv. Chem. Phys.*, **105**, 405-441.

Kong, C. L. (1973). Combining rules for intermolecular potential parameters. II. Rules for the Lennard-Jones (12-6) potential and the Morse potential. *J. Chem. Phys.*, **59**, 2464-2467.

Kristof, T. and J. Liszi (1997). Effective intermolecular potential for fluid hydrogen sulfide. *J. Phys. Chem. B*, **101**, 5480-3.

Kristof ,T., Liszi, J., and I. Szalai (1996). *Molec. Phys.* **89**, 931-942

Kumar, S. K. (2000). Quantitatively accurate lattice models for hydrocarbon polymers. preprint.

Laso, M., de Pablo, J. J., and U. W. Suter (1992). Simulation of phase equilibria for chain molecules. *J. Chem. Phys.*, **97**, 2817-19.

Lyubartsev, A., Martsinovski, A., and S. Shevkunov (1992). New approach to Monte Carlo calculation of the free energy: Method of expanded ensembles. *J. Chem. Phys.*, **96**, 1776-83.

Mackerell, A. D., Wiorkiewiczkuczera, J., and M. Karplus (1995). *J. Am. Chem. Soc.*, **117**, 11946.

Martin, M. G. and J. I. Siepmann (1998). Transferable potentials for phase equilibria. 1. United-atom description of n-alkanes. *J. Phys. Chem. B*, **102**, 2569-77.

Martin, M. G., Chen, B., and J. I. Siepmann (1998). A novel Monte Carlo algorithm for polarizable force fields: Application to a fluctuating charge model for water. *J. Chem. Phys.*, **108**, 3383-5.

Martin, M. G. and J. I. Siepmann (1999). Novel configurational-bias Monte Carlo method for branched molecules. Transferable potentials for phase equilibria. 2. United-atom description of branched alkanes. *J. Phys. Chem B*, **103**, 4508-17.

McDonald, I. R. and K. Singer (1967). Machine calculation of thermodynamic properties of a simple fluid at supercritical temperatures. *J. Chem. Phys.*, **47**, 4766-72.

Nath, S. K., Escobedo, F. A., and J. J. de Pablo (1998). On the simulation of vapor-liquid equilibria for alkanes. *J. Chem. Phys.*, **108**, 9905-11.

Panagiotopoulos, A.Z. (1987). Direct determination of phase coexistence properties of fluids by Monte Carlo simulation in a new statistical ensemble. *Molec. Phys.*, **61**, 813-826.

Panagiotopoulos, A. Z., Quirke, N., Stapleton, M., and D. J. Tildesley (1988). Phase equilibria by simulation in the Gibbs ensemble. Alternative derivation, generalization and extension to mixture and membrane equilibria. *Molec. Phys.*, **63**, 527-545.

Panagiotopoulos, A. Z. (1995). chapter in Baus, M., Rull, L. F., and J.-P. Ryckaert (editors) *Observation, Prediction and Simulation of Phase Transitions in Complex Fluids*. NATO ASI Ser. C, vol. 460, pp. 463-501.

Panagiotopoulos, A.Z., Wong, V., and M. A. Floriano (1998). Phase equilibria of lattice polymers from histogram reweighting Monte Carlo simulations. *Macromolecules*, **31**, 912-918.

Panagiotopoulos, A. Z. and S. K. Kumar (1999). Large lattice discretization effects on the phase coexistence of ionic fluids. *Phys. Rev. Lett.*, **83**, 2981-84.

Panagiotopoulos, A. Z. (2000a). Monte Carlo methods for phase equlibria of fluids. *J. Phys.: Condens. Matter*, **12**, R25-R52.

Panagiotopoulos, A. Z. (2000b). On the equivalence of continuum and lattice models for fluids. *J. Chem. Phys.*, **112**, 7132-7.

Potoff, J. J. and A. Z. Panagiotopoulos (1998). Critical point and phase behavior of the pure fluid and a Lennard-Jones mixture. *J. Chem. Phys.*, **109**, 10914-20.

Rosenbluth, M. N. and A. W. Rosenbluth, (1955). Monte Carlo calculations of the average extension of molecular chains. *J. Chem. Phys.*, **23**, 356-59.

Service, R. F. (1997). Can chip devices keep shrinking? *Science*, **275**, 1401-4.

Siepmann, J. I. and D. Frenkel, (1992). Configurational bias Monte Carlo: a new sampling method for flexible chains. *Molec. Phys.*, **75**, 59-70.

Siepmann, J. I. (1999). *Monte Carlo Methods in Chemical Physics*, **105**, 443-460.

Smit, B., De Smedt, Ph., and D. Frenkel (1989). Computer simulations in the Gibbs ensemble. *Molec. Phys.*, **68**, 931-50.

Spyriouni T., Economou, I. G., and D. N. Theodorou (1999). Molecular simulation of alpha-olefins using a new united-atom potential model: Vapor-liquid equilibria of pure compounds and mixtures. *J. Am. Chem. Soc.*, **121**, 3407-13.

Stern, H. A., Kaminski, G. A., Banks, J. L, Zhou, R., Berne, B. J., and R. A. Friesner (1999). Fluctuating charge, polarizable dipole, and combined models: Parameterization from *ab initio* quantum chemistry. *J. Phys. Chem. B*, **103**, 4730-7.

Sun, H. (1998). COMPASS: An *ab initio* force-field optimized for condensed-phase applications-Overview with details on alkane and benzene compounds. *J. Phys. Chem. B*, **102**, 7338-64.

Thompson, T. B., editor (1999). *Chemical Industry of the Future: Technology Roadmap for Computational Chemistry*, available from http://www.ccrhq.org/vision/index/roadmaps/complete.html.

Valleau, J.P. (1991). Density-scaling: a new Monte Carlo technique in statistical mechanics. *J. Comp. Phys.*, **96**, 193-216.

Van Leeuwen, M. (1996). Prediction of the vapour-liquid coexistence curve of alkanols by molecular simulation. *Molec. Phys.*, **87**, 87-101.

Visco, D. P. Jr. and D. A. Kofke (1998). Vapor-liquid equilibria and heat effects of hydrogen fluoride from molecular simulation. *J. Chem. Phys.*, **109**, 4015-27.

Vlot, M. J., Huinink, J., and J. P. van der Eerden (1999). Free energy calculations on systems of rigid molecules: An application to the TIP4P model of H2O. *J. Chem. Phys.*, **110**, 55-61.

Vorontsov-Velyaminov P., Broukhno, A., and T. Kuznetsova (1996). Free energy calculation by expanded ensemble method for lattice and continuous polymers. *J. Phys. Chem.*, **100**, 1153-8.

Wilding, N. B. and M. Müller (1994). Accurate measurements of the chemical potential of polymeric systems by Monte-Carlo simulation. *J. Chem. Phys.*, **101**, 4324-30.

Wilding, N. B. (1995). Critical point and coexistence-curve properties of the Lennard-Jones fluid: A finite-size scaling study. *Phys. Rev. E*, **52**, 602-11.

Wilding, N. B. (1997). Critical end point behavior in a binary fluid mixture. *Phys. Rev. E*, **55**, 6624-6631.

Wood, W. W. (1968). Monte Carlo calculations for hard disks in the isothermal-isobaric ensemble. *J. Chem. Phys.*, **48**, 415-434.

MOLECULAR-SCALE MODELING
OF REACTIONS AND SOLVATION

Donald G. Truhlar
Department of Chemistry and Supercomputer Institut
University of Minnesota
Minneapolis, MN 55455-0431

Abstract

This paper presents an overview of recent developments in molecular modeling of reactions and solvation. It includes both electronic structure and dynamics, and it considers both electronically adiabatic processes and electronically nonadiabatic ones.

Keywords

Continuous surface switching, Density functional theory, Diabatic representations, Direct dynamics, Ehrenfest method, Electron correlation, Hybrid Hartree-Fock-density-functional theory, Hybrid quantum mechanical molecular mechanical methods, Integrated molecular orbital methods, Multi-coefficient correlation methods, Multi-configuration molecular mechanics, Nonequilibrium effects, Specific reaction parameters, Tight binding theory, Trajectory surface hopping, Variational transition state theory.

Introduction

Molecular modeling of gas-phase molecular structure has achieved tremendous success in recent years. In many cases, theoretical structures and energies are more accurate than experimental results. Perhaps even more important though is the availability of reasonably reliable theoretical results where no experiments are practical or available. Theoretical chemistry can also be used to predict dynamics and spectroscopy, and it can model systems in the condensed phase as well as the gas-phase. This paper summarizes some recent advances in these areas, especially for molecular modeling of reaction energies, activation barriers, and solvation.

Chemical reactions may be divided into two types, electronically adiabatic and electronically nonadiabatic. Even when one calculates the electronic structure step-by-step along with the nuclear motion, the questions of interaction potentials and dynamics are conceptually separable. Interaction potentials will be discussed first, then dynamics. The discussion of dynamics focuses on rate constants and overall rate processes when state-to-state information is not required.

Interactions and Forces

For electronically adiabatic reactions, i.e., those in which the adiabatically defined electronic state does not change during the collision, the forces on the nuclei may be obtained as the gradient of a single-valued adiabatic potential energy function (PEF), also called the potential surface (Truhlar et al. 1987, Schatz 1989, Truhlar 1992). A potential surface is the expectation value of the electronic Hamiltonian (including nuclear repulsion) for a given electronic wave function with fixed nuclei; the surface is a function of the positions of the nuclei. Thus the calculations of forces between nuclei (i.e., between atoms and atomic ions) eventually reduce to calculating electronic wave functions as functions of coordinates. Technically, however, some electronic structure methods work directly with electronic densities rather than electronic wave functions, or—in a hybrid way—they work with wave functions and densities.

For electronically nonadiabatic reactions involving a small number N of electronic states, which is the usual situation for gas-phase photochemistry or condensed-phase photochemistry in non-electronically–conducting media, the forces on the nuclei may be derived from N adiabatic potential surfaces, V_1, V_2, ... V_N, and $(N^2 - N)/2$ matrix elements of the nuclear momentum operator, which is a $3N_{atoms}$-dimensional vector (thus one requires $3N_{atoms}(N^2 - N)/2$ elements). (Note: N_{atoms} is the number of atoms.) When, as is usually the case for interesting applications, the electronically nonadiabatic coupling may be assumed to arise from changes in orbital occupation rather than from the inability of electrons in a given orbital to completely follow the nuclei due to their finite speed, one may change the electronic representation from an adiabatic one to a diabatic one (Garrett and Truhlar 1981, Sidis 1992). In the diabatic representation, the forces may be derived to a good approximation—but not exactly (Mead and Truhlar 1982)—from N diabatic potential surfaces, U_{11}, U_{22}, ..., U_{NN}, which are again expectation values, but now calculated with diabatic electronic wave functions, plus $(N^2 - N)/2$ off-diagonal matrix elements of the electronic Hamiltonian, which is a scalar operator. This is a great simplification. The final situation is the case of chemical reactions in or on an electronically conducting medium, such as reactions at a metal surface. This case is hardest to treat although there is some justification for continuing to use potential surfaces, possibly with the addition of electron-hole pairs to the model (Chester 1961, Marcus, 1964, Wonchoba et al. 1994, Head-Gordon and Tully 1995, Billing 1998).

Electronically Adiabatic Reactions

Electronic structure calculations have so far had the most success in the electronically adiabatic case, in part because the adiabatic states are uniquely defined, and the variational principle provides a convenient way to approximate the lowest adiabatic state. Two developments in this area have been the backbone of most other advances:

- Coupled cluster-type methods including the "parentheses T" terms, which is an explicit-correlation approach, and which includes coupled clusters with single and double excitations and quasiperturbative triple excitations, CCSD(T), and also quadratic configuration interaction with single and double excitations and quasiperturbative triple excitations, (QCISD(T)) (Raghavachari and Anderson 1996), and

- A density-based approach using hybrid Hartree-Fock-density functional theory (HF-DFT), which is justified on the basis of the adiabatic connection method (ACM) (Harris and Jones 1974, Harris 1984, Becke 1993).

Whereas CCSD(T) and QCISD(T) methods and Møller-Plesset perturbation theory, which may be thought of as an approximation to these methods, employ explicit forms for the many-electron wave function, DFT is based—in principle—directly on the one-electron density. But even so-called pure DFT methods are usually based on the Kohn-Sham formulation, and therefore they use a wave function to calculate the kinetic energy; whereas so-called hybrid HF-DFT methods use a hybrid formulation for the exchange part of the potential energy, calculating a fraction of it by Hartree-Fock exchange integrals and the rest by a functional of the density or a functional of the density and its gradient.

Explicitly Correlated Wave Functions

The quasiperturbative triple excitations in the CCSD(T) and QCISD(T) methods include both a fourth-order term and a fifth-order term in a well balanced way (Scuseria and Lee 1990, Stanton 1997), and they are generally believed to give a result that is often close to full configuration interaction for a given one-electron basis (Lee and Scuseria 1995). However, (i) the methods are expensive even for moderate (unconverged) basis sets, and (ii) convergence with respect to the one-electron basis is slow. There are four strategies for coping with these problems.

The first is the Gaussian-x strategy (Curtiss and Raghavachari 1998, Curtiss et al. 1999). To overcome difficulty (i) by this strategy, one performs less complete correlated calculations with a large basis and QCISD(T) or CCSD(T) calculations with a small basis and assumes that higher-order correlation effects and extended-basis-set effects are additive. To make up for systematic errors and help overcome difficulty (ii), one also adds in a so-called higher-level correction, which contains 2 to 4 parameters and changes discontinuously when the number of paired electrons change, e.g., during the process R· + H· $\rightarrow$ RH.

The second strategy is basis set extrapolation. Originally, this strategy was aimed primarily at very small molecules by testing how well one could extrapolate from very large basis sets (e.g., correlation-consistent polarized valence quadruple zeta or higher) to the infinite basis set limit (Montgomery et al. 1994, Martin 1998). More recently, it has been developed for the more affordable choice of extrapolating from polarized double and triple zeta to the infinite basis set limit (Ochterski et al. 1996, Fast et al. 1999).

The third strategy is to extrapolate the many-electron level of the treatment correlation energy, either *ab initio* or semiempirically. Examples of the latter are the scaling external correlation (SEC) and scaling all correlation (SAC) methods (reviewed elsewhere: Corchado and Truhlar 1998).

The fourth strategy is the multi-coefficient (MC) approach that attempts to combine the advantages of all three of these approaches (Tratz et al. 1999, Fast and Truhlar 2000). The most general class of MC methods is called multi-coefficient correlation methods (MCCMs),

Table 1. Mean Unsigned Errors Per Bond for 82 Atomization Reactions[a].

Scaling	Method	MUE/bond (kcal)	Cost[b] (relative)
N^5 methods	*single-level*		
	MP2/cc-pVDZ	7.8	8
	MP2/MG3	2.4	100
	MP2/cc-pVTZ	2.0	160
	multi-level		
	SAC-MP2/cc-pVTZ	3.1	8
	SAC-MP2/cc-pVTZ	1.9	160
	MCCM-CO-MP2	1.5	170
	MCCO-CO-MP2; MG3; 6-31+G*	1.3	150
N^6 methods	*single-level*		
	MP4SDQ/cc-pVDZ	11.5	17
	CCSD/cc-pVDZ	11.9	120
	MP4SDQ/cc-pVTZ	5.6	250
	CCSD/cc-pVTZ	6.1	2300
	multi-level		
	SAC-MP4SDQ/cc-pVDZ	2.1	17
	MCSAC-QCISD/6-31G*	2.6	25
	MCCM-UT-QCISD/6-31G(2df,p); 6-31G*	1.0	74
	MCCM-UT-MP4SDQ	0.8	180
	MC-QCISD	0.7	120
N^7 methods	*single-level*		
	QCISD(T)/6-31G*	10.2	42
	MP4/cc-pVDZ	9.6	92
	QCISD(T)/6-311G(d,p)	8.0	130
	CCSD(T)/cc-pVDZ	10.3	260
	MP4/6-311G(2df,p)	3.5	540
	MP4/cc-pVTZ	2.6	2250
	CCSD(T)/cc-pVTZ	3.7	3600
	multi-level		
	Gaussian-3	0.36	720
	MCG3	0.31	210

[a] based on Tratz et al. (1999) and Fast and Truhlar (2000).

[b] Cost: an average based on single-point energies for molecules with 5–11 atoms. MP2/MG3 ≡ 100.

and it includes the MCSAC, EIB, MCCM-CO, MCCM-UT, MCCM-NM, MCG2, MCG3, G3S, and MC-QCISD approaches. In the two papers just referenced, we recommended eleven of these methods as providing the best compromise of accuracy and economy for calculating bond energies (or, more generally, for calculating any gas-phase thermochemical quantities for stable species). Table 1 compares the performance (mean unsigned errors) of some of these methods (which, like the Gaussian-x methods, may all be called multi-level or dual-level methods) to the more traditional single-level electronic structure methods. The computational efficiency (defined as the cost for achieving a given average accuracy) is striking, even though the cost function is based on molecules with inly 5–11 atoms. As system size is increased, the costs of the methods eventually scale as $(N_{atoms})^n$, where n is 5, 6, or 7. Because of these scaling properties, Table 1 has subsections for each of these values of n. For *very* large systems, only the smallest n calculations will be affordable. However, for any n, the multi-level methods are much more accurate for a given cost than the single-level methods. An important area of current research is linear scaling methods in which the algorithms are reformulated such that $n = 1$. Only recently has there been progress on developing linear scaling algorithms for methods with $n = 5$ and 6 (Ayala and Scuseria 1999, Schultz et al. 1999, Scuseria and Ayala 1999).

All electronic structure methods become less accurate when attention changes from stable species to transition states. A special problem occurs for doublet and triplet transition states; for these cases the most popular explicitly correlated treatments, which are based on unrestricted Møller-Plesset perturbation theory, often lead to wave functions that have significant contributions from states with an incorrect expectation value of S^2, where S is electronic spin. This so-called spin contamination problem is corrected by restricted open-shell Møller-Plesset theory, but it appears that the fix-up is not physical enough to guarantee more accurate results (Chuang et al. 2000). The CCSD and QCISD approaches, and especially CCSD(T) and QCISD(T), appear to be more accurate in this respect, but a truly satisfactory procedure seems to require a multi-configuration wave function as a zero-order state for perturbation theory or as a reference state for configuration interaction. Conventional configuration interaction methods, whether based on single-configuration references or multi-configuration references, are not size-cßonsistent, and this is a serious problem in predicting potential energy surfaces. One promising method that includes both size-consistency and multi-configuration reference states is the MRPT2 method of Hirao and Nakano (Nakano et al. 1997, Roberto-Neto et al. 1999, Kobayashi et al. 2000). It will be interesting to test this method further when computer code for analytic gradients (Nakano et al. 1998) becomes available. Other multireference perturbation theory methods may also be recommended (e.g., Finley and Freed 1995, Kozlowski et al. 1995, Staroverov and

Table 2. Mean Unsigned Errors in Barrier Heights and Reaction Energies for 20 Reactions[a].

Basis	Method	Mean |error| (kcal)		Cost[b] (relative)
		$V^{\ddagger}$	ΔE	
MG3	MP2	4.3	2.9	100
	B3LYP	3.9	2.1	50
	BH&HLYP	2.3	3.0	50
	MPW1PW91	3.6	1.5	50
	MPW1K	1.6	1.7	50
	QCISD(T)// MPW1K	1.3	1.4	100
6-31+G(d,p)	MPW1K	1.6	2.0	7
	QCISD(T)// MPW1K	3.1	3.1	12
multi-basis	MC-QCISD// MPW1K	1.5	1.2	10

[a] based on Lynch et al. (2000).

[b] Cost: an estimate for 10 energies, 10 gradients, and 1 Hessian based primarily on the reaction of amino radical with ethane. MP2/MG3 ≡ 100.

Davidson 1998, Celani and Werner 2000; see also additional references in Sect. 1.2).

With these caveats about single-configuration reference methods in mind, the MCCM methods may still be recommended for transition states if used with appropriate caution. We shall return to this subject in the section below.

Density Functional Methods

Local density functional methods, with or without gradient corrections, systematically overestimate binding energies (Van Leuwen and Baerends 1994, Yxkinten et al. 1997, Perdew 1999) and systematically underestimate barrier heights (Porezag and Pederson 1995, Baker et al. 1996). (Sometimes gradient-corrected functionals are called nonlocal, but here we reserve the word "nonlocal" for integral operators like the Hartree-Fock exchange operator.) Although the situation may improve in the future with better density functionals or by including multireference character in the computational scheme (see next section), at the present time pure DFT cannot be recommended for quantitative work on molecules (it is better for solids).

Hybrid HF-DFT

The poor performance of local density functional methods for molecular energies has been explained theoretically (Becke 2000) in terms of the implied locality of the exchange hole, and this explanation, combined with the adiabatic connection method (Harris and Jones 1974, Harris 1984), points to the need for incorporating nonlocal operators, such as Hartree-Fock exchange. Apparently, at least 12% Hartree-Fock exchange is required to incorporate the necessary nonlocality (Becke 2000). Methods that mix local density functionals (usually gradient-corrected) with nonlocal Hartree-Fock exchange (Becke 1993) are called hybrid HF-DFT methods. When compared to using explicitly correlated wave functions, hybrid HF-DFT methods provide remarkably high accuracy on a price-performance basis, and for moderate-sized and large systems, they will often give the best results of all affordable methods. The most widely used hybrid method is B3LYP (Stephens et al. 1994), with 20% Hartree-Fock exchange, although a method employing the same functionals but with 50% Hartree-Fock exchange has often been reported to be more accurate for chemical reaction barrier heights and for applications in surface science. The newer MPW1K hybrid method, with a modified exchange functional and 42.8% Hartree-Fock exchange, is even more accurate; see Table 2 (Lynch et al. 2000). Nevertheless, one must still be very cautious about the reliability of DFT and hybrid HF-DF theory for molecular modeling, as indicated by the poor performance of DFT for hydration effects (Hall et al. 2000). The methods can be very useful when used in situations where they have been validated, but they cannot be used universally or uncritically.

Even with improved functionals of the one-electron density, DFT or hybrid HF-DFT based on the Kohn-Sham equations and a non-interacting single-determinant reference state cannot handle all situations equally well. That is because such a DFT formulation includes dynamical (external) correlation effects but not static (internal) ones; inclusion of static correlation requires a multi-reference treatment, i.e., the use of a multiconfiguration reference state, or taking account of the two-electron density. In recent work good progress has been made in sorting out the relevant issues and taking initial steps toward practical multi-reference DFT (Sancho-Garciá et al. 2000, Gräfenstein and Cremer 2000).

A separation of the kinetic or potential energy into non-exchange and exchange parts is not unique, and one must be careful not to overinterpret individual components of the energy. Nevertheless such analyses may be useful for designing better DFT methods. The exchange kinetic energy, for example, is known to play a critical role in realistic descriptions of bonding (Goddard and Wilson 1972).

Another idea illustrated in Table 2 is the use of DFT to optimize geometries. The notation "X//Y" refers to optimizing a stationary point or calculating a reaction path at level Y and then carrying out energy calculations at these geometries (so-called single-point energies) by method X. Table 2 shows that single-point energies with the MC-QCISD method at MPW1K stationary-point geometries provides an economical way to calculate reasonably accurate barriers for chemical reactions. The // scheme can also be applied along reaction paths, where it is called ISPE or IRCmax, but it is less successful than a method called /// that involves limited geometry optimizations at the higher level (Chuang et al. 1999a).

Large Systems

Very large systems typically require the use of lower-level methods. Even DFT calculations may be too expensive for some applications, or they may be insufficiently accurate. One therefore turns to simpler, less expensive methods, especially those with parameters that may be adjusted semiempirically. Molecular mechanics (i.e., the use of valence force fields) is very popular for conformational analysis of stable species, but is inapplicable in its original form to reactions. Empirical valence bond (Åqvist and Warshel 1993) and multi-configuration molecular mechanics (Kim et al. 2000) provide two ways to extend molecular mechanics to reactive systems. Another approach is constituted by the so-called quantum-mechanical-molecular-mechanical (QM/MM) methods (Gao and Thompson 1998) in which quantum mechanical methods, often semiempirical ones, are used for a subsystem, and molecular mechanics is used for the rest of the system. When the subsystem is an entire solute not connected by bonds to the rest of the system, a method based on link atoms (Field et al. 1990) appears to be very useful. This approach has also been used successfully (Ridder et al. 1998) when the subsystem boundary is in the middle of a bond (which is essential for modeling the critical residues of enzymes), although in this case the generalized hybrid orbital method (Gao et al. 1998, Alhambra et al. 1999), which involves boundary atoms, and the use of adjusted connection atoms (Antes and Thiel 1999) appear more robust. Another QM/MM method that allows the boundary of the subsystem to be in the center of a bond is the integrated-molecular-orbital-molecular-orbital (IMOMO) method (Humbel et al. 1996, Svensson et al. 1996, Corchado and Truhlar 1998b) or the IMOHC extension for optimizing geometries (Corchado and Truhlar 1998a). The IMOMO method is very flexible and allows one to use explicitly correlated wave functions for the subsystems, which is called the correlated capped subsystem (CSS) method (Coitiño et al. 1996, Noland et al. 1997, Coitiño et al. 1997); it is particularly powerful for calculating substituent effects. Another approach, which is similar in some respects to IMOMO but both less flexible (a disadvantage) and simpler (a definite advantage) is the use of locally dense basis sets (LDBSs), as demonstrated by DiLabio and Wright (1998).

Atoms and molecules at nanosurfaces show different behavior and reactivity from that of ordinary materials. Furthermore, the properties of nanoparticles may be tuned by size as well as constitution. Modeling the reactivity of nanoparticles and large clusters efficiently may require specialized models. When one is modeling a specific system, one can often obtain more accurate results by using specific reaction parameters (SRPs) (Gonzalez-Lafont et al. 1991, Rossi and Truhlar 1995, Bash et al. 1996). For example, in work mentioned above, general parameters have been optimized to give improved thermochemistry via hybrid Hartree-Fock/density-functional theories (HF-DFT) (Becke 1993, Lynch et al. 2000). Although these parameters have been optimized to make the methods robust for a diverse set of applications, for any one application, other values of the parameters may prove more effective (Chuang et al. 1999b). This approach can be implemented by means of various bootstrap strategies. For example, if one can afford to calculate the barrier height and/or transition state geometry for a given reaction system by a more accurate method, one can adjust an SRP (or SRPs) to reproduce this critical region of the potential energy surface. Then the specifically recalibrated HF-DFT-SRP method provides an affordable engine for direct dynamics calculations that should yield useful results for that reaction system. Other strategies are also possible, e.g., one can optimize parameters on a smaller but similar system or on a subsystem of a larger system.

As we move up in size in the intermediate-size cluster systems, at some point we may find that even hybrid HF-DFT-SRP methods present a prohibitive cost. In that case there are two general strategies: (i) use an intrinsically less accurate level of theory, but use SRPs to calibrate it for the system of interest; (ii) use multilevel techniques based on treating a subset of the atoms in the cluster at a higher level than the full cluster.

Ab initio and hybrid HF-DF electronic structure techniques can be used to understand which lower-level methods are most reliable for various kinds of systems (e.g., systems without ions or polar bonds) and, when appropriate, to validate lower-level methods for such properties. Even more powerful is the use of high-level methods to parameterize semiempirical models that can be applied to larger systems. There is a hierarchy of such semiempirical methods that may be used. With respect to using lower level theory with SRPs we note that we have had considerable success using neglect-of-diatomic-differential-overlap (Pople and Beveridge 1970) theories as the lower level (Gonzalez-Lafont et al. 1991, Liu et al. 1993, Corchado et al. 1995, Chuang et al. 1999b). One can use genetic algorithms to optimize the parameters, a procedure that we have demonstrated successfully for other systems (Rossi and Truhlar 1995, Bash et al. 1996, Liu and Truhlar 2000). For even larger clusters, one can use tight binding theory (Sutton et al. 1988, Ducastelle 1991, Galli and Mauri 1994, Wang and Mak 1995, Wasserman et al. 1996, Elstner et al. 1998, Horsfield 1997, Galli et al. 1998, Liu and Truhlar 2000) to generate the potentials. Tight-binding methods are especially alluring when reformulated so the computational cost scales linearly with N_{atoms} (Goedecker and Teter 1995, Jayanthi et al. 1996, Horsfeld et al. 1998, Sternberg et al.

1999, Bowler and Gillan 1999). As an example of the size limitations, nonorthogonal, non-self-consistent tight binding could be used for long simulations of up to 5000 atoms, but nonorthogonal, self-consistent tight binding is probably most useful for long simulations in which the system size is restricted to 2000 or less atoms. As a starting point, it is important to understand the accuracies of various general levels as well as to have universal parameterizations available as starting points for specific reaction parameterization. Tight-binding methods have been most successful for homonuclear systems (e.g., silicon) and systems where all atoms have similar electronegativity (e.g., hydrocarbons).

Potential surfaces for liquid-phase condensed-phase systems may be modeled in the full space of solute and solvent (this is called the explicit-solvent approach) (for collections of articles, see Jensen and Truhlar 1987, Gao and Thompson 1998), or the effect of solvent may be taken into account implicitly by replacing the solute potential energy by a potential of mean force (for a recent review, see Cramer and Truhlar 1999). The explicit approach includes both nonequilibrum solvation effects and equilibrium ones, whereas the implicit approach usually includes only equilibrium solvation. In the implicit approach, nonequilibrium effects, if significant, can be added by using collective solvent coordinates representing the deviation from equilibrium (Chuang and Truhlar 1999). Explicit methods also have the advantage of providing a seemingly more natural way to include specific interactions in the first solvation shell, although these can be included in an average way in implicit models via atomic surface tensions. However, explicit-solvent methods have several severe drawbacks, most especially that they are expensive. As a result of this chief drawback, such treatments usually neglect solute and/or solvent polarizability effects or treat these in a parameterized way.

Whereas explicit-solvent models require new parameters for every new solvent to be considered, implicit-solvent models have been parameterized in a "universal" way for general organic solvents by representing the solvents in terms of a small number (about a half dozen) of solvent descriptors such as dielectric constant, hydrogen bond acidity, and so fourth (Zhu et al. 1998, Hawkins et al. 1999). The dielectric constant accounts for bulk electric polarization of the solvent, and the other descriptors account for deviations from this due to cavitation, short-range forces, such as dispersion forces and hydrogen bonding, and solvophobic effects, if any. A critical element in calculating the electric polarization effects is a realistic model of liquid-phase electrostatics. Class IV charges calculated by the self-consistent reaction field method in the liquid phase provide a semiempirical, low-cost means of including solute polarizability in such calculations (Zhu et al. 1998).

Electronically Nonadiabatic Reactions and Spectroscopy

The best current methods for calculating excited-state potential curves or sets of coupled ground and excited surfaces are multiconfiguration SCF (MCSCF) calculations combined with perturbation theory for dynamical electron correlation (both diagonalize-then-perturb, e.g., CASPT2 (Roos et al. 1994, Roos et al. 1996, Merchan et al. 1999), and perturb-then-diagonalize, e.g., QCMRPT2 (Nakano 1993, Hoffmann and Khait 1999), procedures are available) and MCSCF followed by multi-reference configuration interaction (MRCI) (Harrison and Shepard 1994, Werner 1995, Dachsel et al. 1977). Diabatic states are usually defined by using valence bond configurations, or state-averaged natural orbitals (García et al. 1997, Klüner et al. 1999) but more systematic (and systematically improvable) methods are available (Atchity and Ruedenberg 1997), although they are not still completely general.

Recent work has shown progress in practical extensions of DFT methods to excited states, and there is now considerable activity in this area.

A problem on the borderline between structure and dynamics is the calculation of vertical excitation spectra in the condensed phase. By the Franck-Condon principle, the excitation is sudden on the time scale of nuclear motion, but electronic polarization occurs on a faster time scale. Recent progress has been made by combining a two-time-scale treatment of electric polarization of the solvent (treated as a continuous dielectric medium) with additional terms for dispersion effects and hydrogen bonding (Aguilar et al. 1993, Li et al. 2000).

Dynamics

Electronically Adiabatic Processes

A variety of molecular dynamics methods can be used for the dynamics of the interacting atoms once accurate forces are known. The most straightforward methods are based on classical trajectories. For statistically averaged (i.e., microcanonical or canonical) chemical reaction rate constants, generalized transition state theory has some advantages. One advantage of transition state theory is that although the original fundamental justification of the theory was classical, it has been shown that quantal effects may be incorporated by a two-step process (Garrett et al. 1980, Truhlar et al. 1982, Isaacson and Truhlar 1982). First, one replaces the classical phase space integrals that define the partition functions of transition state theory by quantum mechanical sums over states. This quantizes all degrees of freedom of the reactant and all degrees of freedom of the transition state except the reaction coordinate (which is missing in the transition state partition coefficient). This is done prior to optimizing the location of the variational transition state so as to include quantum effects on the location of the variational transition state as well as on the reactive flux that passes through it. In the second step, one multiplies by a transmission coefficient that accounts for the competition between tunneling and overbarrier processes; this includes the quantum effects on the reaction-coordinate motion. Variational transition state theory with semiclassical multidimensional transmission coefficients (VTST/SMT) has been well validated against accurate quantum dynamics for a diverse set of few-body systems (Allison and Truhlar 1998), but accurate quantum dynamics rapidly becomes prohibitively expensive as system size grows.

At a deeper level, we now recognize that the transition state partition function, usually interpreted as a Boltzmann average of sharp energy levels of the transition state, can be also interpreted as a Boltzmann average of transition state resonances with finite energy widths (due to their finite lifetimes) (Truhlar and Garrett 1992) or as a Boltzmann average of the transmission probability for flux passing the quantized dynamical bottlenecks (Chatfield et al. 1992, Chatfield et al. 1996)— either of these methods automatically includes tunneling and other quantum effects on the reaction coordinate, and they provide an alternative to the two-step procedure.

Other (closely related) approaches to including quantum effects in transition state theory are called path integral quantum transition state theory, PI-QTST (Gillan 1988, Voth et al. 1989; for a comparison to VTST/MT, see Truhlar et al. 1996) and the instanton approach (Miller 1975, Bendarskii et al. 1993, Smedarchina et al. 1995). The PI-QTST method has similar accuracy to VTST/MT for symmetric bimolecular reactions in regions of its validity (McRae et al. 1992), and work is in progress on extending it to asymmetric systems and metastable potentials (Jang et al. 1999). The instanton approach has had considerable success for proton tunneling rearrangements (Fernández-Ramos et al. 1999), but in its current form it is not applicable to bimolecular reactions or to unimolecular reactions with reaction paths involving large-amplitude motions.

An alternative approach that is being developed for proton and hydride transfer reactions is to treat the transferred atom quantum mechanically with the rest of the system treated by classical mechanics (Hammes-Schiffer 1998).

It is useful to compare transition state theory to classical trajectories even in a classical mechanical world. If a rate process is dominated by a single dynamical bottleneck region, a very efficient way to calculate the reactive flux by trajectories would be to sample at the dynamical bottleneck region and integrate the trajectories just far enough forward and backward to count how many times they recross the bottleneck region. If one has identified a true dynamical bottleneck, the trajectories with positive momentum at the saddle point will proceed directly to products and those with negative momentum along the reaction coordinate will proceed directly to

reactants, where "directly" means without recrossing the dynamical bottleneck. This no-recrossing limit is transition state theory, and it is often an excellent approximation (Truhlar and Garrett 1980, Truhlar et al. 1996). Thus, in a classical world, transition state theory may be thought of as an efficient way to carry out trajectory simulations, i.e., as a form of rare event sampling.

When one considers condensed-phase systems, the number of possible dynamical bottlenecks can become large. Nevertheless in some cases one can still calculate all the rate constants for elementary steps and predict the complex-systems kinetics by a master equation or by kinetic Monte Carlo algorithms   However, when the number of possible elementary processes becomes too large to catalog, a classical mechanical simulation of the entire complex system may still be used. At first it might seem that transition state theory becomes less useful here, but that is not true. Voter has shown that combining transition state theory ideas with full-system classical dynamics provides an efficient means of rare event sampling even when the dynamical bottlenecks are not pre-identified; the resulting method is called hyperdynamics (Voter 1997, Voter and Sorensen 1999, Rudd and Voter 1999). In a rough description, hyperdynamics consists in flooding (filling in) the minima to make it easier to surmount the barriers. This may be compared to an alternative strategy (Ota and Brünger 1997) of scaling down the barriers when the goal is the computation of free energies of flexible molecules rather than dynamics.

If energy transfer processes that replenish the reactant Boltzmann distribution for reactive states are slow compared to reaction, or if energy transfer events that stabilize highly reactive states of reactants formed by reverse reaction are slow, then one must add corrections for nonequilibrium effects to either trajectory calculations or to transition state theory, because the assumption that reaction is sampling an equilibrium ensemble breaks down (Lim and Truhlar 1986). A second circumstance in which one must consider nonequilibrium effects is when one or more of the reactants is a solute, and it is assumed that the solvent is at equilibrium with it. This is often a good first-order assumption, but we must consider quantitative nonequilibrium corrections (Chuang and Truhlar 1999).

Direct dynamics is the use of electronic structure theory to calculate energies, gradients, and Hessians on an "as needed" basis, or perhaps by local interpolation, but without fitting this data to a global or semiglobal potential energy function (Wang and Karplus 1973, Leforestier 1978, Truhlar et al. 1982). In other words, the dynamics are generated directly from the electronic structure calculations. Car and Parrinello (1985) pioneered an extended Lagrangian formulation for the interface between the two elements in direct dynamics calculations combining density functional theory (DFT) for the electronic structure element with purely classical trajectories for the dynamics. In their method, the electronic degrees of freedom are treated by fictitious dynamics while the nuclei are treated with classical dynamics. Direct dynamics has also been widely used with other computational strategies and with other dynamics methods. For example, there has been a considerable amount of work on direct dynamics methods for generalized transition state theory in the gas phase (Truhlar and Gordon 1990, Truhlar 1995, Corchado et al. 1998c, Chuang et al. 1999a) and in the liquid state (Chuang et al. 1999b).

An especially difficult challenge for condensed-phase dynamical studies is the incorporation of quantal dynamical effects into enzyme catalysis, but recently progress has been reported in this area by two especially promising approaches (Hwang and Warshel 1996, Alhambra et al. 1999, Alhambra et al. 2000).

Electronically Nonadiabatic Dynamics

The treatment of electronically nonadiabatic collisions is much less developed than the treatment of reactions that proceed on a single potential energy surface. In some cases when there is a dense manifold of closely coupled electronic states, one may use statistical theories even for electronic degrees of freedom (Truhlar and Dixon 1979). For the coupling of the ground state to this manifold or for couplings among a few low-lying, widely separated electronic states, there are three main approaches under current study: trajectory surface hopping (Tully 1998), self-consistent potential methods (Meyer and Miller 1979, Topaler et al. 1998), and density matrix evolution (Martens and Fang 1997). Recently we have developed a method, called continuous surface switching (CSS), that combines some of the more desirable features of the TSH and Ehrenfest methods, without having their worst drawbacks (Volobuev et al. 2000, Hack and Truhlar 2000). An alternative method that accomplishes some of the same objectives by using a stochastic Schrödinger equation has also been proposed (Prezhdo 1999).

Methods are also available for the treatment of electronically nonadiabatic processes in condensed-phase systems (Prezhdo and Rossky 1997, Rabani et al. 1999), and a review is available (Egorov et al. 1999).

Concluding Remarks

It is now well established that molecular-scale modeling competes with experiment for accuracy and convenience on small systems, and the size of the system that can be treated with useful accuracy is rapidly increasing. For complex systems, molecular-scale modeling is not only a technique for predicting accurate numerical values of system parameters, but also it serves as a techniques of gaining insight and understanding.

Acknowledgments

I am grateful to my collaborators (see references) for their contributions to the work discussed here. The author's research on gas-phase variational transition state theory and its interface with electronic structure calculations is supported by the U. S. Department of Energy, Office of Basic Energy Sciences. Other research described in this report is supported by the National Science Foundation.

References

Aguilar, M. A., F. J. Olivares del Valle, and J. Tomasi (1993). Nonequilibrium solvation: An ab initio quantum-mechanical method in the continuum-cavity-model approximation. *J. Phys. Chem.*, **99**, 7375–7384.

Alhambra, C., J. Gao, J. C. Corchado, J. Villà, and D. G. Truhlar (1999). Quantum mechanical dynamical effects in an enzyme-catalyzed proton-transfer reaction, *J. Am. Chem. Soc.*, **121**, 2253–2258.

Alhambra, C., J. C. Corchado, M. L. Sánchez, J. Gao, and D. G. Truhlar (2000). Quantum dynamics of hydride transfer in enzyme catalysis. to be published.

Allison, T. C. and D. G. Truhlar (1998). Testing the accuracy of practical semiclassical methods: variational transition state theory with optimized multidimensional tunneling. In *Modern Methods for Multidimensional Dynamics Computations in Chemistry*, D. L. Thompson (Ed.), World Scientific, Singapore, 1998, 618–712.

Antes, I. and W. Thiel (1999). Adjusted connection atoms for combined quantum mechanical and molecular mechanical methods. *J. Phys. Chem. A*, **103**, 9290–9265.

Atchity, G. J. and K. Ruedenberg (1997). Determination of diabatic states through enforcement of configurational uniformity. *Theor. Chem. Acc.*, **97**, 47–58.

Ayala, P. Y. and G. E. Scuseria (1999). Linear scaling second-order Møller-Plesset theory in the atomic orbital basis for large molecular systems. *J. Chem. Phys.*, **110**, 3660–3671.

Baker, J., M. Muir, J. Andzelm, and A. Scheiner (1996). Hybrid Hartree-Fock density-functional theory functionals: The adiabatic connection method. In *Chemical Applications of Density Functional Theory* (ACS Symposium Series, Vol. 629), B. B. Laird, R. B. Ross, and T. Ziegler (Eds.), American Chemical Society, Washington, 342–367.

Bash, P. A., L. L. Ho, A. D. MacKerell, D. Levine, and P. Hallstrom (1996). Progress toward chemical accuracy in the computer simulation of condensed phase reactions. *Proc. Natl. Acad. Sci. U. S. A.*, **93**, 3698–703.

Becke, A. D. (1993). Density-functional thermochemistry. III. The role of exact exchange. *J. Chem. Phys.* **98**, 5648–5652.

Becke, A. D. (2000). Simulation of delocalized exchange by local density functionals. *J. Chem. Phys.*, **112**, 4020–4026.

Bendarskii, V. A., D. E. Makarov, and C. H. Wight (1994). Chemical dynamics at low temperatures. *Adv. Chem. Phys.*, **88**, 1–385.

Billing, G. D. (1998). Vibrational relaxation of adsorbed molecules by coupling to electron-hole pair excitation. *Chem. Phys. Lett.*, **290**, 150–154.

Bowler, D. R. and M. J. Gillan (1999). Density matrices in $O(N)$ electronic structure calculations: theory and applications. *Comp. Phys. Comm.*, **120**, 95–108.

Car, R. and M. Parrinello (1985). Unified approach for molecular dynamics and density functional theory. *Phys. Rev. Lett.* **55**, 2471–2474.

Celani, P. and H.-J. Werner (2000). Multireference perturbation theory for large restricted and selected active space reference wave functions. *J. Chem. Phys.*, **112**, 5546–5557.

Chatfield, D. C., R. S. Friedman, D. W. Schwenke, and D. G. Truhlar (1992). Control of chemical reactivity by quantized transition states. *J. Phys. Chem.*, **96**, 2414–2421.

Chatfield, D. C., R. S. Friedman, S. L. Mielke, G. C. Lynch, T. C. Allison, D. G. Truhlar, and D. W. Schwenke (1996). Computational spectroscopy of the transition state. In *Dynamics of Molecules and Chemical Reactions*, R. E. Wyatt and J. Z. H. Zhang, Eds., Marcel Dekker, New York, 323–386.

Chester, G. V. (1961). The theory of the interaction of electrons with lattice vibrations in metals. *Adv. Phys.*, **10**, 357–400.

Chuang, Y.-Y., J. C. Corchado, and D. G. Truhlar (1999a). Mapped interpolation scheme for single-point energy corrections in reaction rate calculations and a critical evaluation of dual-level reaction-path dynamics methods. *J. Phys. Chem. A*, **103**, 1140–1149.

Chuang, Y.-Y., M. L. Radhakrishnan, P. L. Fast, C. J. Cramer, and D. G. Truhlar (1999b). Direct dynamics for free radical kinetics in solution: Solvent effect on the rate constant for the reaction of methanol with atomic hydrogen. *J. Phys. Chem.*, **103**, 4893–4909.

Chuang, Y.-Y. and D. G. Truhlar (1999). Nonequilibrium solvation effects for a polyatomic reaction in solution. *J. Amer. Chem. Soc.*, **121**, 10157–10167.

Chuang, Y.-Y., E. L. Coitiño, and D. G. Truhlar (2000). How should we calculate transition state geometries for radical reactions? The effect of spin contamination on the prediction of geometries for open-shell saddle points. *J. Phys. Chem. A*, **104**, 446–450.

Coitiño, E. L., D. G. Truhlar, and K. Morokuma (1996). Correlated capped subsystem calculations as a way to include electron correlation locally: A test for substituent effects on bond energies, *Chem. Phys. Lett.*, **259**, 159–164.

Coitiño, E. L. and D. G. Truhlar (1997). Systematic analysis of bond energies calculated by the integrated molecular orbital-molecular orbital method. *J. Phys. Chem. A*, **101**, 4641–4645.

Corchado, J. C., J. Espinosa-Garcia, W.-P. Hu, I. Rossi, and D. G. Truhlar (1995). Dual-level reaction-path dynamics (The /// approach to VTST with semiclassical tunneling). Application to $OH + NH_3 \rightarrow H_2O + NH_2$. *J. Phys. Chem.*, **99**, 687–694.

Corchado, J. and D. G. Truhlar (1998a). Integrated molecular orbital method with harmonic cap for molecular forces and its application to geometry optimization and the calculation of vibrational frequencies. *J. Phys. Chem. A*, **102**, 1895–1898.

Corchado, J. and D. G. Truhlar (1998b). Dual-level methods for electronic structure calculations that use quantum mechanics as the lower level. In *Combined Quantum Mechanical and Molecular Mechanical Methods* (ACS Symposium Series, Vol. 712), J. Gao and M. A. Thompson (Eds.), American Chemical Society, Washington, 106–127.

Corchado, J. and D. G. Truhlar (1998c). Interpolated variational transition state theory by mapping. *J. Phys. Chem. A*, **102**, 2424–2438.

Cramer, C. J. and D. G. Truhlar (1999). Implicit solvation models: equilibria, structure, spectra, and dynamics, *Chem. Rev.*, **99**, 2161–2200.

Curtiss, L. A. and K. Raghavachari (1998). Computational methods for calculating accurate enthalpies of formation, ionization potentials, and electron affinities. In *Computational Thermochemistry* (ACS Symposium Series, Vol. 677), K. K. Irikura and D. J. Frurip (Eds.), American Chemical Society, Washington, 176–196.

Curtiss, L. A., K. Raghavachari, P. C. Redfern, A. G. Baboul, and J. A. Pople (1999). Gaussian-3 theory using coupled cluster energies. *Chem. Phys. Lett.*, **314**, 101–107.

Dachsel, H., H. Lischka, R. Shepard, J. Nieplocha, and R. J. Harrison (1997). A massively parallel multireference configuration interaction program: the parallel COLUMBUS program, *J. Comp. Chem.*, **18**, 430–448.

DiLabio, G. A. and J. S. Wright (1998). Calculation of bond dissociation energies for large molecules using locally dense basis sets, *Chem. Phys. Lett.*, **297**, 181–186.

Ducastelle, F. (1991). Tight-binding potentials. In M. Meyer and V. Pontikis (Eds.), *Computer Simulations in Materials Science*. Kluwer, Dordrecht, 233–253. [NATO ASI Ser., Ser. E, Vol. 205].

Egorov, S. A., E. Rabani, and B. J. Berne (1999). On the adequacy of mixed quantum-classical dynamics in condensed-phase systems. *J. Phys. Chem. B*, **103**, 10978–10991.

Elstner, M., D. Porezag, G. Jungnickel, J. Elsner, M. Haugk, T. Frauenheimer, S. Suhai, and G. Seifert, Self-consistent-charge density-functional tight-binding method for simulations of complex materials properties. *Phys. Rev. B*, **58**, 7260–7268.

Fast, P. L., M. L. Sánchez, and D. G. Truhlar (1999). Infinite basis limits in electronic structure theory. *J. Chem. Phys.*, **111**, 2921–2926.

Fast, P. L. and D. G. Truhlar (2000). MC-QCISD: Multi-coefficient correlation method based on quadratic configuration interaction with single and double excitations. *J. Phys. Chem. A*, to be published.

Fernández-Ramos, A., Z. Smedarchina, W. Siebrand, M. Z. Zgierski, and M. A. Rios (1999). Direct-dynamics approaches to proton tunneling rate constants. A comparative test for molecular inversions and an application to 7-azaindole tautomerization, *J. Am. Chem. Soc.*, **121**, 6280–6289.

Field, M. J., P. A. Bash, and M. Karplus (1990). A combined quantum mechanical and molecular mechanical potential for molecular dynamics simulations. *J. Comp. Chem.*, **11**, 700–33.

Finley, J. P. and K. F. Freed (1995). Application of complete space multireference many-body perturbation theory to N2: dependence on reference space and H0. *J. Chem. Phys.*, **102**, 1306–1333.

Galli, G. and F. Mauri (1994). Large scale quantum simulations: C_{60} impacts on a semiconducting surface. *Phys. Rev. Lett.* **73**, 3471–3474.

Galli, G. and J. Kim (1998). Large scale quantum simulations using tight-binding Hamiltonians and linear scaling methods. *Mater. Res. Soc. Symp. Proc.*, **491**, 425–438.

Gao, J., P. Amara, C. Alhambra, and M. J. Field (1998). A generalized hybrid orbital (GHO) method for the treatment of boundary atoms in combined QM/MM calculations. *J. Phys. Chem. A*, **102**, 4714–4721.

Gao, J. and M. A. Thompson, Eds. (1998). *Combined Quantum Mechanical and Molecular Mechanical Models* (ACS Symposium Series, **712**). American Chemical Society, Washington.

García, V. M., M. Reguero, R. Caballol, and J. P. Malrieu (1997). On the quasidiabatic character of average natural orbitals. *Chem. Phys. Lett.*, **281**, 161–167.

Garrett, B. C., D. G. Truhlar, R. S. Grev, and A. W. Magnuson (1980, 1983). Improved treatment of threshold contributions in variational transition state theory. *J. Phys. Chem.*, **84**, 1730–1748, **87**, 4554(E).

Garrett, B. C. and D. G. Truhlar (1981). The coupling of electronically diabatic states in atomic and molecular collisions. In D. Henderson (Ed.), *Theoretical Chemistry: Theory of Scattering* (Theor. Chem. Advances and Perspectives Series, Vol. 6, Part A), Academic Press, New York, 215–289.

Gillan, M. J. (1988). The quantum simulation of hydrogen in metals. *Phil. Mag. A*, **58**, 287–283.

Goddard, W. A. III and C. W. Wilson Jr. (1972). Role of kinetic energy in chemical binding. II. Contragradience. *Theor. Chim. Acta*, **26**, 211–230.

Goedeker, S. and M. Teter (1995). Tight-binding electronic-structure calculations and tight-binding molecular dynamics with localized orbitals. *Phys. Rev. B*, **51**, 9455–9566.

Gonzalez-Lafont, A., T. N. Truong, and D. G. Truhlar (1991). Direct dynamics calculations with neglect of diatomic differential overlap molecular orbital theory with specific reaction parameters. *J. Phys. Chem.*, **95**, 4618–4627.

Gräfenstein, J. and D. Cremer (2000). Can density functional theory describe multi-reference systems? investigation of carbenes and organic biradicals. *Phys. Chem. Chem. Phys.*, **2**, 2091–2103.

Hack, M. and D. G. Truhlar (2000). Nonadiabatic trajectories at an exhibition. *J. Phys. Chem. A*, in press.

Hall, R. J., I. H. Hillier, and M. A. Vincent (2000). Which density functional should be used to model hydration? *Chem. Phys. Lett.*, **320**, 139–143.

Hammes-Schiffer, S. (1998). Mixed quantum/classical dynamics of hydrogen transfer reactions. *J. Phys. Chem. A*, **102**, 10443–10454.

Harris, J. and R. O. Jones (1974). The surface energy of a bounded electron gas. *J. Phys. F*, **4**, 1170–1186.

Harris, J. (1984). Adiabatic-connection approach to Kohn-Sham theory. *Phys. Rev. A*, **29**, 1648–1659.

Harrison, R. J. and R. Shepard (1994). *Ab initio* molecular electronic structure on parallel computers. *Ann. Rev. Phys. Chem.*, **45**, 623–658.

Hawkins, G. D., T. Zhu, J. Li, C. C. Chambers, D. J. Giesen, D. A. Liotard, C. J. Cramer, and D. G. Truhlar (1999).

Universal solvation models. In *Combined Quantum Mechanical and Molecular Mechanical Methods* (ACS Symposium Series, **712**), J. Gao and M. A. Thompson, Eds. American Chemical Society, Washington, pp. 106–127.

Head-Gordon, M. and J. C. Tully (1995). Molecular dynamics with electronic frictions. *J. Chem. Phys.*, **103**, 10137–10145.

Hoffmann, M. R. and Y. G. Khait (1999). A self-consistent quasidegenerate coupled-cluster theory. *Chem. Phys. Lett.*, **311**, 372–378.

Horsfeld, A. P. (1997). Efficient tight-binding. *Phys. Rev. B*, **56**, 6594–6602.

Horsfield, A. P., D. R. Bowler, C. M. Goringe, D. G. Pettifor, D. G., and M. Aoki, M. (1998). A comparison of linear scaling tight binding methods. *Mater. Res. Soc. Symp. Proc.*, **491**, 417–424.

Humbel, S., S. Sieber, and K. Morokuma (1996). The IMOMO (integrated MO MO) method: integration of different levels of molecular orbital approximations for geometry optimization of large systems: Test for *n*-butane conformation and $S_N 2$ reaction: RCl^+ Cl^-. *J. Chem. Phys.*, **105**, 1959–1967.

Hwang, J.-K. and A. Warshel (1996). How important are quantum mechanical nuclear motions in enzyme catalysis? *J. Am. Chem. Soc.* **118**, 11745–11751.

Isaacson, A. D. and D. G. Truhlar (1982). Polyatomic canonical variational theory for chemical reaction rates. Separable-mode formalism with application to OH + $H_2 \rightarrow H_2O$ + H. *J. Chem. Phys.*, **76**, 1380–1391.

Jang, S., C. D. Schweiters, and G. A. Voth (1999). A modification of path integral quantum transition state theory for asymmetric and metastable potentials. *J. Phys. Chem. A*, **103**, 9527–9538.

Jayanthi, C. S., S. Y. Wu, J. Cocks, N. S. Luo, Z. L. Xie, M. Menon, and G. Yang (1998). Order-*N* method for a nonorthogonal tight-binding Hamiltonian. *Phys. Rev. B*, **57**, 3799–3802.

Jensen, K. F. and D. G. Truhlar (1987). *Supercomputer Research in Chemistry and Chemical Engineering* (ACS Symposium Series, **353**). American Chemical Society, Washington.

Kim, Y., J. C. Corchado, J. Villà, J. Xing, and D. G. Truhlar (2000). Multiconfiguration molecular mechanics algorithm for potential energy surfaces of chemical reactions. *J. Chem. Phys.*, **112**, 2718–2735.

Klüner, T., S. Thiel, and V. Staemaler (1999). Ab initio calculation of proton scattering from He(1s2s, ^{1}S): A first-principles wavepacket study beyond the Born-Oppenheimer approximation. *J. Phys. B*, **32**, 4931–4946.

Kobayashi, Y., M. Kamiya, and K. Hirao (2000). The hydrogen abstraction reactions: a multireference Møller-Plesset perturbation (MRMP) theory study. *Chem. Phys. Lett.*, **319**, 695–700.

Kozlowski, P. M., M. Dupuis, and E. R. Davidson (1995). The Cope rearrangement revisited with multireference perturbation theory. *J. Am. Chem. Soc.*, **117**, 774–778.

Lee, T. J. and G. E. Scuseria (1995). Achieving chemical accuracy with coupled-cluster theory. In *Quantum Mechanical Electronic Structure Calculations with Chemical Accuracy*, S. R. Langhoff, Ed. Kluwer Academic, Dordrecht, 47–108.

Leforestier, C. (1978). Classical trajectories using the full abv initio potential energy surface H⁻ + CH₄ → CH₄ + H⁻. *J. Chem. Phys.*, **68**, 4406–4410.

Li, J., C. J. Cramer, and D. G. Truhlar (2000). Two-response-time model based on cm2/indo/s2 electrostatic potentials for the dielectric polarization component of solvatochromic shifts on vertical excitation energies. *Int. J. Quantum Chem.*, **77**, 264–280.

Lim, C. and D. G. Truhlar (1986). The effect of vibrational-rotational disequilibrium on the rate constant for an atom-transfer reaction. *J. Phys. Chem.*, **90**, 2616–2634.

Liu, Y.-P., D.-h. Lu, A. Gonzàlez-Lafont, D. G. Truhlar, and B. C. Garrett (1993). Direct dynamics calculation of the kinetic isotope effect for an organic hydrogen-transfer reaction, including corner-cutting tunneling in 21 dimensions. *J. Am. Chem. Soc.*, **115**, 7806–7817.

Liu, T. and D. G. Truhlar (2000). A tight-binding model for the energetics of hydrocarbon fragments on metal surfaces. unpublished.

Lynch, B. J., P. L. Fast, M. Harris, and D. G. Truhlar (2000). Adiabatic connection for kinetics. *J. Phys. Chem. A*, **104**, 4811–4815.

Marcus, R. A. (1965). On the theory of electron-transfer reactions. VI. Unified treatment for homogeneous and electrode reactions. *J. Chem. Phys.*, **43**, 679–701.

Martens, C. C. and J.-Y. Fang (1997). Semiclassical-limit molecular dynamics on multiple electronic surfaces. *J. Chem. Phys.*, **106**, 4918–4930.

Martin, J. M. L. (1998). Calibration study of atomization energies of small molecules. In *Computational Thermochemistry* (ACS Symposium Series, **677**), K. K. Irikura and D. J. Frurip, Eds. American Chemical Society, Washington, 212–236.

McRae, R. P., G. K. Schenter, B. C. Garrett, G. R. Haynes, G. A. Voth, and G. C. Schatz (1992). Critical comparison of approximate and accurate quantum-mechanical calculations of rate constants for a model activated reaction in solution. *J. Chem. Phys.*, **97**, 7392–7404.

Mead, C. A. and D. G. Truhlar (1982). Conditions for the definition of a strictly diabatic electronic basis for molecular systems. *J. Chem. Phys.*, **77**, 6090–6098.

Merchan, M., L. Serrano-Andres, R. Gonzalez-Luque, B. O. Roos, and M. Rubio (1999). Theoretical spectroscopy of organic systems. *THEOCHEM*, **463**, 201–210.

Meyer, H. D. and W. H. Miller (1979). A classical analog for electornic degrees of freedom in nonadiabatic collision processes. *J. Chem. Phys.*, **70**, 3214–3223.

Miller, W. H. (1975). Semiclassical limit of quantum mechanical transition state theory for nonseparable systems. *J. Chem. Phys.*, **62**, 1899–1906.

Montgomery, J. A. Jr., J. W. Ochterski, and G. A. Petersson (1994). A complete basis set model chemistry. IV. an improved atomic pair natural orbital method. *J. Chem. Phys.*, **101**, 5990–5909.

Nakano, H. (1993). Quasidegenerate perturbation theory with multiconfigurational self-consistent-field reference functions. *J. Chem. Phys.*, **99**, 7983–7992.

Nakano, H., M. Yamanishi, and K. Hirao (1997). Multireference Møller-Plesset method: accurate description of electronic states and their chemical interpretation. *Trends Chem. Phys.*, **6**, 167–214.

Nakano, H., K. Hirao, and M. S. Gordon (1998). Analytic energy gradients for multiconfiguration self-consistent-field second-roder quasidegenerate

perturbation theory (MC-QDPT). *J. Chem. Phys.* **108**, 5660–5669.

Noland, M., E. L. Coitiño, and D. G. Truhlar (1997). Correlated capped subsystem method for the calculation of substituent effects on bond energies. *J. Phys. Chem. A*, **101**, 1193–1197.

Ochterski, J. W., G. A. Petersson, and J. A. Montgomery Jr. (1996). A complete basis set model chemistry. V. extensions to six or more heavy atoms. *J. Chem. Phys.*, **104**, 2698–2619.

Ota, N. and A. T. Brünger (1997). Overcoming barriers in macromolecular simulations: non-Boltzmann thermodynamic integration. *Theor. Chem. Acc.* **98**, 171–181.

Perdew, J. (1999). The functional zoo. In *Density functional theory: A bridge between chemistry and physics*, P. Geerlings, F. De Proft, and W. Langenaeker, Eds. VUB University Press, Brussels, 87–109.

Pople, J. A. and D. L. Beveridge (1970). *Approximate Molecular Orbital Theory*. McGraw-Hill, New York.

Porezag, D. and M. R. Pederson (1995). Density functional based studies of transition states and barriers for hydrogen exchange and abstraction reactions. *J. Chem. Phys.*, **102**, 9345–9349.

Prezhdo, O. V., and P. J. Rossky (1997). Evaluation of quantum transition rates from quantum-classical molecular dynamics simulations. *J. Chem. Phys.* **107**, 5863–5878.

Prezhdo, O. V. (1999). Mean field approximation for the stochastic Schrödinger equation. *J. Chem. Phys.* **111**, 8366–8377.

Rabani, E., S. A. Egorov, and B. J. Berne (1999). On the classical approximation to nonradiative electronic relaxation in condensed-phase systems. *J. Phys. Chem. A*, **103**, 9539–9544, 9564(E).

Raghavachari, K. and J. B. Anderson (1996). Electron correlation effects in molecules. *J. Phys. Chem.*, **100**, 12960–12973.

Ridder, L., A. J. Mulholland, J. Vervoort, and I. M. C. M. Reitjens (1998). Correlation of calculated activation energies with experimental rate constants for an enzyme catalyzed aromatic hydroxylation. *J. Am. Chem. Soc.*, **120**, 7641–7642.

Roberto-Neto, O., F. B. C. Machado, and D. G. Truhlar (1999). Energetic and structural features of the CH₄ + O(³P) → CH₃ + OH abstraction reaction. Does perturbation theory from a multiconfiguration reference state (finally) provide a balanced treatment of the reaction? *J. Chem. Phys.*, **111**, 10046–10052.

Roos, B. O., M. Fülscher, P.-Å. Malmqvist, M. Merchán, and L. Serrano-Andrés (1995). Theoretical studies of the electronic spectra of organic molecules. In *Quantum Mechanical Electronic Structure Calculations with Chemical Accuracy*, S. R. Langhoff, Ed. Kluwer, Dordrecht, 357–438.

Roos, B. O., K. Andersson, M. Fülscher, P.-Å. Malmqvist, L. Serrano-Andrés, K. Pierloot, and M. Merchán (1996). Multiconfigurational perturbation theory: applications in electronic spectroscopy. *Adv. Chem. Phys.*, **93**, 219–331.

Rossi, I. and D. G. Truhlar (1995). Parameterization of NDDO wavefunctions using genetic algorithms: An evolutionary approach to parameterizing potential energy surfaces and direct dynamics calculations for organic reactions. *Chem. Phys. Lett.* **233**, 231–236.

Sancho-Garciá, J. C., A. J. Pérez-Jiménez, and F. Moscardó (2000). A comparison between DFT and other ab initio schemes on the activation energy in the automerization of cyclobutadiene. *Chem. Phys. Lett.*, **317**, 345–251.

Schatz, G. C. (1989). The analytical representation of electronic potential-energy surfaces. *Rev. Mod. Phys.*, **61**, 669–688.

Schuetz, M., G. Hetzer, and H.-J. Werner (1999). Low-order scaling local electron correlation methods. I. Linear scaling MP2. *J. Chem. Phys.* **111**, 5691–5705.

Scuseria, G. E. and T. J. Lee (1990). Linear scaling coupled-cluster methods which include the effect of connected triple excitations. *J. Chem. Phys.*, **93**, 5851–5855.

Scuseria, G. E. and P. Y. Ayala (1999). Comparison of coupled cluster and perturbation theories in the atomic orbital basis. *J. Chem. Phys.*, **111**, 8330–8343.

Sidis, V. (1992). Diabatic Potential energy surfaces for charge-transfer processes. *Adv. Chem. Phys.*, **82**, 73–134.

Smedarchina, Z., W. Siebrand, M. Z. Zgierski, and F. Zerbetto (1995). Dynamics of molecular inversion: an instantan approach. *J. Chem. Phys.*, **102**, 7024–7034.

Stanton, J. F. (1997). Why CCSD(T) works: A different perspective. *Chem. Phys. Lett.*, **281**, 130–134.

Staroverov, V. N. and E. R. Davidson (1998). The reduced model space method in multireference second-order perturbation theory. *Chem. Phys. Lett.*, **296**, 435-444.

Stephens, P. J., F. J. Devlin, C. F. Chabalowski, and M. J. Frisch (1994). Ab initio calculation of vibrational absorption and circular dichroism spectra using density functional force fields. *J. Phys. Chem.*, **98**, 11623–11627.

Sternberg, M., G. Galli, and T. Frauenheim (1999). NOON—a non-orthogonal localized orbital order-*N* method. *Comp. Phys. Comm.*, **118**, 200–212.

Sutton, A. P., M. W. Finnis, D. G. Pettitor, and Y. Ohta (1988). The tight-binding model. *J. Phys. C*, **21**, 35–66.

Svensson, M., S. Humbel, and K. Morokuma (1996). Energetics using the single point IMOMO (integrated molecular orbital + molecular orbital) calculations: Choices of computational levels and model system. *J. Chem. Phys.*, **105**, 3654–3661.

Topaler, M. S., T. C. Allison, D. W. Schwenke, and D. G. Truhlar (1998, 1999). What is the best semiclassical method for photochemical dynamics in systems with conical intersections? *J. Chem. Phys.*, **109**, 3321-3345, **110**, 687–688(E).

Tratz, C. M., P. L. Fast, and D. G. Truhlar (1999). Improved coefficients for the scaling all correlation and multi-coefficient correlation methods. *Phys. Chem. Comm.*, **2**, 14: 1–10.

Truhlar, D. G. (1992). Potential energy surfaces. In R. A. Meyers (Ed.), *The Encyclopedia of Physical Science and Technology*, Vol. 13, 2nd ed. Academic Press, New York, 385–393.

Truhlar, D. G. (1995). Direct dynamics method for the calculation of reaction rates. In D. Heidrich (Ed.), *The Reaction Path in Chemistry: Current Approaches and Perspectives*. Kluwer, Dordrecht, 229–255.

Truhlar, D. G. and D. A. Dixon (1979). Direct-mode chemical reactions II: classical theories. In *Atom-Molecule*

Collision Theory, R. B. Bernstein, Ed. Plenum, New York, 595–646.

Truhlar, D. G., J. W. Duff, N. C. Blais, J. C. Tully, and B. C. Garrett (1982). The quenching of Na (3^2P) by H_2: interactions and dynamics. *J. Chem. Phys.*, **77**, 764–776

Truhlar, D. G. and B. C. Garrett (1980). Variational transition-state theory. *Acc. Chem. Res.*, **13**, 440–448.

Truhlar, D. G. and B. C. Garrett (1992). Resonance state approach to quantum mechanical variational transition state theory. *J. Phys. Chem.*, **96**, 6515–6518.

Truhlar, D. G., B. C. Garrett, and S. J. Klippenstein (1996). Current status of transition state theory. *J. Phys. Chem.*, **100**, 12771–12800.

Truhlar, D. G. and M. S. Gordon (1990). From force fields to dynamics: classical and quantal paths. *Science*, **249**, 491–498.

Truhlar, D. G., A. D. Isaacson, R. T. Skodje, and B. C. Garrett (1982, 1983). The incorporation of quantum effects in generalized transition state theory. *J. Phys. Chem.*, **86**, 2252–2261, **87**, 4554(E).

Truhlar, D. G., R. Steckler, and M. S. Gordon (1987). Potential energy surfaces for polyatomic reaction dynamics. *Chem. Rev.*, **87**, 217–236.

Tully, J. C. (1998). Nonadiabatic dynamics. In *Modern Methods for Multidimensional Dynamics Computations in Chemistry*, D. L. Thompson, Ed. World Scientific, Singapore, 34–72.

Van Leeuwen, R. and E. J. Baerends (1994). An analysis of nonlocal density functionals in chemical bonding. *Int. J. Quant. Chem.* **52**, 711–730.

Volobuev, Y. L., M. D. Hack, M. S. Topaler, and D. G. Truhlar (2000). Continuous surface switching: An improved time-dependent self-consistent-field method for nonadiabatic dynamics. *J. Chem. Phys.*, in press.

Voter, A. F. (1997). Hyperdynamics: Accelerated molecular dynamics of infrequent events. *Phys. Rev. Lett.*, **78**, 3908–3911.

Voter, A. F. and M. R. Sorensen (1999). Accelerating atomistic simulations of defect dynamics: hyperdynamics, parallel replica dynamics, and temperature-accelerated dynamics. *Mater. Res. Soc. Symp. Proc.*, **538**, 427–439.

Voter, A. F. and W. G. Rudd (1999). Bias potential for hyperdynamics simulations. *Mater. Res. Soc. Symp. Proc.*, **538**, 485–490.

Voth, G. A., D. Chandler, and W. H. Miller (1989). Rigorous formulation of quantum transition state theory and its dynamical corrections. *J. Chem. Phys.*, **91**, 7749–7760.

Wang, I. S. Y. and M. Karplus (1973). Dynamics of organic reactions. *J. Am. Chem. Soc.*, **95**, 8160.

Wang, Y. and C. H. Mak (1995) Transferable tight-binding potential for hydrocarbons. *Chem. Phys. Lett.*, **235**, 37–46.

Wasserman, E., L. Stixrude, and R. E. Cohen (1996). Thermal properties of iron at high pressures and temperatures. *Phys. Rev. B*, **53**, 8296–8309.

Werner, H.-J. (1995). Problem decomposition in quantum chemistry. In *Domain-Based Parallelism and Problem Decomposition Methods in Computational Science and Engineering*, D. E. Keyes, Y. Saad, and D. G. Truhlar, Eds. Society for Industrial and Applied Mathematics, Philadelphia, 239–261.

Wonchoba, S. E., W.-P. Hu, and D. G. Truhlar. Reaction path approach to dynamics at a gas-solid interface: Quantum tunneling effects for an adatom on a non-rigid metallic surface. In H. L. Sellers and J. T. Golab, Eds. *Theoretical and Compuational Approaches to Interface Phenomena.* Plenum, New York, 1–34.

Yxkinten, U., J. Hartford, and T. Holmquist (1997). Atoms embedded in an electron gas: the generalized gradient approximation. *Phys. Sci.*, **55**, 499–506.

DEVELOPMENT AND APPLICATION OF DETAILED KINETIC MECHANISMS FOR FREE RADICAL SYSTEMS

Anthony M. Dean
Corporate Strategic Research
ExxonMobil Research and Engineering Co.
Annandale, NJ 08801

Abstract

In this paper we review some of the approaches used to estimate self-consistent rate coefficient assignments and then illustrate the use of the resulting kinetic models that we have developed to explore two important systems. Methods to estimate rate coefficients for a variety of hydrogen abstraction reactions are described. Methods to account for the pressure and temperature dependence of chemically-activated reactions are also discussed. The first system analysis illustrates the use of a recently developed nitrogen mechanism to capture the major features of the Thermal DeNOx process, whereby NO is reduced to N_2 by addition of NH_3 to power plant effluents. The other describes calculations using a combined hydrocarbon and nitrogen mechanism to explore the utility of secondary injection of fuel in a diesel engine as a means to enable NO reduction. The potential for improved modeling approaches in the future is briefly discussed.

Keywords

Kinetic mechanisms, Free radical kinetics, Thermochemical kinetics, Hydrogen transfer reactions, Chemical activation reactions, Applications to combustion systems, Automatic mechanism generation.

Introduction

Recent advances within the kinetics community make it possible to describe the kinetics of complex free radical systems in terms of elementary reactions. The most important result of this research is a significantly improved understanding of the details of these reactions at the molecular level. This understanding has resulted from concerted efforts of both experimentalists and theorists, particularly in those instances where both approaches have focused upon a specific reaction. This improved understanding has led to more accurate assignment of both elementary reaction rate coefficients and thermodynamic properties. Accurate molecular-level descriptions are particularly advantageous for the following reasons: (1) It provides a much better framework for extrapolation. (2) It provides insights into the most effective ways to improve system performance. (3) It permits the use of "computer experiments" to effectively explore various process options. (4) It identifies those particular sets of initial conditions in which to focus experimental efforts.

Development of a detailed chemical mechanism to describe a process such as combustion of a hydrocarbon fuel is easy to describe, but difficult to do. One needs to systematically identify all the reaction pathways and then to assign accurate rate coefficients to each reaction. Many mechanisms are hierarchical in nature, where extension to a more complex system is accomplished by adding reactions to a previously developed simpler model. Although this approach appears obvious, it may also be a recipe for disaster. For example, the previously developed model might contain rate coefficients that have been "tweaked" to better fit some combination of observations. Without benefit of hindsight, it can be very difficult to tell if such changes are a legitimate method to account for the inherent uncertainty in the rate coefficient measurements, or whether the changes are needed to patch up an inherent defect in the system description. If the latter, these mechanistic defects have a disconcerting tendency to reappear in the larger mechanisms, sometimes leading to further tweaking that only makes matters worse.

One might think of two different approaches to mechanism development. If the goal is to provide the best description of a particular system where there are substantial data available, one might start by assembling a mechanism using the best available information for the individual reactions and then systematically optimizing it by adjusting rate coefficients, preferably within the error limits, to better describe the experimental database. This approach has been employed for the development of the widely-used GRI mechanism for natural gas combustion (Frenklach et al., 1999). This mechanism provides a good description of a variety of natural gas combustion and ignition experiments. However, the authors are very explicit in their warning to users not to make any changes or substitutions to the mechanism. They recognize the possibility that such "improvements" could lead to a marked degradation in predictive ability, due to the complexity of the interactions within the model.

Another approach is to assemble the mechanism based on the best available kinetic information, both experimental and theoretical, and then to compare the predictions of this model to a wide variety of systems. Here the goal is not necessarily to achieve a quantitative description of any particular system, but rather to provide reasonable descriptions of the range of systems. We have used this approach to develop a kinetic description of nitrogen chemistry in combustion and to describe hydrocarbon-nitrogen interactions (Dean and Bozzelli, 2000). We analyzed over 350 reactions and assigned rate coefficients. These assignments were based on experimental measurements where reliable data were available. Otherwise, we used thermochemical kinetic principles and theoretical guidance to estimate the rate coefficients. The predictions of the mechanism assembled from these reactions then were compared to a variety of nitrogen systems, ranging from simple flow reactor studies of ammonia oxidation to reactive intermediate formation in rich ammonia flames. The observed agreement suggested that the overall chemistry is reasonably well understood.

Of course, it is possible, even likely, that such comparisons on other systems might not be so favorable. If a kinetic model has difficulty describing the major features of a system, several choices are available. The most obvious approach, and frequently the worst choice, is to adjust the rate coefficients of the most sensitive reactions to improve the fit. From a pragmatic perspective, continuing advances in both experimental and theoretical methods have reduced the uncertainties in the rate coefficient assignments, thereby limiting the range of legitimate adjustment of these values. Moreover, such adjustments, without taking into the consideration the effect of such changes on predictions in other systems, might obscure underlying fundamental defects in the

mechanism. This is the reason that the adjustments made in the GRI mechanism involve a wide-ranging set of experiments. We think that a much more prudent strategy is to recognize that the poor fit might reflect the fact that some critical chemistry is missing from the model. At this point, it is usually best to retreat to the lab to collect additional data, especially on reactive intermediates, that might shed light on the missing chemistry. The continued iterations between modeling, experiment, and theory is unquestionably tedious, but recent advances in all of these areas suggest the effort is worthwhile and there is reasonable hope for convergence.

With either approach, critical components of kinetic model development include identification of the types of reactions that should be included in the mechanism and development of systematic "rules" for rate coefficient assignments. These rules assume even more importance for the cases where the mechanisms are computer-generated (Susnow et al., 1997; Glaude et al., 1998; Grenda et al., 2000).

In this paper we will review some of the approaches that allow such self-consistent rate coefficient assignments and then illustrate the use of the resulting kinetic models that we have developed to explore two important systems. The first of these illustrates the utility of a nitrogen mechanism to describe the Thermal DeNOx process (Lyon, 1975) whereby NO is reduced to N_2 by addition of NH_3 to power plant effluents. The other describes how a combination of hydrocarbon and nitrogen mechanisms were used to identify secondary injection of fuel as a means to enable NO reduction in diesel engines (Weissman et al., 1999; Weissman et al., 2000).

Estimation of Thermodynamic Quantities

It is difficult to overestimate the importance of reliable thermodynamic property estimates for quantitative applications of chemical kinetic models. Accurate heats of formation for reactants and products are needed to identify plausible pathways and to estimate activation energies. Free energies of reaction are required to calculate the equilibrium constants, which are used to obtain reverse rate coefficients. This is especially important in high-temperature environments where many reactions can be partially-equilibrated and the ratios of various free radical species are governed by thermodynamic, not kinetic, parameters. Group additivity approaches provide an accurate method to estimate the necessary properties (Benson, 1976; Reid et al., 1987; Cohen, 1993). Several computer programs are available to calculate thermodynamic properties based upon the group-additivity principle (Ritter and Bozzelli, 1991; Stein, 1994). An extension of this approach, where the thermodynamic properties of hydrocarbon radicals are estimated by calculating the expected changes that occur upon breaking the C–H bond of the parent, has also been developed (Lay et al., 1995).

An automated computational approach (Grenda et al., 2000) has been developed to estimate thermodynamic and physical properties of gas phase radical and molecular species using group additivity methods based on earlier work by Ritter and Bozzelli (1991). Species are entered using a variety of methods ranging in complexity from straightforward character-string based nomenclature to detailed bond connectivity tables. These are used to computationally construct the species atomic structure and then automatically identify the appropriate group additivity families. The complexity of species may vary from small alkanes or oxygen-containing hydrocarbons to multi-ring fused aromatics. Properties for radical and biradical species are calculated by applying bond dissociation increments to a stable parent molecule to reflect the loss of an H atom. The thermodynamic properties estimated include enthalpy and entropy of formation and heat capacity at several discrete temperatures. Polynomial coefficients convenient for modeling calculations are generated and output in NASA polynomial format. The polynomial coefficients, generated using fitting and extrapolation methods based upon the harmonic oscillator model, an exponential model, or Wilhoit polynomials (Ritter, 1989), are valid from 300-5000K. This approach to fitting the temperature-dependent heat capacity estimates also yields reduced frequencies (and degeneracies) for use with QRRK pressure-dependent analyses. Physical properties such as boiling point, critical properties, and Lennard-Jones parameters are also estimated using a variety of literature group additivity procedures.

Estimation of Hydrogen Atom Abstraction Reaction Rate Coefficients

Elementary reactions in which a hydrogen atom is abstracted from a stable species by an atom or radical are ubiquitous in free radical kinetics. Unfortunately, experimental data, especially over an extended temperature range, are limited and this has forced us to estimate abstraction rate coefficients. We developed an estimation procedure (Dean and Bozzelli, 2000) to deal with these reactions in a self-consistent way.

Rate coefficients are estimated by relating the parameters for a specific reaction type, e.g., abstractions by OH, to that for a well-characterized hydrocarbon analog, e.g., OH + C_2H_6. Consistent use of such reference reactions avoids propagating whatever inconsistencies are present in published compilations. Moreover, this approach accounts for the different temperature dependencies found for various types of hydrogen abstractions. It is now well-recognized (Tsang and Hampson, 1986; Baulch et al., 1992) that many hydrogen abstraction reactions show upward curvature in their Arrhenius plots, leading to appreciably higher rate coefficients at high temperatures than one would estimate from a linear extrapolation of low-temperature measurements. Such temperature dependencies have to be included in order to predict reliable rate coefficients under high-temperature conditions. A convenient way to include this dependence is to express the rate constant in modified Arrhenius form, $A(T^n)\exp(-E/RT)$, rather than to use the

Arrhenius form, $A'\exp(E'/RT)$. The modified Arrhenius form is consistent with Transition State Theory (TST).

Our approach assigns values of **A** and **n** for all reactions in a homologous series based on a single reference reaction. For abstraction by H, O, and OH, we use the recommendations of Baulch et al. (1992) for the corresponding abstractions from ethane and assume that **A** scales directly with the number of equivalent abstractable hydrogens. The preexponential factors thus take the form $A_{C2H6}*(n_H / 6)$, where n_H is the number of equivalent hydrogens that can be abstracted from the molecule of interest. We treat other abstractions differently. For example, for CH_3 abstractions we believe that it is more consistent to base our assignment of **A** and **n** on its reaction with H_2O implied by the rate constant for OH + $CH_4 = CH_3 + H_2O$ and the equilibrium constant for this reaction. With **A** and **n** now assigned for all methyl abstractions, we adjust **E** for $CH_3 + C_2H_6$ until k(T) agrees with that reported by Baulch et al. Although the resulting fit is not perfect, it is adequate for our purposes, and it provides the necessary thermodynamic consistency between abstraction by OH and CH_3. This connection of rate coefficient assignments to thermodynamics is critical for construction of reliable kinetic models. In an analogous way, the rate coefficients for larger alkyl radicals were assigned on the basis of the values for OH abstraction from the respective parents. Values for unsaturated radicals such as C_2H_3 and C_2H were also obtained using this approach. For HO_2, the reverse of OH + H_2O_2 was used.

The value of **E** is estimated using the "Evans-Polanyi relationship", which states that the activation energy E for reactions of a similar type but with different enthalpy change is proportional to the enthalpy difference. This can be written as:

$$E = E_{ref} - f \{ \Delta H_{ref} - \Delta H \}$$

The proportionality constant **f** is known as the Evans-Polanyi factor. For hydrocarbon systems, we have estimated **f** using ethane as the reference molecule. For each reaction class, we selected values for **f** which seemed to give the overall best fit to selected data taken from the NIST compilation (Mallard et al., 1993) and other available sources. **f** is generally observed to be less than unity, reflecting the fact that all of the change in enthalpy is not reflected in the change in activation energy. A complication of this form is that **f** must usually be different (unless f = 0.5) for endothermic and exothermic reactions in order that $E_{endo} - E_{exo} = \Delta E_{rxn}$. To avoid this problem, we prefer to use different values for **f** for abstraction from different types of hybridized carbon, reflecting the different enthalpy changes. (The values shown in Table 1 are for abstraction from sp^3 carbon.)

Another complication that can arise when using the Evans-Polanyi relationship for reactions with very large exothermicities is that the resulting activation energy might become negative, which is physically meaningless for an abstraction reaction. To avoid this, we suggest assigning a lower limit to the activation energy. We assign this limit to be -nR(300). [In this expression, **n** is the exponent of T in the modified Arrhenius form.] Using this limit, the value of **E'** in the simple Arrhenius form cannot become negative at 300K.

Table 1 lists the parameters that we use to estimate some of the most common abstraction reactions.

Table 1. Rate Coefficient Parameters for Hydrogen Atom Abstraction Reactions.

$k = n_H \, A \, T^n \exp(-\{E_0 - f(\Delta H_0 - (\Delta H)\}/RT) \; cm^3 \, mole^{-1} \, s^{-1}$					
<u>R</u>	<u>A</u>	<u>n</u>	<u>E_0 (kcal)</u>	<u>ΔH_0 (kcal)</u>	<u>f</u>
H	2.4E8	1.5	7.4	-3.1	0.65
O	1.7E8	1.5	5.8	-1.1	0.75
OH	1.2E6	2.0	0.9	-18.3	0.50
HO_2	1.4E4	2.69	18.9	12.7	0.60
CH_3	8.1E5	1.87	10.6	-3.7	0.65

Fig. 1 compares estimates of the rate coefficients for a variety of OH abstraction reactions using this approach to the measurements of Koffend, et al. (1996) near 1100K. (The actual temperature varied from 962K to 1186 K, and each estimate was computed at the appropriate temperature.) All of these species contain 6 primary C–H bonds and from 0 (ethane) to 16 (decane) secondary C–H bonds. The agreement obtained in this comparison suggests that the assumption that the rate coefficient simply scales with the number of equivalent abstractable hydrogen atoms is reasonable.

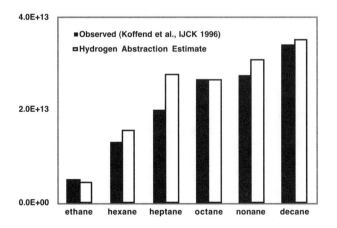

Figure 1. Comparison of estimated and measured OH abstraction rate coefficients near 1100K.

Abstractions that produce resonantly-stabilized carbon-centered radicals must be treated differently. Rate coefficients for these reactions are often reported in terms of "normal" Arrhenius parameters, i.e., with no temperature-dependent preexponential factor. (The reason for this is radical addition can compete with abstraction here, making it more difficult to assign an unambiguous rate coefficient for abstraction.) Use of such two-parameter Arrhenius expressions for these abstractions can significantly underestimate the high-temperature rate coefficients for these reactions. This problem is further exacerbated if the three-parameter modified-Arrhenius fits, accounting for curvature, are used for other types of abstraction reactions. It is necessary to include a non-Arrhenius dependence for the reactions that produce resonantly-stabilized radicals. We choose to use the same **n** used for the non-resonant analogs.

Although these abstractions are typically ~12 kcal/mole more exothermic, the limited data available (Tsang, 1991) suggest that these are only slightly faster. Use of the usual Evans-Polanyi factors would substantially overestimate these rate coefficients. One could assign lower **f** parameters to these reactions, or one could assume that these abstractions will have an activation energy ~1 kcal/mole lower than its non-resonant analog. Additional complications arise in assignment of **A** for these systems. Formation of the resonantly-stabilized system restricts rotation about the radical center, thus lowering the entropy of the transition state. This results in a decrease in the A-factor. Abstractions of aldehydic hydrogens as well as hydrogen from carbon atoms that are adjacent to carbonyl groups also require special treatment.

Even with the residual uncertainties, kinetic models based on these estimates can be quite useful. One reason for this is that many complex kinetic systems are often sensitive to a surprisingly small number of rate coefficients. Thus, even semi-quantitative accuracy in rate coefficient assignments might be adequate for many kinetic models. However, it is important to check this conclusion on a case-by-case basis by performing sensitivity analysis. (Since the typical sensitivity analysis explores perturbations about the calculated solution, it is important that the assignments are sufficiently accurate to produce a result close to the "answer". Otherwise, the sensitivity analysis could be misleading.) One can reexamine the assignments for the reactions that are highly sensitive on a reaction-by-reaction basis.

Despite the fact that we can sometimes deal with the residual uncertainties from the simple estimation techniques outlined above, there is no question that improved estimation methods are desired. One likely source of such improvements is the use of a more fundamental analysis based on electronic structure theory and molecular dynamics. Fortunately, such approaches are becoming feasible, e.g., Hu, et al. (1997). The computationally efficient Density Functional Theory (DFT) shows promise for accurate transition state energetics (Jursic, 1997), but remaining uncertainties in predicted barrier heights may limit the predictive capabilities. An alternative approach is to use DFT to compute accurate transition state geometries and frequencies, thus allowing calculation of the temperature-dependent Arrhenius A-factor via a straight-forward application of Transition State Theory. The barrier height can be fit by comparison to experimental results at a specific temperature, typically near 300K where more data are often available. An example of this approach is shown in Fig.2 (Susnow et al., 1999). This combination of a computed preexponential and a fitted **E** permits a more reliable extrapolation to higher temperatures.

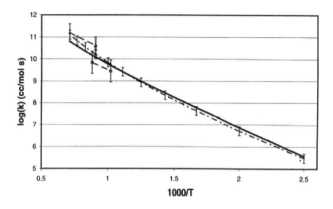

Figure 2. Predicted vs. experimental rate coefficients for $CH_3 + C_2H_6$. The solid line is calculated using a DFT barrier height adjusted from the calculated value of 9.4 kcal/mole to 12.4 kcal/mole (Susnow et al., 1999).

This is particularly advantageous for analysis of kinetic systems where application of more computationally intensive *ab initio* techniques, e.g., (Petersson, 1997; Mayer et al., 1998), to obtain barrier heights for each reaction would not be practical. However, these more computationally intensive analyses of model reactions serve the very important purpose of delineating subtleties of the potential energy surfaces and reaction dynamics that can then perhaps be generalized to more complex reactions. The significant activity in this area suggests improved estimates of hydrogen-transfer rate coefficients will emerge, with the potential for improved accuracy of kinetic models.

Pressure-dependent Reactions

Kineticists have long known that it is necessary to consider the pressure-dependence of the rate coefficients of dissociation and recombination reactions. The pressure effect can be qualitatively understood by recognizing that these reactions are really not one-step processes. Considering the recombination reaction,

$$A + B = AB$$

one has to take into account that formation of the new chemical bond A–B initially produces AB*, an energy-rich adduct that can either redissociate to A and B or be stabilized by an energy-transferring collision with a "bath gas" molecule M to form products:

$$A + B = AB*$$

$$AB* + M \rightarrow AB + M$$

This competition between the unimolecular reaction of the energized adduct back to reactants and its collisional stabilization can be analyzed by assuming d[AB*]/dt ~ 0 (the steady-state assumption) which leads to the result that the recombination rate coefficient is independent of pressure at high P and scales linearly with pressure at low P (Robinson and Holbrook, 1972).

Similar considerations apply for any reaction which proceeds via formation of a new chemical bond. In the context of free radical chemistry, this means that one must account for the pressure dependence of radical addition to unsaturated molecules (and the reverse reaction, beta-scission to form an unsaturated molecule and a radical) and insertion reactions as well as recombinations. Another complication in analyzing reactions proceeding via energized adducts is the possibility that additional dissociation channels to new products will be available. Thus, radical addition, recombination, and insertion reactions can manifest very complex temperature and pressure behavior as the stabilization channel competes with these multiple decomposition channels. It is not uncommon for a combustion mechanism to contain a large number of such chemically-activated reactions, typically one-third to one-half of all the reactions. Substantial errors can result when experimental measurements on these rate coefficients are extrapolated to other regions of temperature and pressure without accounting for the competition between unimolecular reactions of the energized adducts and their bimolecular collisional stabilization.

A straightforward approach to treat chemically-activated reactions is available. This approach (Dean, 1985) was initially based on the quantum version of Kassel theory (QRRK) to compute the energy-dependent rate coefficient k(E) for reactions of the energized adduct. It used a single geometric-mean frequency to characterize the adduct. It has since been improved in several ways, including development of a three-frequency model. The three frequencies and the associated degeneracies are computed from the fits to heat capacity estimates (Ritter, 1989). The heat capacity data can be reliably estimated using group-additivity. This approach offers the advantage of avoiding the specification of the complete frequency distribution of the adduct; for many of the molecules of interest, such information is unavailable. This three-frequency model provides a suitable approach to analyze both unimolecular falloff and bimolecular chemical activation reactions. It provides a simple framework by which the effects of temperature and pressure can be readily understood and evaluated. It is especially useful for those cases that require such an analysis for many reactions. This approach is described in greater detail in Dean and Bozzelli (2000), where it is applied to a number of reactions involving nitrogen compounds.

The most important parameters required to calculate the temperature and pressure dependence of these reactions are the high-pressure rate coefficients for reactions of the various adducts. Of course, if electronic structure calculations are available that provide the detailed structure and energetics of the transition state for any of the adduct reactions, it is possible to use this information to compute the isomerization rate coefficient directly using Transition State Theory. However, for most systems of interest, we currently do not have the luxury of such detailed information, so various estimation approaches are used. For reactions where the adduct dissociates to products, it is often possible to obtain a reliable estimate by using the reverse reaction (radical addition or recombination) together with the equilibrium constant for the reaction. For adduct isomerizations, one can estimate the rate coefficients using thermochemical arguments (Benson, 1976). For example, the A-factor can be estimated by considering how many rotational degrees of freedom are tied up in the cyclic transition state. The barrier can be estimated by considering the combination of the inherent barrier for the hydrogen transfer (obtained from analysis of the analogous bimolecular hydrogen transfer) and the ring strain associated with the cyclic transition state, with this ring strain estimated based on the analogous cyclic alkane. Calculation of the unimolecular rate coefficient for the adducts also requires the density of states of the adduct. It is possible to estimate this microscopic information quite accurately from the macroscopic heat capacity of the adduct. (Bozzelli et al., 1997) This thermodynamic property can be estimated very well using group additivity, as discussed earlier. As a result of these various estimation techniques, one can calculate the detailed pressure and temperature dependence of the various pathways of chemically-activated reactions without need of assigning arbitrary (adjustable) parameters.

Given that one can estimate k(T,P) for a chemically-activated reaction, another important issue is how to incorporate this pressure-dependence into a mechanism. For simple one-well systems, the approach of Troe and co-workers (Gilbert et al., 1983) works well, and this parametrization can be directly incorporated into CHEMKIN (Kee et al., 1989), a widely used kinetic modeling package. For more complex chemically-activated systems with several energized adduct isomers, alternative approaches are more useful–cf. Venkatesh et al. (1997), Venkatesh (2000) and references therein.

Selected Applications

Thermal DeNOx

The Thermal DeNOx process, whereby ammonia injection into the post-combustion zone selectively reduces NO, was invented in Exxon's Corporate Labs by Lyon (1975) and has been applied in over a hundred commercial units. The kinetics are very interesting in that there is a temperature window, typically between 1100 and 1400K,

where NO is reduced to N_2 by NH_3. At lower temperatures, the kinetics are too slow whereas at higher temperatures the NO concentration increases. Some representative data (Lyon, 1987) are shown in Fig. 3. At ~1225K and ~0.2 s residence time at 1.1 atm, addition of 380 ppm NH_3 reduced the NO concentration in a mixture containing 4% O_2 and 10% steam from an original 230 ppm to ~35 ppm.

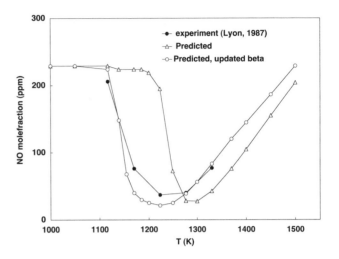

Figure 3. Comparison of predictions for NO reduction at various temperatures to the experimental data of Lyon for typical Thermal DeNOx conditions. The line marked "Predicted, updated beta" uses the recently measured branching ratio for the $NH_2 + NO$ reaction, which is only slightly larger than that used in the base case (cf. text).

The predictions, also shown in Fig. 3, were done using a detailed mechanism developed in our analysis of the combustion chemistry of nitrogen (Dean and Bozzelli, 2000). This included 371 reactions of nitrogen species as well as 151 reactions describing the hydrogen/oxygen system taken from Miller and Bowman (1989). The predictions using this mechanism, with no adjustments in rate coefficients, clearly capture the qualitative behavior of this system, but the temperature window is shifted to slightly higher temperature, suggesting that the kinetics in the model are somewhat slow. Analysis showed that the predictions are unusually sensitive to the branching ratio of the reaction:

$$NH_2 + NO = N_2 + H_2O$$

$$= NNH + OH$$

Our model assigned a value of 0.28 to the fraction of this reaction that passes through the NNH + OH route (branching ratio $\beta = 0.28$) at 1200K. This value was based on a simple average of several experimental measurements with substantial scatter, suggesting significant uncertainty in this assignment. Subsequent experimental work on this reaction (Votsmeier et al., 1999) measured this branching ratio using improved diagnostics at higher temperatures (1350-1750K). The data show a linear increase of branching ratio with temperature. Extrapolation down to 1200K yields a value of $\beta = 0.35$, reasonably close to the value used in the original model. However, this small change has a substantial effect, as shown in Fig. 3. (The value for the total rate coefficient was kept constant.) The agreement is very encouraging, with a good prediction of the location and width of the temperature window, and only a slight overestimate of the maximum amount of NO reduction.

Another feature of the kinetics is shown in Fig. 4, where the effect of removing steam is considered. The kinetic model, using the updated experimental measurement of the branching ratio, properly predicts the shift toward lower temperatures as steam is removed. This is an example of a system where steam is chemically involved and is not simply an inert diluent.

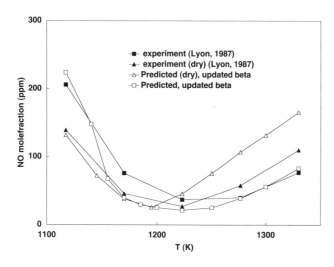

Figure 4. Effect of removing steam on Thermal DeNox kinetics. Both experiment and model show a shift to lower temperature as the system becomes more reactive.

The reasonable descriptions of the Thermal DeNox process are encouraging. This mechanism is significantly different in some ways from earlier ones, e. g., the Miller-Bowman mechanism (1989). The changes include a much shorter lifetime for NNH to be consistent with relatively recent theoretical analysis (Koizumi et al., 1991) and a lower branching ratio for the NH_2+NO reaction to reflect more recent experimental measurements. (Miller and Bowman had used $\beta = 0.50$.) The current model also includes a number of species that were not considered in the earlier models. As it turns out, the addition of these species and their associated reactions resulted in a reasonable overall description of the Thermal DeNOx kinetics, as well as other systems (Dean and Bozzelli, 2000). It is especially encouraging that this agreement was obtained with the improved assignments regarding NNH and the $NH_2 + NO$ branching ratio. The earlier models represent an example of the dangers discussed earlier of

tweaking critical rate coefficients to fit data. At the time those models were developed, the experimental data on the branching ratio were not yet available and the theoretical analysis of the NNH lifetime had not yet been done, so the assignments seemed reasonable at the time. However, as new information emerged, it forced a reexamination of this system, resulting in an improved understanding.

It turns out that there are significant practical implications of a shorter NNH lifetime (Bozzelli and Dean, 1995). The shorter lifetime (faster dissociation) implies that the reverse reaction

$$H + N_2 = NNH$$

is also faster, resulting in establishment of a low, but finite, concentration of NNH when N_2 and H are present. In a typical combustion environment, the presence of both N_2 from the air and the relatively high concentration of H atoms shifts this equilibrium to the right, increasing the concentration of NNH. In turn, NNH can react with oxygen atoms in a very fast reaction

$$NNH + O = NO + NH$$

to produce NO. (Some fraction of the NH formed will also be oxidized to NO.) The NNH pathway to NO is not as dependent on temperature as the Thermal NO route (Zeldovich, 1946). Furthermore, calculations that included this pathway showed that 40% of the total NO formed in a low-pressure stoichiometric propane-air flame was predicted to form via the NNH route (Harrington et al., 1996). These results emphasize the need to include the NNH pathway into models used to explore methods to minimize NO production. Inclusion of only the Thermal NO pathway could overestimate the amount of NO reduction achieved by reducing the temperature.

NO Reduction in a Diesel Engine

The high thermal efficiency of diesels makes them logical choices to improve fuel economy in vehicles. However, their NO_x and soot emissions must be carefully controlled. NO_x control is especially difficult since the usual methods of catalytic NO_x reduction require a stoichiometric mixture of fuel and air at the catalyst, while the diesel engine always operates with excess air. The pursuit of "lean DeNO$_x$," catalysis is currently a major research effort around the world, as this represents enabling technology for more widespread use of diesels in passenger cars (Klein et al., 1999). However, even with a suitable catalyst, the amount of residual unburned hydrocarbons from normal diesel combustion is significantly lower than required to reduce the NO. One approach to provide the necessary hydrocarbons for the lean DeNOx catalyst is illustrated in Fig. 5 (Weissman et al., 1999; Weissman et al., 2000).

Small amounts of fuel (typically 2-5%) can be injected during the expansion stroke to provide the necessary hydrocarbons. The timing of this "secondary injection" is critical. If injected too early, when the combustion gases have not been sufficiently cooled by the adiabatic expansion of the piston, the added fuel would simply burn to produce CO_2 and H_2O, thus not providing the desired reductant at the catalyst. If injected too late, the fuel would not react in the cooler zone, and the unreacted diesel fuel would arrive at the catalyst. However, diesel fuel is not the optimum hydrocarbon for the catalyst. Partially reacted fuel components, such as oxygenates and olefins, are better reductants (Masters and Chadwich, 1998). Injection at the point where the post-combustion gases are at just the proper temperature allows the fuel to be partially reacted to produce these desired oxygenates and olefins.

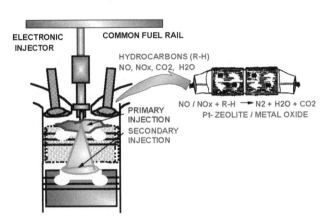

Figure 5. Use of secondary injection to introduce the hydrocarbon reductant for lean DeNOx catalysis.

To explore the feasibility of secondary injection and to identify the optimum timing, we developed a homogeneous variable volume reactor model to simulate the heating and cooling induced by the piston movement. The model also accounts for the heat release during combustion of the primary fuel. Since the amount of heat released is a function of the speed and load of the engine, it is necessary to run the simulation for a variety of conditions. Although this approach is a drastic oversimplification of the conditions in a real diesel, the results were extremely useful for our exploration of the validity of the concept of secondary injection.

The reactions of diesel fuel were approximated using a detailed hydrocarbon oxidation mechanism that described the high-temperature oxidation of n-hexane, 2,3-dimethyl butane, and cyclohexane. This mechanism was constructed following the principles outlined earlier. It consisted of ~250 species and ~2000 reactions. In addition to the hydrocarbon chemistry, it included the nitrogen kinetics described in the earlier application. This mechanism is clearly an oversimplification of a detailed description of "real diesel fuel", but it should be reasonably representative of the aliphatic portion of diesel fuel, typically 60-70% of the total. It does not attempt to describe the kinetics of the aromatic components of diesel.

For given speed/load conditions, and thus a given temperature vs. crank angle distribution, we calculated the effect of adding the secondary fuel at different times (post injection times). The results of a typical calculation are shown in Fig. 6.

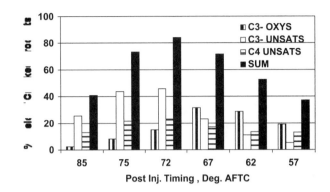

Figure 6. Predicted yields of partially oxidized products vs. timing for secondary injection. Engine parameters: 3.1 bar BMEP, 1.4 bar inlet pressure, 2250 rpm.

The results suggest that it is indeed feasible to produce the desired partially oxidized products in this manner. For this particular engine condition, the preferred timing is near 72 degrees after top-dead-center. Earlier injection leads to a decrease in yield since the temperature is too high and more of the secondary fuel is burned to CO and CO_2. At later times, the expansion has cooled the primary combustion gases to the point where the secondary injection fuel fails to react, allowing unreacted fuel to reach the catalyst.

The effect of changing engine conditions is illustrated in Fig. 7. Higher values of BMEP (Brake Mean Effective Pressure) correspond to higher load. The higher load conditions require more fuel and generate higher in-cylinder temperatures, which shifts the optimum injection to later times. (Note the change in direction of increasing values of post injection timing between Figs. 6 and 7.) Injection timing is more sensitive to engine load than to engine speed or intake pressure.

Another interesting feature of the calculations is the similar shape of the yield vs. crankangle curves. Plotting these same results with temperature rather than crankangle as the independent variable (Fig. 8) moves the various engine conditions closer together, showing the importance of temperature. The remaining offset, with somewhat lower temperatures for higher load, is due to different pressures for the various engine conditions. The peak pressure varies with load, and the pressure, relative to peak pressure, is lower at high load where the injection comes later during the expansion stroke.

Fig. 9 reveals the effect of secondary injection timing upon the nature of the partially oxidized species formed. Note that all engine conditions can be represented by one curve once the abscissa is redefined to reflect the difference

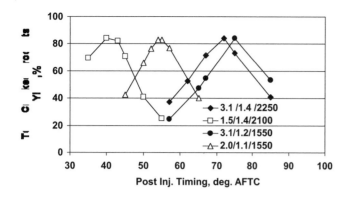

Figure 7. Predicted effect of changing engine conditions on the optimum timing for secondary injection. Engine parameters: BMEP(bar)/inlet pressure(bar)/rpm.

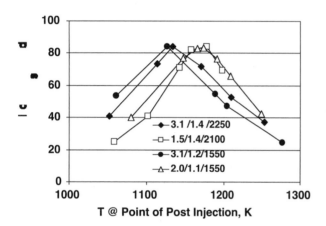

Figure 8. Predicted effect of variations in temperature at time of secondary injection for various engine conditions. Engine parameters: BMEP(bar)/inlet pressure (bar)/rpm.

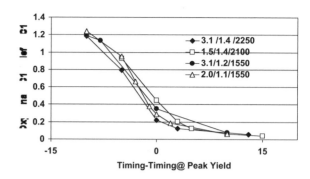

Figure 9. Predicted effect of variations in injection timing upon the relative amount of the oxygenates and olefins for various engine conditions. Engine parameters: BMEP(bar)/inlet pressure (bar)/rpm.

in crankangle from the peak conversion. In essence, this shift accounts for the temperature differences for the various conditions. As can be seen from the figure, at the point of maximum yield of partially oxidized products, olefins predominate over oxygenates. At higher injection temperatures (earlier injection timing) the relative amount of oxygenate increases. This is consistent with the basic mechanistic concept that the initial reaction products are olefins, produced by beta-scission of the radicals formed by abstraction of hydrogen atoms from the parent fuel. At the higher temperatures, reaction has proceeded beyond this point, the relative amount of oxygenates increases as the fuel fragments become progressively more oxidized, but the total yield of partially oxidized products decreases. Thus one has the ability to tune the proportion of oxygenates to olefins (perhaps at some expense in overall yield) to better fit the appetite of the catalyst used for the lean DeNOx.

The modeling also revealed another advantage for secondary injection. In the main combustion zone in the diesel, a very important reaction is chain-branching via:

$$H + O_2 = OH + O$$

However, as the temperature decreases during the expansion stroke to reach that suitable for partial conversion of the injected fuel, the following reaction

$$H + O_2 + M = HO_2 + M$$

becomes more important. This HO_2 quickly converts NO to NO_2:

$$HO_2 + NO = NO_2 + OH$$

It is substantially easier for the catalyst to reduce NO_2 than NO to N_2 (Penetrante et al., 1999). Thus the modeling predicts that one can use secondary injection to simultaneously produce the desired reductants and also convert the NO to the more easily reduced NO_2.

Fig. 10 shows the predicted conversion of NO to NO_2 for the same combination of speed/load conditions considered earlier. For the assumed conditions where both the secondary fuel and NO are uniformly dispersed in the cylinder, it is possible to achieve almost complete conversion of NO to NO_2 in the available contacting time. The higher load (higher T) conditions have the maximum conversion at later injection times; earlier injections encounter a temperature zone that is too hot to produce the HO_2 required for the conversion. Instead, chain-branching dominates and the fuel is completely burned. The dropoff in conversion efficiency at later injection timings at all loads reflects the decreasing rate of oxidation of the fuel at the lower temperatures resulting from the adiabatic expansion.

These predictions were sufficiently promising that we initiated a series of engine tests to explore the potential of secondary injection to facilitate lean DeNOx catalysis in partnership with P. S. A. (Peugeot)[1]. The results of these tests were consistent with our predictions.

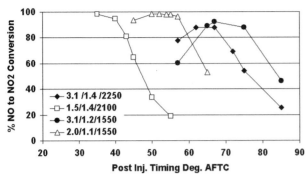

Figure 10. Predicted conversion of NO to NO_2 as a result of secondary injection. Engine parameters: BMEP(bar)/inlet pressure (bar)/rpm.

Fig. 11 shows that it is possible to co-optimize production of the desired partially reacted fuel and NO_2 production to maximize the delivery of both the preferred reductant and NO_2 to the catalyst.

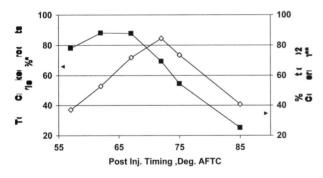

Figure 11. Predictions for both cracked product yield and NO to NO_2 conversion for the 3.1 bar BMEP, 1.4 bar inlet presure/2250 rpm case.

Other Applications

The specific applications discussed above are based on our research at ExxonMobil. However, there are a substantial number of similar studies emerging from other research groups. Many detailed mechanisms describing various aspects of hydrocarbon oxidation have been developed. A recent review provides an annotated listing (Lissianski et al., 2000).

A theme emerging from current research in kinetics is an increasingly sophisticated approach to couple chemistry and transport. For some time it has been feasible to compute the propagation of 1D laminar flames through a homogeneous mixture. More recently, calculations describing flame propagation through a stratified mixture where the fuel/air ratio varies with distance have been reported (Ra, 1999; Pires da Cruz et al., 2000). However, the direct inclusion of detailed chemical mechanisms into CFD codes used to describe the fluid dynamics in engines is not yet computationally feasible. Nonetheless, it is possible to gain significant insight into the chemical behavior of such systems. A good example is the recent analysis of Homogeneous Charge Compression Ignition (HCCI) (Aceves et al., 2000). These workers use a hybrid procedure in which the first step is use of the CFD code KIVA to calculate the temperature distribution inside the cylinder without any combustion. This information is then used to define the initial conditions for a series of kinetic calculations, each calculation representing a different temperature zone within the engine. A 10-zone model was shown to provide accurate predictions of maximum pressure, burn duration, and efficiency. For this case, where the fuel is uniformly dispersed in the cylinder, this post-processing methodology provided a feasible approach to account for the temperature gradients.

The Future

In this paper, I have used some examples of our research at ExxonMobil to try to illustrate some of the possibilities that can emerge from construction and application of detailed chemical mechanisms. Those of us in the kinetic modeling community face a bright future, given our collective increase in understanding of the fundamentals of free radical reactions and the prospect of even faster computers, allowing more explicit coupling of kinetics and transport. The ever-growing utility of electronic structure calculations provide a tremendous opportunity for us to refine our chemical intuition by careful analysis of increasingly more complex elementary reactions. This should provide us the means to improved "rules" for prediction of elementary reactions. These improved rules should substantially improve the quality of automatic mechanism generation codes, both in terms of the rate coefficients that are included, as well as in improved screening techniques to help keep these mechanisms to a more manageable size. Under any circumstances, the role of mechanism reduction becomes even more important (Tomlin et al., 1997). It is especially encouraging to see new avenues to mechanism reduction showing promise. (Iepapetritou and Androulakis, 1999; Petzold and Zhu, 1999; Tonse et al., 1999; Androulakis, 2000).

---

[1] Our P. S. A. collaborators included G. Belot, M. Ehresmann, J. C. Momique, and O. Salvat.

Acknowledgments

The author has had the luxury of working with many talented colleagues, including J. W. Bozzelli, I. P. Androulakis, H. H. Carstensen, A. Y. Chang, J. T. Farrell, W. H. Green, J. M. Grenda, F. Hershkowitz, J. E. Johnston, H. S. Pink, R. G. Susnow, P. K. Venkatesh, M. D. Weisel, and W. Weissman. Many thanks to all of them for their efforts as well as stimulating discussions.

References

Aceves, S. M., Flowers, D. L., Westbrook, C. K., Smith, J. R., Pitz, W. J., Dibble, R., Christensen, M. and B. Johansson (2000). A multi-zone model for prediction of HCCI combustion and emissions. *SAE Conference*, **2000-01-0327**.

Androulakis, I. P. (2000). Kinetic mechanism reduction based on an integer programming approach. *AIChE J.*, **46**, 361-371.

Baulch, D. L., Cobos, C. J., Cox, R. A., Esser, C., Frank, P., Just, T., Kerr, J. A., Pilling, M. J., Troe, J., Walker, R. W. and J. Warnatz (1992). Evaluated kinetic data For combustion modeling. *J. Phys. Chem. Ref. Data*, **21**, 411-734.

Benson, S. W. (1976). *Thermo. Kinetics*. New York, John Wiley and Sons.

Bozzelli, J. W., Chang, A. Y. and A. M. Dean (1997). Molecular density of states from estimated vapor phase heat capacities. *Int. J. Chem. Kin.*, **29**, 161-170.

Bozzelli, J. W. and A. M. Dean (1995). O + NNH: a possible new route for NOx formation in flames. *Int. J. Chem. Kin.*, **27**, 1097-109.

Cohen, N. (1993). The thermochemistry of alkyl free radicals. *J. Phys. Chem.*, **96**, 9052-9058.

Dean, A. M. (1985). Predictions of pressure and temperature effects upon radical addition and recombination reactions. *J. Phys. Chem.*, **89**, 4600-4608.

Dean, A. M. and Bozzelli, J. W. (2000). Combustion chemistry of nitrogen. In *Gas-Phase Combustion Chemistry*, W. C. Gardiner, Jr., Ed. Springer, 125-341.

Frenklach, M., Bowman, C. T., Smith, G. P. and W. C. Gardiner Jr. (1999). GRI-Mech, Version 3.0 http://www.me.berkeley.edu/gri_mech/

Gilbert, R. G., Luther, K. and J. Troe (1983). Theory of thermal unimolecular reactions in the fall-off range. II. Weak collision rate constants. *Ber. Bunsenges. Phys. Chem.*, **87**, 169-177.

Glaude, P. A., Warth, V., Fournet, R., Battin-Leclerc, F., Scacchi, G. and G. M. Come (1998). Modeling of the oxidation of n-octane and n-decane using an automatic generation of mechanisms. *Int. J. Chem. Kin.*, **30**, 949-959.

Grenda, J. M., Bozzelli, J. W. and A. M. Dean (2000). Automated methods of treating chemically activated reactions in kinetic mechanism generation. SIAM Eighth International Conference on Numerical Combustion, Amelia Island, FL.

Harrington, J. E., Smith, G. P., Berg, P. A., Noble, A. R., Jeffries, J. B. and D. R. Crosley (1996). Evidence for a new NO production mechanism in flames. *Proc. Combust. Inst.*, **26**, 2133-38.

Hu, W.-P., Rossi, I., Corchado, J. C. and D. G. Truhlar (1997). Molecular modeling of combustion kinetics. The abstraction of primary and secondary hydrogens by

hydroxyl radical. *J. Phys. Chem. A*, **101**, 6911-6921.

Iepapetritou, M. G. and I. P. Androulakis (1999). Uncertainty considerations in the reduction of chemical reaction mechanisms. 5th International Conference on Foundations of Computer-Aided Process Design, Breckenridge, CO.

Jursic, B. S. (1997). An accurate evaluation of activation barriers for hydrogen abstraction reactions with Becke's 88 density functional theory and high-level G1 and G2 ab initio methods. *Int. J. Quant. Chem.*, **65**, 75-82.

Kee, R. J., Rupley, F. M. and J. A. Miller (1989). *Chemkin-II - A Fortran Chemical Kinetics Package for the Analysis of Gas-Phase Chemical Kinetics*. Sandia National Laboratories, SAND89-8009 UC-401.

Klein, H., Lopp, S., Lox, E., Kawanami, M. and M. Horiuchi (1999). Hydrocarbon deNOx catalysis–system development for diesel passenger cars and trucks. SAE International Conference and Exposition, Detroit, Michigan, 1999-01-0109.

Koffend, J. B. and N. Cohen (1996). Shock tube study of OH reactions with linear hydrocarbons near 1100 K. *Int. J. Chem. Kin.*, **28**, 79-87.

Koizumi, H., Schatz, G. C. and S. P. Walch (1991). A coupled channel study of HN_2 unimolecular decay based on a global ab-initio surface. *J. Chem. Phys.*, **95**, 4130-4135.

Lay, T. H., Bozzelli, J. W., Dean, A. M. and E. R. Ritter (1995). Hydrogen atom bond increments for calculation of thermodynamic properties of hydrocarbon radical species. *J. Phys. Chem.*, **99**, 14514-27.

Lissianski, V. V., Zamansky, V. M. and Gardiner, W. C., Jr. (2000). Combustion chemistry of nitrogen. In *Gas-Phase Combustion Chemistry*, W. C. Gardiner, Jr., Ed. Springer, 1-123.

Lyon, R. K. (1975). *Method for the Reduction of the Concentration of NO in Combustion Effluent Using Ammonia*. U. S. Patent #3,900,554.

Lyon, R. K. (1987). *Div. Fuel Chem,* **32**, 433.

Mallard, W. G., Westley, F., Herron, J. T., Hampson, R. F. and D. H. Frizzell (1993). NIST Chemical Kinetics Database Version 5.0.

Masters, S. G. and D. Chadwich (1998). Selective catalytic reduction of nitric oxide from stationary diesel sources by methanol over promoted alumina catalysts. *Catalysis Today*, **42**, 137-143.

Mayer, P. M., Parkinson, C. J., Smith, D. M. and L. Radom (1998). An assessment of theoretical procedures for the calculation of reliable free-radical thermochemistry, A recommended new procedure. *J. Chem. Phys.*, **108**, 604-615.

Miller, J. A. and C. T. Bowman (1989). Mechanism and modeling of nitrogen chemistry in combustion. *Prog. Energy Combust. Sci.*, **15**, 287-338.

Penetrante, B. M., Brusasco, R. M., Merritt, B. T. and G. E. Vogtlin (1999). Sulfur tolerance of selective partial oxidation of NO to NO_2 in a plasma. SAE, 1999-01-3687.

Petersson, G. A. (1997). Complete basis set thermochemistry and kinetics. In *Computational Thermochemistry*. K. Irikura and D. J. Fpurip, Ed. American Chemical Society, Washington.

Petzold, L. and W. J. Zhu (1999). Model reduction for chemical kinetics, An optimization approach. *AIChE J.*, **45**, 869-886.

Pires da Cruz, A., Dean, A. M. and J. M. Grenda (2000). A numerical study of the laminar flame speed of stratified methane/air flames. *Proc. Combust. Inst.*, **28**, (in press).

Ra, Y. (1999). Ph.D. Thesis. Massachusetts Institute of Technology.

Reid, R. C., Prausnitz, J. M. and B. E. Poling (1987). *Properties of Gases and Liquids*. McGraw Hill, New York.

Ritter, E. R. (1989). Ph.D. Thesis. New Jersey Institute of Technology.

Ritter, E. R. and J. W. Bozzelli (1991). THERM: Thermodynamic property estimation for gas phase radicals and molecules. *Int. J. Chem. Kin.* **23**, 767-778.

Robinson, P. J. and K. A. Holbrook (1972). *Unimolecular Reactions*. Wiley Interscience, London.

Stein, S. E. (1994). NIST Structures and Properties Database and Estimation Program Version 2.0.

Susnow, R. G., Dean, A. M. and W. H. Green (1999). Hydrogen abstraction rates via density functional theory. *Chem. Phys. Lett.*, **312**, 262-268.

Susnow, R. G., Dean, A. M., Green, W. H., Peczak, P. and L. J. Broadbelt (1997). Rate-based construction of kinetic models for complex systems. *J. Phys. Chem. A*, **101**, 3731-3740.

Tomlin, A. S., Turnayi, T. and M. J. Pilling (1997). Mathematical methods for the construction, investigation and reduction of combustion mechanisms. In *Low-Temperature Combustion and Autoignition*. M. J. Pilling, Ed. Chemical Kinetics, Vol. 35, Elsevier.

Tonse, S. R., Moriarty, N. W., Brown, N. J. and M. Frenklach (1999). PRISM: piecewise reusable implementation of solution mapping. An economical strategy for chemical kinetics. *Isr. J. Chem*, **39**, 97-106.

Tsang, W. (1991). Chemical kinetic data base for combustion chemistry. Part 5. Propene. *J. Phys. Chem. Ref. Data*, **20**, 221-273.

Tsang, W. and R. F. Hampson (1986). Chemical kinetic data base for combustion chemistry. Part 1. Methane and related compounds. *J. Phys. Chem. Ref. Data*, **15**, 1087-1279.

Venkatesh, P. K. (2000). Damped psuudospectral functional forms of the falloff behavior of unimolecular reactions. *J. Phys. Chem. A*, **104**, 280-287.

Venkatesh, P. K., Dean, A. M., Cohen, M. H. and R. W. Carr (1997). Approximating the pressure- and temperature dependent behavior of chemically-activated reactions: A review of existing semi-empirical formalisms and the utility of chebyshev expansions and sensitivity analysis. *Rev. in Chem. Eng.*, **13**, 1-67.

Votsmeier, M., Song, S., Hanson, R. K. and C. T. Bowman (1999). A shock tube study of the product branching ratio for the reaction $NH_2 + NO$ using frequency-modulation detection of NH_2. *J. Phys. Chem. A*, **103**, 1566-1571.

Weissman, W., Dean, A. M. and H. S. Pink (1999). *NO to NO2 Conversion Control in a Compression Injection Engine by Hydrocarbon Injection During the Expansion Stroke*. U. S. Patent #5,947,080.

Weissman, W., Hershkowitz, F., Dean, A. M. and H. S. Pink (2000). *DeNOx Reductant Generation in a Compression-Ignition Engine by Hydrocarbon Injection During the Expansion Stroke*. U. S. Patent #6,029,623.

Zeldovich, Y. B. (1946). *Acta Physicochim. URSS*, **21**, 577.

HIGH-PERFORMANCE COMPUTING TODAY

Jack Dongarra
Computer Science Department
University of Tennessee
Knoxville, TN 37996-1301

Hans Meuer
Direktor des Rechenzentrums
Universität Mannheim
68131 Mannheim, Germany

Horst Simon
National Energy Research Supercomputing Center
Lawrence Berkeley National Laboratory
Berkeley, CA 94720

Erich Strohmaier
Computer Science Department
University of Tennessee
Knoxville, TN 37996-1301

Abstract

In 1993 for the first time a list of the top 500-supercomputer sites worldwide has been made available. The Top500 list allows a much more detailed and well-founded analysis of the state of high-performance computing (HPC). This paper summarizes some of the most important observations about HPC as seen through the Top500 statistics. The major trends we document here are the continued dominance of the world market in HPC by the U.S., the completion of a technology transition to commodity microprocessor based highly parallel systems, and the increased industrial use of supercomputers in areas previously not represented on the Top500 list. The scientific community has long used the Internet for communication of email, software, and papers. Until recently there has been little use of the network for actual computations. The situation is changing rapidly with a number of systems available for performing grid based computing.

Keywords

High performance computing, Parallel supercomputers, Vector computers, Cluster computing, Computational grids, Top500 list.

Introduction

In the last 50 years, the field of scientific computing has seen a rapid change of vendors, architectures, technologies and the usage of systems. Despite all these changes, the evolution of performance on a large scale, however, seems to be a very steady and continuous process. Moore's Law is often cited in this context. If we plot the peak performance of various computers of the last 5 decades in Fig. 1 that could have been called the "supercomputers' of their time we indeed see how well this law holds for almost the complete lifespan of modern computing. On average we see an increase in performance of two orders of magnitude every decade.

In the second half of the seventies the introduction of vector computer systems marked the beginning of modern Supercomputing. These systems offered a performance advantage of at least one order of magnitude over conventional systems of that time. Raw performance was the main if not the only selling argument. In the first half of the eighties the integration of the vector system in conventional computing environments became more important. Only the manufacturers that provided standard programming environments, operating systems and key applications were successful in getting industrial customers and survived. Performance was mainly increased by improved chip technologies and by producing shared memory multiprocessor systems.

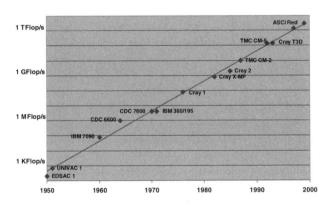

Figure 1. Moore's Law and peak performance of various computers over time.

Fostered by several Government programs, massive parallel computing with scalable systems using distributed memory got in the focus of interest at the end of the eighties. Overcoming the hardware scalability limitations of shared memory systems was the main goal. The increase in performance of standard microprocessors after the RISC revolution, together with the cost advantage of large-scale production, formed the basis for the "Attack of the Killer Micro". The transition from ECL to CMOS chip technology and the usage of off-the-shelf microprocessor instead of custom designed processors for MPPs was the consequence.

The acceptance of the MPP system not only for engineering applications but also for new commercial applications especially for database applications emphasized different criteria for market success such as stability of system, continuity of the manufacturer and price/performance. Success in commercial environments is now a new important requirement for a successful Supercomputer business. Due to these factors and the consolidation in the number of vendors in the market hierarchical systems built with components designed for the broader commercial market are currently replacing homogeneous systems at the very high end of

performance. Clusters built with off-the-shelf components also gain more and more attention.

At the beginning of the nineties, while the MP vector systems reached their widest distribution, a new generation of MPP system came on the market with the claim to be able to substitute or even surpass the vector MPs. To provide a better basis for statistics on high-performance computers, the Top500 list (Dongarra et al., 1999) was begun.

This report lists the sites that have the 500 most powerful computer systems installed. The best Linpack benchmark performance (Dongarra, 1989) achieved is used as a performance measure in ranking the computers. The Top500 list has been updated twice a year since June 1993. In the first Top500 list in June 1993 there were already 156 MPP and SIMD systems present (31% of the total 500 systems).

The year 1995 saw some remarkable changes in the distribution of the systems in the Top500 for the different types of customer (academic sites, research labs, industrial/commercial users, vendor installations, and confidential sites). Until June 1995, the major trend seen in the Top500 data was a steady decrease of industrial customers, matched by an increase in the number of government-funded research sites. This trend reflects the influence of the different governmental HPC programs that enabled research sites to buy parallel systems,

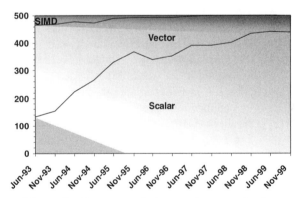

Figure 2. Processor design used as seen in the Top500.

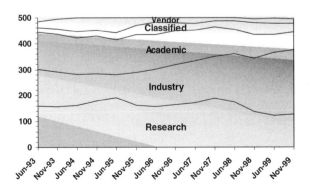

Figure 3. The number of systems of the different types of customers over time.

especially systems with distributed memory. Industry was understandably reluctant to follow this step, since systems with distributed memory have often been far from mature or stable. Hence, industrial customers stayed with their older vector systems, which gradually dropped off the Top500 list because of low performance. Beginning in 1994, however, companies such as SGI, Digital, and Sun started to sell symmetrical multiprocessor (SMP) models of their major workstation families. From the very beginning, these systems were popular with industrial customers because of the maturity of these architectures and their superior price/performance ratio. At the same time, IBM SP2 systems started to appear at a reasonable number of industrial sites. While the SP initially was sold for numerically intensive applications, the system began selling successfully to a larger market, including database applications, in the second half of 1995.

Subsequently, the number of industrial customers listed in the Top500 increased from 85, or 17%, in June 1995 to about 241, or 48.2%, in June 1999. This appears to be a trend because of the following reasons:

- The architectures installed at industrial sites changed from vector systems to a substantial number of MPP systems. This change reflects the fact that parallel systems are ready for commercial use and environments.
- The most successful companies (Sun, IBM and SGI) are selling well to industrial customers. Their success is built on the fact that they are using standard workstation technologies for their MPP nodes. This approach provides a smooth migration path for applications from workstations up to parallel machines.
- The maturity of these advanced systems and the availability of key applications for them make the systems appealing to commercial customers. Especially important are database applications, since these can use highly parallel systems with more than 128 processors.

While many aspects of the HPC market change quite dynamically over time, the evolution of performance seems to follow quite well some empirical laws such as Moore's law mentioned at the beginning of this chapter. The Top500 provides an ideal data basis to verify an observation like this. Looking at the computing power of the individual machines present in the Top500 and the evolution of the total installed performance, we plot the performance of the systems at positions 1, 10, 100 and 500 in the list as well as the total accumulated performance. In Figure 4 the curve of position 500 shows on the average an increase of a factor of two within one year. All other curves show a growth rate of 1.8 +- 0.07 per year.

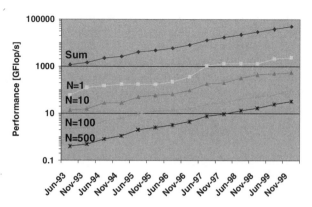

Figure 4. Overall growth of accumulated and individual performance as seen in the Top500.

To compare these growth rates with Moore's Law, we now separate the influence of increasing processor performance and of the increasing number of processors per system on the total accumulated performance. To get meaningful numbers we exclude the SIMD systems for this analysis, as they tend to have extreme high processor numbers and extreme low processor performance. In Fig. 4 we plot the relative growth of the total processor number and of the average processor performance defined as the quotient of total accumulated performance by the total processor number. We find that these two factors contribute almost equally to the annual total performance growth factor of 1.82. The processor number grows per year on the average by a factor of 1.30 and the processor performance by 1.40 compared to 1.58 for Moore's Law.

Based on the current Top500 data which cover the last 6 years and the assumption that the current performance development continue for some time to come we can now extrapolate the observed performance and compare these values with the goals of the mentioned government programs. In Fig. 5 we extrapolate the observed performance values using linear regression on the logarithmic scale. This means that we fit exponential growth to all levels of performance in the Top500. These simple fittings of the data show surprisingly consistent results. Based on the extrapolation from these fits we can expect to have the first 100~TFlop/s system by 2005 which is about 1--2 years later than the ASCI path forward plans. By 2005 also no system smaller then 1~TFlop/s should be able to make the Top500 any more.

Looking even further in the future we could speculate that based on the current doubling of performance every year, the first Petaflop system should be available around 2009. Due to the rapid changes in the technologies used in HPC systems there is however at this point in time no reasonable projection possible for the architecture of such a system at the end of the next decade. Even as the HPC market has changed its face quite substantially since the introduction of the Cray 1 three decades ago, there is no end in sight for these rapid cycles of re-definition.

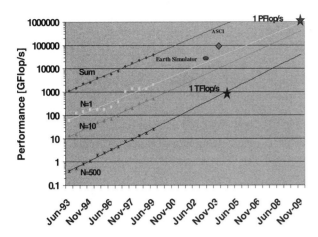

*Figure 5.   Extrapolation of recent growth rates
of performance seen in the Top500.*

Computational Grids

Two things remain consistent in the realm of computer science: i) there is always a need for more computational power than we have at any given point, and ii) we always want the simplest, yet most complete and easy to use interface to our resources. In recent years, much attention has been given to the area of Grid Computing. The analogy is to that of the electrical power grid. The ultimate goal is that one day we be able to plug any and all of our resources into this Computational Grid to access other resources without need for worry, as we do our appliances into electrical sockets today. We are developing an approach to Grid Computing called NetSolve (Casanova and Dongarra, 1998). NetSolve allows for the easy access to computational resources distributed in both geography and ownership. We also describe a parallel simulator with support for visualization that runs on workstation clusters and show how we have used NetSolve to provide an interface that allows one to use the simulator without obtaining the simulator software or the tools needed for visualization.

Network Enabled Solvers

The NetSolve project, under development at the University of Tennessee and the Oak Ridge National Laboratory, has been a successful venture to actualize the concept of Computational Grids. Its original motivation was to alleviate the difficulties that domain scientists usually encounter when trying to locate/install/use numerical software, especially on multiple platforms. NetSolve is of a client/agent/server design in which the client issues requests to agents who allocate servers to service those requests; the server(s) then receives inputs for the problem, does the computation and returns the output parameters to the client. The NetSolve client user gains access to limitless software resources without the tedium of installation and maintenance. Furthermore,

NetSolve facilitates remote access to computer hardware, possibly high-performance supercomputers with complete opacity. That is to say that the user does not have to possess knowledge of computer networking and the like to use NetSolve. In fact, he/she does not even have to know remote resources are involved. Features like fault tolerance and load balancing further enhance the NetSolve system.

At this point, we offer a brief discussion of the three aforementioned components. The NetSolve agent represents the gateway to the NetSolve system. It maintains a database of servers along with their capabilities (hardware performance and allocated software) and usage statistics. It uses this information to allocate server resources for client requests. The agent, in its resource allocation mechanism, balances load amongst its servers; it is also the primary component that is concerned with fault tolerance. The NetSolve server is the computational backbone of the system. It is a daemon process that awaits client requests. The server can be run on all popular strains of the UNIX operating system and has been ported to run on almost any architecture. The server can run on single workstations, clusters of workstations, or shared memory multiprocessors. It provides the client with access to software resources and also provides mechanisms that allow one to integrate any software with NetSolve servers.

The NetSolve client user submits requests (possibly simultaneously) and retrieves results to/from the system via the API provided for the language of implementation. NetSolve currently supports the C, FORTRAN, Matlab, and Mathematica programming interfaces. The functional interface completely hides all networking activity from the user. NetSolve-1.2 can be downloaded from the project web site at www.cs.utk.edu/netsolve.

There are many research projects in the area of grid-based computing. Recently, a book with a number of contributors was published on computational grids. Ian Foster and Carl Kesselman, the originators of Globus, were the editors for this book. It was just published by Morgan Kaufman and is titled *The Grid: Blueprint for a New Computing Infrastructure* (Foster and Kesselman, 1998). It contains ideas and concepts by many people associated with this new area of computational grids, trying to build up an infrastructure that will allow users to get access to the remote users in a way that makes sense.

Conclusions

The scientific community has long used the Internet for communication of email, software, and papers. Until recently there has been little use of the network for actual computations. This situation is changing rapidly and will have an enormous impact on the future. NetSolve is an environment for networked computing whose goal is to deliver the power of computational grid environments to users who have need of processing power, but are not expert computer scientists. It achieves this goal with its three-part client-agent-server architecture.

References

Casanova, H. and J. Dongarra (1998). Applying netSolve's network enabled server. *IEEE Comp. Sci. & Eng.*, **5**(3), 57-66.

Dongarra, J.J., H.W. Meuer, and E. Strohmaier (1999). *Top500 Supercomputer Sites*. University of Tennessee Computer Science Tech. Report, UT-CS 99-434, Knoxville, Tennessee.

Dongarra, J. (2000) *Performance of Various Computers Using Standard Linear Equations Software*. University of Tennessee Computer Science Tech. Report UT-CS 89-85. Knoxville.

Foster, I. and C. Kesselman, eds. (1998). *The Grid: Blueprint for a New Computing Infrastructure*. Morgan Kaufman Publishers, San Francisco, CA, 677.

RIDING THE COMPUTING TIDAL WAVE:
A VIEW FROM COMPUTATIONAL
MOLECULAR SCIENCE

Thom H. Dunning, Jr. and Robert J. Harrison
Environmental Molecular Sciences Laboratory
Pacific Northwest National Laboratory[1]
Richland, Washington 99352

Abstract

A new generation of supercomputers based on clusters of shared-memory multiprocessors will provide unprecedented computing capabilities. Computer systems with this architecture capable of 1 TFLOPS (trillion arithmetic operations per second) are already in place and systems capable of 100 TFLOPS are expected to be available by 2004-5. This new capability, if it can be realized in scientific simulation software, will have a dramatic impact on our ability to model a broad range of complex natural and engineered systems. In chemistry, it will extend quantitative predictions of molecular properties to systems with as many as a hundred atoms and semi-quantitative calculations to thousands of atoms. However, the development of scientific and engineering codes for these parallel supercomputers poses a number of technical challenges. These challenges range from the development of efficient, scalable algorithms for microprocessor-based supercomputers to the development of efficient communications and input/output software for communications-intensive applications. In the development NWChem, a new molecular modeling code written specifically for parallel supercomputers, these challenges were addressed by establishing close collaborations between theoretical and computational chemists, computer scientists, and applied mathematicians. This new "community code" provides the basis for a major advance in our ability to predict the properties and behavior of molecules by bringing the full power of terascale supercomputers to bear.

Keywords

Molecular modeling, Computational chemistry, Electronic structure, Schrödinger equation, Software development, Parallel supercomputers, Massively parallel computers, Non-uniform memory access, NUMA, Northwest Chem, NWChem.

Introduction

Understanding chemical processes in real-world systems requires a broad range of information on the molecules involved: their structure and interactions, their electronic and optical properties, the heats and rates of chemical reactions, and so on. Problems of importance to the U.S. Department of Energy include combustion and atmospheric chemistry, chemical vapor deposition,

protein structure and enzymatic chemistry, and industrial chemical processing. Molecules of interest range from simple species important in the atmosphere (*e.g.*, methane and hydroxyl radical) to the complex species responsible for life (*e.g.*, proteins and polysaccarides). Although much has been gained by qualitative applications of the concepts of molecular quantum mechanics to chemical problems, quantitative predictions are often required to resolve outstanding questions. This can only be achieved

---

[1] The Pacific Northwest National Laboratory is operated by Battelle Memorial Institute for the U.S. Department of Energy under Contract No. DE-AC06-76RLO 1830.

by obtaining accurate solutions of the electronic Schrödinger equation.

Although solution of the Schrödinger equation is still a daunting task, the development of sophisticated mathematical models and computational algorithms coupled with an exponential growth in computing power over the past two decades has made it possible to solve the electronic Schrödinger equation for a large class of molecules. For small molecules, the accuracy of the calculations can rival that obtained from experiment. For large molecules, less computationally intense, but also less accurate methods must be used. This can introduce a level of uncertainty into the predictions that may be unacceptable. Thus, there is a continuing need to push accurate quantum chemical approaches to larger and larger molecules.

In this article, we discuss the issues involved in the development of molecular modeling codes for solving the electronic Schrödinger equation on the new generation of parallel supercomputers. Using clusters of shared memory multiprocessors (also called symmetric multiprocessors, SMPs), computers capable of 1 trillion arithmetic operations per second (TFLOPS) are already in place, e.g., at the San Diego Supercomputer Center and Oak Ridge National Laboratory, and computers capable of 100 TFLOPS are expected within the next five years. In fact, IBM recently announced plans for the development of a 1,000 TFLOPS computer, called "Blue Gene," in this same time frame,[2] although it will be limited to a specific application, namely, protein folding. The availability of computers capable of 100 TFLOPS or more will have a profound impact on our ability to simulate the fundamental molecular processes that underlie complex natural and engineered systems.

Unfortunately, the benefits of these new computer systems will not come easily. First, the scientific codes written and optimized for the vector supercomputers of the past must be revised for the new parallel supercomputers. Software development, especially of the magnitude required here, is a very manpower-intensive effort and not one that most computational scientists would undertake lightly, especially those groups responsible for such large molecular science codes as Gaussian (Frisch *et al.*, 1998), Aces II (Stanton *et al.*), CADPAC (Amos *et al.*, 1995), GAMESS (Schmidt *et al.*, 1993), MOLPRO (Werner *et al.*, 2000), DALTON (Helgaker *et al.*, 1997) and others. Second, disciplinary computational scientists must become familiar with the tools that are being developed—largely by computer scientists—to enable the efficient use of parallel supercomputers. Since many of these tools are in an active state of development, this can be a neverending task. In addition, many of these tools require "hardening"

before they will be truly usable for developing the very complex software packages found in molecular modeling applications.

In this paper, we provide a brief glimpse into the (near) future of supercomputing, explore the major issues involved in developing software for parallel supercomputers, and illustrate how these problems were addressed in the development of NWChem, a new "community" code for molecular simulations. The emphasis will be on solving the electronic Schrödinger equation, although NWChem provides a broad range of molecular modeling capabilities, including molecular dynamics and Monte Carlo calculations using traditional force fields, *ab initio* forces, or mixed quantum mechanical forces/ molecular mechanical force fields.

In our discussions, we will adopt the usual definition of a supercomputer as "the fastest computer system(s) available at a given time for numerical computations." To qualify as a true supercomputer parallel computers must now have many hundreds, if not thousands, of processors.

Near Term Evolution of Supercomputing

For fifty years, the history of computing has been one of continuous progress, with advances over the past twenty years or more being in line with Moore's Law, which (paraphrasing) states that the computing power of (micro)processors will increase by a factor of two every 18-24 months. This exponential increase in computing power has led to spectacular new computing capabilities. For example, in 1980, the fastest computer in NERSC, DOE's National Energy Research Scientific Computing Center, was a Cray 1 with a peak speed of 140 MFLOPS (or millions of arithmetic operations per second). By

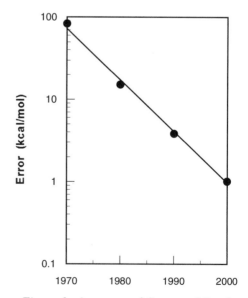

Figure 1. Accuracy of Computed Bond Energies.

---

[2] General information on "Blue Gene" may be found at http://www.research.ibm.com/bluegene/.

1990, a Cray 2 with a peak speed of 4,000 MFLOPS had been installed at NERSC. In 2000, a Cray T3E, with a peak speed of 620,000 MFLOPS is in routine production and NERSC is awaiting the delivery of a new IBM RS/6000 Scalable Parallel computer system with a peak speed of 3,600,000 MFLOPS.

The impact of these advances in computing technology on computational science and engineering has been stunning. In Fig. 1 we plot the average error in the prediction of bond energies from 1970 to the present. Knowledge of bond energies is critical for understanding many chemical processes, including combustion, atmospheric chemistry, and industrial chemical processing. In 1970, the average error in calculated bond energies was so large that predictions were totally useless for real-world chemical applications. By 2000, however, through a combination of methodological and algorithmic advances plus increases in computing power, the error has been reduced by a factor of almost 100. The calculations are now more accurate than most experimental measurements! There are, of course, limitations on the size of molecules that can be modeled using these sophisticated computational techniques, but advances in computational algorithms and computer technology continue to push this limit to larger and larger molecules.

Although predicting the future of any technology that is evolving as rapidly as computer technology is difficult (if not foolhardy), a number of trends are already in place that will determine the basic architecture of super-computers delivered within the next five years or so. Over this period, U.S. computer vendors will focus on building parallel supercomputers using clusters of shared memory multiprocessors (also called symmetric multiprocessors) or SMPs. Thus, the IBM machine scheduled for delivery to Lawrence Livermore National Laboratory in the Fall 2000 ("ASCI White") will have 512 nodes with each node containing 16 processors and 8 or 16 GBYTES of memory[3] (note the switch from "mega" to "giga," or millions to billions). All told, this machine will have a total of 8,192 processors and more than 6,000 GBYTES of memory. In addition, over 160,000 GBYTES of disk storage will be available to the user. Since the peak speed of each of the processors in ASCI White is 1.5 GFLOPS and there are 484 compute nodes, the peak speed of this parallel supercomputer will be nearly 12,000 GFLOPS (the remaining nodes are devoted to interactive and debugging use as well as managing GPFS, IBM's General Parallel File System). *These numbers—in computer speed, in memory size, and in disk storage—suggest more than a modest change in how we use computers to model molecules and molecular processes is well underway.*

Given these trends, what can we expect in the near future? IBM's future plans[4] call for faster and faster processors (to several GFLOPS per processor), increased numbers of processors per node (by 4x or so), and lower latency, higher bandwidth switches connecting the nodes. The maximum number of nodes is expected to remain about the same. This is similar to the roadmap provided by Compaq,[5] which envisions increasing the peak performance of its Alpha processor to several GFLOPS, the number of processors in the node from 4 to 64, the number of nodes from 128 to 256, and (again) lower latency, higher bandwidth switches. These advances will lead to parallel supercomputers running at 100,000 GFLOPS or 100 teraflops (TFLOPS) or more by 2004-5! In fact, DOE's National Nuclear Security Agency is actively pursuing the development of these computers as a part of its Stockpile Stewardship Program. As a result of the investments by NNSA, computers with a substantial fraction of these capabilities are expected to be available in the NSF and civilian DOE supercomputing centers in a comparable, if slightly delayed, time frame.

Despite the overall similarity in the IBM and Compaq roadmaps, their detailed implementation plans differ. For example, IBM intends to use its prowess in silicon chip fabrication to place multiple processors and large caches on each chip. IBM's next generation processor, the Power-4, will have two processors, a shared 1½ MBYTE L2 cache on each chip, and a large off-chip shared L3 cache. Future processors in this family will have an increased number of processors per chip as well as larger on-chip L2 caches. As an extreme example, the chips to be used in "Blue Gene" are expected to have 32 processors (each capable of one GFLOPS) and 16 MBYTES of memory on a (less than) 1-inch square chip, although the processors will have a reduced instruction set. These trends must be kept in mind in designing software to run on parallel supercomputers. Otherwise, at the pace that computer technology is changing, the effort devoted to revising software could swamp that for doing scientific research.

Much of the progress in computer technology is being driven by increases in the computing power of microprocessors (a result of advances in microtechnology). These advances are not coming just from the traditional computer manufacturers like Sun, IBM, Hewlett-Packard, and Compaq (through its acquisition of DEC). For example, Sony recently introduced the PlayStation 2 in Japan (it will become available in the U.S. in the Fall 2000). The PlayStation 2, which is priced

---

[3] Additional information on "ASCI White" may be found at: http://www.llnl.gov/asci/platforms/white/index.html.

[4] General information on IBM's RS/6000 SP computers can be found at: http://www.rs6000.ibm.com/hardware/largescale/.

[5] General information on Compaq's AlphaServer computers can be found at: http://www.compaq.com/hpc/.

in the $300-400 price range, contains a microprocessor capable of 6 GFLOPS (although in 32-bit mode) and 32 MBYTES of memory. Since a Linux development environment is available for the PlayStation 2, it is quite possible that future scientific studies will be carried out on a computer that evolved from this game console. When this advance in microprocessor technology is combined with the consequences of a true commodity market, the results are truly awesome. In the next four years, Sony expects to sell 100 million PlayStation 2s. This corresponds to the production of more than 1 quadrillion (million billion) arithmetic operations (petaflops, PFLOPS) of computing capability every 2½ days!

Even the chips designed for personal computers are rapidly approaching what would have been supercomputer performance levels just 10 years ago. The currently available 1GHz AMD Athlon can perform two arithmetic operations per cycle for a peak performance of 2 GFLOPS. Later this year, Intel will release the first processor based on its new IA-64 architecture—the Itanium.[6] The initial clock frequency of the Itanium will be 800 MHz. Since the Itanium can execute up to four arithmetic operations per cycle, its peak performance will be 3.2 GFLOPS. Faster IA-64 processors are already in various advanced stages of development and will become available in the next few years. Again, this new capability will be impelled rapidly forward by the mass market served by Intel.

A word of caution though. For the informed buyer, the commodity market provides a wonderful opportunity for cost-effective scientific simulation. For many research groups, small clusters of PCs will very effectively provide the resources needed to carry out their research in computational modeling and simulation. For clusters containing more than a hundred PCs, on the other hand, we say *caveat emptor*! Not only does the bus and memory architecture of personal computer systems limit the range of applications that can realize the performance promised by the microprocessor and the speed of the switch that can be used to connect the processors, but PCs are not designed to have the fault tolerance or very low failure-rates required by truly massively parallel supercomputers. In an organization of 8,000+ people, who cares if a few PC's have to be rebooted each day due to "hardware glitches"? But, you can be sure that the users of "ASCI White" will care.

Scientific Computing on Parallel Supercomputers— Software Issues

In this and the following section, we discuss a number of issues related to the performance of scientific modeling and simulation codes on parallel supercomputers. Although much of the discussion will be general, at times the issues being discussed will be illustrated by examples from quantum chemistry. In this section, we address software issues; hardware issues will be discussed in the following section.

Algorithms for Electronic Structure Calculations

It has long been recognized that advances in mathematical models and algorithms have contributed as much to advances in computational modeling and simulation capabilities as have advances in computer technology. Thus, we no longer use grids of energies to calculate the structure of a molecule. Rather, we use far more efficient quasi-Newton procedures, analytically computing the derivative of the energy with respect to nuclear coordinates and following the steepest descent path to the minimum energy structure. Even for a medium size molecule, this reduces the computing requirements by an order of magnitude. As we shall see, algorithmic advances will continue to be important in the future.

During the past decade it has been established that the coupled cluster approach provides an accurate description of the electronic structure of atoms and molecules (Lee and Scuseria, 1997; Peterson and Dunning, 1997; Martin, 1998; Bak *et al.*, 2000; Dunning, 2000). A coupled cluster approximation that includes all single and double excitations plus a perturbative estimate for the *connected* triple excitations—CCSD(T)—provides a good balance between accuracy and computational cost (Raghavachari *et al.*, 1989). For molecules well described by a Hartree-Fock wave function, the CCSD(T) method predicts bond energies, ionization potentials, and electron affinities to an accuracy of approximately ±0.5 kcal/mol, bond lengths accurate to ±0.0005 Å, and vibrational frequencies accurate to ±5 cm^{-1}. This level of accuracy is more than adequate to answer the questions which arise in studies of the complex chemical systems noted in the Introduction.

Basis set expansion techniques are usually used to solve the electronic Schrödinger equation. This converts the original partial differential equation into a set of algebraic equations suitable for solution on electronic computers. Analysis of the algebraic equations allows us to determine the computational order of the various methods used to solve the electronic Schrödinger equation. Thus, the CCSD(T) algorithm has an iterative N^6 step (the CCSD algorithm) and an N^7 step (the perturbation theory algorithm for the triple excitations), where N is the number of functions in the basis set. Using the results from these analyses, computational chemists have developed very sophisticated algorithms for CCSD(T) calculations. As a result, it is now possible to perform accurate CCSD(T) calculations on molecules as large as benzene on well configured workstations.

---

[6] General information on the Itanium and the IA-64 family may be found at: http://developer.intel.com/design/ia-64/.

Nonetheless, the steep dependence of the CCSD(T) method on the size of the basis set dramatically restricts the size of molecules that can be studied using this method.

Hartree-Fock calculations, which are based on a simple independent-electron description of the electronic structure of molecules, scale as N^4. It has long been known that the use of screening techniques in the HF algorithm can reduce the exponent in the scaling law to less than four. The simplest screening technique is based on the fact that certain classes of integrals become negligibly small when the atomic centers on which the functions are located are well separated. Use of a simple estimation procedure dramatically decreases the number of integrals that must be computed to form the HF Hamiltonian—the source of the N^4 scaling. For moderate-size molecules, it is found that the scaling exponent is around three for calculations that use screening, and, in the large molecule limit, the screened HF algorithm scales as $O(N^2)$. Other efficiencies are also possible. For example, the charge density may be fitted to an atom-centered basis as in density functional theory, DFT (Dunlap, Connolly and Sabin, 1979), thereby reducing the formal cost of evaluating the Coulomb matrix to just $O(N^3)$. Using sparsity reduces this to $O(N^2)$ but with a much smaller pre-factor than the conventional approach. Using more advanced mathematical techniques, such as the fast multipole method (FMM) to handle the long-range Coulomb interaction (Greengard and Rohklin, 1997), and a separate treatment of the exchange interaction, it is possible to develop a HF algorithm that exhibits linear scaling in the very large molecule limit (Ochsenfield, White and Head-Gordon, 1998). Combining these techniques with Fourier methods provides improved scaling even for small systems in high-quality basis sets.

Similar reductions in the scaling laws are possible for correlated calculations, including CCSD(T) calculations. This is now an important consideration, because the availability of parallel supercomputers has extended correlated calculations to molecules large enough for the reduction in the scaling law exponent to be significant. Although a detailed review of this work is beyond the scope of this paper, it is instructive to briefly summarize current activities because they clearly demonstrate the power of algorithmic advancements.

The scaling laws referred to above assume that the correlated calculations are carried out in terms of the canonical, delocalized Hartree-Fock orbitals. For small molecules, this is, in fact, the most efficient approach. However, electron correlation effects, which are a result of the difference between the full inter-electronic interaction ($\Sigma\Sigma'/r_{ij}$) and the averaged HF interaction, are local. They asymptotically decay as $^1/r^6$—and, therefore, there is no physical reason why their computation should scale as a high power of the number of basis functions. In fact, as in the HF method, for a sufficiently large molecule, it should be possible to develop an algorithm to compute the correlation energy that scales linearly with N.

Pulay and Saebø (Pulay and Saebø, 1993) were the first to advance a systematic approach for exploiting the locality of electron correlation effects. Recently, Werner and co-workers (Hampel and Werner, 1996; Schütz and Werner, 2000) put forward a local CCSD(T) method. Their algorithm combines localized HF occupied MOs with a projected set of AOs representing the virtual space. Use of this orbital set allows a systematic truncation of the excitations included in the wave function based upon physical criteria, e.g., the distance between the localized orbitals. This method leads to much improved scaling with molecular size. There is considerable merit in this approach, but it unavoidably introduces hard-to-quantify approximations and has difficulty achieving linear scaling since in most models the number of wave function amplitudes scales as $O(N^2)$. It also requires an iterative solution of the equations even for perturbation theory. Some of the local correlation methods (Maslen et al., 1998 and 2000) improve upon this approach because they avoid solving large sets of equations, can be systematically extended, and can, in the limit of a complete basis set for each atom/fragment, approach the exact result. However, they still do not realize linear scaling.

As in the HF method, it is possible to exploit screening in calculating the correlation energy. However, to take advantage of screening, the equations for the various correlated methods must be re-written in terms of AOs rather than MOs. This has recently been done for the coupled cluster method by Scuseria and Ayala (Scuseria and Ayala, 1999), who showed that, *with screening alone*, the CCD equations could be solved just as efficiently in the AO basis as in the MO basis, *for sufficiently large molecules*. Since the effectiveness of screening and multipole summation techniques increase with molecule size, the question is not whether the use of AOs is more efficient than the use of MOs, but where the crossover point occurs. The preliminary studies of Scuseria and Ayala suggest that the crossover point for extended molecules may occur for molecules containing as few as 10-20 atoms. The crossover point for compact three-dimensional molecules in high-quality basis sets will clearly be higher, but this is nevertheless an exciting development. What is needed now is a systematic comparison of the accuracy and cost of local correlation methods on systems of such size that their computational advantages can be evaluated and any inaccuracies exposed by comparison to full CCSD(T) calculations.

Save for the HF (and DFT) algorithms, the impact of reduced scaling algorithms has yet to be felt in chemistry. As the above discussion shows, reduced scaling algorithms for correlated molecular calculations are imminent. When these algorithmic advances are combined with the advances in computing technology noted here, the impact will be truly revolutionary. Problems that currently seem intractable will not only become doable, they will become routine. A whole new set of challenges will then await the adventurous soul.

Programming Models for Parallel Supercomputers

Parallel supercomputers based on clusters of SMPs pose problems very different than the parallel supercomputers of the 1990s, where there was only one processor in each node. Although one could use a message-passing model for SMP clusters, this approach does not take advantage of the communications efficiencies possible within the shared-memory nodes. The lack of a well defined, efficient programming model for clusters of SMPs has been a major impediment in their use to date.

Parallel programming tools for distributed- and shared-memory systems are quite distinct. The message-passing interface (MPI) provides a portable mechanism to exchange data between processes by passing messages between the two processors involved (Message Passing Interface Forum, 1995). The processes may be running on either a shared- or distributed-memory computer, but the programming model itself is intrinsically distributed-memory since a process' data is defined to be private. This distributed model is both a strength and a weakness of message passing. Distributing the data forces the programmer to pay attention to data locality, which is essential for efficient parallel (and sequential!) execution. Message-passing programs are also portable and, for algorithms with straightforward data parallelism and static load balance, are easy to write. However, it is difficult to use message passing to write scalable parallel algorithms that manipulate complex data structures or use dynamic load balancing.

In contrast, Open-MP specifies a portable mechanism for writing shared-memory parallel programs based upon the data-parallel decomposition of DO-loops (OpenMP, 1999). In principle, the processes may be running on either shared- or distributed-memory computers, but most implementations are for shared-memory systems. An Open-MP program only expresses the data parallelism; the compiler does the rest. Just as in previous incarnations of this programming model, an Open-MP model can work well on tightly coupled shared-memory computers if the algorithms have lots of loop-level data-parallel operations. Unfortunately, most of quantum chemistry algorithms do not fit this description. Furthermore, Open-MP does not give the programmer control over data locality or alignment.

Some vendors of parallel supercomputers are recommending that their customers pursue a mixed Open-MP and MPI programming model in order to use clusters of SMPs. This level of complexity may be necessary for "bleeding-edge" applications, but represents an intolerable increase in effort (and cost!) for most projects.

We recommend a programming model that unites both sequential and parallel algorithms as well as distributed- and shared-memory computers. The central concept is that of *non-uniform memory access* (NUMA). The familiar structure of a sequential computer has a memory hierarchy—registers, caches, memory, disk—with each level requiring successively more time to access (latency) and having slower transfer speeds (bandwidth). A parallel computer, no matter how its memory is organized, just adds additional levels to this memory hierarchy. Thus, the same methods that we use to improve the performance of sequential algorithms, *e.g.*, data blocking, may be directly applied to parallel algorithms.

From the NUMA perspective, parallel algorithms only differ from sequential algorithms in that they must also express and control concurrency, and distributed and shared-memory computers only differ in the manner in which memory must be accessed. NUMA algorithms are much easier to design and model. However, a new capability is required to make NUMA programs easier to write and more efficient than message passing programs—one-sided memory access.

One-sided memory access is a familiar property of shared-memory programming. In SMPs any process can *independently* access any piece of data. This is why shared-memory programs are nominally easier to write than message-passing programs, which require other processes to explicitly participate in the access of "remote" data. It turns out that one-sided memory access can be implemented at least as efficiently, if not more efficiently, than message passing even on distributed memory computers. At its lowest level, the network connecting two nodes just moves data from the memory of one node into that of another. Message passing is built on top of that capability using a not-inexpensive protocol that forces "receives" to block until the messages arrive, or buffers incoming messages. In contrast, a simple one-sided remote memory copy needs little, if any additional overhead. One-sided access enables straightforward manipulation of distributed data structures, and makes dynamic load balancing easy, since each process can function independently.

Putting the pieces together, a NUMA algorithm combined with a portable tool to provide shared data structures with one-sided operations, such as Global Arrays (to be discussed later), and augmented with standard message-passing tools (*if needed*), is a very powerful approach for developing parallel algorithms that are both scalable and efficient on either shared- or distributed-memory computers, or any combination of the two.

Operating Systems for Parallel Supercomputers

At the base of high-performance computing software is the operating system for parallel supercomputers. At

Table 1. IBM and Compaq Parallel Supercomputers.

	IBM	COMPAQ
Processor	**Power3-II**	**EV67**
Clock Frequency (MHz)	375	667
Max. FLOPS/cycle	4	2
Peak Performance (GFLOPS)	1.50	1.33
L1 Data Cache		
Size (KBYTES)	64	64
Bandwidth (GBYTES/S)	6.0	10.6
L1B/F (BYTES/FLOP)	4.0	8.0
Computational Node	**Nighthawk-2**	**ES40**
Number of Processors	16	4
Peak Performance (GFLOPS)	24	5.3
L2 Cache/Processor		
Size (MBYTES)	8	8
Bandwidth (GBYTES/S)	8.0	7.1
L2B/F (BYTES/FLOP)	5.3	5.3
Memory System		
Size (GBYTES)	32	64
Bandwidth (GBYTES/S)	16	5.2
B/F (BYTES/FLOP)	0.67-1.33	0.98
Communication	**Colony**[a,b]	**Quadrics**[c]
Bandwidth (MBYTES/S)	500	210
Latency (μs)	15	<6
NETB/F (BYTES/FLOP)	0.02-0.33	0.04-0.16
Computer System	**ASCI White**	**AlphaServer SC**
Max. Number of Nodes	512	128
Max. Number of Processors	8,192	512
Peak Performance (GFLOPS)	12,288	748

[a] Refers to maximum uni-directional bandwidth; bi-directional bandwidths will be 2x higher. Rates achieved by MPI will be lower.

[b] The data given here refers to the Colony (single/single) switch. Higher bandwidths (up to 4x) will be possible in future versions of the switch.

[c] The data given here is for one "rail" of the Quadrics switch. Additional "rails" may be added to increase the bandwidth of the switch.

this point, none of the computer vendors can deliver an operating system for parallel supercomputers that provides the functionality of those provided earlier for vector supercomputers. The operating system that is delivered with parallel supercomputers is immature and missing one or more important, if not critical, components. It then becomes the responsibility of staff at the computer site, working closely with the vendor's staff, to "harden" the operating system and fill in the missing pieces. This is a technically challenging task and, as a result, the time between delivery of the computer system and the availability of that system to the general user has grown to several months (it is usually available to "friendly" users much earlier). A movement is now underway in the computer science community to use open-source software, such as Linux, as the basis for an operating system for future parallel computers (PITAC, 2000). The operating system would be developed by a consortium of computer scientists in the national laboratories, universities, and computer companies and maintained and extended for the benefit of all who use parallel supercomputers for technical computing. This is a formidable task, but one that may need to be undertaken to ensure that the computers being built for business applications can also be used as tools for scientific research.

Scientific Computing on Parallel Supercomputers—Hardware Issues

In this section, we will examine a number of performance issues related to the architecture and implementation of current parallel supercomputers. Use of parallel supercomputers poses a number of problems unlike those encountered in the previous generation of vector supercomputers. One cannot obtain supercomputer performance from massively parallel computers without understanding the intimate connection between the computer hardware and the algorithms used in the scientific simulation codes.

Architecture and Implementation of the Computational Nodes in Parallel Supercomputers

The architecture as well as the implementation of the computational nodes in a parallel supercomputer has a major influence on the performance of an algorithm. In fact, to achieve maximum efficiency, algorithms must be adjusted to, if not tailored for, the node. Important features to consider for the computational node are the speed and architecture of the microprocessor (number of functional units, size of on-chip caches, speed of the communication ports, *etc.*), the number of microprocessors used in the computational node, the architecture of the memory subsystem (size, speed and nature of the off-chip caches; nature and speed of the microprocessor-memory interconnect; *etc.*).

Detailed specifications for two current supercomputer systems—the IBM "ASCI White" system[3,4] and the Compaq AlphaServer SC system[5]—are summarized in Table 1. The *not-to-be-exceeded* peak speed of a processor is commonly determined as the product of the clock frequency and the maximum number of floating point operations that can be performed per cycle. For example, the 375 MHz IBM Power3-II can perform two multiply-add operations per cycle (a total of four arithmetic operations) and the 667 MHz Compaq Alpha EV67 microprocessor can perform a multiplication and an addition on each cycle (two arithmetic operations). This gives about the same peak speed for both processors—1.50 versus 1.33 GFLOPS. So why is it commonly seen that the Alpha EV67 outperforms the Power3-II on most scientific applications (Worley, 2000)?

The fact that the clock frequency of the Alpha microprocessor is nearly twice that of the Power3-II processor gives it an inherent advantage since all of the core operations, including address computation, occur more quickly. The Alpha's superior clock speed also implies that the bandwidth from its L1 cache is nearly twice that of the Power3. Furthermore, the Power3 only delivers its peak performance if two multiply-add operations are scheduled every cycle, which is unlikely for most code processed by a compiler. In contrast, the Alpha only requires that independent multiplication and addition operations be performed to achieve maximum speed. So, it is not surprising to find that, in real benchmarks, the delivered speed of the IBM Power3-II is significantly less than that of the Compaq Alpha EV67.

External to the processors, all computer designers must confront the increasing gulf between the speed of the processors and that of the memory subsystem. Since the early days of Cray Research, Inc., it has been recognized that the performance of many scientific applications is determined by the performance of the memory subsystem as much as by the speed of the floating-point unit. In fact, beginning with the Cray X-MP, the balance between memory bandwidth (B, in BYTES/SEC) and floating point performance (F, in FLOPS) became an explicit design element in Cray supercomputers. The later Cray models boasted exemplary memory bandwidths. For example, in the Cray T-90, the last vector supercomputer from CRI, the B/F ratio was over 15! This is to be compared to a B/F ratio of 0.67-1.33 for the Nighthawk-II nodes used in the ASCI White system and 0.98 for the ES40 nodes used in the AlphaServer SC systems. For the IBM machine, the lower figure corresponds to the case where all of the processors are simultaneously requesting data from memory, whereas the higher figure refers to the bandwidth when only one processor is requesting data. To put this in context, for a DAXPY operation

$$y_i = ax_i + y_i,$$

a common arithmetic operation in scientific codes, to execute at full speed *out of memory* requires that $B/F \geq 12$.

Modern microprocessors attempt to solve the memory performance problem by using caches. Caches are built from fast, but expensive memory. Both the IBM Nighthawk-II and Compaq ES40 nodes have an 8 Mbyte L2 cache for each microprocessor. The B/F ratio for the IBM L2 cache is 5.3, an improvement of as much as a factor of 8 over retrieving data from main memory. Clearly, if an algorithm can be structured so that the data are in cache when needed, the performance of the microprocessor will improve substantially. For a small number of scientific applications, it may be possible to keep the data in cache. For others, the key to a successful cache strategy is data reuse. In addition, high efficiency can be achieved if sufficient operations are performed for each memory reference, *e.g.*, evaluation of an exponential or square root.

Modern microprocessors also have a number of other features to enhance memory performance, including instruction and data pre-fetch and the ability to schedule multiple outstanding memory requests. In spite of all of these enhancements, however, the *sequential* performance of many scientific codes on microprocessor-based computational nodes is still a major problem.[7] Both the ASCI Path Forward program and the National Security Agency are currently funding research aimed at increasing the performance of memory subsystems in parallel supercomputers. We expect to see substantial improvements in the B/F ratio of microprocessor-based parallel supercomputers in the next few years.

It is worthwhile to note that Japanese vector-parallel supercomputer systems do not have the serious memory bandwidth problems noted above. They have been built on the principles enunciated by Cray and, thus, have memory subsystems well designed for scientific applications. For example, the latest supercomputers offered by NEC, the SX-5,[8] have a B/F ratio of 8. Although the U.S. Government has added a 400% tariff to Japanese supercomputers in the U.S. (because of "dumping" concerns) and many senior computer technologists have declared this architecture a "dead end," it is clear that a large number of scientific applications would benefit greatly from the increased memory bandwidth provided by these systems.

Architecture and Implementation of the Communications Fabric in Parallel Supercomputers

For the processors in a parallel computer to collaborate in solving all but the most trivial problems, they must have an efficient mechanism for sharing data. The most attractive option is a fast and uniform shared memory, but it is not practical to build such systems with hundreds, let alone thousands of processors. Shared-memory systems with non-uniform memories are affordable and were once in vogue (*e.g.*, the BBN Butterfly and the Kendall Square Research KSR-1 and

---

[7] Current climate modeling codes require very high memory bandwidths to achieve optimal performance. Tests with version 2 of the Community Climate Model show that on microprocessor-based computers CCM-2 typically obtains 10% or less of peak performance, whereas on vector supercomputers it achieved nearly 40% of peak (Climate Research Committee, 1998).

[8] Additional information on the NEC line of supercomputers can be found at: http://www.hstc.necsyl.com/index.html.

KSR-2) but did not offer good price-performance due to the price and performance penalty for the hardware that handles the additional address manipulations, cache coherency, and data routing. Distributed-memory systems, in which nodes only have direct access to their own memory and must access data in other nodes over a network, inevitably have a better "raw" price-performance. We use the term "raw" here because this simple analysis does not take into account the greater ease of use, administration, and programming of shared-memory systems.

Large shared-memory systems are now enjoying a renaissance with systems up to 512 nodes available from SGI, and smaller systems from other vendors (*e.g.*, Sun). This renaissance is a result of three major factors. First, the speed of current microprocessors demands the tighter coupling between processors and memory provided by shared-memory systems. Second, advances in technology have made non-uniform, shared-memory systems more affordable. And, third, there are now many commercial users (airlines, banks, investment brokerages, dot coms, *etc.*) who need the level of performance and ease of programming provided by these systems. These users are far less tolerant of manpower inefficiencies than are scientists and engineers, who are used to being on the cutting (often bleeding) edge of new computing technologies. The increasing availability of shared memory multiprocessors is, in fact, the driver behind the current trend to construct parallel supercomputers from clusters of SMPs tied together by high-speed switches.

This trend has mixed implications for the performance of applications on parallel supercomputers. Larger SMP nodes allow more expensive, and hence higher performance, network cards or interconnects to be used since their cost can be amortized over all of the processors in the node. And, as long the processes on a node do not saturate the network interface, a performance gain will be realized. However, communication-intensive applications will fare worse on parallel supercomputers based on large SMP clusters. Consider the latency and bandwidth of the interconnect on the IBM SP series. The IBM SP-1 switch had a bandwidth of just 8 MBYTES/S and a message-passing latency of about 80 μs. The switch on the next generation of parallel supercomputer from IBM, the SP-2, delivered about 80 MBYTES/S with a latency of 30 μs. The most recent switch, the "Colony" switch in ASCI White, delivers a uni-directional bandwidth of 500 MBYTES/S (twice this in two directions) and about 15 μs latency. So, bandwidth and latency have improved dramatically. However, over the same period of time, processors have also been getting much faster and, now, with the move to SMP clusters more processors are being bundled into a node.

Above we used the ratio of memory bandwidth to floating-point speed (B/F) to characterize the balance in computational nodes. The same may be used for the network, denoted as NETB/F. The five-years-old IBM SP-

2 at PNNL/EMSL has a single processor node running at 120 MHz with a peak performance of 480 MFLOPS. Thus, the NETB/F ratio is 0.17. The 16-processor 375 MHz Nighthawk-II nodes have a peak performance of 24 GFLOPS, which with the Colony switch yields a NETB/F of 0.02 assuming that all 16 processes are sending data, or 0.33 if only one processor is sending data. Thus, despite the substantial increase in the communications bandwidth of the "Colony" switch, *on average*, the communications-to-compute balance in "ASCI White" is worse than it was in the IBM SP-2. The implication with respect to the parallel scalability of jobs within and between nodes is clear—the scientific code developer must maximize computation within the node and minimize communication among the nodes. This trend shows no sign of abating.

Other proprietary and commodity interconnects, some of which have been very popular with PC clusters, have followed a similar, though less aggressive, path. For instance, state-of-the-art workstation clusters are now connected with switched gigabit (note the use of "bit," not "byte") ethernet (or Myrinet) instead of the older 100-baseT (100 MBITS/S) or 10-baseT (10 MBITS/S) products. However, a gigabit network is still not a commodity item. 100-baseT connections are built into most PCs and switches cost a few hundred dollars. Gigabit (or Myrinet) network cards still cost about $1,000 dollars and switches $10,000 and over.

The network bandwidth problem can be solved, in principle, by increasing the number of network interface cards per node (although packaging eventually becomes a problem). The latency problem is not so readily addressed. The 30 μs latency of the IBM SP-2 is equivalent to 14,400 FLOPS. The equivalent number for the Nighthawk-II is between 22,500 and 360,000 FLOPS depending on the number of processors waiting for communication to be initiated. The speed of signal transmission provides a lower bound on the latency and this is one of the driving forces for putting multiple processors on a chip or in tightly integrated SMP modules (other reasons include reducing power consumption and costs). However, most of the latency in current systems arises from design limitations (*e.g.*, latencies inherent in buses), overhead due to switching control between different contexts or threads or processors or procedures, instruction and data cache flushing, and plain bad software. Many of these problems cannot be eliminated, so we must develop algorithms that tolerate or hide latency.

Transferring data in large blocks minimizes the sensitivity to latency. The value of

$$n_{\frac{1}{2}} = latency*bandwidth$$

indicates how much data must be transferred so that 50% of the peak transfer rate is obtained, or equivalently so that 50% of the cost is in latency. For the Nighthawk-II node, this is 7.5 KBYTES. For some applications, it may not be possible to generate messages this large. Various software techniques, such as asynchronous communication, may hide some latency, but these approaches are complex and not broadly applicable.

Cray (formerly Tera) are pursuing an interesting mixed hardware-software solution in their multi-threaded architecture (MTA).[9] Each processor handles up to 128 executing threads that can be scheduled with very low overhead so that the latency of memory operations can be completely hidden by rescheduling threads. Many algorithms could have sufficient fine grain parallelism to generate this many threads. However, the actual performance as well as the price-performance of this system is yet to be determined.

In years gone-by there was much discussion about appropriate topologies for the high-performance network of a parallel computer. The emerging victor in the market place is a switched network in which any node can send data directly to any other without contention except for the links directly to/from each node. Why has this happened? The most compelling reasons are that the scalability of a fully connected switched network is greater than any of the alternatives, and incremental growth of the system is much easier. Fast switches are still expensive but their price can be expected to fall steadily. However, an attractive feature of lower dimension interconnects is that the switches or routers are simpler and there are fewer of them, so it is affordable to make them faster. ASCI Red is a good example of this strategy with a link bandwidth of over 300 MBYTES/S; the bandwidth of the CRAY-T3E is similar.

In the foregoing, the statistics for various IBM systems have been used as illustration only. Similar conclusions could have been drawn from the systems of any other vendor. Our comments should not be taken as a criticism of IBM in particular.

Architecture and Implementation of the IO Subsystem in Massively Parallel Supercomputers

Disk drives, just like memory, are increasing dramatically in capacity but not as rapidly in speed. In the past, out-of-core solvers were common in scientific calculations, and, in electronic structure calculations, it was once routine to compute all of the integrals once and store them on disk for repeated use later in the calculation. The first parallel supercomputers had just a few processors in each node and, thus, there was room to include several local disk drives. Such nodes were good building blocks for scalable disk subsystems, *i.e.*, disk subsystems whose bandwidth increased in direct proportion to the number of processors. Examples of

machines like this were the IBM SP-2 and Compaq Alpha clusters, and traditional out-of-core techniques were very efficient on these machines. General parallel file systems were often layered on top of these distributed disk systems, although fault-tolerance and interference with running jobs were problems.

The trend towards large SMP nodes means there is relatively little room in the nodes for disks. So, general purpose subsystems must be used for I/O. Advantages of this approach are that fault-tolerance can be built in, and the parallel file system is separated from the compute nodes so both deliver more reliable performance. The downsides are that such a system can never deliver the raw bandwidth of disks local to each node, I/O consumes network bandwidth, and there is the additional overhead of the parallel file system software. Experience over the years with general parallel file systems from many vendors has not been positive. They have been both slow and unstable. However, the latest version of the General Parallel File System (GPFS) from IBM seems to be both fast and robust when supported by the appropriate hardware and used by large parallel applications.

The cost of large file systems is no longer much of a problem, at least for electronic structure calculations. The capacity and prices of commodity disk drives, *e.g.*, 36 Gbyte ultra-160 SCSI for $500, makes it possible to build very large file systems for modest investments. For instance, a 100 TBYTE file system requires 2,778 disks at a cost of only $1.4M. The actual cost of the file system is, of course, much higher due to considerations of power supply, hot-swapping, sufficient controllers for sustainable bandwidth, file system structure, back-up, and fault tolerance.

The last point mentioned above is important. Modern disk drives have theoretical mean-time-between-failures (MTBF) of up to 1,000,000 hours (114 years), but this amazing statistic does not account for the enormous stress that scientific applications can place on a disk drive. In addition, this analysis neglects other external factors such as inadequate conditioning of the power supply. A more realistic MTBF for such disks in a parallel supercomputer is 100,000 hours and we can roughly estimate that, in our 100 TBYTE file system, we will have to replace a drive every two days.

NWChem—A Molecular Modeling Code for Parallel Supercomputers

Parallel supercomputers require a substantial change in the process by which scientific algorithms and codes are designed and implemented. Before beginning the NWChem Project in 1992, two years were spent investigating the issues involved in parallel computing, including the performance issues involved in mapping computational chemistry data structures and algorithms to a distributed set of computational nodes. Among other things, this work indicated that it would be better to

---

[9] Additional information in the Cray MTA may be found at: http://www.cray.com/products/systems/craymta/.

design the new code as a parallel code from the beginning rather than try to port a code such as Gaussian (Frisch *et al.*, 1998) to a parallel supercomputer—the sequential nature of most legacy codes are well embedded in their structure. It also revealed the benefits of a NUMA model and one-sided memory access.

The major challenges associated with the development of a new quantum chemistry code can be classified into two general categories. First, electronic structure codes are very large by scientific code standards (>500,000 lines of code) with many specialized algorithms (calculation of integrals over one- and two-electron operators, transformation of massive four-index arrays, construction and solution of matrices with dimensions from millions to billions, *etc.*) being used in the time critical parts of the code. Thus, an enormous amount of general infrastructure as well as a large number of new parallel algorithms must be developed. Second, the existing parallel computing systems software, *e.g.*, the interprocessor communication software, mathematical algorithm libraries, and IO software, often did not provide the functionality or level of performance needed for a new parallel computational chemistry code.

In this section, we discuss the major issues involved in the development of NWChem (Kendall *et al.*, 2000 and references therein). NWChem implements many of the standard electronic structure methods currently used to compute the properties of molecules and periodic solids. In addition, NWChem has the capability to perform classical molecular dynamics and free energy simulations with the forces for these simulations being obtainable from a variety of sources, including *ab initio* calculations. Examples of NWChem's capabilities include:

- Direct, semi-direct, and conventional Hartree-Fock (RHF, UHF, ROHF) calculations using up to 10,000 gaussian basis functions with analytic first and numerical second derivatives (analytic second derivatives are being tested).

- Direct, semi-direct, and conventional density functional theory (DFT) calculations with a wide variety of local and non-local exchange-correlation potentials, using up to 10,000 basis functions with analytic first and numerical second derivatives (analytic second derivatives are being tested).

- Complete active space (CASSCF) with analytic first and numerical second derivatives.

- Semi-direct and RI-based second-order perturbation theory (MP2) calculations for RHF and UHF wave functions using up to 3,000 basis functions; fully direct calculations based on RHF wave functions; analytic first derivatives and numerical second derivatives of the MP2 energy.

- Coupled cluster, CCSD and CCSD(T), calculations based on RHF wave functions using up to 1000 basis

functions with numerical first and second derivatives of the coupled cluster energy. (A new CCSD(T) code is described below.).

Some of these calculations can make use of the new $O(N)$ or "fast" Coulomb, exchange, and quadrature techniques. For a more complete listing of the capabilities of NWChem, including its capabilities for periodic solids and molecular dynamics simulations, the reader is referred to the NWChem web site:

http://www.emsl.pnl.gov:2080/docs/nwchem/nwchem.html

In addition to the above, a powerful scripting interface, based on the widely-used Python object-oriented language, is available for NWChem.

The PNNL group is in the final stages of completing a new version of the coupled cluster code that will implement an algorithm for massively parallel supercomputers with the minimum number of floating-point operations for the expensive triples term. The new code is designed to perform CCSD(T) calculations with up to 3,000 basis functions and scale well up to 10,000 processors. It will also incorporate some of the new approaches being proposed to reduce the scaling of coupled cluster calculations, *e.g.*, local correlation approaches and fast summation techniques.

NWChem runs on essentially all parallel super-computing platforms, including IBM SPs, Cray T3Ds and T3Es, and SGI Origin 2000s. Despite the fact that NWChem was designed for parallel supercomputers, it also runs on workstations, PCs running Linux or Windows, as well as clusters of desktop computers or workgroup servers. The design features behind this capability are discussed below.

NWChem has already been used for a number of computational studies that go well beyond the *state-of-the-art* possible with the last generation of supercomputers. It is not possible for us to provide an exhaustive list here. Rather, we will briefly discuss two examples. All calculations were carried out on subsets of processors of the IBM SP-2 in the Molecular Science Computing Facility at PNNL.

Water clusters have long been subjects of study at PNNL and elsewhere. Water clusters are important in understanding the fundamental nature of liquid water. In addition, they are of intrinsic interest as intermediates between the gas and liquid phase. Past theoretical studies identified four low-lying isomers of the water hexamer, $(H_2O)_6$. Because of basis sets limitations, these calculations were unable to accurately predict the relative energies of the isomers. Xantheas and Harrison (2000) recently reported MP2 calculations on $(H_2O)_6$ using up to 1722 basis functions. By using basis sets that systematically approach the complete basis set limit (Dunning, 1989; Kendall, Dunning and Harrison, 1992), they determined that, *at the MP2 limit*, the binding energies of the prism, cage, and book isomers all lie within 0.3 kcal/mol of each other and the fourth isomer, the ring isomer, lies less than 1 kcal/mol higher. The nature of the basis sets used in these calculations, combined with the large number of basis functions in a small volume, make these calculations far more demanding than, for instance, an equivalent-size calculation in a 6-31G* basis on a larger molecule.

Calculations often serve as the only source of information on the weak interactions between molecules, information that is needed for molecular dynamics or Monte Carlo simulations of the condensed phase behavior of these species. Unfortunately, past calculations rarely provided interaction potentials of the required accuracy. With the new capabilities provided by NWChem, this situation has changed. Recent MP2 calculations by Feller (1999), using basis sets that approach the complete basis set limit (some with over 1,000 basis functions), predicted that $\Delta H(0\ K)$ for H_2O-benzene was 2.7 ± 0.2 kcal/mol, in excellent agreement with the experimental value of 2.5 ± 0.1 kcal/mol. To support computational studies of the aqueous chemistry of "buckytubes" and graphite

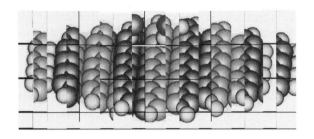

Figure 2.   Water on a Graphite Fragment
($C_{24}H_{12}$).

sheets, Feller and Jordan (2000) have now extended these studies to the interaction of H_2O with seven (7) fused benzene rings ($C_{24}H_{12}$); see Fig. 2. For H_2O-$C_{24}H_{12}$, MP2 calculations with 1649 basis functions provide detailed information on the distance dependence of the interaction potential and predict that the electronic binding energy is nearly 50% larger than that for H_2O-C_6H_6.

The NWChem Team

Let us now turn to issues associated with the development of NWChem. The NWChem Project was organized around a core team of theoretical and computational chemists, computer scientists, and applied mathematicians at PNNL, augmented by a worldwide group of collaborators (25+ scientists). The core team was responsible for designing the overall architecture of NWChem, writing a major portion of the code, assigning tasks to collaborators, resolving major issues related to the parallel implementation of mathematical algorithms, and integrating the software provided by collaborators into NWChem. The collaborators were chosen to provide technical expertise missing or over-subscribed in the core team.

The core team consisted of six theoretical and computational chemists plus an equal number of postdoctoral fellows and three computer scientists and applied mathematicians. It was a tightly knit group, and this level of integration was essential to achieving the goals set for NWChem. Many of the problems encountered in developing NWChem were technically complex and their resolution required an in-depth understanding of both the scientific algorithms and the computer hardware and systems software. The solutions to these problems were often unique. Global Arrays (GA) and ChemIO (now called ParIO), which arose because of the need for efficient solutions to the data distribution/access and I/O problems, were solutions to two such problems.

Although the combination of an on-site core team plus off-site collaborators provided the range of technical capabilities needed to develop NWChem, there are lessons to be learned about managing such a highly distributed project. For example:

- The time and effort required for integration of sequential or parallel codes into the new code framework was always larger than estimated.

- The preparation of documentation, for both users and programmers, should have been initiated early in the project. The programmer's manual is especially important because this document provides the guidelines needed to ensure that the software produced by the distributed team will work together.

- The software components that are on the critical path should be developed in-house, since the time schedules and priorities of collaborators inevitably differ from those of the core team.

Our experience suggests that a distributed technical software development team can be successful *if* the core team is large enough to develop all of the software components on the critical path and *if* sufficient guidance is provided to the collaborators on the format and content for their contributions and their progress is carefully monitored.

Elements of the Design of NWChem

Development of scientific codes for parallel supercomputers poses demanding technical challenges. The mathematical algorithms must achieve high-performance levels on cache-based microprocessors and scale to thousands of processors (or more). In contrast, many existing algorithms achieve only 5% to 10% of the "peak" performance of the processors, and few algorithms scale well beyond a few tens of processors. In addition, scientific codes must be carefully designed so that they can easily accommodate new mathematical models and algorithms as knowledge advances. If scientific codes can not evolve as new knowledge is gained, they will rapidly become outdated. It must also be possible to move the codes from one generation of computers to the next without undue difficulty as computer technology advances. The lifetime of scientific

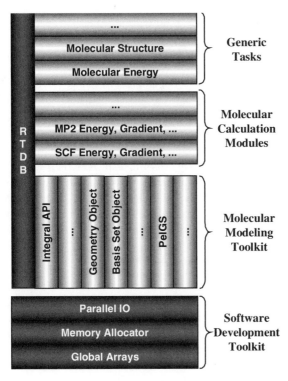

Figure 3. Layered, Modular Architecture of NWChem

codes is measured in decades, the lifetime of computers in years. This has been a major challenge for parallel supercomputers since the architectures of parallel systems have evolved substantially during the past decade. Although we currently seem to be on an evolutionary plateau, this situation is unlikely to persist to the end of the decade.

The key to achieving the above goals is a carefully designed architecture that emphasizes layering and modularity; see Fig. 3 for a simplified version of the structure of NWChem. At each layer of NWChem, subroutine interfaces or styles were specified in order to control critical characteristics of the code, such as ease of restart, the interaction between different tasks in the same job, and reliable parallel execution. Object-oriented design concepts were used extensively within NWChem. Basis sets, molecular geometries, chunks of dynamically allocated local memory, and shared parallel arrays are all examples of "objects" within NWChem. NWChem is implemented in a mix of C and Fortran-77, since neither C++ nor Fortran-90/95 was suitable at the start of the project though we anticipate supporting these languages in the future. Since we did not employ a true object-oriented language, and in particular did not support

inheritance, NWChem does not have "objects" in the strict sense of the word. However, careful design with consideration of both the data and the actions performed upon the data, and the use of data hiding and abstraction, permits us to realize many of the benefits of an object-oriented design.

In the very bottom layer of NWChem, are the Memory Allocator (MA), Global Arrays (GA), and Parallel IO (ParIO). MA is a type-safe memory allocator that allows a multi-language application to manage dynamically allocated memory, flexibly (providing both a stack and a heap) and reliably (providing several debug features including detection of over-/under-run). Global Arrays (GA) will be described in more detail below. It is our parallel-programming tool that provides one-sided access to shared, distributed multi-dimension arrays. ParIO is a library that supports several models of input/output (IO) on a wide range of parallel computers. The models include independent files for each process, or a file shared between processes which may independently perform operations with independent file-pointers. It also supports an extension of Global Arrays to disk, called Disk Arrays, through which processes collectively move complete or partial global arrays to/from external storage. The "Software Development Toolkit" was (and still is) primarily the responsibility of the computer scientists involved in NWChem. It essentially defines a "hardware abstraction" layer that provides a machine independent interface to the upper layers of NWChem. When NWChem is ported from one computer system to another, nearly all changes occur in this layer, with most of the changes elsewhere being for tuning or to accommodate machine specific problems such as compiler flaws. The "Software Development Toolkit" contains only a small fraction of the code in NWChem, less than 2%, and only a small fraction of the code in it is machine dependent (notably the address-translation and transport mechanisms for the one-sided memory operations). This is the reason that NWChem can be ported easily from one computer system to another.

The next layer, the "Molecular Modeling Toolkit," provides the functionality commonly required by computational chemistry algorithms. This functionality is provided through "objects" and APIs (application programmer interfaces). Two of the "objects" defined here support the simultaneous use of multiple basis sets, *e.g.*, as required in some density functional calculations, and multiple molecular geometries, *e.g.*, as may be required for manipulating molecular fragments or molecular recognition. Examples of the APIs include those for the one- and two-electron integrals, that for the quadratures used by the DFT method, and a number of basic mathematical routines including those for parallel linear algebra and Fock-matrix construction. Nearly everything that might be used by more than one type of computational method is exposed through an application interface. Common blocks are not used for passing data across APIs, but are used to support data hiding behind APIs.

The runtime database (RTDB) is a key component of NWChem, tying together all of the layers of NWChem. This database plays a role similar to the GAUSSIAN checkpoint file or the GAMESS dumpfile, but is much easier and safer to use. Arrays of typed data are stored in the database using simple ASCI strings for keys (or names). The database may be accessed either sequentially or in parallel.

The next layer within NWChem, the "Molecular Calculation Modules," is comprised of independent modules that communicate with other modules only via the RTDB or other persistent forms of information. This design ensures that, when a module completes, all persistent information is in a consistent state. Some of the inputs and outputs of modules (via the database) are also prescribed. Thus, all modules that compute an energy store it in a consistently named database entry - in this case "<module>:energy", substituting the name of the module for "<module>". Examples of modules include computation of the energy for SCF, DFT, and MCSCF wave functions. Surprising to some, the code to read the user input is also a module. This makes the behavior of the code more predictable, e.g., when restarting a job with multiple tasks or steps, by forcing the state of persistent information to be consistent with the input already processed. Modules often invoke other modules, e.g., the MP2 module invokes the module to compute the SCF wave function, and the QM/MM module invokes one of the modules to compute the energy and gradient.

The highest layer within NWChem is the "task" layer, sometimes called the "generic-task" layer. Functions at this level are also modules—all of their inputs and outputs are communicated via the database, and they have prescribed inputs and outputs. However, these capabilities are no longer tied to specific types of wave functions or other computational details. Thus, regardless of the type of wave function requested by the user, the energy may always be computed by invoking "task_energy()" and retrieving the energy from the database entry named "task:energy". This greatly simplifies the use of generic capabilities such as optimization, numeric differentiation of energies or gradients, and molecular dynamics. It is the responsibility of the "task"-layer routines to determine the appropriate module to invoke.

NWChem was designed to be extensible in several senses. First, the clearly defined task and module layers make it easy to add substantial new capabilities to NWChem. This can be done for both parallel and sequential codes. An example of this would be the recent integration of POLYRATE, which performs direct dynamics calculations (Chuang et al., 2000), into NWChem. Although tighter integration is possible by using lower-level APIs, this simple approach can provide much of the needed functionality. Second, the wide selection of lower-level APIs, e.g., for manipulating basis sets and evaluating integrals, makes it easier to develop new capabilities within NWChem than within codes in which these capabilities are not easy to access. Finally, having a standard API means that a change to an implementation will affect the whole code. For example, to add scalar relativistic corrections to all supported wave function or density functional models, we only had to modify the routines that compute the one or two-electron integrals, e.g., adding routines to calculate the extra integrals needed by the Douglas-Kroll-Hess (Hess, 1986) or Dyall NESC (Dyall, 2000) formalisms. Adding a faster integrals module also speeds up the whole code. Sometimes, however, new APIs must be added. This was done when the Texas integral package was integrated into NWChem several years ago, and an API that supported the more efficient evaluation of integrals in batches was needed. This can be done straightforwardly, but obviously requires more extensive code or algorithmic changes to realize the benefit.

Numerical Algorithms in NWChem

As noted in previous sections, the algorithms to be used on parallel supercomputers must be carefully chosen if maximum performance is to be realized. Two major issues must be addressed. The first is efficient use of the computational node and the second is efficient use of the communications network. For the present, use of multiple disks on each node solves the disk IO problem. Parallel supercomputers are based on high-performance microprocessor nodes which have reasonable B/F ratios for the L2 cache, but poor B/F ratios for main memory. Thus, maximum performance will be achieved only if the calculations can be run out of cache. For this reason the subroutines for electronic structure calculations in NWChem make heavy use of matrix and matrix-vector operations as instantiated in the BLAS, or Basic Linear Algebra Subroutines (Dongarra et al., 1990). The BLAS are optimized for each processor-memory system and make efficient use of the memory subsystem by maximizing reuse of data in the cache. The gain in performance can be dramatic—often a speed-up of more than an order of magnitude can be realized over compiler-generated code. Most vendors have developed efficient implementations of the BLAS.

Use of the BLAS in electronic structure calculations is not new. With the introduction of vector supercomputers in the late 1970s, it was realized that the use of matrix and matrix-vector operations enabled the calculations to run at maximum speed. This led to the systematic revision of electronic structure codes to emphasize the use of matrix and matrix-vector operations and, eventually, the development of an *implicit* pact with the applied mathematics community supporting the BLAS that they would develop efficient implementations of these subroutines on new computer systems. This pact, with the support of computer vendors, has been honored. In fact, Whaley and Dongarra (1999) have recently developed an approach that automatically generates and optimizes the BLAS for microprocessor-based computer systems.

Computational chemistry calculations have traditionally relied very heavily on out-of-core techniques for solving the electronic Schrödinger equation. Because the size of the major intermediate data sets, the two-electron integrals over AOs and MOs, increases as N^4, it has not been possible to use in-core techniques except for very small molecules in very small basis sets. In fact, for very large basis sets and/or very large molecules, direct techniques, which recompute the integrals on the fly rather than computing them once and storing them on disk, must be used. However, this requires that the integrals be computed every time they are needed—a time-consuming task. With the large memories and disk stores available on parallel supercomputers, traditional techniques can be used for much larger molecules than heretofore possible. In addition, the large memories and disk stores offer opportunities for semi-direct procedures that compute and store the most expensive integrals in memory/on disks and recompute the remaining integrals on the fly. NWChem takes advantage of these various possibilities by first determining the amount of memory/disk store available on the compute nodes that it has been allocated and then selecting the optimum algorithm given these resources. This can result in super-linear speed-ups as the number of compute nodes is increased. In the future, with computers that have 10-100 times the memory on our IBM SP-2, we expect such adaptive algorithms to become commonplace.

The NETB/F ratio is small for all current parallel computers and the situation is unlikely to improve in the near future. Thus, it is important to maximize the amount of computing done on the data within a node before data must be exchanged with another processor. This requires algorithms in which the amount of work done on the node can be optimized, at least over a selected range. This has been done in NWChem whenever possible but has proved to be a major challenge. In fact, this is expected to be one of the major challenges faced by future scientific software developers.

Global Arrays

The Global Arrays (GA) library implements a shared-memory programming model in which the programmer explicitly manages data locality (Nieplocha, Harrison, and Littlefield, 1996). This is achieved by explicit calls to functions that transfer data between a global address space (distributed multi-dimension arrays) and local storage. In this respect, the GA model has similarities to distributed shared-memory models that provide an explicit acquire/release protocol. However, the GA model acknowledges that remote data is slower to access than local data and allows data locality to be explicitly specified and managed. The GA model exposes to the programmer the NUMA characteristics of modern high-performance computer systems, and, by recognizing the communication overhead for remote data transfer, it promotes data reuse and locality of reference. The GA library allows each process in a parallel program to access, asynchronously, logical blocks of physically distributed matrices, without the need for explicit cooperation by other processes. This functionality has proved very useful in numerous computational chemistry applications. Today, computational codes in other disciplines, e.g., atmospheric modeling, fluid dynamics, and even financial forecasting, are beginning to make use of GA.

GA provides extensive support for controlling array distribution and accessing locality information. In addition, it provides interfaces to high-performance basic linear algebra operations. Global arrays can be created by:

- Allowing the library to determine array distribution;
- Specifying the decomposition for one or more array dimensions and allowing the library to determine the others; and
- Specifying irregular distribution as a cartesian product of irregular distributions for each axis.

The distribution and locality information is available through library operations that:

- Specify the array section held by a given process;
- Specify which process owns a particular array element; and
- Returns a list of processes and blocks of data for a given section of an array.

GA fully inter-operates with MPI.

In order to encourage the wide use of Global Arrays, as well as Parallel IO and the Memory Allocator, the source codes for these programs have been placed in the public domain and the full source codes, documentation, and related articles can be downloaded from:

http://www.emsl.pnl.gov:2080/docs/parsoft/

Software Engineering Practices in the Development of NWChem

The process for developing NWChem was far more formal than that associated with most scientific code development projects. The overall framework for the NWChem Project was guided by a report from a workshop held at PNNL in February 1990 and organized by one of the authors (THDJr). The participants in that workshop included senior theoretical and computational chemists from national laboratories, universities, and industry. Following this workshop, a formal software development plan was written by the core team at PNNL and peer reviewed. Once work started, periodic peer reviews were held to assess the quality and pace of the work. We are indebted to the community for both the support and time that they gave to this project. The NWChem Project profited substantially from the advice given in the workshop and reviews.

Software engineering practices for the development of scientific codes rarely have the rigor associated with commercial software engineering practices. There are many good reasons for this. Often the codes are part of a larger research project and are designed to answer a specific question. These codes tend to be small and it is unlikely that they will ever be used once the project is over. Even the early stages of "community codes" usually involve an exploratory phase where the use of more formal software engineering approaches would slow down progress with little real benefit. However, the use of good software engineering practices is essential for the design and implementation of large software systems like NWChem. These practices are still not as formal as standard commercial software engineering practices (McConnell, 1998), or the practices used by NASA for its mission critical software.[10] Nonetheless, they are essential for creating scientific codes with the needed level of performance, extensibility, portability, and maintainability. Good software engineering practices are also critical for coordinating the input of a widely distributed software development team.

The essential elements of the software engineering practices used by the NWChem Project were:

- A management structure consisting of a project manager and a chief software architect. Although it is tempting to consolidate these two positions into one, this must be resisted. It is simply not possible for one individual to do both jobs well.

- A timeline with well-defined milestones and deliverables. This was established at the beginning of the project and revised thereafter as needed (the revisions were subject to change control). The deliverables for some milestones were reports, for others it was pieces or versions of the code.

- Two major stages were defined: design and implementation of a prototype code, followed by design and implementation of the final production code. Prototyping reduces risk by testing new ideas before committing to their implementation. It also provides the users with early versions of the code for both evaluation and research use.

- Within each of the stages above, we carried out a) a requirements and analysis step to define the functionality of the code[11]; b) a code design step, which defined the architecture of the code, its "objects" and APIs, and the algorithms to be used; c) identified the major unresolved issues and established research efforts to address these problems; and d) created the code in well-defined versions, using revision control to track all changes.

- Finally, we tested the final code against the requirements resulting from a) above.

One innovation, at least as far as scientific computing is concerned, was an automatic "build" of the code for several different computing platforms. This helped eliminate incompatibilities as well as more than a few errors.

A number of documents were prepared in the development of NWChem. First was a relatively brief document describing the project along with the major timelines and deliverables. This document described the NWChem Project at a high level, but provided sufficient detail so that its overall organization, objectives, approach, and timelines could be evaluated by experts in the field. Additional documents were prepared in the course of developing the prototype and production versions of the software. These included a requirements and analysis document, a code design document, a programmers manual, and a users manual. In addition, a report was prepared that described the testing and verification of the final version of the software.

Conclusion

The era of terascale computing is here! Computer systems capable of 1 trillion arithmetic operations per second (TFLOPS) are already in place at the San Diego Supercomputer Center and Oak Ridge National Laboratory, a 10+ TFLOPS system will be delivered to Lawrence Livermore National Laboratory at the end of the summer (although security restrictions will limit external access to this system), and computer systems capable of 30-100 TFLOPS are in various planning stages. These terascale systems are based on clusters of shared-memory multiprocessor systems (SMPs) with cache-based microprocessors as the building blocks of the SMPs.

---

[10] Information on NASA's software engineering practices can be found in the documents listed on the web page: http://sel.gsfc.nasa.gov/website/documents/online-doc.htm

[11] The reader may be surprised to learn that the requested functionality actually did change from 1991 to 1994, a result of the increasing interest in density functional theory during that period.

In the literature and many private discussions, the performance of a parallel supercomputer is often computed as the product of the peak performance of the processors and the number of processors in the computer. We know better than to do this, of course, but it is a simple calculation to perform. And, of course, we are all tempted at one point or another to participate in "hype" of high-performance computing (as we did in the opening paragraph of this section). Of course, what we are really interested in is the delivered (or sustained) performance of the parallel supercomputer system on real scientific applications. Unfortunately, because of the effort involved in benchmarking parallel supercomputers, this is usually determined *after the fact*, *i.e.*, after the computer is purchased. What is needed is a manageable set of realistic benchmarks targeted for parallel supercomputers as well as performance models that allow us to predict the behavior of parallel supercomputers for selected scientific applications (for an example of the former, see Wong *et al.*, unpublished).

The advances promised by parallel supercomputers will not come gratis, as have many other computing advances in the past, *e.g.*, the transition from the Cray X-MP to the Y-MP to the C-90. Many of the problems encountered in developing scientific codes for parallel supercomputers are technically complex and their resolution requires an in-depth understanding of both the scientific algorithms and the computer hardware and systems software. Hardware problems to be overcome range from the memory bandwidth limitations of microprocessor-based compute nodes to the bandwidth and latency limitations of the interprocessor communications switch. Software problems to be overcome range from the choice of programming model to the development of numerical algorithms that scale to thousands of processors. And, in the end, we want a code that is extensible, portable, and maintainable. Solving these problems is best done by establishing close collaborations between theoretical and computational scientists, computer scientists, and applied mathematicians. As the NWChem Project illustrated, scientific codes that achieve these goals can be met by combining a core team located at one institution with a worldwide group of collaborators. Teams such as this will not only allow us to take full advantage of terascale computing but will also position us to meet the challenges of petascale computing toward the end of this decade.

As daunting as the above problems seem, it will be worth it! Combining the computing advances described above with advances in mathematical models and computational algorithms will lead to revolutionary new modeling and simulation capabilities. Problems that currently seem intractable will not only become doable, they will become routine. In chemistry, computational studies will become an integral and irreplaceable part of studies aimed at understanding the chemical processes involved in the production of smog, the burning of hydrocarbon fuels, and the industrial production of materials. The fidelity of modeling complex biomolecules will also take a major step forward, greatly increasing the contributions of computational chemistry to the life sciences. To realize these opportunities, however, it is critical that the federal agencies that fund molecular and computational science make the investments in scientific simulation software, computing system software, and mathematical libraries necessary to fully capitalize on terascale and petascale computing (For a discussion of these issues, see the NSF-DOE Workshop Report, 1998, and the PITAC report, 1999). Focused efforts such as those outlined here for NWChem need to be supported in a number of scientific disciplines, including computer science. This is the goal of the new investments proposed by the Office of Science in the U.S. Department of Energy in its FY2001 request for "Scientific Discovery through Advanced Computing" (Office of Science, 2000).

Acknowledgments

We thank Drs. Mark Seager, Jeff Nichols, Ricky Kendall, George Fann, and Jarek Nieplocha for numerous helpful and provocative discussions. We also thank Dr. Ricky Kendall for his detailed comments on the manuscript and Dr. David Feller for use of Figure 2. This work was supported by the Division of Chemical Sciences in the Office of Basis Energy Sciences of the U.S. Department of Energy.

References

Amos, R. D.; I. L. Alberts, J. S. Andrews, S. M. Colwell, N. C. Handy, D. Jayatilaka, P. J. Knowles, R. Kobayashi, K. E. Laidig, G. Laming, A. M. Lee, P. E. Maslen, C. W. Murray, J. E. Rice, E. D. Simandiras, A. J. Stone, M.-D. Su, and D. J. Tozer (1995). *CADPAC: The Cambridge Analytic Derivatives Package Issue 6, Cambridge.* For further information, see: http://ket.ch.cam.ac.uk/software/cadpac.html

Bak, K. L., P. Jørgensen, J. Olsen, and W. Klopper (2000). Accuracy of atomization energies and reaction enthalpies in standard and extrapolated electronic wave function/basis set calculations. *J. Chem. Phys.*, **112**, 9229-9242.

Chuang, Y.-Y, J. C. Corchado, P. L. Fast, J. Villà, W.-P Hu, Y.-P. Liu, G. C. Lynch, C. F. Jackels, K. A. Nguyen, M. Z. Gu, I. Rossi, E. L. Coitiño, S. Clayton, V. S. Melissas, B. J. Lynch, R. Steckler, B. C. Garrett, A. D. Isaacson, and D. G. Truhlar (2000). *POLYRATE-version 8.4.1,* University of Minnesota, Minneapolis. For further information, see: http://comp.chem.umn.edu/polyrate/

Climate Research Committee. (1998). Capacity of U.S. climate modeling to support climate change assessment activities. National Academy Press: Washington, D.C.

Dunlap, B. I., J. W. D. Connolly and J. R. Sabin (1979). On some approximations in applications of Xalpha theory. *J. Chem. Phys.*, **71**, 3396-3402.

Dunning, T. H. Dunning, Jr. (1989). Gaussian Basis Sets for Use in Correlated Molecular Calculations. I. The Atoms Boron through Neon and Hydrogen. *J. Chem. Phys.*, **90**, 1007-1023.

Dunning, T. H., Jr., K. A. Peterson, and D. E. Woon (1998). In P. v. R. Schleyer, P. R. Schreiner, N. L. Allinger, T. Clark, J. Gasteiger, P. Kollman, H. F. Schaefer III (Eds.), *Encyclopedia of Computational Chemistry*. Wiley & Sons, Berne (Switzerland). Vol. 1, 88-114.

Dunning, T. H., Jr. (2000). A roadmap for the calculation of molecular binding energies. *J. Phys. Chem. A*, in press.

Dongarra, J., J. Du Croz, I. Duff, and S. Hammarling. (1990) A set of Level 3 Basic Linear Algebra Subprograms. *ACM Trans. Math. Soft.*, **16**, 1-17.

Dyall, K. G. (2000). Interfacing relativistic and non-relativistic methods. III. Atomic 4-spinor expansions and integral approximations. *J. Chem. Phys.* accepted for publication.

Feller, D. (1999). Strength of the benzene-water hydrogen bond. *J. Phys. Chem. A*, **103**, 7558-7561.

Feller, D. and K. Jordan. (2000). Submitted for publication.

Frisch, M. J., G. W. Trucks, H. B. Schlegel, G. E. Scuseria, M. A. Robb, J. R. Cheeseman, V. G. Zakrzewski, J. A. Montgomery, R. E. Stratmann, J. C. Burant, S. Dapprich, J. M. Millam, A. D. Daniels, K. N. Kudin, M. C. Strain, O. Farkas, J. Tomasi, V. Barone, M. Cossi, R. Cammi, B. Mennucci, C. Pomelli, C. Adamo, S. Clifford, J. Ochterski, G. A. Petersson, P. Y. Ayala, Q. Cui, K. Morokuma, D. K. Malick, A. D. Rabuck, K. Raghavachari, J. B. Foresman, J. Cioslowski, J. V. Ortiz, B. B. Stefanov, G. Liu, A. Liashenko, P. Piskorz, I. Komaromi, R. Gomperts, R. L. Martin, D. J. Fox, T. Keith, M. A. Al-Laham, C. Y. Peng, A. Nanayakkara, C. Gonzalez, M. Challacombe, P. M. W. Gill, B. G. Johnson, W. Chen, M. W. Wong, J. L. Andres, M. Head-Gordon, E. S. Replogle, and J. A. Pople (1998). *Gaussian 98 (Revision A.9)*. Gaussian, Inc., Pittsburgh, PA. For further information, see: http://www.gaussian.com/index.htm

Greengard, L. and V. Rohklin (1997). *Acta Numerica*, **6**, 229, and references therein.

Halkier, A., T. Helgaker, P. Jørgensen, W. Klopper, H. Koch, J. Olsen, and A. K. Wilson (1998). Basis-set convergence in correlated calculations on Ne, N_2, and H_2O. *Chem. Phys. Lett.*, **286**, 243-252..

Hampel, C. and H.-J. Werner (1996). Local treatment of electron correlation in coupled cluster theory. *J. Chem. Phys.*, **104**, 6286-6297.

Helgaker, T.; H. J. Aa. Jensen, P. Jørgensen, J. Olsen, K. Ruud, H. Ågren, T. Andersen, K. L. Bak, V. Bakken, O. Christiansen, P. Dahle, E. K. Dalskov, T. Enevoldsen, B. Fernandez, H. Heiberg, H. Hettema, D. Jonsson, S. Kirpekar, R. Kobayashi, H. Koch, K. V. Mikkelsen, P. Norman, M. J. Packer, T. Saue, P. R. Taylor, and O. Vahtras (1997). *DALTON, An Ab Initio Electronic Structure Program, Release 1.0*. For further information, see: www.kjemi.uio.no/software/dalton/dalton.html

Hess, B.A (1986). Relativistic electronic-structure calculations employing a two-component no-pair formalism with external-field projection operators. *Phys. Rev. A*, **33**, 3742-3748.

Kendall, R. A., T. H. Dunning, Jr., R. J. Harrison (1992). Electron Affinities of the First-Row Atoms Revisited.

Systematic Basis Sets and Wave Functions. *J. Chem. Phys.*, **96**, 6796-6806.

Kendall, R. A., E. Aprà, D. E. Bernholdt, E. J. Bylaska, M. Dupuis, G. I. Fann, R. J. Harrison, J. Ju, J. A. Nichols, J. Nieplocha, T. P. Straatsma, T. L. Windus, A. T. Wong (2000). High-performance computational chemistry: An overview of NWChem, a distributed parallel application. *Comput. Phys. Comm.*, **128**, 260-283. For further information, see: www.emsl.pnl.gov:2080/docs/nwchem/nwchem.html

Lee, T. J. and G. E. Scuseria (1997). Achieving chemical accuracy with coupled cluster theory. In S. R. Langhoff (Ed.), *Quantum Mechanical Electronic Structure Calculations with Chemical Accuracy*. Kluwer, Dordrecht, 47-108.

Martin, J. M. L. (1996). Ab initio total atomization energies of small molecules—towards the basis set limit. *Chem. Phys. Lett.*, **259**, 669-678.

Martin, J. M. L. (1998). In P. v. R. Schleyer, P. R. Schreiner, N. L. Allinger, T. Clark, J. Gasteiger, P. Kollman, and H. F. Schaefer III (Eds.), *Encyclopedia of Computational Chemistry*. Wiley & Sons, Berne (Switzerland). Vol. 1, 115-128.

Maslen, P. E., M. Head-Gordon (1998). Noniterative local second order Møller-Plesset theory: Convergence with local correlation space. *J. Chem. Phys.*, 109, 7093-7099.

Maslen, P. E., M. Head-Gordon (2000). An accurate local model for triple substitutions in fourth order Møller-Plesset theory and in perturbative corrections to singles and doubles coupled cluster methods. *Chem. Phys. Lett.*, **319**, 205-212.

McConnell, S. (1998). *Software Project Survival Guide*. Microsoft Press, Redmond, WA.

Message Passing Interface Forum (1995). MPI: A Message-Passing Interface Standard (note that version 1.1 fixes errata in the original version).

Nieplocha, J. R. J. Harrison, and R. J. Littlefield (1996). Global Arrays: A nonuniform memory access programming model for high-performance computers. *J. Supercomputing*, **10**, 197-220.

NSF-DOE Workshop Report, National Workshop on Advanced Scientific Computing, July 30-31, 1998. A copy of this report can be downloaded from: http://www.er.doe.gov/production/octr/mics/mics_documents.htm

Ochsenfeld, C., C. A. White, and M. Head-Gordon (1998). Linear and sublinear scaling formation of Hartree-Fock-type exchange matrices. *J. Chem. Phys.*, **109**, 1663-1669.

Office of Science, U.S. Department of Energy (2000). Scientific discovery through advanced computing. A copy of this report can be downloaded from: http://www.er.doe.gov/four/Features.asp

OpenMP (1999). *OpenMP FORTRAN Application Program Interface, Version 1.1*. Additional information may be found at: http://www.openmp.org.

Peterson, K. A. and T. H. Dunning, Jr. (1997). The CO molecule: Role of basis set and correlation treatment in the calculation of molecular properties. *J. Molec. Struct. (Theochem)*, **400**, 93-117.

PITAC, President's Advisory Committee on Information Technology, *Information Technology Research: Investing in Our Future*, February 1999. A copy of the report (in pdf format) may be downloaded from:

http://www.ccic.gov/ac/report/.

PITAC, President's Advisory Committee on Information Technology, *Recommendations of the Panel on Open Source Software for High End Computing*, September 11, 2000. A copy of the recommendations can be downloaded from:
http://www.ccic.gov/ac/pres-oss-11sep00.pdf

Pulay, P. and S. Saebø (1993). Local treatment of electron correlation. *Ann. Rev. Phys. Chem.*, **44**, 213-36.

Raghavachari, K., G. W. Trucks, J. A. Pople, and M. Head-Gordon (1989). A fifth-order perturbation comparison of electron correlation theories. *Chem. Phys. Lett.*, **157**, 479-483.

Schmidt, M. A.; K. K. Baldridge, J. A. Boatz, S. T. Elbert, M. S. Gordon, J. H. Jensen, S. Koseki, N. Matsunaga, K. A. Nguyen, S. Su, T. L. Windus, M. Dupuis, and J. A. Montgomery (1993). General atomic and molecular electronic structure system. *J. Comp. Chem.*, **14**, 1347-1363. For further information, see:
www.msg.ameslab.gov/GAMESS/GAMESS.html

Schütz, M. and H.-J. Werner (2000). Local perturbative triples correction (T) with linear cost scaling. *Chem. Phys. Lett.*, **318**, 370-378.

Scuseria, G. E. and P. Y. Ayala (1999). Linear scaling coupled cluster and perturbation theories in the atomic orbital basis. *J. Chem. Phys.*, **111**, 8330-8343.

Stanton, J.F.; J. Gauss, J. D. Watts, M. Nooijen, N. Oliphant, S. A. Perera, P. G. Szalay, W. J. Lauderdale, S. A. Kucharski, S. R. Gwaltney, S. Beck, A. Balková, D. E. Bernholdt, K. K. Baeck, P. Rozyczko, H. Sekino, C. Hober, R. J. Bartlett, and R. J. Integral packages included are *VMOL* (Almlöf, J.; Taylor, P. R.); *VPROPS* (Taylor, P. R.), *ABACUS*; (Helgaker,, T.; Aa. Jensen, H. J.; Jørgensen, P.; Olsen, J.; Taylor, P. R.). *ACES II*. For further information, see: www.qtp.ufl.edu/Aces2/.

Straatsma, T. P. and R. D. Lins, to be published.

Werner, H.-J.; P. J. Knowles, R. D. Amos, A. Berning, D. L. Cooper, M. J. O. Deegan, A. J. Dobbyn, F. Eckert, C. Hampel, G. Hetzer, T. Leininger, R. Lindh, A. W. Lloyd, W. Meyer, M. E. Mura, A. Nicklaß, P. Palmieri, K. A. Peterson, R. M. Pitzer, P. Pulay, G. Rauhut, M. Schütz, H. Stoll, A. J. Stone, and T. Thorsteinsson (2000). *MOLPRO Quantum Chemistry Package 2000.1*. For further information, see: www.tc.bham.ac.uk/molpro/.

Whaley, R. C. and J. Dongarra. (1999). Automatically tuned linear algebra software (ATLAS). A copy of this paper can be found at:
http://www.cs.utk.edu/~rwhaley/ATL/INDEX.HTM

Wong, A. T., L. Oliker, W. T. C. Kramer, T. L. Kaltz, and D. H. Bailey (unpublished). Evaluating system effectiveness in high performance computing systems. A copy of this article can be downloaded from:
http://www.nersc.gov/aboutnersc/pubs.html

Worley, P. H. (2000). Performance studies using CCM/MP-2D. A report on these studies may be found at:
http://www.epm.ornl.gov/~worley/studies/ccm-mp-2d.html

Xantheas, S.S. and R. J. Harrison (2000). Critical issues in the development of transferable interaction models for water: II. Accurate energetics of the first few water clusters from first principles, *J. Chem. Phys.*, submitted.

CONCEPTUAL DESIGN OF CHEMICAL PROCESSES: OPPORTUNITIES FOR MOLECULAR MODELING

Michael F. Doherty
Department of Chemical Engineering
University of California
Santa Barbara, California 93106

Abstract

In recent years, process development has changed quite significantly in many companies. There is an increased emphasis on tighter justification through integration of business development with process development. This is a very positive step. There is also an emphasis on reducing the development time and cost by omitting some phases of the activity that have traditionally been considered vital (e.g., omitting the pilot plant). This has not always been such a positive step. In this article we consider some of the ways that molecular modeling can contribute to improved procedures for process development.

Keywords

Conceptual design, Attainable region, Reactive separations, Crystallization.

Introduction

Conceptual process design is the activity of inventing a process flowsheet based on preliminary estimates of the business opportunity for the product and initial exploratory data and knowledge of the process chemistry. It takes place in an environment of great uncertainty where engineers are required to make decisions quickly based on limited technical and business data. During this phase of process development about 5-10% of the project costs are spent, but as much as 70-80% of the total project costs are locked-in, see Fig. 1. If the project survives this phase of development (most do not) then the structure of the flowsheet starts to become frozen so that detailed equipment design, followed by operability and control studies can be performed.

A guiding principle of conceptual design is that the economic performance of an unoptimized flowsheet using the best technology is superior to a highly optimized flowsheet using inferior technology. It has been estimated that for every dollar it costs to correct a problem at the conceptual design stage it will cost $10 at the flowsheeting stage, $100 at the detailed design stage,

$1000 after the plant is built, and over $10,000 to clean up the mess after a failure (Kletz, 1989).

The most important outcomes from the conceptual design stage are: (1) identification of one or more good process flowsheet structures and their key design

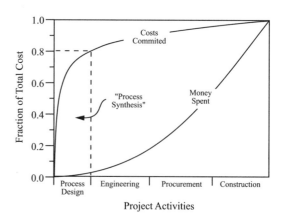

Figure 1. Costs commited and money spent during process development.

120

variables, (2) initial estimates of the equipment sizes and operating conditions, (3) an initial estimate of the process costs (both investment costs and operating costs) and profitability, (4) based on this information, a plan for critical experiments and modeling needed for the next phase of development, (5) a decision to either go forward to the next phase of process development or to abandon the project.

Process chemistry has a major impact on the selection of the best process technology. Moreover, the structure of the process flowsheet has a major impact on the key design variables and on the best conditions for conducting process chemistry experiments. Therefore, a vital component of successful process development is close integration of process chemists with conceptual design engineers.

There is never a single solution to an engineering problem. There are many decisions to make and justify in order to find good alternative designs, and therefore a systematic approach to decision-making is needed. One very effective approach that has been used successfully on many projects is Douglas's Design Procedure (Douglas, 1988). In this approach, decisions are structured hierarchically and based on economics. The first few levels of this procedure are:

Level 1: Determine the number of plants and their interconnections, i.e., determine the structure of the plant complex.

Level 2: Perform overall material balances on each plant to determine input and output flows as a function of the reaction selectivities and any other key design variables.

Level 3: Select the reactor and recycle structures. Perform recycle material balances.

Level 4: Select the separation system structure.

Later levels deal with energy management, process safety, operability, control, etc. One of the key initial steps at each level is to decide quickly what's feasible and what's not. For example, it is fruitless to evaluate the profitability of a process with a reaction selectivity to the desired product of 95% at a reactor conversion of 65% if such conditions can never be achieved in any kind of reactor configuration.

In the last two decades geometric methods have proved to be effective tools for feasibility analysis. These include attainable region methods for reactor systems (Glasser et al., 1987), residue curve maps plus related ideas for separation systems (Doherty and Malone, 2001), and pinch analysis for heat exchanger systems (Linnhoff et al., 1982).

Molecular modeling has an important role to play in support of these activities by providing estimates of critical pieces of data that are missing in the early stages of process development, and new models that are

uniquely suited to providing information that classical models have failed to provide.

Some of these opportunities are briefly described in this article.

Attainable Regions for Reactor Systems

For a given system of reactions with given reaction kinetics, the *attainable region* is defined as, "a domain in the concentration space that can be achieved by using any system of steady-flow chemical reactors, i.e., by using the processes of *mixing* and *reaction* (Glasser et al., 1987). The concept was originally proposed by Horn (1964) and developed to its present level by Feinberg, Glasser, Hildebrandt and co-workers (e.g., Feinberg and Hildebrandt, 1997).

Necessary conditions for the attainable region (AR) include: (1) the region is convex, i.e., it cannot be enlarged by mixing, (2) no reaction vector points outward from the boundary (reaction vectors on the boundary can point inward, be tangential, or have zero magnitude), (3) the negative of any reaction vector outside the attainable region cannot point inwards, see Fig. 2.

For isothermal consecutive reactions A→B→C, the attainable region is 2-dimensional (i.e., it can be represented as a region on a graph where the ordinate represents the concentration of A and the abscissa the

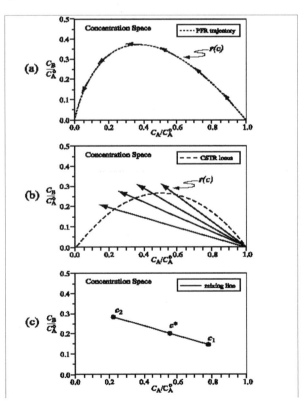

Figure 2. Geometry of reaction and mixing illustrated for the reaction system A→B→C. (a) tangency, (b) collinearity, (c) lever rule.

concentration of B). It's boundary is defined by a plug flow reactor trajectory, therefore, all other reactors or combinations of reactors give outlet compositions that lie inside or on the boundary of this region.

A typical plot of the AR for this chemistry with $k_1 = k_2 = k$ is given in Fig. 3 on a yield (this measures the concentration of B) vs. conversion (this measures the concentration of A) diagram. Since yield (Y), conversion (x) and selectivity (s) are related by $Y=sx$, the selectivity of the reactor system at any point in the attainable region is given by the slope of the straight line from the origin to that point. Therefore, a lot of useful information can be read from such a diagram. For conceptual design it is extremely important to see that for this chemistry and for the estimated values of k_1 and k_2 it is not possible to get high selectivities at conversions above 25-30% in any kind of reactor configuration. This implies that unless novel technology can be discovered (better catalyst, multiphase reactors, combined reactor-separator, etc.) then the process is doomed to operate at low conversions and high recycle flow rates.

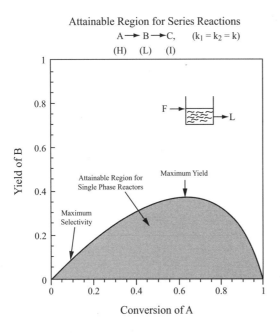

Figure 3. Attainable region for series reaction in single-phase reactors.

In order to estimate the reactor volume we must know the absolute rate of each reaction (i.e., values for both k_1 and k_2). However, to estimate the AR and the selectivity we need only an estimate of the relative rates (i.e., a value for k_1/k_2). Commonly, the cost of the reactor shell is not a significant investment cost but the impact of selectivity on the process economics is often very substantial.

Once the reaction species are identified (which is not as simple as it sounds) there are good methods for identifying sets of independent reactions that are consistent with the known reactants and products (Aris and Mah, 1963). For gas phase chemistries there are good molecular modeling approaches for refining the procedure to identify the actual reaction pathway (or at least to narrow it down), but little is known about how to do this for liquid phase chemistries.

Two important contributions that molecular modeling could make to process design are:

(MM 1) To couple gas phase reaction pathways plus estimates of the relative reaction rates for each channel predicted by molecular modeling with gas phase reactor analysis to estimate the attainable region.

(MM 2) To develop effective methods for estimating liquid phase reaction pathways and relative rates of reaction.

The first of these should be possible to attempt right away and could provide valuable "proof-of-concept" insights. The second is a more challenging problem in the theory of liquids.

Simultaneous Reaction and Separation

It is well known that combining separation with chemical reaction can be an effective way of improving the performance of a process flowsheet (Agreda et al., 1990; Siirola, 1996). However, improved process technology does not always result from combining these operations, which is why methods are needed to support decision-making. Methods include: attainable regions, residue curve maps, difference-point methods, bifurcation and continuation approach to feasibility, and optimization-based approaches.

We again consider the A→B→C chemistry, but this time we treat a case where A is the highest boiling component, B is the lowest boiler and C boils at an intermediate temperature. If the reaction occurs only in the liquid phase (e.g., due to a mineral acid catalyst or solid catalyst placed in the liquid), then intuitively we expect higher yield and selectivity if B is preferentially removed from the reactor, say by evaporation. Removing B reduces the rate at which C is made (therefore improves selectivity), and increases the rate at which B is formed because of the increased concentration of A (therefore improves yield). This is captured very effectively by plotting the trajectory for a two-phase vapor-liquid CSTR, which lies *outside* the AR for single phase reactors, see Fig. 4. Although it is not yet known how to calculate AR's in multiphase reactor systems, this example demonstrates that the single-phase AR can be enlarged in some cases by combining separation with chemical reaction in a multiphase reactor. One embodiment of such a reactor system is a reactive distillation column.

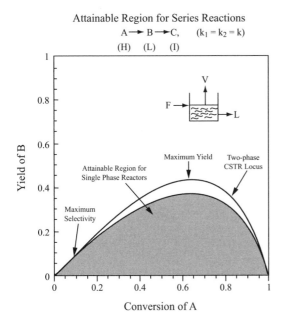

Figure 4. Feasible region for vapor-liquid CSTR.

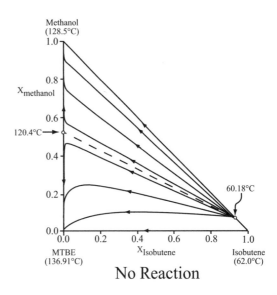

No Reaction

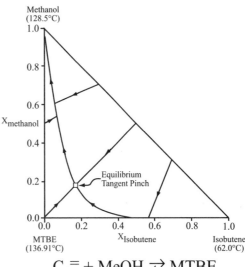

$$C_4^{=} + MeOH \rightleftarrows MTBE$$

Figure 5. Residue curve maps for MTBE chemistry in the limits of no chemical reaction (top), and equilibrium chemical reaction (bottom).

Feasible products from reactive distillation columns can be estimated using residue curve maps and related ideas (Doherty and Malone, 2001, Ch. 10). Residue curve maps for the mixture isobutene, methanol and methyl tertiary butyl ether (MTBE) are shown in Fig. 5. Fig 5(a) shows the RCM with no chemical reaction and Fig. 5(b) shows it near the limit of combined phase and reaction equilibrium. The main point to notice is that they are different, therefore, the products from non-reactive and reactive distillation columns will also be different.

Classical models of phase equilibrium, reaction equilibrium and chemical kinetics are often adequate for estimating residue curve maps and related tools for conceptual design, except in the following situations where molecular modeling is expected to play an important role:

- When one (or more) of the chemicals in the process is dangerous (e.g., toxic, explosive) or extremely expensive then physical properties are difficult to measure and are usually missing at the conceptual design stage of process development. Molecular modeling provides a means of estimating missing properties, including phase equilibria. The resulting estimates can then be used as surrogate data for estimating parameters in classical models, or the molecular models can be directly incorporated into conceptual design calculations as demonstrated recently by Escobedo (2000) for the calculation of residue curve maps.

- The thermodynamic reaction equilibrium constant, K_{eq}, [defined as $\exp(-\Delta G^0(T)/RT)$] is often difficult to estimate or measure accurately. For example, the reported values for K_{eq} for the MTBE reaction at 347 K range from 30 to 50. The scatter in reported values for methyl acetate chemistry is even larger. Scientists are trained to ask, "which value is correct," while engineers are trained to ask, "does it matter which value I use." It would very useful to have methods available for refining these estimates because sometimes (though not always) it does matter. In the case of the reactions cited, good estimates are much more critical in MTBE chemistry than in methyl acetate chemistry (Okasinski and Doherty, 1997).

- Self-catalyzed reactions are a big challenge in multiphase reaction-separation systems design. Examples include: aqueous formaldehyde chemistry, chlorination chemistries, aqueous nitric acid chemistry, and many more. Since the chemistry cannot be stopped it is difficult to estimate parameters in conventional models for phase equilibrium, reaction equilibrium, and chemical kinetics because all the effects occur simultaneously. Although progress has been made in this area for classical models in recent years (Hasse et al., 1990) there are very promising results from the application of reaction ensemble Monte Carlo simulations and related molecular modeling techniques (Lisal et al., 2000; Gubbins et al., 1994). This has the potential to be a very fruitful source of new models and methods that will find application in conceptual design.

Crystallization of Organic Materials

Crystallization is used here as a typical instance of processes that lead to microstructured products. Other examples include emulsions, gels, foods, home and personal care products, etc. Such systems may include reaction and/or separation steps like conventional chemical systems but, in addition, they always include structure-forming steps and often formulation steps. For example, the active ingredient in a pharmaceutical process not only needs to be made and structured (e.g., as a crystal) it must normally be formulated with other ingredients before it can be sold as a product.

Molecular modeling is especially suited to treating these types of materials because of the natural way that microstructure enters the modeling schemes. In crystallization, for example, the molecules are placed on a lattice by defining an asymmetric unit and a set of symmetry operations that act on this unit to create both the unit cell and a periodic array of lattice points. The shape of a crystals produced by a process often has a major impact on product quality (functionality) as well as processability. Estimating the shape of crystals during the discovery and conceptual design phases of product & process development is of major value in many cases.

The Bravais-Friedel-Donnay-Harker, and Attachment Energy models are first order approaches for predicting crystal morphology and shape. They are effective at estimating the likely faces on a crystal. Their computer implementations are fast and easy to use, and they have proved to be accurate for predicting vapor-grown crystal

shapes. These methods also yield geometric and energetic data that may be useful as input to more detailed kinetic models. The main drawback of these approaches is their inability to account for effects of solvent and other process conditions (i.e., impurities, supersaturation). For example, the experimentally observed shape of succinic acid crystals grown from water is hexagonal plates, whereas the crystals grow as needles out of isopropanol.

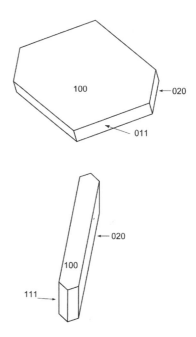

Figure 6. Predicted shape of succinic acid crystals grown from water (top), and isopropanol (bottom).

crystal. Their computer implementations are fast and easy to use, and they have proved to be accurate for predicting vapor-grown crystal shapes. These methods also yield geometric and energetic data that may be useful as input to more detailed kinetic models. The main drawback of these approaches is their inability to account for effects of solvent and other process conditions (i.e., impurities, supersaturation). For example, the experimentally observed shape of succinic acid crystals grown from water is hexagonal plates, whereas the crystals grow as needles out of isopropanol.

The recent approaches of Liu and Bennema (1996), and Winn and Doherty (1998) are the first attempts at using detailed kinetic theory for crystal shape prediction. Figure 6 shows the shape of succinic acid crystals grown from water, and from isopropanol predicted by this new modeling approach. The results are in good agreement with experiment. Both models recognize the significance of interfacial phenomena in crystal shape modeling, and lead the way for future developments, such as new simulation and/or group contribution methods for interfacial free energy prediction.

Some of the key areas for future experimental and modeling research are:

(MM6) *Mixed Solvents*: Crystals grown from a mixture of two or more solvents can have different characteristics than those grown from any one of the solvents alone. This effect is especially significant if the solute has very different solubility in each solvent. There is great potential for performing modeling studies for mixed solvents.

(MM7) *Hydrogen Bonds & Growth Unit*: Identification of the nature of the growth unit that incorporates in the growing crystal faces is an important factor in morphology. Several researchers have discussed pre-condensation in the solution phase to form dimers or other precursors, and the need to account for this effect in morphological modeling Hydrogen bonds in the solute and/or solvent molecules play an important role. However, there is still a need for better theories and models for predicting growth units in solution crystallization.

(MM8) *Polymorphs*: Polymorphs have always been of interest in crystallization, but have become a critically important factor in pharmaceutical production and registration because of recent FDA requirements. Different polymorphs have different crystal structures, optical properties, dissolution rates, shapes and interfacial properties. Thus, models of solution-crystal interactions might be able to predict polymorph selection and/or transition. Systematic studies along these lines would be of great practical interest.

(MM9) *Chiral Separations*: Single-enantiomer product molecules are a rapidly growing sector of the pharmaceutical industry, and crystallization is one of the key technologies for chiral selection. Crystals of the racemate have different structures than the individual enantiomers. As with polymorphism, interfacial phenomena may influence enantiomeric selectivity, and the challenge is to develop technology, and protocols (aided by modeling) to produce single enantiomer products.

(MM10) *Process Modeling*: An important and challenging area for chemical engineering research is to link interfacial models, capable of capturing the above effects, to process models. Such models would allow for novel designs and operating protocols to be developed systematically before they are tested experimentally. This is one of the ways that engineering, combined with molecular modeling, can contribute to faster process development.

Concluding Remarks

In the last two decades there has been an explosion of new methods and models in support of conceptual process design. Most of this activity has focused on classical models for chemical and petrochemical process systems. As the envelope of these systems is pushed, and as greater interest develops in conceptual design methods for systems that manufacture microstructured products, there will be a growing reliance on molecular models in support of process development. As molecular modeling improves we should expect significant benefits for conceptual process design leading to the development of better processes in shorter time.

Molecular modeling can contribute in a great variety of areas, just a few of which have been discussed in this article.

Acknowledgements

Many thanks to my students and colleagues for valuable assistance during the preparation of this article, especially: Mike Malone, Jim Douglas, Monty Alger, Sagar Gadewar (who provided Figures 3 and 4), and Daniel Winn. Financial support is acknowledged from the NSF as well as the Design & Control Center, and the Fortune Project at the University of Massachusetts.

References

Agreda, V. H., Partin, L. R., and W. H. Heise (1990). High purity methyl acetate via reactive distillation. *Chem. Eng. Prog.*, **86**(2), 40-46.

Doherty, M. F., and M. F. Malone (2001). *Conceptual Design of Distillation Systems*, McGraw-Hill, NY.

Douglas, J. M. (1988). *Conceptual Design of Chemical Processes*, McGraw-Hill, NY.

Escobedo, F. A. (2000). Molecular and macroscopic modeling of phase separation. *AIChE J*, **46**, 2086-2096.

Feinberg. M., and D. Hildebrandt (1997). Optimal reactor design from a geometric viewpoint-I. Universal properties of the attainable region. *Chem. Eng. Sci.*, **52**, 1637.

Glasser, D., D. Hildebrandt, and C. Crowe (1987). A geometric approach to steady flow reactors: The attainable region and optimization in concentration space. *Ind. Eng. Chem. Res.*, **26**, 1803 .

Horn, F. (1964). Attainable and non-attainable regions in chemical reaction technique. *Euro. Symp. on Reaction Engineering*, Pergamon Press, NY, 293.

Johnson, J. K., Panagiotopoulos, and K. E. Gubbins (1994). *Mol. Phys.*, **81**, 717-733.

Kletz, T. (1989). Friendly plants. *Chem. Eng. Prog*, **85**(7), 18-26.

Linnhoff, B, D. W. Townsend, D. Boland, and G. F. Hewitt (1982). *A User Guide on Process Integration for the Efficient Use of Energy*, Institution of Chemical Engineers, Rugby (Warwickshire).

Lisal, M., W. R. Smith, and I. Nezbeda (2000). Molecular simulation of multicomponent reaction and phase equilibria in MTBE ternary system. *AIChE J.*, **46**, 866-875.

Liu, X, Y, and P. Bennema (1996). Growth morphology of crystals and the influence of fluid phase. *Crystal Growth of Organic Materials*, A. S. Myerson, D. A. Green, and P. Meenan, eds., ACS, Washington DC.

Okasinski, M. J., and Doherty, M. F. (1997). Thermodynamic behavior of reactive azeotropes. *AIChE J.*, **43**, 2227-2238.

Siirola, J. J. (1996). Industrial applications of chemical process synthesis. *Adv. Chem. Eng.*, **23**, 1-62.

Winn, D. and M. F. Doherty (1998). A new technique for predicting the shape of solution-grown organic crystals. *AIChE J.*, **44**, 2501-2514.

PERICYCLIC REACTIONS IN THERMAL PYROLYSIS OF POLYCARBODIIMIDE

Karin Rotem and Phillip R. Westmoreland
Department of Chemical Engineering
University of Massachusetts Amherst
Amherst, MA 01003-3110

Abstract

Experimental data from polycarbodiimide pyrolysis are interpreted as proceeding by pericyclic steps that generate the monomer. Electronic structure calculations at Hartree-Fock and semiempirical levels were used for predicting reaction thermochemistry and kinetics. Although bond dissociation energies were too strong to explain the pyrolysis kinetics, the activation energy for a six-electron pericyclic transition state was lower and consistent with the data.

Keywords

Pyrolysis, Polymer, Concerted, Pericyclic, Kinetics, Polycarbodiimide.

Introduction

Polymers – thermoplastics, thermosets, and elastomers – are ubiquitous in our daily life, and polymer synthesis at a large scale has been a triumph of chemistry and chemical engineering. At the same time, incineration and flammability of polymers have been important parts of their life cycles. Numerous schemes have been proposed to recycle polymers by pyrolytically decomposing them into monomers or other valuable light hydrocarbons, and burning the polymers for fuel value becomes very attractive as landfill space becomes scarce and costly. Polymer flammability likewise depends on pyrolysis, as little direction oxidation occurs. Instead, the evolved decomposition gases burn, radiating heat back to the surface.

A conventional description of pyrolysis is that bonds first rupture homolytically and then by radical beta scission, aided by radical abstraction processes. Van Krevelen (1961) and other authors further describe polymer pyrolysis as dominated either by random scission (e.g., polyethylene) or by chain-unzipping reactions (e.g., polypropylene). Experimentally, polyethylene decomposes to a wide molecular-weight distribution of products. This distribution has been interpreted as resulting from random homolytic fission of bonds within an idealized $poly(CH_2)$ chain. When enough bonds break, short chains escape into the vapor, generating considerable ethylene, propylene, and other olefins. In polypropylene, random fission would occur, but beta scissions would unzip the chain as soon as a break occurred.

That seems not to be the case for the polymer examined here, polycarbodiimide (polyguanidine). Non-radical pericyclic reactions (Woodward and Hoffmann, 1970) can best account for the experimentally observed kinetics.

The present work examines reaction pathways and kinetics for polymer pyrolysis using experiments and computational chemistry. Decomposition is explored by thermogravimetric analysis / mass spectrometry and by calculations of thermochemistry and kinetics. A pathway is proposed that involves a six-electron pericyclic transition state, regenerating the monomer in a non-radical chain unzipping. This hypothesis is consistent with the pyrolysis data.

Approach and Procedure

Experimental

Mass and heat fluxes to a sample of decomposing polymer were measured in a commercial TGA/DSC (Rheometrics STA-1500) during a linear 10°C/min ramp of temperature. A few mg of polymer were used in a typical run. Evolved gases were sampled just above the sample pan, drawn through a heated sample line, and analyzed at three- to four-min intervals (30 to 40°C) with a Hewlett-Packard 5890 Series II Gas Chromatograph with 5972 Mass Sensitive Detector. Species identities were assigned using Hewlett-Packard GC/MS software, and mole fractions were estimated by proportionality of total-ion signal to carbon number. DSC data were not used in the study of this polymer.

Computation

Structures and energies of species and transition states were calculated mainly using *Gaussian 94*, *Gaussian 98*, and *Gaussian 98 for Windows* (Frisch et al., 1998). Visualization was with a variety of codes, primarily *Cerius²* (MSI). Thermochemical properties and transition-state-theory rate constants were calculated with the BAC-MP4 method (Melius and Binkley, 1986; Melius, 1990). That method uses calculations of UHF/6-31(d) geometry optimization and frequency, MP4/6-31G(d,p) single-point energy, frequency scaling, and energy corrections based on the bonded elements and their interatomic bond lengths. For bond dissociation energies, enthalpies ($\Delta_f H^°_{298}$) were predicted for model compounds and their radical fragments.

Results and Discussion

Experimental

Poly(carbodiimide) polymers or polyguanidines have the general repeating unit -(-C(=NR)-NR´-]-, where R and R´ are substituent groups. They were thought to decompose cleanly to their monomers, which have cumulative allene-type pi bonds in the form RN=C=NR´.

Experimentally, four different polycarbodiimides were tested: poly (methylbenzyl methyl) carbodiimide, poly (di-*n*-hexyl) polycarbodiimide, poly (methyl-TEMPO) carbodiimide, and poly (N-pentafluorophenyl N'-*n*-hexyl) polycarbodiimide. Note that the first polymer, MBM-polycarbodiimide, would not be a good candidate to be a fire suppressant because of the high heat of combustion of its ethylbenzene. Rather, it was tested as an initial proof of principle that polycarbodiimide decomposition could be exploited.

Two types of decomposition temperatures were obtained. The conventional value is obtained from the mass vs. time (or temperature) by extrapolating a linear region of initial decay back to intersect the baseline mass of undecomposed polymer. For the polymers above, the

range was 178-208°C. Using the somewhat different criterion of 1% mass loss, the range was 178-207°C.

Consider a simple first-order model

$$\frac{dP}{dt} = -A \cdot \exp\left(-\frac{E_{act}}{RT}\right) \cdot P \qquad (1)$$

where P is the polymer mass and $T = b \cdot t$. We would like to know the activation energy that would yield 1% decomposition (i.e., a detectable amount of decomposition) at 178-207°C. There is no closed-form analytical solution of Eq. (1), but numerical integration using a reasonable range of Arrhenius pre-exponential factors A gives a range of 33-42 kcal/mol (Fig. 1) for the four polymers. For the MBM-polycarbodiimide ($T_{dec,1\%} = 185$ K), the range is 33-39 kcal/mol.

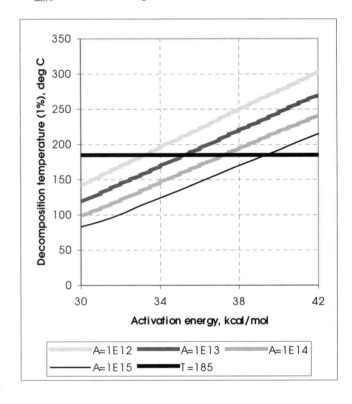

Figure 1. Decomposition temperatures corresponding to first-order kinetics in a sample heated at 10°C/min.

From TGA/GC-MS experiments on the MBM-polycarbodiimide, the dominant product was the monomer, methylbenzyl methyl carbodiimide (R = 1-methyl benzyl or more properly 1-phenylethyl; R´ = methyl). Secondary products were styrene and a peak tentatively identified as (1-isocyanatoethyl)-,(S)-benzene, which may have been a residue of the polymerization or of partial oxidation.

Computational

The first hypothesis was that some bond in the polymer must be weak enough to cause the low decomposition temperature; that is, on the order of 35-40 kcal/mol. Homolytic fission typically has a much smaller rate constant and higher activation energy (70-80 kcal/mol) than beta-scission (25-35 kcal/mol).

Calculations of bond strengths with model compounds gave the weakest bond as 58 kcal/mol between the imide nitrogen and the phenylethyl group (Fig. 2). This is indeed weak, but still much stronger than 40 kcal/mol. It is weakened by being benzylic from the hydrocarbon side.

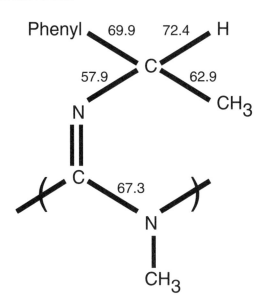

Figure 2. Key bond energies (kcal/mol) in methylbenzyl methyl polycarbodiimide.

Other bond energies were even stronger, so pericyclic reactions were examined as alternatives to homolytic bond fission. Pericyclic reactions have transition states involving concerted making and breaking of bonds around a cyclic perimeter of bonds. This bonding ring is structured such that the number of electrons conveys aromatic stability to the ring. A six-electron cycle (Hückel $4n+2$) is one classical form. The bonds make and break concertedly (or nearly so), so the energy cost of breaking bonds is somewhat offset by the energy return of making new bonds. For example, in the reaction of 1,3,5-hexatriene to 1,3-cyclohexadiene, three C=C π bonds are converted into C-C σ bonds, two C-C bonds are converted into C=C bonds, and a new C-C bond is created. Typical activation energies are a π-bond energy or less, 30-55 kcal/mol and even lower.

These transition states are tight, though, with Arrhenius pre-exponential factors $A = 10^9$ to 10^{11} s^{-1}. These values are much lower than the 10^{14} to 10^{16} s^{-1} values for gas-phase homolytic fissions. The latter A-factors are so

high because the transition state is centrifugally loosened relative to the gas-phase parent (Benson, 1976). In a polymer, the A-factor would be lower but still higher than for the pericyclic reactions. Thus, there is a competition for fastest rate constant between low-A, low-E_{act} pericyclic decomposition reactions and the high-A, high-E_{act} bond fissions. If temperatures are high enough, bond fission easily wins because of the nonlinear importance of the activation energy, and subsequent low-E_{act} radical reactions like abstraction and addition to π bonds become dominant. At low temperatures, activation energy makes either reaction slow. It is within the window between these two regimes that pericyclic decomposition reactions can be important.

The compound chosen to model the polymer for these calculations had the structure HN=CH-NH-CH=NH, using H's in place of the rest of the polymer, the chain methyl group, and the methylbenzyl group. The most likely pericyclic route was expected to involve transfer of an imide R group (H, here) to a nearby imide nitrogen. The process is depicted in Fig. 3. Simultaneously, its results would be to change the first imide to a cyano group, to change the second imide into an amine, to convert the adjacent chain C-N into an imide, and to break the polymer at the other C-N bond.

Such a transition state was indeed located, falling into the class of retro-ene cycloreversions (Woodward and Hoffman, 1970). The heat of reaction was calculated

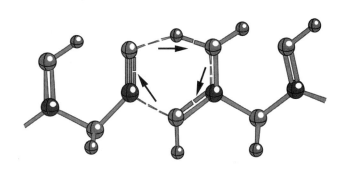

Figure 3. Depiction of pericyclic transition state proposed for decomposition of simple polycarbodiimide. Large dark spheres in chain are carbon; large light spheres are nitrogen; and small spheres are hydrogen.

to be endothermic by 8.9 kcal/mol. The heats of formation at 298 K for the initial stable structure, transition state, cyano group, and pseudo-monomer were calculated to be 54.0, 86.7, 26.3, and 36.6 kcal/mol, respectively. The activation energy for ideal-gas unimolecular decomposition was 33.2 kcal/mol with an A-factor of $9.8 \cdot 10^{12}$ s^{-1}. In the modeling described in Fig. 1, these parameters would give 1% decomposition at 159°C, slightly lower than the low end of the experimental range of decomposition temperatures for the polymers.

Calculations using larger model compounds are being pursued, but the essential observation is that these rate parameters are much more consistent with the observed kinetics than are the activation energies for homolytic bond fission. Note that a cyano group would be formed in a mid-polymer break, and few such groups were observed experimentally. On the other hand, this reaction could completely unravel the polymer into monomer by starting at the imide end after a chain break or at the end of the original polycarbodiimide.

These results are not sufficient to prove the mechanism, but they provide strong support for the validity of the hypothesis.

Conclusions

Experimental data from polycarbodiimide pyrolysis are interpreted as proceeding by pericyclic steps that generate the monomer. Electronic structure calculations at Hartree-Fock and semi-empirical levels were used for predicting reaction thermochemistry and kinetics. Bond dissociation energies, calculated at BAC-MP4/6-31G(d,p)//UHF/6-31G(D) level, were too strong to explain the pyrolysis kinetics. However, the calculated activation energy for a six-electron pericyclic transition state proved to be consistent with the data.

Acknowledgments

We gratefully acknowledge support of this work through Cluster F of the Center for UMass-Industry Research on Polymers by the following organizations: Amoco, Army Research Laboratories, Boeing, CIBA, DuPont, the U.S. Federal Aviation Administration, General Electric, Solutia, Titeflex, and Union Carbide.

References

Benson, S. W. (1976). *Thermochemical Kinetics, 2nd Ed.* Wiley, New York.

M. J. Frisch, G. W. Trucks, H. B. Schlegel, G. E. Scuseria, M. A. Robb, J. R. Cheeseman, V. G. Zakrzewski, J. A. Montgomery, Jr., R. E. Stratmann, J. C. Burant, S. Dapprich, J. M. Millam, A. D. Daniels, K. N. Kudin, M. C. Strain, O. Farkas, J. Tomasi, V. Barone, M. Cossi, R. Cammi, B. Mennucci, C. Pomelli, C. Adamo, S. Clifford, J. Ochterski, G. A. Petersson, P. Y. Ayala, Q. Cui, K. Morokuma, D. K. Malick, A. D. Rabuck, K. Raghavachari, J. B. Foresman, J. Cioslowski, J. V. Ortiz, A. G. Baboul, B. B. Stefanov, G. Liu, A. Liashenko, P. Piskorz, I. Komaromi, R. Gomperts, R. L. Martin, D. J. Fox, T. Keith, M. A. Al-Laham, C. Y. Peng, A. Nanayakkara, C. Gonzalez, M. Challacombe, P. M. W. Gill, B. Johnson, W. Chen, M. W. Wong, J. L. Andres, C. Gonzalez, M. Head-Gordon, E. S. Replogle, and J. A. Pople (1998). *Gaussian 94*; *Gaussian 98*; *Gaussian 98 for Windows*. Gaussian, Inc., Pittsburgh PA.

Melius, C. F. (1990). In S. U. Bulusu, S. (Ed.), *Chemistry and Physics of Energetic Materials*. Kluwer Academic Publishers, Dorderecht, 21-52.

Melius, C.F., and J. S. Binkley (1986). *Proc. Combustion Institute*, **21**, 1953.

Van Krevelen, D.W. (1961). *Properties of Polymers*. Elsevier, Amsterdam.

Woodward, R.B., and R. Hoffman (1970). *The Conservation of Orbital Symmetry*. VCH Publishers / Verlag Chemie, Weinheim.

CATALYSIS FROM FIRST PRINCIPLES

M. Mavrikakis[1] and J. K. Nørskov[2]

[1]Department of Chemical Engineering
University of Wisconsin-Madison
Madison, WI 53706

[2]Center for Atomic-Scale Materials Physics
Department of Physics
Technical University of Denmark
DK-2800, Lyngby, Denmark

Abstract

Recent progress in the theoretical description of elementary reactions on transition metal surfaces is discussed. Calculations based on density functional theory and a non-local description of exchange and correlation effects can now be used to predict changes in reactivity from one system to the next. On the basis of the calculations, models can be developed elucidating the "electronic" and "geometric" factors in catalysis.

Keywords

Catalysis, Surface reactions, Transition metals, DFT.

Introduction

In the present contribution some of the recent progress in a "first principles" understanding of elementary steps in simple catalytic processes will be discussed. This "first principles" understanding is based on a detailed quantum mechanical description of the interactions in the reacting system.

The theoretical description of interatomic interactions at surfaces has developed greatly over the last few years. The main reason is that computational methods based on Density Functional Theory (DFT) have reached a point where several of the complex systems of interest to catalysis can be treated with reasonable accuracy. These calculations are free of parameters and have elucidated the detailed reaction mechanisms and reaction energetics, including both thermochemistry and kinetics, for a number of elementary reaction steps on metal surfaces. This in itself opens up new opportunities for an improved understanding of chemical processes at the molecular level.

An equally important development is that along with the new computational possibilities, the models describing variations in adsorption energies or activation energy barriers for reaction has been developed further. This means that we are now approaching the point where we can make predictions about the direction in which the rate of a specific reaction should vary from one surface to the next. As a result, we are beginning to understand which properties of the clean metal surface govern its reactivity.

In the following, a large number of calculations for the adsorption energies and activation energy barriers for reactions on transition metal and noble metal surfaces is summarized. It will be shown that the variations in adsorption energies and reaction barriers can be understood largely as governed by a few physical parameters characterising the clean metal surface. In particular, these parameters are related to the energy distribution of the local d states of the metal surface, the size of the coupling matrix elements of the surface with the relevant adsorbate states, and the local geometry of the reaction site.

Calculational Details

All the calculations presented here are based on slab models of the surface. Ionic cores are described by soft (or ultra-soft) pseudopotentials (Vanderbilt, 1990) and the Kohn-Sham one-electron valence states are expanded in a basis of plane waves with kinetic energies below 50 (or 25) Ry, respectively. The exchange-correlation energy and potential are described by the Generalized Gradient Approximation(GGA-PW91) (Perdew,1992; White, 1994). The self-consistent PW91 density is determined by iterative diagonalization of the Kohn-Sham Hamiltonian, Fermi-population of the Kohn-Sham states ($k_BT=0.1$ eV), and Pulay mixing of the resulting electronic density (Kresse, 1996; Hammer, 1999}. All total energies have been extrapolated to $k_BT=0$ eV.

In the calculations the adsorbate(s) and the uppermost atomic layers of the metal slab have been allowed to relax to their energetically most favorable position. Transition states for reactions are found by varying a single coordinate (the reaction coordinate) and minimizing all other degrees of freedom of the reactants. In cases where it was not evident which coordinate would be the reaction coordinate, the latter was found by the iterative method of Ulitsky and Elber (Ulitsky, 1990).

Electronic Factors

In Fig. 1 a large number of data from the literature (Hammer, 1999; Kratzer, 1996a; Kratzer, 1996b; Hammer, 1997; Hammer, 1996; Holmblad, 1996; Mavrikakis, 1998), all extracted from DFT calculations, has been collected. These data represent calculated adsorption energies of atomic and molecular adsorbates as well as activation energy barriers for surface reactions. They all describe a situation where the adsorbate interacts with the same kind of metal atom(s) in the same local geometry, but the environment varies. The environment has been changed via either one of the following four different ways:

1. By considering stretched or compressed surfaces.
2. By considering different single crystal facets, including surfaces with and without steps.
3. By considering one metal as a pseudomorphic overlayer on top of another metal.
4. By considering surface alloys.

The calculated values of adsorption energies and barriers for dissociation show large variations. It can, for instance, be seen that CO bound ontop of a Pt atom can have adsorption energies varying by about 1 eV (ca. 100 kJ/mol) depending on the surroundings of the Pt atom in question. The largest adsorption energies are found for adsorption on steps and kinks and the lowest on the compressed hexagonal overlayer structures on Pt(100). Such effects have been observed experimentally.

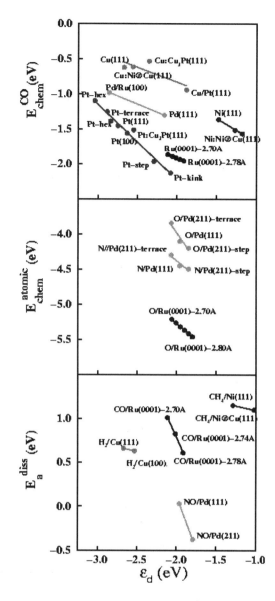

Figure 1. Molecular E_{chem}^{mol} and atomic E_{chem}^{atom} binding energy as a function of the d-band center (ε_d) of the metal surface (top and middle panel, respectively). The barrier for dissociation of small molecules, referenced to gas phase zero, as a function of ε_d is shown in the bottom panel. Common colors are used for data corresponding to the same metal throughout the three panels. Lines drawn represent best linear fits. Adapted from (Mavrikakis, 1998).

An increased adsorption energy for CO at steps has been observed using thermal desorption spectroscopy (McClellan, 1981; Hayden, 1985; Yates, 1995), and a similar effect has been observed in supported small metal clusters (Henry, 1998). The relative inactivity of the compressed hexagonal overlayer on Pt(100) has also been observed experimentally (Yeo, 1995). Steps on Pd surfaces (Pd(211) consists of (111) facets and steps) have also been found to show a lower barrier for dissociation of NO than the flat (111) surface, again in agreement with experiment (Gao, 1994). Changes in adsorption energy and dissociation barriers for overlayer structures and surface alloys have also been observed in many cases (Rodriguez, 1992; Kampshoff, 1994; Larsen, 1998). Most recently the effect of strain on adsorption properties has been observed directly (Gsell, 1998).

In Fig. 1 we show the data as a function of the center of mass of the density of states projected onto the atomic d states of the clean surface. For convenience, we use all the d states here, instead of the ones with the correct symmetry for bonding with the various adsorbates. This makes no major difference, when the adsorption geometry remains similar.

Fig. 1 shows clearly that for a given adsorbate or reaction and for a given metal, the variations in adsorption energies or activation energies are governed largely by the variations in the energy of the surface d-states. There are good reasons for this (Hammer, 1997). The interaction between the adsorbate states and the metal d states is an important part of the interaction energy, and while the sp bands of the metal are broad and structure-less, the d bands are narrow, and small changes in the environment can change the d states and their interaction with adsorbate states significantly. The d-band center (ε_d) is the simplest possible measure for the position of the d states.

The correlation between interaction strength (adsorption energy or activation energy barrier) and ε_d holds for many different adsorbates and metals. Calculations for H_2 dissociation on transition and noble metals have shown a similar relationship to hold also when different metals are compared (Hammer, 1997; Eichler, 1998

The correlation between the interaction strength and the d-band center, Fig. 1, is found for all the transition metals considered. The identity of the metal involved shows up in the *strength* of the effect, that is, the slope of $E(\varepsilon_d)$ through the size of the coupling matrix element. The relative ordering in the coupling strength is 5d>4d>3d following the relative sizes of the d-wave functions (Hammer, 1997).

Having established that the (local) transition metal d band center, ε_d, is an important parameter determining the ability of a metal atom to interact with a reacting atom or molecule, the question arises what determines the variations in ε_d from one system to the next.

First, consider the case where the metal is simply strained. When the lattice is expanded parallel to the surface, the overlap between the d-electrons on neighboring metal atoms becomes smaller, the band width decreases and to keep the d-occupancy fixed, the d states have to move up in energy (for a more than half-filled band). When the structure of the surface changes, for instance by introducing a step, the local d-projected density of states is not changed due to strain, but due to a change in the number of metal neighbors, the metal coordination number. The general rule is that the lower the coordination number, the smaller the local band width and the higher the ε_d (for metals with more than half-filled d-bands). For alloys and overlayers, a large portion of the change in ε_d can be attributed to changes in the metal-metal distances in the surface (Ruban, 1997), as it is the case for the strained slab.

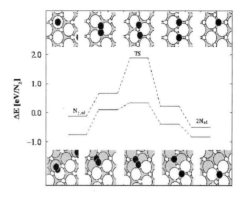

Figure 2.   Results from density functional calculations comparing N_2 dissociation on a terrace and at a step on Ru(0001). The upper curve shows the adsorption and transition state (TS) energies for the dissociation on the terrace, whereas the lower curve shows the same energies at the step. The configuration is shown in each case. The energy zero is taken to be the energy of the N_2 molecule in the gas phase. (Dahl, 2000)

Geometric Factors

The effect of local geometric arrangement of the surface metal atoms on the surface reactivity has been clearly demonstrated, with large scale DFT calculations, at least for three specific cases:

1. Dissociation of N_2 on a flat Ru(0001) surface requires an activation energy barrier which is higher than that needed on a stepped Ru(0001) surface by ca. 1.5 eV (Dahl, 1999), c.f. Fig. 2.
2. Dissociation of NO at a corrugated Ru(0001) surface (Hammer, 1999b), where monatomic steps provide reaction paths with activation energies smaller than the activation energy required on the flat Ru(0001) surface by more than 1 eV.
3. Dissociation of CO on a Rh(111) has an activation energy barrier higher than that necessary on the stepped Rh(211) surface by ca. 1 eV (Mavrikakis, 2000).

In all these cases, a thorough analysis of the transition and final states of the interesting reaction paths indicate that the favorable barriers at the step edges originate from the minimized degree of metal atom sharing between the two atoms of the adsorbate molecule at the transition state and the final state. In particular, five metal atoms are associated with the transition state complex at the step site, rather than four at the terrace. As a result, the transition state configuration at the step avoids the indirect repulsive interactions that are responsible for the higher barrier on the terrace.

Conclusion

In conclusion, it has been illustrated that large scale density functional calculations can now be used to predict adsorption energies and barriers for reactions at metals surfaces. In addition, we can begin to understand the physical origin of variations in adsorption energies and barriers from one system to the next. There is a general correlation between the reactivity of a surface and shifts in the center of the metal d-bands (ε_d). The coupling strength between the adsorbate states and the d states is another important parameter. Together with the number of d-electrons, it determines the variations in interaction strengths from one metal to the next (Hammer, 1997).

When combined with experimental investigations of surface structure and reactivity, the new developments in the theory of surface reactivity make it possible to suggest new catalytically interesting systems. An example of such a catalyst design from first principles has been reported recently (Besenbacher, 1998).

The present work was in part financed by The Danish Research Councils through The Center for Surface Reactivity and grant 9501775. The Center for Atomic-scale Materials Physics is sponsored by the Danish National Research Foundation. MM thanks the Graduate School at UW-Madison for partial financial support.

References

Besenbacher, F., Chorkendorff, I., Clausen, B. S., Hammer, B., Moelenbroek, A. M., Norskov, J. K., and I. Stensgaard (1998). *Science*, **279**, 1913.

Dahl, S., *et al.* (1999). *Phys. Rev. Lett.*, **83**, 1814.

Eichler, A., Kresse, G., and J. Hafner (1998). *Surf. Sci.*, **397**, 116.

Gao, Q., Ramsier, R. D., Neergaard Waltenburg, H., and J. T. Yates, Jr. (1994). *J. Am. Chem. Soc.*, **116**, 3901.

Gsell, M., Jakob, P., and D. Menzel (1998). *Science*, **280**, 717.

Hammer, B., Morikawa, Y., and J. K. Norskov (1996). *Phys. Rev. Lett.*, **76**, 2141.

Hammer, B., Nielsen, O. H., and J. K. Norskov (1997a). *Catal. Lett.*, **46**, 31.

Hammer, B., and J. K. Norskov (1997b). in: *Chemisorption and Reactivity on Supported Clusters and Thin Films*, R. M. Lambert and G. Pacchioni, Eds. Kluwer Academic Publishers, The Netherlands, 285-351

Hammer, B. (1998). *Faraday Dis.*, **110**, 323.

Hammer, B., Hansen, L. B., and J. K. Norskov (1999a). *Phys. Rev. B*, **59**, 7413.

Hammer, B. (1999b). *Phys. Rev. Lett.*, **83**, 3681.

Hayden, B. E., Kretzschmar, K., Bradshaw, A. M., and R. G. Greenler (1985). *Surf. Sci.*, **149**, 394.

Henry, C. R. (1998). *Surf. Sci. Rep.*, **31**, 231.

Holmblad, P. M., *et al.*, (1996). *Catal. Lett.*, **40**, 131.

Kampshoff, E., Hahn, E., and K. Kern (1994). *Phys. Rev. Lett.*, **73**, 704.

Kratzer, P., Hammer, B., and J. K. Norskov (1996a). *Surf. Sci.*, **359**, 45.

Kratzer, P., Hammer, B., and J. K. Norskov (1996b). *J. Chem. Phys.*, **105**, 5595.

Kresse, G. and J. Forthmuller (1996). *Comp. Mat. Sci.*, **6**, 15.

Larsen, J. H., and I. Chorkendorff (1998). *Surf. Sci.*, **405**, 62.

Mavrikakis, M., Hammer, B., and J. K. Norskov (1998). *Phys. Rev. Lett.*, **81**, 2819.

Mavrikakis, M., Stoltze, P., and J. K. Norskov (2000). to be submitted.

McClellan, M. R., Gland, J. L., and F. R. McFreeley (1981). *Surf. Sci.*, **112**, 63.

Perdew, J. P., et al. (1992). *Phys. Rev. B*, **46**, 6671.

Rodriguez, J. A., and D. W. Goodman (1992), *Science*, **257**, 897.

Ruban, A., *et al.* (1997). *J. Mol. Catal. A*, **115**, 421.

Ulitsky, A. and R. Elber (1990). *J. Chem. Phys.*, **92**, 1510.

Vanderbilt, D .H. (1990). *Phys. Rev. B*, **41**, 7892.

Yates, Jr., J. T. (1995). *J. Vac. Sci. Tech. A*, **13**, 1350.

Yeo, Y. Y, Wartnaby C. E., and D. A. King (1995). *Science*, **268**, 1731.

White, J. A. and D. M. Bird (1994). *Phys. Rev. B*, **50**, 4954.

MOLECULAR SIMULATION FOR METHANOL SYNTHESIS IN SUPERCRITICAL N-HEXANE

Xiaogang Zhang and Buxing Han
Center for Molecular Sciences, Institute of Chemistry
Chinese Academy of Sciences
Beijing 100080, China

Yongwang Li, Bing Zhong, and Shaoyi Peng
The State Key Laboratory of Coal Conversion, Institute of Coal Chemistry
Chinese Academy of Sciences
Taiyuan 030001, China

Abstract

A computer model based on the Monte Carlo method for methanol synthesis from CO and H_2 in supercritical n-hexane was investigated. The model took into account the effect of methanol aggregation on the synthesis reaction and converted Monte Carlo steps to real time according to Boltzman equation. The results indicated that TOF (turnover frequency) and conversion number (defined as the ratio of the total number of methanol molecules produced by reaction over the total number of carbon monoxide impinging on the surface) increased with methanol aggregation number (MAN) at the same reaction condition, and MAN mainly influenced methanol re-adsorption. We also simulated the influence of mole fraction of n-hexane on the reaction under different MAN and found that conversion number increased monotonously with the mole fraction of n-hexane. Each TOF versus n-hexane mole fraction curve had a maximum, and the maximum shifted towards higher n-hexane concentration as the MAN increased.

Keywords

Monte Carlo simulation, Methanol synthesis, Supercritical n-hexane, Methanol aggregation number.

Introduction

Supercritical fluids (SCFs) are attractive media for chemical reactions because of their unique properties. Many of the physical and transport properties of SCFs are intermediate between those of a liquid and a gas. The State Key Laboratory of Coal Conversion used SC n-hexane as the reaction medium for methanol synthesis using CO and H_2, and found that the conversion and selectivity could be improved.

The main advantage of the MC simulation is that the configurations sampled by the moves can be restricted to those that are of direct interest to the process being modeled. MC simulation for the analysis of molecular processes occurring on solid surface is now becoming a powerful tool for understanding elementary steps and mechanism of the kinetics of heterogeneous reaction.

In the previous study [Zhang, 1998], we simulated the microscopic structure of binary mixtures of methanol-n-hexane at different densities and showed that there existed significant aggregation between methanol molecules in supercritical n-hexane. The aim of this paper is to investigate the effects of methanol aggregation number (MAN) on the reaction behavior of methanol synthesis in SCF n-hexane through MC simulation.

135

Model and Algorithm

Mechanistic investigations of methanol synthesis using CO and H_2 have been carried out mainly on Cu/ZnO and Cu/ZnO/Al$_2$O$_3$. Evidence from infrared spectroscopy [Saussey et al., 1982; Hindermann et al., 1985; Deluzarche et al., 1985; Edwards and Schrader, 1985] and chemical trapping [Deluzarche et al., 1979; Deluzarche et al., 1981; Vedage et al., 1984] suggested that formyl group (HCO) is the key intermediate in CO hydrogenation. Shustorovich and Bell [1991] analyzed the mechanism of methanol synthesis using BOC-MP approach, and showed that the formation of methanol from CO included the following steps:

$$H_2(g) \rightarrow 2H(a) \tag{1}$$
$$CO(g) \rightarrow CO(a) \tag{2}$$
$$H(a) + CO(a) \rightarrow HCO(a) \tag{3}$$
$$H(a) + HCO(a) \rightarrow H_2CO(a) \tag{4}$$
$$H(a) + H_2CO(a) \rightarrow H_3CO(a) \tag{5}$$
$$H(a) + H_3CO(a) \rightarrow H_3COH(a) \tag{6}$$
$$H_3COH(a) \rightarrow CH_3OH(g) \tag{7}$$

where (a) indicates the species adsorbed on the surface. Upon adsorption, the H_2 dissociates into two-H atoms, each residing on a separate surface site, while CO requires only a single site.

The catalyst surface is represented by a lattice consisting of a regular square array, where each lattice site has four nearest neighbors, referred to as adjacent sites. Adsorbed atoms are allowed to migrate to adjacent sites only, diagonal movement is not allowed.

For the reaction, possible events include adsorption of H_2 or CO; surface movement of adsorbed species; reaction between two adsorbed species (if the reaction could occur); and desorption of CH_3OH or CO. The reaction between two species might depend on the local environment of these species. In our model, the rates of these processes are written in terms of the local environment of the individual surface species. For each process the probability P_i of a successful event is given by:

$$P_i = A \exp(-E_i/RT) \, \Delta t \tag{8}$$

where A is the pre-exponential factor, E_i is the activation energy, Δt is a time interval.

One of the important features of a supercritical fluid mixture is that local density differs from bulk density or average value. This phenomenon has been referred as "clustering" or "aggregation". Our simulation studies [Zhang, 1998] have shown that there existed aggregation between methanol molecules in supercritical n-hexane. In this case, it is possible that the opportunity of methanol molecules impinging on catalyst surface decreases. In our model, we considered the effects of methanol aggregation on reaction behavior.

In our model, we defined conversion number η as the ratio of the total number of methanol molecules produced by reaction divided by the total number of carbon monoxide impinging on the surface.

Catalytic reaction on the surface is a dynamic process with the evolution of time. How to convert Monte Carlo steps to real time is the key of this model. The number of hits per unit time on a given surface at fixed pressure and temperature can be calculated from the Boltzmann equation. For hydrogen molecule,

$$F_{H_2} = P_{H_2}/(2\pi m_{H_2} k_B T)^{1/2} \tag{9}$$

where F_{H_2} is the number of hits of hydrogen molecules per unit time on unit surface, P_{H_2} is the pressure of hydrogen, k_B is Boltzmann constant, m_{H_2} is the mass of hydrogen, and T is absolute temperature.

We can obtain the time interval of hits of hydrogen molecules at a given pressure and temperature.

Parameter Estimation

In our model, the data of activation energy of every elementary step were estimated according to the BOC method [Shustovorich, 1990], and pre-exponential factors were estimated based on transition-state theory or obtained from literature [Zhdanov, 1988; Dumesic et al., 1991]. Table 1 summarizes the values of these parameters, which were used in our model.

Table 1. Pre-Exponential Factors and Activation Energies of Elementary Steps for Simulation Over Cu Catalyst.

Step	Pre-exponential factors		Activation energies (kcal/mol)	
	Forward	Reverse	Fwd	Rev
(1)	8.34 ← 10^{2}Pa^{-1}s^{-1}	1.0 ← 10^{-3}cm^{2}s^{-1}	7	15
(2)	2.23 ← 10^{3}Pa^{-1}s^{-1}	1.0 ← 10^{13}s^{-1}	0	12
(3)	1.0 ← 10^{-1}cm^{2}s^{-1}	1.0 ← 10^{13}s^{-1}	24	0
(4)	1.0 ← 10^{-2}cm^{2}s^{-1}	1.0 ← 10^{13}s^{-1}	0	20
(5)	1.0 ← 10^{-2}cm^{2}s^{-1}	1.0 ← 10^{13}s^{-1}	4	9
(6)	1.0 ← 10^{-2}cm^{2}s^{-1}	1.0 ← 10^{13}s^{-1}	10	18
(7)	1.0 ← 10^{13}s^{-1}	2.08 ← 10^{3}Pa^{-1}s^{-1}		

Results and Discussion

Effect of the Methanol Aggregation Number (MAN) on Synthesis Reaction

Our simulation shows that the MAN (ranges from 1 to 4) could influence the reaction behavior because the synthesis reaction is reversible. The effect of MAN on

TOF, η, and concentrations of species absorbed on the surface was studied at 498 K and 8.0 MPa with feedstock composition of n-C_6H_{14} = 50 mol %, H_2 = 33.5 mol %, CO = 16.5 mol %, and the results are demonstrated in Figures 1 to 3.

η and TOF increase with MAN in the MAN range from 1 to 4, as can be seen from Figures 1 and 2. Figure 3 shows that the steady-state concentration of H, CO and active empty sites are not sensitive to MAN, but that of CH_3OH decreases notably with MAN. We can conclude that MAN mainly influences the re-adsorption of methanol molecules produced by reaction.

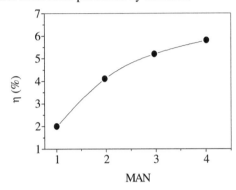

Figure 1. Effect of MAN on η.

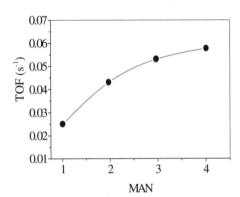

Figure 2. Effect of MAN on TOF.

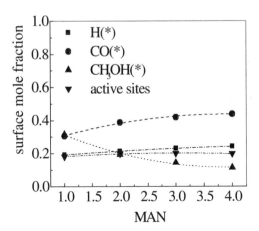

Figure 3. Plot of the surface concentration for different adsorbed species versus MAN.

Effect of the Mole Fraction of n-Hexane on the Synthesis Reaction

The influence of n-hexane mole fraction plays an important role in the methanol synthesis. Here we analyze the changes of TOF, η with mole fraction of n-hexane. Figure 4 shows the simulation results at 498 K and 8.0 MPa, and the feed gas H_2: CO=2:1. It can be seen that η increases monotonically with the mole fraction of n-hexane for different MAN.

Figure 5 shows that each TOF versus n-hexane mole fraction curve has a maximum, and the maximum shifts towards higher n-hexane concentration as the MAN increases.

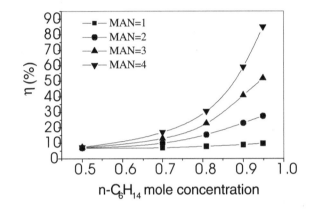

Figure 4. Plot of η versus C6H14 mole concentration for different MAN.

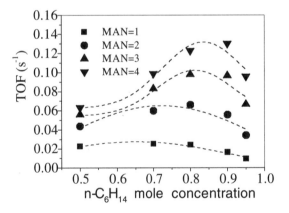

Figure 5. Plot of TOF versus C_6H_{14} mole concentration for different MAN.

Conclusion

With the help of the reaction of methanol synthesis in supercritical n-hexane, we simulate the influence of MAN and the mole fraction of n-hexane on catalyst activity. For the reaction, there is still an uncertainty about the sticking probability and about the other model parameters. The real strength is that it can be used to understand variations in the catalytic reaction system. This Monte Carlo method is not limited to modeling existing experimental data. Good estimations can be obtained for the conditions at which experiments cannot be conducted. By changing the model parameters, it is possible to understand the effect of these parameters on the reaction, which might help in catalyst development.

In this paper, we use methanol synthesis on Cu as an example, but the same method can be applied to some other catalytic processes.

References

Zhang, X. G. (1998). Ph.D. Thesis, Institute of Coal Chemistry, Chinese Academy of Sciences, China.

Saussey, J.,Lavalley, J. C., J. Lamotte, et al. (1982). *J. Chem. Soc. Chem. Comm.*, 278.

Hindermann, J. P., Schlieffer, E., H. Idris, et al. (1985). *J. Mol. Catal.*, **33**, 133.

Deluzarche, A., Hindermann, J. P., A. Kiennemann, et al. (1985). *J. Mol. Catal.*, **31**, 225.

Edwards, J. F., and G. I. Schrader (1985). *J.Catal.*, **94**, 175.

Deluzarche, A., Cressley, J., and R. Kieffer (1979). *J. Chem. Res. (s)*, **136**, 1657.

Deluzarche, A., Hindermann, J. P., and R. Kieffer (1981). *J. Chem. Res. (s)*, **72**, 934.

Vedage, G. A., Pitchai, R., R. G. Herman, et al. (1984). Am. Chem. Soc., Div. Fuel Chem. Prepr., Paper 29, 196.

Shustorovich, E., and A. T. Bell (1991). *Surf. Sci.*, **253**, 386.

Shustorovich, E. (1990). *Adv. Catal.*, **37**, 101.

Zhdanov, V. P. (1988). *Catal. Rev. – Sci. Eng.*, **30**, 501.

Dumesic, J. A., Rudd, D. F, L. M. Aparicio, et al. (1991). *The Microkinetics of Heterogeneous Catalysis*, American Chemical Society, Washington DC.

MOLECULAR DYNAMICS SIMULATION OF THE REACTION OF H AND SiH₃ RADICALS WITH HYDROGENATED SILICON SURFACE

Masahiko Hirao, Shinya Muramatsu, and Masatoshi Shimada
Department of Chemical System Engineering
The University of Tokyo
Tokyo 113-8656, Japan

Abstract

We report on the result of molecular dynamics simulations of H and SiH$_3$ radicals on hydrogenated silicon surface, and discuss the effect of hydrogen coverage on elementary surface reactions. We prepared a model which describe monohydride and dihydride Si(100) surfaces and modified the hydrogen coverage of a local region in the range from 0.5 to 2.0 monolayers (ML) so as to describe growing surface. A gas phase radical, H or SiH$_3$, fell onto modified hydrogenated silicon surface. By repeating such a simulation against several substrates, which differ in their hydrogen coverage, the effect of hydrogen coverage was investigated. As a result of our simulations, H radical mainly causes hydrogen adsorption and abstraction reactions and altered hydrogen coverage of the surface. The reactions which generate the 0.5 ML or 1.5 ML structure occurred frequently. By density functional calculations, it is considered that the local 0.5 ML and 1.5 ML structures are metastable staructures. Thus it is expected that H radical has an effect of increasing the 0.5 ML and 1.5 ML structures. While the reactions of SiH$_3$ radicals with monohydride or dihydride surface rarely occurred, these radicals reacted more frequently with the 0.5 ML and 1.5 ML structures. These results indicate that the change of local hydrogen coverage caused by H radical may induce subsequent surface reactions with SiH$_3$ radical.

Keywords

Hydrogenated silicon surface, Radical reaction, Molecular dynamics simulation, Density functional calculation, Hydrogen coverage.

Introduction

Chemical vapor deposition (CVD) process is widely used for thin film growth in semiconductor industry. Many reaction conditions, such as composition of source gas, substrate temperature and surface properties influence the growth. Among those conditions, the hydrogen coverage on the silicon substrate has a great influence. Surface hydrogen is considered to inhibit silicon crystal growth (Copel, 1999) and the creation of dangling bonds by hydrogen abstraction process is considered to be the rate-limiting step in film growth (Abelson, 1993). On the other hand, Matsuda (1983) reported that sufficient hydrogen coverage of a silicon surface helps the surface diffusion of silicon hydride species and results in microcrystal growth. Hydrogen coverage depends on other reaction conditions. When a hydrogen-terminated silicon substrate is heated, hydrogen molecules desorb from the surface and surface structure is reconstructed according to the temperature. The hydrogen coverage is also modified by surface reactions. Recently, the dynamics of the abstraction of hydrogen were analyzed by molecular dynamics (MD) simulations (Ramalingam, 1998). Hydrogen radicals induce both hydrogen abstraction (Srinivasan, 1998) and adsorption reactions (Boland, 1992), and equilibrium hydrogen coverage increases when a silicon substrate is exposed to sufficient atomic hydrogen. As described in these examples, hydrogen coverage is closely related to various conditions, thus it is difficult to understand its direct effect. Therefore, the detailed analysis of surface phenomena by molecular modeling techniques including MD simulations is needed for undestanding and development of CVD processes.

In this paper, we report on the results of MD simulations in which gas-phase radicals fell onto a hydrogen-terminated Si(100) surface, and discuss the effect of hydrogen coverage on surface phenomena.

Model of Simulation

We performed a MD simulation in which a radical fell onto a properly modeled hydrogen-terminated Si(100) surface. By repeating such a simulation against several model surfaces, the effect of hydrogen coverage was investigated. It is considered that hydrogen coverage of a silicon surface changes continuously during silicon thin film growth owing to surface reactions with radicals. In order to model such a growing surface, the hydrogen coverage of a local structure was separated from that of a periodic structure, as shown in Fig.1. The model substrate has a periodic surface structure, and has a different local surface structure where radicals fall. The four local structures, whose hydrogen coverage is defferent from each other, were used. Those structures are stable and representative ones for 0.5 to 2.0 monolayers (ML) of hydrogen coverage. As the periodic hydrogenated silicon surface, the monohydride and dihydride surfaces were prepared. An initial structure was modeled by combining these local and periodic structures. The unit cell consisted of 20 layers of 100 silicon atoms and additional surface hydrogen atoms to satisfy its hydrogen coverage. Two-dimensional periodic boundary conditions were applied so as to model an infinite surface.

As gas-phase radicals, H and SiH_3 were selected in this study. Many hydrogen radicals are produced when the source gas is diluted with hydrogen, and SiH_3 is the main species for plasma-enhanced CVD (Abelson, 1993). Those radicals were placed above the local surface structure as an initial condition, then fell onto the local structure of substrate giving kinetic energies according to the velocity distribution at 300 K. Twelve times MD simulations were carried out by using NVE ensenble for the same hydrogen coverage at the same temperature. As the interatomic potential for a system which consists of silicon and hydrogen atoms, an extended form of the Tersoff potential presented by Murty (1995) was used. SiH_3 radical is also modeled by this interatomic potential, thus decomposition and reaction on the surface can be simulated.

It has not been confirmed that the extended Tersoff potential is applicable to the simulation of surface phenomena. Therefore, density functional (DFT) calculations were carried out using DMol3 program (MSI, 1999) and energetics were investigated to verify the results of MD simulations. As shown in Fig. 2, a slab model which consisted of eight layers of two silicon atoms is used. The top layer of silicon atoms was hydrogen-terminated so as to satisfy the 0.5 to 2.0 ML hydrogen coverage, and all the dangling bonds at the bottom were terminated with hydrogen atoms. The most stable structure for each surface model was obtained by geometry optimization. During the optimization, the four bottom layers of silicon atoms and the bottom hydrogen atoms were fixed.

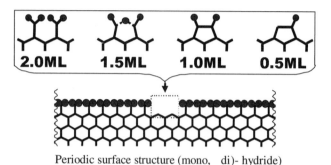

Periodic surface structure (mono, di)- hydride)

Figure 1. Surface structure model for MD simulation. Local structures correspond to 0.5 to 2.0 ML hydrogen coverage.

Figure 2. Surface structure model for DFT calculation. The 0.5 ML local structure is shown.

Results and Discussion

When H radical fell onto the hydrogenated silicon furface, H radicals mainly caused hydrogen adsorption and abstraction reactions, and altered hydrogen coverage of the surface. Figure 3 shows the transition of hydrogen coverage caused by H radicals at 300K. The right arrows and left arrows designate abstraction and adsorption reactions, respectively. The numbers beside the arrows indicate the frequency of that reaction over twelve times of simulations. Other types of reaction are not counted in this figure. The reactions which generate the 0.5 ML or 1.5 ML structures occurred more frequently than the reactions which generate the 1.0 or 2.0 ML structures. It is expected that H radicals have the effect to increase the number of 0.5 ML and 1.5 ML structures. The same tendency was observed by the simulations at 700 K.

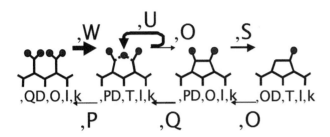

Figure 3. The frequency of H adsorption and abstraction reactions by H radicals at 300 K. The right- and left- directed arrows designate H abstraction and absorption, respectively. The numbers indicate the frequency of reaction over twelve times of simulations.

On the other hand, SiH₃ radicals caused many kinds of reactions such as adsorption, hydrogen abstraction and self-decomposition. Among them, adsorption reactions occurred most frequently. Table 1 shows the effect of hydrogen coverage and substrate temperature on the probability of SiH₃ adsorption. This table indicates the frequency of the adsorption reaction over twelve times of simulations. Though perature of substrate did not affect on the adsorption probability, the effect of local hydrogen coverage was clearly observed. While SiH₃ radicals rarely reacted with the 1.0 ML or 2.0 ML local surface structure, they reacted more frequently with the 0.5 ML or 1.5 ML local surface structure. As depicted in Fig. 3, those 0.5 ML and 1.5 ML structures are mainly generated by reactions with H radicals. Therefore, we can conclude that easier adsorption of SiH₃ radicals is realized by the change of the local hydrogen coverage caused by H.

Table 1. The Probability of SiH₃ Adsorption. The Numbers Indicate the Frequency of Adsorption Reaction Over Twelve Times of Simulations.

Temp [K]	Periodic structure						
	Monohydride			Dihydride			
	Local structure (ML)						
	0.5	1.0	1.5	0.5	1.0	1.5	2.0
300	5	0	3	5	1	5	0
700	4	2	2	6	0	6	0

However, the 0.5 ML and 1.5 ML structures has not been clearly observed by experiments. Thus, we have to analyze whether such structures are stable or not. Table 2 shows the energy changes accompanying reactions with H radicals obtained by DFT calculations. The values are the energy differences before and after the reaction. For example, the values on the most right row are calculated by following equations.

$$E_{adsorption} = (V_{1.5ML} + V_H) - V_{2.0ML} \qquad (1)$$

$$E_{abstraction} = (V_{2.0ML} + V_H) - (V_{1.5ML} + V_{H_2}) \qquad (2)$$

where E_x is the energy changes accompanying reaction x, V_y is the potential energy of structure y.

These values indicate that all the reactions are exothermi and that such reactions may occur more frequently than their reverse reactions. If the 1.5 ML structure is generated by H radicals, it will not spontaneously revert to the original structure without the help of other radicals or the reconstruction with its surrounding environment. However, the heat of reaction accompanying the production of the 1.5 ML structure is higher than others. Therefore the 1.5 ML structure is a metastable structure.

Table 2. Energy Changes Accompanying Reactions with H Radicals, Obtained by DFT Calculation. The Energy Unit in kJ/mol.

| Type of Reaction | Reaction path | | |
	0.5ML - 1.0ML	1.0ML - 1.5ML	1.5ML - 2.0ML
H adsorption	+252.4	+151.4	+315.0
H abstraction	+213.5	+314.5	+150.9

Although several periodic structures for hydrogenated Si(100) surface are observed and they correspond to hydrogen coverages of 1.0 ML, 1.33 ML and 2.0 ML, the existence of the 1.5 ML structure has not been reported. However, the 1.5 ML structure presented in this paper is not a periodic structure but a local structure. Surrounded by the 1.0 ML or 2.0 ML periodic structures. Furthermore, it is worth noting that the 1.5 ML structure may not be a static structure but a dynamic one. Although only one kind of 1.5 ML structure is shown in Fig. 1, there are other 1.5 ML structures as stable as that shown in Fig. 1 and transition between those structures occurred continuously during the MD simulation. For these reasons, we believe that an local 1.5 ML structure can exist, and that it is an instantaneous structure which continuously changes its shape.

The local 1.5 ML structure may play an important role in the film growth by CVD process. We think that the 1.5 ML structure will be produced more efficiently by H abstraction caused by gas-phase radicals than by thermal desorption of surface H atoms. Microscopically, thermal desorption requires two neighboring surface H atoms. In other words, even if the 1.5 ML structure is produced by thermal desorption, it will be accompanied by the production of another 1.5 ML structure in its neighborhood. Thus the liftime of 1.5 ML structure produced by thermal desorption is very short because two 1.5 ML structures neighboring each other are not stable. On the other hand, gas-phase radicals such as H and SiH_3 abstract one surface hydrogen atom and produce a stable local 1.5 ML structure.

Figure 3 depicts that hydrogen coverage decrease rather than increase upon reactions with H radicals. This is somewhat different from experimental results where equilibrium hydrogen coverage on silicon surface increases by exposing to H radicals. This difference is due to the limitation of our simulation model. Our model substrates, whose local hydrogen coverage was modified, can not represent the equilibrium growing surface with the gas phase composition at a given temperature. Therefore, we cannot conclude equilibrium hydrogen coverage from our simulation results. However, whichever direction hydrogen coverage may go toward an equilibrium state, it is considered that both adsorption and abstraction reactions are induced by H radicals, microscopically.

It is said that when a silicon substrate surface is sufficiently covered with hydrogen, silicon hydride species diffuse on the surface until they reach an active site. In our simulation, however, many SiH_3 radicals were repelled at the surface and return to gas phase, and surface diffusion of those radicals was rarely observed. This may be in conseqence of the relatively short cutoff radius of the potential function. Though the Tersoff potential has been applied to the system of a silicon crystal and a silicon surface and has successfully reproduced their properties, silicon hydride species on the surface may easily go outside the cutoff radius and fly away by small repulsive force acts on it from surface H atoms. For this reason, it might be reasonable to expect that some of the radicals repelled in the simulation should diffuse on the surface originally. Further investigation should be done, however, to judge whether or not the cutoff radius is sufficient for simulating such surface phenomena.

Conclusions

The effect of hydrogen coverage on elementary surface reactions has been examined by MD simulations. We developed a model with a local surface structure correspond to 0.5 to 2.0 ML hydrogen coverage embedded in periodic monohydride or dihydride Si(100) surface. One H or SiH_3 radical is placed above the modeled surface and fall onto the local structure. It has been shown that H radicals modify the hydrogen coverage and generate local surface structures which correspond to 0.5 ML or 1.5 ML hydrogen coverage. While the adsorption probability of SiH_3 radicals changes according to hydrogen coverage, the probability is high for the 0.5 ML or 1.5 ML structures. The DFT calculation also suggested that those local surface structures are metastable at least they are isolated. These results indicate that the change in the local hydrogen coverage caused by H radicals may induce subsequent surface reactions.

References

Abelson, J. R. (1993). Plasma deposition of hydrogenated amorphous silicon: Studies of the growth surface. *Appl. Phys. A,* **56**, 493-512.

Boland, J. J. (1992). Role of bond-strain in the chemistry of hydrogen on the Si(100) surface. *Surf. Sci.,* **261**, 17-28.

Copel, M. and M. Tromp (1994). H coverage dependence of Si(001) homoepitaxy. *Phys. Rev. Lett.* **72**, 1236-1239.

Matsuda, A. (1983). Formation kinetics and control of microcrystallite. *J. Non-Cryst. Solids.* **59-60**, 767-774.

MSI (1999), *Cerius2 4.0.0 Manual.*

Murty, M. V. R., and H. A. Atwater (1995). Empirical interatomic potential for {Si-H} interactions. *Phys. Rev. B,* **51**, 4889-4893.

Perdew, J. P. and Y. Wang (1992). Accurate and simple analytic representation of the electron-gas correlation energy. *Phys. Rev. B,* **45**, 13244-13249.

Ramalingam, S., D. Maroudas, E. S. Aydil, and S. P. Walch (1998). Abstraction of hydrogen by SiH_3 from hydrogen-terminated Si(001)-(2x1) surfaces. *Surf. Sci.,* **418**, L8-L13.

Srinivasan, E., H. Yang, and G. N. Parsons (1998). Ab initio calculation of hydrogen abstraction energetics from silicon hydrides. *J. Chem. Phys.,* **105**, 5467-5471.

MODELING RESID HYDROPROCESSING WITH TETRACHLOROALUMINATE CATALYSTS USING COMPUTATIONAL CHEMISTRY

Mark A. Plummer
Marathon Oil Company
7400 South Broadway
Littleton, CO 80122

Abstract

Tetrachloroaluminates with differing metal cations were previously tested experimentally for their capability to catalyze resid upgrading chemistries of hydrogenation, alkylation, hydrocracking and desulfurization. In the current effort, a computational chemistry procedure was developed to correlate these experimental results with calculated charge differences between the tetrachloroaluminate catalysts and pure hydrocarbons representative of those in resids.

Keywords

Computational chemistry, Electron charge differences, Tetrachloroaluminate catalysts, Hydroprocessing of resids, Pure compound models.

Introduction

Recently, several tetrachloroaluminates with differing metal cations(Table 1) were tested experimentally for their capability to catalyze resid upgrading chemistries of hydrogenation, alkylation, hydrocracking and desulfurization. This work used commercial resids and a pure hydrocarbon feed containing molecular types representative of those in the resid feeds(Table 1)

Table 1. Tetrachloroaluminates, Pure Hydrocarbons and Resid Feeds.

Tetrachloroaluminates

Monovalent – Molten	Divalent – Molten
Li Na K	Zn
Divalent – Solid	Trivalent – Solid
Ni Mo Pd	V Co

Pure Compounds (Mole % In Feed)
Eicosane(52.6) Dibenzothiophene(16.3) Pyrene(21.1)
Resid Feeds
Vacuum Resid   ROSE Pitch

In the current effort, these experimental results were analyzed using computational chemistry methods. The purpose was to develop correlations for optimizing the selection of tetrachloroaluminates or other catalysts for use in hydroprocessing of resids.

Pure hydrocarbon conversion and product property results were first found to correlate with calculated electron charge differences between adsorbed hydrogen plus hydrocarbon and the catalyst evaluated. Then, charge differences for the pure hydrocarbons were assigned to representative resid fractions. Resid conversion and product property results were found to correlate very well with assigned charge differences from the pure hydrocarbon calculations. Electron charge differences were calculated using the semi-empirical quantum mechanics model Zindo/1(HyperChem V5.1). The approach in these calculations is similar to that previously used to model adsorption of hydrocarbons on tetrachloroaluminate catalysts(Plummer, et al, 1998).

Experimental Conditions

The details of the hydroprocessing conditions have been published elsewhere(Saski, et al, 2000). In summary, the pure hydrocarbon feed and resids were contacted with the tetrachloroaluminates at 400 °C for 20 minutes under hydrogen pressures of 5,515 to 10,342 kPa. The amount of catalyst used was 10 mole% of the pure hydrocarbon feed mixture and 10 weight% of the resid feeds.

Molecular Modeling Approach

All catalyst structures were first energy minimized. Then, a hydrogen molecule was placed two Angstroms from an aluminum atom in the catalyst and the combination energy minimized. This yielded a structure for hydrogen adsorbed on the catalyst. Then, each pure hydrocarbon of the feed mixture was separately placed two Angstroms from the adsorbed hydrogen molecule and the structure was again energy minimized. This yielded a structure and an electron distribution for hydrogen and a pure hydrocarbon adsorbed on the catalyst. Next, an electron charge difference of hydrogen charge plus hydrocarbon charge minus catalyst charge was calculated for each energy minimized structure.

For correlating resid results, the electron charge difference calculated for each catalyst with eicosane ($C_{20}H_{42}$) was assigned to the saturate fraction based on the weight% of the saturate fraction in the resid feed. And, the average charge difference obtained with dibenzothiophene and pyrene for each catalyst was similarly assigned to the aromatic fraction in the resid feed.

Detailed Discussion of Results

It was first noted that conversion and product property results via the pure hydrocarbon feed formed separate correlations versus charge difference for the monovalent, divalent and trivalent catalysts. This is not a satisfactory correlation method to optimize catalyst selection. However, it is interesting to note that all charge differences are positive and greater than 1.0 electrostatic units(ESU). This suggests that conversion mechanisms were via electron transfers from adsorbed hydrocarbon and hydrogen molecules to the catalyst.

Next, all charge differences were divided by the number of aluminates in the catalysts. These values were termed "averaged" charge differences.

With one exception, all results obtained with the pure hydrocarbon feed correlate very well with averaged electron charge difference. The general correlation form is two straight lines intersecting in a "V" shape at about 1.1 ESU. Figure 1 shows a typical correlation for total feed conversion to liquid products via hydrocracking, alkylation hydrogenation and desulfurization reactions. Total conversion increased with decreasing charge difference below 1.1 ESU for the divalent and trivalent transition metal catalysts. Above 1.1 ESU, total liquid conversion increased with increasing charge difference for the monovalent alkali metal catalysts.

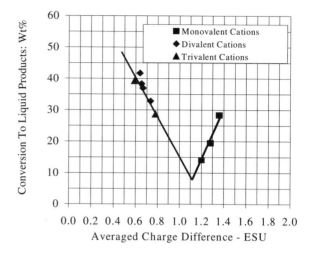

Figure 1.   Conversion of pure compounds to liquid products.

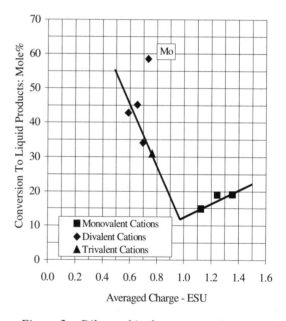

Figure 2.   Dibenzothiophene conversion to liquid products.

The exception to the pure compound results is that only the Mo catalyst yielded any desulfurization of dibenzo-thiophene. Hence, the Mo catalyst result do not fit the general "V" shape curve for total dibenzothiophene conversion to liquid products (Fig. 2). However as detailed below, all catalysts studied gave significant levels of desulfurization with the two resids evaluated,. Hence, dibenzothiophene is not a complete model for the resids studied.

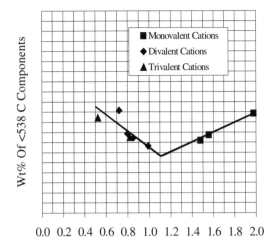

Figure 3.   Concentration of <538 C components in products via vacuum resid.

It is not known why the "V" shape correlation occurs since no major shift was noted in preferences between hydrocracking, alkylation and hydrogenation reactions in the transition and alkali metal regions.   One possible explanation may be that the molten monovalent catalysts tend to agglomerate as shown by computational chemistry modeling.   Di and trivalent catalysts do not agglomerate. More work in the agglomeration area is underway.

It was also found that conversion and product property results via the resids studied correlate with averaged charge difference.   And, the generalized correlation is again a "V" shape with the intersection near 1.1 ESU.   As an example, Fig. 3 shows the correlation obtained for the concentration of components boiling below 538 °C in products via the vacuum resid.

Also, the level of desulfurization obtained with the ROSE pitch forms a typical "V" shaped correlation with averaged charge difference(Figure 4).   This correlation shows that the dual Ni/Mo catalyst yielded essentially the same level of desulfurization as all single cation catalysts.

Conclusions

A procedure has been developed to correlate experimental results from resid hydroprocessing. This was accomplished with calculated charge differences between representative pure hydrocarbons and tetrachloro-aluminate catalysts.   This procedure can be used to optimize the tetrachloroaluminates needed for resids other than those used in this study.   It is expected that this procedure can also be used in the evaluation of other types of catalysts.

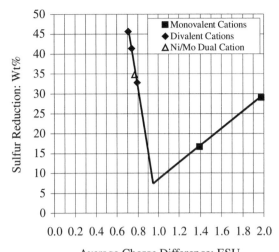

Figure 4.   Sulfur reduction via hydroprocessing ROSE pitch.

Acknowledgements

The author wishes to thank Drs. Chunshan Song, Masahide Sasaki and Boli Wei for supervising and performing the experimental work at The Pennsylvania State University.

References

HyperChem V5.1, HyperCube Inc

Plummer, M. A. and S. A. Plummer (1998).   Hydrocarbon adsorption in tetrachloroaluminate catalysts—Effects of catalyst cation, hydrocarbon polarity and charge transfer via molecular modeling. *Catalyst Today*, **43**, 327-338.

Saski, M., Song, C, and M. A. Plummer (2000).   Transition metal tetrachloroaluminate catalysts for probe reaction simulating petroleum resids upgrading. *Fuel*, **79(4)**, 295-303.

SIMPLE MODEL FOR INSERTION/DELETION ASYMMETRY OF FREE-ENERGY CALCULATIONS

Nandou Lu and David A. Kofke
Department of Chemical Engineering
State University of New York at Buffalo
Buffalo, NY 14260-4200

Abstract

Free-energy perturbation (FEP) calculations in molecular simulation are highly prone to inaccuracy. Moreover, there are two directions in which any FEP calculation can be applied, and the inaccuracies of the two tend to have opposite signs. However, the *magnitudes* of these inaccuracies do not show the same symmetry, and usually one direction gives a much more accurate result than the other. We apply a simple three-state model for the FEP calculation process to demonstrate this asymmetry and uncover the conditions that promote it. For typical FEP calculations it is shown that the smaller inaccuracy accompanies the FEP performed in the direction of decreasing entropy.

Keywords

Free-energy perturbation, Accuracy, Entropy, Widom insertion, Widom deletion.

Introduction

Free-energy calculations are important to the characterization of many interesting phenomena, including phase and reaction equilibria, binding affinity, solvation, thermodynamic stability, and so on. Some of the most widely used methods for computing free energies in molecular simulation are based on the free-energy perturbation (FEP) formalism. The FEP calculation involves two systems, the *reference* and the *target*. It can be performed in either of two directions, depending on which system is designated the reference. In the context of chemical potential calculations, these implementations are known as the Widom insertion (Widom, 1963; Jackson and Klein, 1964) and Widom deletion (Shing and Gubbins, 1982) methods. It is well known that insertion and deletion, although equivalent mathematically, provide estimates of free energies that differ systematically from the correct value (Han, Cushman, and Diestler, 1990; Kofke and Cummings, 1997; Parsonage, 1996a, 1996b; Powles, Evans and Quirke, 1982). A simple example is seen in computing the chemical potential of the hard sphere (HS) system, where the deletion method *always* gives zero residual

chemical potential (defined as the chemical potential minus the ideal-gas value at the same density); except at zero density, this result is obviously wrong, and consistently underestimates the correct value. On the other hand, the insertion method yields an infinite chemical potential until sufficient sampling has been performed to result in a non-overlap insertion attempt. This clearly overestimates the correct value but, unlike the deletion calculation, with enough sampling it can converge to the correct result.

These features of the Widom method are present in other, more general FEP calculations. The complementary over/under-estimation of the two FEP directions is well known, but the fact that one direction is much more accurate than the other is not widely appreciated. Understanding the performance of the FEP calculation is important for the proper interpretation of the results and for choosing the FEP method that provides more reliable results. Some effort has been put toward understanding the asymmetry of insertion and deletion calculations (Parsonage, 1996a, 1996b; Powles, Evans, and Quirke, 1982) and to characterize quantitatively the reliability of results, considering both the precision and accuracy (Kofke and Cummings, 1998; Lu and Kofke, 1999). Nevertheless, the accuracy of the FEP calculation is a very important issue that is still not well studied, is often overlooked, and is routinely misunderstood.

We use the energy distribution formalism to investigate the inaccuracy of FEP calculations. We define the inaccuracy of a calculation having a discrete energy distribution function via a maximum likelihood analysis. We then look at the behavior of the insertion and deletion calculations using a simple 3-state model.

Generalized Insertion and Deletion

The reference and target systems in a FEP calculation typically have different values of entropy. Our previous studies (Kofke and Cummings, 1998; Lu and Kofke, 1999), showed that the entropy difference between two systems is important in determining both the accuracy and precision of free-energy calculations. Therefore it is convenient to identify the perturbation systems by their entropies: the higher entropy system is denoted H and the lower one is L. We define the difference of a quantity between two systems as $(L) - (H)$. Thus the free energy, internal energy, and entropy difference are defined $\Delta A = A_L - A_H$, $\Delta U = U_L - U_H$, and $\Delta S = S_L - S_H$. We adopt units such that Boltzmann's constant and the temperature are unity, so $kT = 1$ and $A = U - S$.

Additionally, we take the configurational potential energy $u(\mathbf{r}^N)$ as the distinguishing feature of the reference and target systems, and we designate the difference between them for any configuration simply as u.

$$u \equiv \Delta u = u_L(\mathbf{r}^N) - u_H(\mathbf{r}^N) \qquad (1)$$

In the Widom chemical potential methods, for example, u describes the energy change associated with the transformation of one of the N molecules from a noninteracting ideal-gas molecule (u_H) to a fully interacting "real" molecule (u_L).

We can use either the H or L system as our reference in the FEP calculation. Corresponding to the Widom chemical potential calculation methods, we define the generalized insertion and generalized deletion directions according to the sign of entropy difference along the perturbation direction. If in a calculation the H system is used as the reference, so the entropy difference along the perturbation direction $S(\text{target})-S(\text{reference})$ is negative, we call it a generalized insertion calculation. Similarly, a generalized deletion calculation uses the L system as the reference, and proceeds in a direction of increasing entropy.

The free-energy difference in a generalized insertion calculation is given by the ensemble average of the energy-change Boltzmann factor sampled according to the distribution on the H system

$$\exp(-\Delta A) = \left\langle \exp(-u) \right\rangle_H \qquad (2)$$

The subscripted angle brackets $\langle...\rangle_H$ indicate the canonical-ensemble average at system H. Similarly, in a generalized deletion calculation, the free-energy difference is given by

$$\exp(+\Delta A) = \left\langle \exp(+u) \right\rangle_L \qquad (3)$$

Energy Distribution Functions

The ensemble averages in Eqns. (2) and (3) can be rewritten in terms of a one dimensional integral weighted by the distribution of energy changes encountered in a simulation. This analysis was first detailed by Shing and Gubbins (1982). Following their notation, we define $f(u)$ as the distribution function of the potential energy change u encountered in a simulation using insertion method. Then Eqn. (2) becomes

$$\exp[-\Delta A] = \int du \exp(-u) f(u) \qquad (4)$$

(we leave implied the integration limits of $(-\infty, +\infty)$) Likewise, the deletion free-energy average is given by

$$\exp[+\Delta A] = \int du \exp(+u) g(u) \qquad (5)$$

with $g(u)$ the distribution function of u encountered in a simulation performing the deletion calculation. Note that both $f(u)$ and $g(u)$ are normalized, and are related (Allen, 1996)

$$f(u) \exp(\Delta A) = g(u) \exp(u) \qquad (6)$$

Also, for the chemical potential calculation it is true that

$$\Delta U = \int u g(u) du \qquad (7)$$

but in general this equation does not hold. However, if it does then specification of f or g is sufficient to evaluate ΔA, ΔU, and thus ΔS.

Accuracy

The accuracy of a FEP calculation depends on how well the important configurations are sampled. For an insertion calculation, the important configurations are those in which u is close to zero or is negative, while for a deletion calculation, the high energy (large u) region contributes most to the Boltzmann average given by Eqn. (3). The barriers to sampling the low-energy region in an insertion method are *entropic* in origin; the high-energy region is not sampled by a deletion calculation due to *energetic* barriers (Kofke and Cummings, 1998). The inaccuracy of the FEP calculation can be quantified by focusing on the inadequate sampling of the important region. In this study we are going to demonstrate an analysis of inaccuracy using a maximum likelihood approach for a discrete distribution model. Methodology for the continuous distribution is somewhat different, although the basic idea is similar; this extension will be discussed in detail in a future publication.

We consider a FEP calculation in which only a finite number n values of energy changes are possible. In the following, we will use the term "energy state" to mean configurations in which the energy difference u has a particular value. The distribution functions for the discrete energy states can be obtained in principle by an infinite length simulation. We denote them as f_i and g_i for $i=1,...,n$. Then the exact free-energy difference by insertion and deletion methods can be computed according to a discrete form of Eqn. (2) and Eqn. (3),

$$\exp\left(-\Delta A\right) = \sum_{i=1}^{n} f_i \exp\left(-u_i\right) \qquad (8)$$

$$\exp\left(+\Delta A\right) = \sum_{i=1}^{n} g_i \exp\left(+u_i\right) \qquad (9)$$

Suppose in a simulation M independent trial free-energy perturbations are performed. We will observe a certain number of trials, m_i, encountered for energy state u_i. For each simulation we record a set $\{m_1, m_2, ..., m_n\}$ which satisfies $M = \sum m_i$. The discrete energy distribution f_i or g_i can be estimated by m_i/M for $i=1,...,n$. The free-energy difference then could be computed using Eqn. (8) or (9) with the estimated f_i or g_i.

Separate simulations each of length M would yield different sets $\{m_i\}$. The probability of observing a specific set in any given simulation is

$$\Omega\left(\{m_i\}\right) = M! \prod p_i^{m_i} / m_i! \qquad (10)$$

where p_i is either f_i or g_i (*i.e.*, the true distributions) Among all possible sets of $\{m_i\}$, there is one that corresponds to the maximum Ω. This set, denoted $\{m_i^*\}$, is the most likely outcome of any single simulation. Ignoring the possibility that several sets $\{m_i\}$ could yield the same observed free-energy difference ΔA, we take the most likely value of ΔA observed in a single simulation to be that obtained from the most-likely distribution of energies; accordingly we designate this value ΔA^*.

We now define the inaccuracy in ΔA as the difference between the most-likely value and the true value, *i.e.*, $\Delta A^* - \Delta A$. This definition of the inaccuracy allows us to proceed with some quantitative analysis of the insertion and deletion calculations. In practice, a simulation yielding a distribution of states $\{m_i\}$ would be analyzed using those results to put confidence limits on the simulation averages. Thus we consider also in our analysis the most-likely confidence limits obtained from a single simulation, computed here as one would extract them from the most-likely distribution of states $\{m_i^*\}$.

Three-State Model

The simplest discrete model is exemplified by the chemical-potential measurement of the hard sphere model, which has only two energy states u: 0 and ∞. The simplicity of this model makes very obvious the inadequacies of the deletion versus insertion FEP calculation as reviewed in the Introduction. No one attempting to measure the hard-sphere chemical potential would come to believe that the residual chemical potential is zero or infinity, as would be indicated by a poorly applied FEP calculation.

We wish to show how only a slightly more complex FEP calculation can lead to results that are very inaccurate but nevertheless believable. Thus we turn to a 3-state model. The square-well model at high dilution (each sphere with at most one neighbor) or at high density (each sphere surrounded by its full coordination of neighbors) would be an example of the 3-state model. A 3-state model is specified by the three available energy states, referred as u_0, u_1, and u_2. For convenience of notation, let $u_0 < u_1 < u_2$.

To look at the inaccuracy in a 3-state FEP calculation, we need knowledge of the exact energy distribution or the true free-energy and entropy difference *a priori*. It is more convenient to specify the free-energy difference between two systems, as well as the entropy difference to enable us to ensure the calculation direction, e.g., $\Delta S < 0$ indicates an insertion calculation. So we preset the values of ΔA, ΔS, as well as the energy states u_0, u_1, and u_2. Using Eqns. (6) to (10) and normalization of f and g, we can easily obtain $\{g_i\}$ and $\{f_i\}$ for $i=0$, 1, and 2. For a simulation with length M, the most-likely set of $\{m_i\}$, and therefore the inaccuracy and most-likely confidence limits, can be computed according to the method described in the preceding section.

Results and Discussion

We performed tests varying ΔA, ΔS, and energies u_i. Note that ΔA and ΔS values should be selected to ensure well-defined distribution functions, i.e., $0<f_i,g_i<1$. For convenience we chose $u_0 = 0$. We do not lose generality with this choice, as it is easy to show that a model with free-energy difference ΔA and energy states u_0, u_1, and u_2 has the exact same behavior of the model with free-energy difference $(\Delta A - u_0)$, energy states 0, $(u_1 - u_0)$, and $(u_2 - u_0)$.

Figure 1 shows the inaccuracy and the most-likely confidence limits as a function of simulation length M for both the insertion and deletion calculations. As expected, the deletion inaccuracy is always negative, and the insertion inaccuracy is positive.

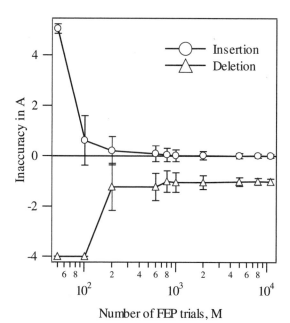

Figure 1. Inaccuracy in the free-energy as computed by the insertion and deletion methods, with error bars indicating most-likely confidence limits. Model conditions are: $\Delta A=4.0$, $\Delta S=-3.95$, $u_1=8$, $u_2=20$, $f_0=0.0182$, $f_1=0.3412$, $f_2=0.6406$, $g_0=0.9937$, $g_1=0.00625$, $g_2=7.2 \times 10^{-8}$.

The figure clearly demonstrates the persistent inaccuracy of the deletion versus the insertion calculation. More damaging than the inaccuracy itself is the reasonableness of the most-likely error bars attached to the measurements. Results of this type do not raise suspicion, such as might arise if the error bars were exactly zero (as in the HS deletion calculation) or were very large (as in HS insertion).

The outcome shown in Fig. 1 was selected to make our point. Other types of inaccuracy behaviors are observed in the model, depending particularly on the selection of the energy states u_i. For larger u_1, the step toward zero in the deletion calculation is not observed in the scale of M shown. Conversely, for smaller u_1 and u_2, the inaccuracy can reach zero for M in this range, and in fact the deletion calculation may become superior to the insertion method. However, these choices for u_i do not correspond to the energy distributions typically encountered in FEP calculations (especially chemical potential calculations). Regardless, for no models that we studied did the deletion pathology seen in Fig. 1 arise in the insertion calculation.

As additional energy states are added to the model, the behavior of the inaccuracy becomes more subtle. In the continuum limit, of most interest practically, the deficiencies in a deletion-oriented FEP calculation would be very difficult to detect without an awareness of the issues we present here.

Conclusions

Free-energy perturbation calculations should always be performed in a way that has the high-entropy system serving as the reference, with the low-entropy system the target. Application of the standard error analysis can be misleading, because the estimated variance of the distribution is subject to the same inaccuracies as the simulation average. One need only examine the sign of the entropy change to know that something may be awry.

Acknowledgments

Acknowledgment is made to the Donors of the Petroleum Research Fund, administered by the American Chemical Society, for support of this research.

References

Allen, M.P. (1996). Simulation and Phase Diagrams. In K. Binder and G. Ciccotti (Eds.), *Proceedings of the Euroconference on Computer Simulation in Condensed Matter Physics and Chemistry*, **49**, 255-284.

Han, K.-K., J. H. Cushman, and D. J. Diestler (1990). A new computational approach to the chemical potential. *J. Chem. Phys.*, **93**, 5167-5171.

Jackson, J.L. and L. S. Klein (1964). The potential distribution method in equilibrium statistical mechanics. *Phys. Fluids*, **7**, 228-231.

Kofke, D.A., and P. T. Cummings (1997). Quantitative comparison and optimization of methods for calculating the chemical potential by molecular simulation. *Mol. Phys.*, **92**, 973-996.

Kofke, D.A., and P. T. Cummings (1998). Precision and accuracy of staged free-energy perturbation methods for computing the chemical potential by molecular simulation. *Fluid Phase Equil.*, **151**, 41-49.

Lu, N., D. A. Kofke (1999). Optimal intermediates in staged free-energy calculations. *J. Chem. Phys.*, **111**, 4414-4423.

Shing, K.S., and K. E. Gubbins (1982). The chemical potential in dense fluids and fluid mixtures via computer simulation. *Mol. Phys.*, **46**, 1109-1128.

Powles, J.G., W. A. B. Evans, and N. Quirke (1982). Non-destructive molecular dynamics calculation of the chemical potential of a fluid. *Mol. Phys.*, **46**, 1347-1370.

Parsonage, N. G. (1996a). Computation of the chemical potential in high-density fluids by a Monte Carlo method. *Mol. Phys.*, **89**, 1133-1144.

Parsonage, N. G (1996b). Chemical potential paradox in molecular simulation. Explanation and Monte Carlo results for the Lennard-Jones fluid *J. Chem. Soc., Faraday Trans.*, **92**, 1129-1134.

Widom, B. (1963). Some topics in the theory of fluids. *J. Chem. Phys.*, **39**, 2808-2812.

MOLECULAR SIMULATIONS FOR ACETONITRILE AND METHANOL BASED ON *AB INITIO* PLUS POLARIZABLE POTENTIAL

Amadeu K. Sum and Stanley I. Sandler
Center for Molecular and Engineering Thermodynamics
Department of Chemical Engineering, University of Delaware
Newark, DE 19716

Abstract

The phase behavior of acetonitrile and methanol are studied using Gibbs Ensemble Monte Carlo simulations with pair potentials derived from *ab initio* calculations and a polarizable model to account for multibody interactions. The properties of acetonitrile are well described by only pairwise interactions, with little contribution from multibody interactions, even though acetonitrile has a large polarizability. On the other hand, the pair potential of methanol is insufficient to accurately predict its phase behavior, and the incorporation of multibody effects with a polarizable model greatly improves the predictions of the vapor-liquid coexistence properties. The addition of polarization comes at a substantial increase in computational time, but it is required for a proper description of the intermolecular interactions of fluids such as methanol in a dense phase.

Keywords

Ab initio pair potential, Multibody interaction, Polarizable model, Acetonitrile, Methanol, Gibbs Ensemble Monte Carlo, Simulation, Vapor-liquid equilibrium.

Introduction

Traditionally, intermolecular potentials have been regressed to a variety of thermodynamic and structural data and then are able to reasonably describe the behavior of systems within the limits of their development. These "effective potentials" commonly include pairwise interactions only, which makes their development and use convenient. Effective potentials have been developed for a large variety of molecules, and these have been used in simulations studies. One of the problems associated with these effective potentials is the implicit accounting of nonpairwise interaction effects through the values of the potential parameters. Therefore, attempts to systematically correct for the deficiencies of effective potentials do not follow a clear strategy, as addition of multibody effects can not be properly distinguished from pairwise effects. Multibody effects are usually incorporated with a polarizable model where, to a first approximation, the induced dipole of one molecule generates an electric field that affects all other molecules in the system (Böttcher, 1973).

There are several examples in the literature in which multibody effects were added to effective potentials to improve and/or extend the range of applicability of properties predictions with molecular simulation methods. Most of studies focused on water and the widely used SCP, SCP/E, and TIP4P effective potentials (Sprik and Klein, 1988; Watanabe and Klein, 1989; Ahlström *et al.*, 1989; Bernardo *et al.*, 1994; Chialvo and Cummings, 1996, 1998; Kiyohara *et al.*, 1998). Improvements in properties prediction with the inclusion of multibody effects using a polarization model were only significant when the parameters for the pairwise portion of the potential were re-adjusted (Dang and Chang, 1997; Yoshii *et al.*, 1998).

One approach to remove the empiricism associated with the deficiencies of effective potentials, is to perform rigorous *ab initio* molecular simulations (Car and Parrinello, 1995), where the energy of the system is calculated at every step of the simulation without any approximation to the type of interactions. Although philosophically this is a very attractive method, it is limited by the type and size of systems that can be studied with present computational resources.

Another route to produce more accurate property predictions that is not as computationally intensive as fully *ab initio* simulations, is to obtain interaction potential from *ab initio* calculations. In this approach: i) one can choose the extent of interactions (pairwise or higher-order), ii) the interaction energies can be calculated with most quantum chemistry packages separate from the simulation, iii) the potential energy surface can then be modeled with reasonably simple functions, iv) the potentials developed can be used with conventional molecular simulations techniques, and v) systematic corrections can be made based on the physics of the intermolecular interactions. This is the subject we address here. We have studied the adequacy of pair intermolecular potentials derived from *ab initio* calculations, for the prediction of vapor-liquid coexistence properties using the Gibbs Ensemble Monte Carlo technique. Based on the simulation results obtained considering only pairwise interactions, we added a polarizable contribution to account for multibody interactions. By following this approach, we not only remove the empiricism associated with potential model development and properties predictions, but we also demonstrate that greater insight can be obtained by determining what are the important intermolecular interactions among different classes of molecules.

Model Description and Simulation Details

Recently, Bukowski *et al.* (1999) performed a comprehensive study of the potential energy surfaces for several molecular pairs using symmetry-adapted perturbation theory (Jeriorski and Szalewicz, 1998), two of which are of particular interest to our study, acetonitrile and methanol. In addition to calculating the interaction energies, Bukowski *et al.* fitted the energies to a site-site potential expression suitable for simulation. In order to represent the *ab initio* energies as accurately as possible, from the minimum energy configuration to highly repulsive states, their potential expression was composed of separate components for the electrostatic, dispersion, and induction contributions to the total interaction energy. Consequently, the potential for acetonitrile was fitted with 100 parameters and that for methanol with 181. We used these potential models for the pairwise interactions of acetonitrile and methanol (see Bukowski *et al.* (1999) for the details and discussion of the potential).

Multibody interactions were added using a polarizable model accounting for the induced dipole interactions among the molecules in the system. Our model had the following form for the polarization energy (Böttcher, 1973),

$$U_{pol} = -\tfrac{1}{2}\sum_i \mu_i \cdot E_i^{\circ} \tag{1}$$

where E_i° is the electric field at site i generated by the fixed charges (q_j) in the system,

$$E_i^{\circ} = \sum_{j \neq i} \frac{q_j r_{ij}}{r_{ij}^3} \tag{2}$$

and if the electric field is not large, the induced dipole μ_i at site i, can be expressed as a linear response to the field,

$$\mu_i = \alpha E_i \tag{3}$$

$$E_i = E_i^{\circ} + \sum_{j \neq i} \tilde{T}_{ij} \cdot \mu_{ij} \tag{4}$$

where α is the polarizability, E_i is the total electric field, $\tilde{T}_{ij}$ is the dipole tensor,

$$\tilde{T}_{ij} = \frac{1}{r_{ij}^3}\left(\frac{3 r_{ij} r_{ij}}{r_{ij}^2} - 1 \right) \tag{5}$$

$r_{ij} = r_i - r_j$ is the separation distance vector, and r_{ij} is the distance between the sites i and j. An iterative procedure was adopted to obtain the induced dipoles required in the calculation of the polarization energy. The induced dipoles were solved self-consistently by satisfying Eqns. (3) and (4) to a specified convergence in the polarization energy (10^{-6} between successive iterations). The initial guess for the induced dipoles was taken as the converged set of the previous Monte Carlo step; the induced dipoles were also used in the calculations of the virial pressure of the system. Only those molecules within a specified radius (one-half the box length) were considered in the polarization energy calculations.

The only additional parameter required for the simulations with the polarizable model was the molecular polarizability. We assumed isotropic polarizability for both acetonitrile and methanol, and we used the experimental values of 4.48 and 3.29 Å³, respectively, as reported in Lide (1995).

Because the pair potential models are true two-body potentials derived from *ab initio* calculations, they include the polarization of any two molecules. Therefore, in order to avoid double counting of the polarization effect due to the two-body interactions, the sum of the polarization energies of all molecular pairs was subtracted from the total polarization energy.

The inclusion of polarizabilities added substantially to the computational cost of the simulations compared to that when only pair interactions were considered, increasing the simulation times by a factor of about 5-10. An additional problem with simulations involving the polarizable model is what is known as polarization catastrophe (Böttcher, 1973; Ahlström et al., 1989), an unphysical phenomenon that occurs at short distances due to the point charge and point polarizability description not present in real molecules. This phenomenon occurs at an intermolecular distance of $r_{ij} = \sqrt[6]{4\alpha^2}$. This distance is 2.08 and 1.87 Å for acetonitrile and methanol, respectively. At distances smaller than the polarization catastrophe, the induced dipoles of the molecules in close proximity diverge and make it impossible to find a proper set of induced dipoles for the molecules in the system. This problem is especially pronounced for acetonitrile and methanol, as their polarizability is relatively large compared to other more common molecules for which most polarization studies have been done (e.g., water that has a polarizability of 1.44 Å^3 and a polarization catastrophe distance of 1.42 Å). In fact, the polarization catastrophe has not been greatly discussed in the literature. The problem has generally been avoided by either introducing damping functions at small distances (Stillinger and David, 1978; Thole, 1981; Bernardo et al., 1994) or by treating the charged centers with a charge distribution instead of simple point charges (Sprik and Klein, 1988; Chialvo and Cummings, 1998). Our approach to this problem was simply to set a lower cut-off radius slightly greater than the polarization catastrophe radius so that this phenomenon was eliminated during the computations. The lower cut-off for the polarization calculations was set to 2.4 Å for both acetonitrile and methanol. We choose this solution to the problem based on the radial distribution functions for the case without any polarizability; there, the distance of closest approach of any two molecules was always greater than the cut-off.

The vapor-liquid phase behavior of acetonitrile and methanol was computed using Gibbs Ensemble Monte Carlo (Panagiatopoulos, 1987) simulations with the Bukowski et al. (1999) pair potentials for acetonitrile and methanol, both with and without the polarizable contribution. The simulations were performed in the *NVT* ensemble with 512 rigid molecules, periodic boundary conditions, and long-range corrections to the energy and pressure. The cut-off radius was set to one-half of the box length. Simulations were performed at several temperatures, and for each state point, the system was initially equilibrated, followed by production runs in which the energy, pressure, particle distribution, and chemical potential were block averaged. The chemical

potential was calculated as a check to ensure equilibration of the system. No other long-range corrections were applied. In addition, the critical properties were estimated from the simulated equilibrium vapor-liquid points by using the renormalization group theory scaling law and the law of rectilinear diameters.

Results and Discussion

The GEMC simulation results for acetonitrile and methanol with the Bukowski et al. (1999) potential without the polarization contribution have been published elsewhere (Hloucha et al., 2000). All the results presented here for the simulations with the polarizable model are at this time preliminary, and final results will be presented at the meeting.

The phase behavior of acetonitrile and methanol from our simulations with and without the polarization contribution are shown in Figs. 1 and 2, respectively, and compared with experimental data. Table 1 lists the estimated critical properties for the different models.

The simulation results in Fig. 1 show that pairwise interactions are sufficient to model acetonitrile both as a vapor and as a dense fluid. The simulations with the polarizable model indicated that the polarization energy is only a small fraction of the total energy, thus suggesting multibody effects are not significant to describe the interaction of acetonitrile molecules. On the other hand, multibody effects are essential to describe the phase behavior of methanol, especially as a dense fluid. The polarization energy for methanol in the dense phase accounted for over 10% of the total energy, and as seen from Fig. 2, this contribution has a dramatic effect in the prediction of the coexistence properties.

Table 1.   Estimated Critical Properties.

System	Model	T_c (K)	ρ_c (g/cm^3)
Acetonitrile	experimental	547.3	0.237
	w/o polariz.	558.5	0.276
	w/ polariz.	(521.2)	(0.267)
Methanol	experimental	512.7	0.278
	w/o polariz.	379.9	0.315
	w/ polariz.	(441.4)	(0.238)

Conclusions

Intermolecular potentials derived from *ab initio* calculations can provide a clearer understanding of the important contributions to the interactions between different type of molecules. By identifying the deficiencies associated with pairwise interactions, corrections for multibody effects can be implemented in the form of a polarizable model with point charges and point polarizabilities to a first approximation. Our GEMC

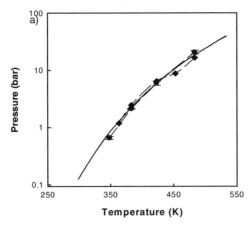

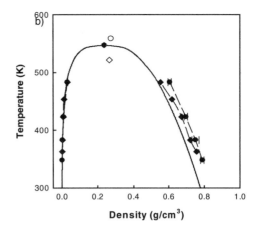

Figure 1.   Phase diagram for acetonitrile: vapor pressure (a) and vapor-liquid coexistence curve (b). Solid line: experimental data; diamonds and circles: GEMC simulations w/ and w/o polarization, respectively; open symbols: estimated critical point.

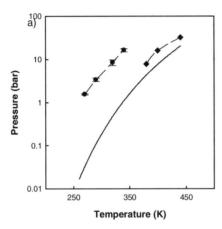

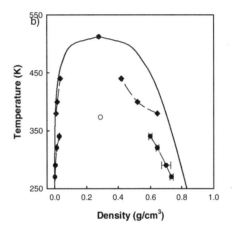

Figure 2.   Phase diagram for methanol: vapor pressure (a) and vapor-liquid coexistence curve (b). Solid line: experimental data; diamonds and circles: GEMC simulations w/ and w/o polarization, respectively; open symbols: estimated critical point.

simulations with and without the polarizable contribution, showed that acetonitrile is well described by pairwise interactions both as a vapor and in a dense phase. Methanol, on the other hand, because of its ability to hydrogen bond, can not be simply modeled with pairwise interactions, and multibody effects play a significant role in properly describing its behavior, especially as a dense fluid.

Acknowledgments

We thank the valuable discussions with Dr. R. Bukowski and Prof. K. Szalewicz with regard to the polarizable model and their *ab initio* data. We would also like to thank Dr. M. Hloucha for the initial development of the GEMC code, and the financial support by the National Science Foundation (CTS-9521406) and the Department of Energy (DE-FG02-85ER13436).

References

Ahlström, P., A. Wallqvist, S. Engström, and B. Jönsson (1989). A molecular dynamics study of polarizable water. *Molec. Phys.*, **68**, 563-581.

Bernardo, D. N., Y. Ding, K. Krogh-Jespersen, and R. M. Levy (1994). An anisotropic polarizable water model: incorporation of all-atom polarizabilities into molecular mechanics force fields. *J. Phys. Chem.*, **98**, 4180-4187.

Böttcher, C. J. F. (1973). Theory of Electric Polarization. 2nd ed., Elsevier, Amsterdam.

Bukowski, R., K. Szalewicz, and C. F. Chabalowski (1999). Ab Initio interaction potentials for simulations of dimethylnitramine solutions in supercritical carbon dioxide and cosolvents. *J. Phys. Chem. A*, **103**, 7322-7340.

Car, R. and M. Parrinello (1985). Unified approach for molecular-dynamics and density-functional theory. *Phys. Rev. Lett.*, **55**, 2471-2474.

Chialvo, A. A., and P. T. Cummings (1996). Engineering a simple polarizable model for the molecular simulation of water applicable over wide ranges of state conditions. *J. Chem. Phys.*, **105**, 8274-8281.

Chialvo, A. A., and P. T. Cummings (1998). Simple transferable intermolecular potential for the molecular simulation of water over wide ranges of state conditions. *Fluid Phase Equil.*, **150-151**, 73-81.

Dang, L. X., and T.-M. Chang (1997). Molecular dynamics study of water clusters, liquid, and liquid-vapor interface of water with many-body potentials. *J. Chem. Phys.*, **106**, 8149-8159.

Hloucha, M., A. K. Sum, and S. I. Sandler (2000). Computer simulation of acetonitrile and methanol with ab initio-based pair potentials. Submitted to *J. Chem. Phys*.

Jeziorski, B., and K. Szalewicz (1998). Intermolecular Interactions by Perturbation Theory, In P. von R. Schleyer (Ed.), Encyclopedia of Computational Chemistry, Wiley, Chichester.

Kiyohara, K., K. E. Gubbins, and A. Z. Panagiotopoulos (1998). Phase coexistence properties of polarizable water Models. *Molec.Phys.*, **94**, 803-808.

Lide, D. R. (Ed.) (1995). *CRC Handbook of Chemistry and Physics*. 76th ed., CRC Press, Boca Raton, FL.

Panagiotopoulos, A. Z. (1987). Direct determination of phase coexistence properties of fluids by Monte-Carlo simulation in a new ensemble. *Mol. Phys.*, **61**, 813-826.

Sprik, M., and M. L. Klein (1988). A polarizable model for water using distributed charge sites. *J. Chem. Phys.*, **89**, 7556-7560.

Stillinger, F. H., and C. W. David (1978). Polarization model for water and its ionic dissociation products. *J. Chem. Phys.*, **69**, 1473-1484.

Thole, B. T. (1981). Molecular polarizabilities calculated with a modified dipole interaction. *Chem. Phys.*, **59**, 341-350.

Watanabe, K., and M. L. Klein (1989). Effective pair potentials and the properties of water. *Chem. Phys.*, **131**, 157-167.

Yoshii, N., H. Yoshie, S. Miura, and S. Okazaki (1998). A molecular dynamics study of sub- and supercritical water using a polarizable potential model. *J. Chem. Phys.*, **109**, 4873-4884.

BRIDGING THE GAP OF MULTIPLE SCALES: FROM MICROSCOPIC, TO MESOSCOPIC, TO MACROSCOPIC MODELS

S. Raimondeau[1], P. Aghalayam[1], M. A. Katsoulakis[2], and D. G. Vlachos[1]

[1]Department of Chemical Engineering, University of Massachusetts

[2]Department of Mathematics and Statistics, University of Massachusetts
Amherst, MA 01003

Abstract

Multiscale models for homogeneous-heterogeneous processes are proposed, using domain decomposition, to capture surface spatiotemporal inhomogeneities by integrating density functional theory with the semi-empirical bond-order conservation method, molecular simulations for the interface, and reactor scale models for the fluid phase. Mesoscopic theories, suitable for larger length and time scales than molecular simulations, are also discussed. As examples, the CO oxidation on Pt and the diffusion of interacting molecules through a membrane are illustrated.

Keywords

Multiscale, Mesoscopic, Monte Carlo, Gradient, Bond order conservation, CO oxidation, Pt, Membranes.

Introduction

Homogeneous-heterogeneous (HH) reaction systems entail a large number of reactors ranging from catalytic to membrane, to deposition, to electrochemical. In such processes, a fluid surrounds a solid surface on which the desirable product usually forms. The disparity in length and time scales encountered in these processes is tremendous. Phenomena occurring at the fluid-surface interface at quantum and molecular length scales, ranging from Å to several nanometers, render classical continuum equations (CE), often referred to as mean field equations, inapplicable. Examples of such phenomena include intermolecular forces, nucleation, spinodal decomposition, and quantum reaction effects. While a first-principles simulation of these phenomena could significantly impact product quality and reactor performance, application of quantum calculations (QC) and molecular simulations, such as Monte Carlo (MC) or molecular dynamics (MD), to the entire spectrum of scales is impractical.

In this paper, we describe multiscale integration hybrid (MIH) simulations for such systems using a domain decomposition approach (Vlachos, 1997). The CO oxidation on a Pt surface embedded in a CSTR and a mesoscopic theory for diffusion through a membrane are presented as illustrative examples of the overall approach.

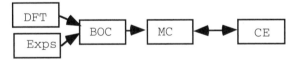

Figure 1. Schematic of multiscale modeling.

Multiscale Algorithms for HH Systems

A flowchart of the proposed methodology for multiscale modeling of HH processes is shown in Fig. 1. The conservation CE for the fluid phase are mature and are not further discussed here. Since QC, such as density functional theory (DFT) (Andzelm and Wimmer, 1991), are computationally very intensive for heterogeneous problems due to the large number of orbitals involved, their application to an entire large surface reaction network is not currently feasible. Semi-empirical techniques can instead be used with input taken either from suitable experiments or DFT. Given that the range of interaction between surface species is often relative short, a limited number of distinct spatial arrangements (nano-configurations) of reactive species can be encountered, termed as classes. Each class of a reaction event has a unique transition probability per unit time. Consequently, a database of kinetic parameters describing all classes is created that is fed to a lattice MC simulation. Mathematically, the QC/MC coupling is one-way, an assumption that can be rationalized because quantum mechanical effects occur at much shorter time scales compared to molecular (rare) events, so that the former can be taken in quasi-steady state.

One such semi-empirical technique developed for catalytic reactions is the unity bond index, quadratic exponential potential (UBI-QEP) or bond-order conservation (BOC) of Shustorovich and co-workers (Shustorovich and Sellers, 1998). The input to this approach is the heats of chemisorption of the species participating in the mechanism, and the output is activation energies with a typical accuracy of 2-4 kcal/mol. Empirical corrections to the predicted activation energies can also be done (Hansen and Neurock, 1999). The maximum number of DFT input parameters to the BOC formalism is equal to the number of surface species rather than the number of surface reactions that is typically much larger. As an example, for the H_2 oxidation on Pt, there are only four species compared to eighteen steps in the reaction mechanism (Park et al., 1999). While the BOC method was used to identify the rate limiting step in several catalytic mechanisms and explain experimental trends, its use in reactor scale simulations has not been done till recently (Park et al., 1999). In the lack of other information, transition state theory (TST) for generic surface reaction types can be employed to estimate pre-exponentials, which in conjunction with the energetics of the BOC, provide an initial set of reaction rate constants for each local arrangement (class) of adsorbates. Refinement of such reaction rate constants is possible by comparison to suitable experimental data and use of appropriate optimization tools discussed in (Aghalayam et al., 2000b).

Lattice MC simulations are now routine for surface processes and diffusion in nanopores, and have extensively been used to elucidate the underlying phenomena under ultra high vacuum (UHV) conditions for which many surface science experiments are conducted. A lattice representation of sites is suitable for chemisorbed species and allows for execution of slower (rare) events rather than studying vibrations in the potential energy surface, the latter done using MD. These MC simulations solve directly a master equation as a Markov process and can be classified as null event and continuous time types of algorithms. In the former algorithms, there is a lack of exact correspondence between real time and MC events (a

problem which can be circumvented at the expense of frequently scanning the entire lattice ((Reese et al., 2000)). This is an impediment in multiscale modeling because rates of surface processes are needed for MC/CE coupling. In contrast, the latter algorithms involve only successful events and can reach longer time scales, an important consideration in MC/QE coupling (Vlachos, 1997). It is these latter algorithms we employ in our simulations.

In MC/CE coupling, fluxes of the CE are matched with the spatially averaged MC rates. This two-way coupling provides temporal information from molecular to macroscopic times. Steady-state solutions are obtained asymptotically from long integration of such time dependent, hybrid simulations. Alternatively, a parametrization of the surface boundary condition(s) used in Newton's method of the entire problem or an iterative steady-state solver can be employed to obtain steady-state solutions more economically (Vlachos, 1997). In all these algorithms, the MC serves as a subroutine for function evaluation needed in the boundary condition(s) of the reactor scale model.

A Catalytic Reactor Example: CO Oxidation on Pt

The reaction mechanism of CO oxidation on Pt is (* denotes a vacant site):

$$CO + * \leftrightarrow CO* \qquad (1)$$
$$O_2 + 2* \leftrightarrow 2O* \qquad (2)$$
$$CO* + O* \rightarrow CO_2 + *. \qquad (3)$$

Repulsive first-nearest neighbor interactions between O species and between CO species are assumed, which have been shown to strongly affect activation energies of mean field models (Aghalayam et al., 2000a). As an example, using the BOC method on a Pt(100) square lattice, there is a total of sixteen possible local nano-configurations and corresponding activation energies for a reaction event (Eqn. (3)). The activation energy varies significantly with neighboring environment, as illustrated in Fig. 2.

Figure 2. Schematic of three selected nano-configurations of adsorbates on a (100) surface. The number of first-nearest neighbors affects the activation energy of the reaction in a nonlinear manner. The activation energy for the central CO-O reactive pair (in bold) varies from left to right from 10.90 to 0.96, to 9.36 kcal/mol.

The resulting kinetic database is input in MC simulations. A similar database for surface diffusion can be built. The reaction rate and coverage of surface species from BOC/MCsimulations, shown in Fig. 3, are in qualitative agreement with UHV experiments (Zhdanov and Kasemo, 1994).

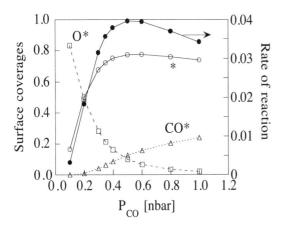

Figure 3. *Coverage of surface species and rate (ML/s) of CO oxidation vs. partial pressure of CO at 475 K and a partial pressure of O_2 of 2.66 nbar on Pt (100) using the BOC/MC model. The pre-exponentials were taken from (Zhdanov and Kasemo, 1994); heats of chemisorption on a clean surface are 34.9 and 51 kcal/mol for CO and O; the interactions were assumed to be 2 and 8 kcal/neighbor for CO-CO and O-O repulsion.*

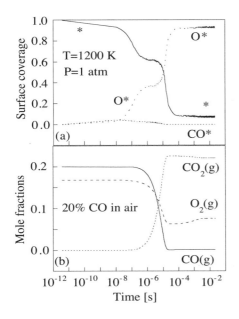

Figure 4. *Surface coverages and gaseous mole fractions during the transient in an isothermal CSTR for an inlet 20% CO in air starting with an empty surface at 1 ms residence time.*

Results from a 1-atm simulation of a Pt surface in a CSTR are shown in Fig. 4. This example entails all coupling routes depicted in Fig. 1. Extension to spatially distributed flows is straightforward and does not impose any additional problems. Transient information from such multiscale problems, that is inherently available, can be significant for on-line control of such reactors. Furthermore, phenomena such as the reconstruction of the 1x1 surface to hex and extension to polycrystalline and supported catalysts can be tackled with the proposed approach in future research. Application to deposition of nanophase materials and thin films is conceptually similar (Vlachos, 1999).

Mesoscopic Scale Modeling

Typical MC simulations presume that spatial heterogeneities in kinetic and transport parameters of surface species are short ranged. As a result, patterns (spatial inhomogeneities) developed are typically at the nanometer length scale, so that a MC simulation box is sufficient to statistically represent reaction rates seen from the fluid phase model. While the above multiscale algorithms perform well, spatiotemporal patterns on surfaces can occur over larger length scales, of the order of many microns (Ertl, 1994). This is typically the case when surface diffusion is fast and diffusion-reaction instabilities occur. A similar situation appears in modeling of diffusion through relatively thick membranes. Under such circumstances, MC simulations are not computationally affordable.

A coarse graining of the master equation on a MC lattice can be done (Hildebrand and Mikhailov, 1996), replacing the local occupation functions (which are either 1 or 0) over a relatively large length scale with a mesoscopic or local coverage by applying the law of large numbers. The resulting mesoscopic equation is a stochastic integro-differential equation and is exact in the limit of an infinite range of intermolecular forces. Such equations can further be scaled and, via asymptotics and homogenization, give macroscopic thermodynamic and transport laws for surface tension, mobility, critical nucleus size, and cluster velocity (Katsoulakis and Vlachos, 2000). Such equations can properly describe nucleation, spinodal decomposition, and uphill diffusion.

As an example, the mesoscopic equation for diffusion of interacting species on a lattice through an Arrhenius hoping mechanism is:

$$c_t - D\nabla \cdot \{e^{-\beta J * c}[\nabla c - \beta c(1-c)\nabla J * c]\} = 0, \quad (4)$$

where c is the concentration, t is the time, D is the diffusion constant, J is the intermolecular potential, β is the inverse of the product of the Boltzmann factor and temperature, and * denotes here convolution.

The solution of Eqn. (4) at steady-state obtained via finite difference discretization and Newton's method is compared with gradient MC simulations in 1D in Fig. 5. These gradient MC simulations are conducted in three control volumes, in analogy to the dual control volume, gradient MD simulations (Heffelfinger and van Swol, 1994). The boundary conditions for this particular simulation are semi-infinite domains with concentration of 1 and 0 at the left and right of the main domain, respectively. Using a pairwise constant potential of attractive interactions, we have found that when the range of interactions is about ≥ 10 in 1D, an asymptotic MC solution is approached (not shown). Using such a range of interactions, quantitative agreement in concentration profiles and fluxes is found for a wide range of temperatures (up to ~5% deviations in fluxes are seen, which are well within the uncertainty of gradient MC simulations), even within the spinodal curve ($\beta w \leq 2$). Such quantitative validation of mesoscopic theories, presented here for the first time, enables future simulations of permeation experiments over realistic length and time scales in a computationally tractable manner.

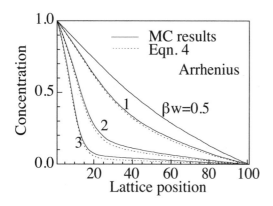

Figure 5. Concentration vs. distance (in lattice units) at steady-state from gradient MC simulations and the mesoscopic theory at various values of the dimensionless intermolecular potential strength βw. Excellent agreement is found when the range of interactions is moderately long.

Conclusions

Multiscale algorithms were presented for modeling of HH processes, which often exhibit spatiotemporal inhomogeneities on the surface. Such algorithms delineate the role of macroscopic, experimentally accessible parameters in microscopics and vice versa, i.e., the effect of molecular and/or quantum information on macroscopic quantities such as rates, ignition temperatures, and so on, that are of direct engineering interest. Finally, mesoscopic theories emerge as an invaluable tool for larger length and time scales, and their incorporation into large reactor or separation scale simulations will be worthy.

Acknowledgments

This work was supported in part by NSF DMS-9626904, DMS-9801769, CTS-9702615, and CTS-9904242.

References

Aghalayam, P., Y. K. Park, and D. G. Vlachos (2000a). A detailed surface reaction mechanism for CO oxidation on Pt. *Symposium (International) on Combustion,* **28**, accepted.

Aghalayam, P., Y. K. Park, and D. G. Vlachos (2000b). Optimization of detailed surface reaction mechanisms. *AIChE J.* (accepted).

Andzelm, J., and E. Wimmer (1991). Dgauss. A density functional method for molecular and electronic structure calculations in the 1990s. *Physica B: Condensed Matter,* **172**, 307-317.

Ertl, G. (1994). Reactions at well-defined surfaces. *Surf. Sci.,* **299/300**, 742-754.

Hansen, E. W., and M. Neurock (1999). Modeling surface kinetics with first-principles-based molecular simulation. *Chem. Eng. Sci.,* **54**, 3411-3421.

Heffelfinger, G. S., and F. van Swol (1994). Diffusion in Lennard-Jones fluids using dual control volume grand canonical molecular dynamics simulations (DCV-GCMD). *J. Chem. Phys.,* **100**, 7548-7552.

Hildebrand, M., and A. S. Mikhailov (1996). Mesoscopic modeling in the kinetic theory of adsorbates. *J. Phys. Chem.,* **100**, 19089-19101.

Katsoulakis, M. A., and D. G. Vlachos (2000). From microscopic interactions to macroscopic laws of cluster evolution. *Phys. Rev. Lett.,* **84**, 1511-1514.

Park, Y. K., P. Aghalayam, and D. G. Vlachos (1999). A generalized approach for predicting coverage-dependent reaction parameters of complex surface reactions: Application to H_2 oxidation over platinum. *J. Phys. Chem. A,* **103**, 8101-8107.

Reese, J., S. Raimondeau, and D. G. Vlachos (2000). Monte Carlo algorithms for complex surface reaction mechanisms. in preparation.

Shustorovich, E., and H. Sellers (1998). The UBI-QEP method: A practical theoretical approach to understanding chemistry on transition metal surfaces. *Surf. Sci. Reports,* **31**, 1-119.

Vlachos, D. G. (1997). Multiscale integration hybrid algorithms for homogeneous-heterogeneous reactors. *AIChE J.,* **43**, 3031-3041.

Vlachos, D. G. (1999). The role of macroscopic transport phenomena in film microstructure during epitaxial growth. *Appl. Phys. Lett.,* **74**, 2797-2799.

Zhdanov, V. P., and B. Kasemo (1994). Kinetic phase transitions in simple reactions on solid surfaces. *Surf. Sci. Reports,* **20**, 111-189.

MULTI-STEP POTENTIAL MODELING
OF METHANE BY DMD/TPT

J. Richard Elliott, Jr. and Jingyu Cui
Chemical Engineering Dept.
The University of Akron
Akron, OH 44325-3906

Abstract

Discontinuous potential models characterize molecular interactions as discrete increments in potential energy. The simplest example is the square-well potential, but multi-step potentials can be easily conceived to resemble continuous potentials like the Lennard-Jones potential. Adaptations based on the work of Wertheim permit characterization of hydrogen bonding interactions. The piecewise nature of the potential model facilitates application of theories like Thermodynamic Perturbation Theory (TPT). The subject of this paper is to demonstrate the accuracy and efficiency that can be achieved when molecular modeling is based on combining Discontinuous Molecular Dynamics (DMD) with second order TPT.   Recent results include force field characterization of n-alkanes, aromatics, alcohols, and water based on accurately matching experimental vapor pressures with those computed by isochoric integration. This paper focuses on the range of methodologies that can be brought to bear through DMD/TPT and their relative merits and limitations. These methodologies include: complete DMD simulation, DMD computation of reference fluid properties and perturbation terms, approximate analytical solutions of TPT, and multiproperty stepwise non-linear regression to characterize the step widths and depths. In this brief article, these methodologies are illustrated for spherical models of methane. The physical properties of vapor pressure, density, and internal energy are correlated to an accuracy of 0.86, 1.58, and 2.72% respectively with a two-step potential model. The optimal step widths are $\lambda=\{1.2, 1.5\}$ with the second well being 80% as deep as the first well.

Keywords

Discontinuous molecular dynamics, Physical properties, Molecular interactions, Thermodynamic perturbation theory.

Introduction

Suppose a theory existed that could predict the results of a molecular simulation to within the accuracy of the simulation in a fraction of the computation time, based purely on the specification of the intermolecular potential function. Noting the limitations of computation speed in performing molecular simulations, such a theory would be very valuable. The search for such a theory has thus far met with limited success, but there are some encouraging recent observations. A close analysis of the vapor pressure comparisons for square-well spheres by Elliott and Hu[1] shows that the predictions of Chang and Sandler's[2] second order Analytical Thermodynamic Perturbation Theory (ATPT2), are in fairly accurate agreement with discontinuous molecular dynamics (DMD)[3, 4] determinations by isochoric integration. Discrepancies between the isochoric integration results and the Gibbs Ensemble results of Vega et al.[5], can now be resolved by comparing to the histogram reweighting analysis of Orkoulas and Panagiotopoulos,[6] indicating that isochoric integration, ATPT2, and histogram reweighting are in close agreement. These observations suggest analyzing second order TPT (TPT2) more generally. In this context, we would like to consider both ATPT2, which has approximations, and TPT2 based on rigorous computation of all terms from DMD simulation. We test the possibility of applying TPT2 as a means of extrapolating and archiving molecular simulation results, at a minimum. More optimistically, it may be possible to apply TPT2 in lieu of molecular simulations under certain conditions. If that is possible, the trial and error process of parameterizing the force fields that form the foundations of molecular modeling can be expedited tremendously. Quantitatively analyzing these possibilities is the subject of this paper.

This paper is organized as follows. DMD simulation is applied to an appropriate reference system and perturbation contributions are compiled in accordance with their definitions.[7] We refer to the results of DMD computations of the TPT2 terms as DMD/TPT2 to distinguish from the ATPT2 of Chang and Sandler. Through comparison of isochoric integration vapor pressures to DMD/TPT2 and ATPT2 we show that DMD/TPT2 is in quantitative agreement with the isochoric integration results while ATPT2 deviates significantly at low temperatures. A comparison of the square-well model to experimental data for methane shows that the accuracy of the square-well model is limited. A particular advantage of Barker and Henderson's[7] formulation of TPT is that the potential can be "dissected" into multiple steps of varying width and depth with little additional computational effort. We treat the step widths and depths as additional force field parameters and analyze the impacts of various segments of the potential function on the modeled physical properties. The optimal potential determined in this way shows some resemblance to conventional potentials like the Lennard-Jones, but retains the advantages of a discontinuous potential.

Comparison of ATPT2 and DMD/TPT2

The essence of perturbation theory[7] is the assumption that the Helmholtz free energy can be decomposed into a reference contribution and an attractive contribution in the form of a power series in $\beta\varepsilon$, where $\beta\equiv1/kT$, $k \equiv$ Boltzmann's constant, and ε is the well-depth.

$$\frac{A^{att}}{NkT} = A1^{*}\beta\varepsilon + A2^{*}(\beta\varepsilon)^{2} + ... \quad (1)$$

For a single well, the $A1^{*}$ and $A2^{*}$ terms become simply:

$$A1^{*} = -\frac{1}{2N}\langle N_{W}\rangle = -\frac{\rho}{2}\int_{1}^{\lambda} g^{ref} 4\pi r^{2} dr \quad (2)$$

$$A2^{*} = -\frac{1}{4N}\left(\langle N_{W}^{2}\rangle - \langle N_{W}\rangle^{2}\right) \quad (3)$$

where N_{w} is the number of atoms within the well of the central atom, ρ is the number density and g^{ref} is the radial distribution function (rdf) of the reference fluid. Our first key emphasis is that the $A1^{*}$ and $A2^{*}$ terms are computed entirely from the reference fluid simulation. Noting that the number of neighboring sites goes as the range of the potential to the third power, and that the computation time is roughly proportional to the number of neighboring sites, the advantage of performing only the reference fluid simulation should be apparent. For example, a typical range for a Lennard-Jones potential is 2.5σ, but the reference fluid could have a range of only σ. The simulation of the reference fluid would require roughly $2.5^{-3} = 6\%$ of the time required for the simulation of the full potential. Note that the DMD method is inherently much faster than conventional MD.[3, 8] The DMD method essentially tracks and schedules collisions similar to the most advanced multiple time step method. So the combination of DMD/TPT has the potential to be two orders of magnitude faster than conventional MD.

Taking the appropriate derivatives shows that,

$$\frac{U - U^{ig}}{N\varepsilon} = A1^{*} + 2A2^{*}\beta\varepsilon + ... \quad (4)$$

$$Z - 1 = Z1^{*}\beta\varepsilon + Z2^{*}(\beta\varepsilon)^{2} + ... \quad (5)$$

where U is the internal energy, Z is the compressibility factor, $Z_{i}^{*} \equiv \eta(\partial A_{i}^{*}/\partial\eta)$, and η is the packing fraction. Our second key point is that TPT2 provides quantitative accuracy when applied at densities higher than $\eta = 0.28$, as evidenced by comparisons to simulations. A plot of internal energy vs. $\beta\varepsilon$ shows that both ATPT2 and DMD/TPT2 provide quantitative accuracy for internal energy at $\eta = 0.3$. Similar comparisons show that the internal energy varies linearly with $\beta\varepsilon$ at all $\eta > 0.28$, obviating the need for higher order perturbation terms. The challenge in applying TPT quantitatively is evaluating the derivative property, Z. A high degree of precision and consistency is required in computing both the $A1^{*}$ and $A2^{*}$. For example, it is not sufficient to estimate $A1^{*}$ from ATPT2 and $A2^{*}$ from DMD/TPT2, even though ATPT2 provides fairly accurate estimates of $A1^{*}$.

The observation that the DMD/TPT2 theory matches the simulations so accurately indicates that DMD/TPT2 can be used in place of the simulation, for $\eta > 0.28$ and at any temperature. In all cases that we have studied to date, this set of conditions encompasses any saturated liquid at a reduced temperature, $T_{r} < 0.9$ where $T_{r} \equiv T/T_{c}$ and T_{c} is the critical temperature. In fact, polyatomic molecules exist as liquids at even higher packing fractions, making the accuracy of DMD/TPT2 even more favorable. This temperature range is sufficient to compute phase equilibria at several temperatures of interest and extrapolate to the critical point through scaling law relations.[9] Since finite size effects necessitate the application of similar extrapolations in the case of molecular simulations,[6] it is fair to say that DMD/TPT2 can act as a substitute for molecular simulation over a broad range of conditions.

Turning our attention to ATPT2, we find that the accuracy is not quite satisfactory. ATPT2 applies the microscopic compressibility approximation in place of Eqn. (3). This degree of approximation is necessary to obtain their analytical solution.[2] The deficiency in Figure 1 may seem small, but it must be appreciated that the temperature where $Z \approx 0$ is closely related to the saturation temperature. Therefore, the deviation apparent in Figure 1 is actually quite significant. These deficiencies become more obvious when considering the saturation pressure in Figure 2. The logarithmic scale in Figure 2 must be noted to appreciate that the error in ATPT2 is roughly 50% for vapor pressure. This behavior can be easily understood in terms of Figure 3. Also shown are the Monte Carlo data of Smith et al.[10] ATPT2 wrongly characterizes $Z2^*$, given by the slope of $A2^*$. ATPT2 gives a significantly positive value for $Z2^*$, whereas Figure 3 shows that $Z2^*$ should be nearer to zero.

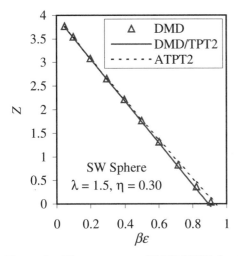

Figure 1. The accuracy of DMD/TPT2 for the compressibility factor.

Perturbation Theory for Multi-Step Potentials

There are two key problems with the square-well potential. First, it offers limited capacity for matching experimental data of physical properties. Second, Hu et al.[11] showed that a single well with $\lambda = 1.5$ cannot provide accurate characterization of the vapor pressure of n-alkanes in the context of a united atom model. Briefly, we feel that the model with a single well must be abandoned if we seek transferable united atom model potentials. Fortunately, Siepmann and coworkers[12] have shown that Lennard-Jones potentials exhibit favorable transferability. If we could use multiple steps to make our discontinuous potential resemble the

Lennard-Jones potential, we might hope to achieve similar transferability.

This observation begs the question of using the Lennard-Jones potential rather than any discontinuous potential. The Lennard-Jones potential has much to recommend it and perturbation theory could be adapted, but the identification of the reference fluid and characterizing the temperature dependence resulting from the softness of the Lennard-Jones potential lead to ambiguities that would complicate the issues of accuracy and consistency that we have encountered. To address this question in another way, a multistep potential with an infinite number of steps could clearly match the Lennard-Jones potential exactly. What is the justification of using an infinite number of steps if two or three would suffice?

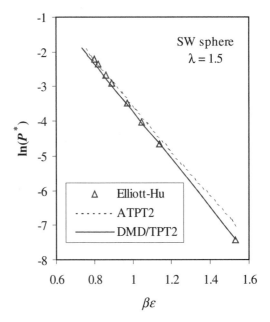

Figure 2. TPT comparisons for the vapor pressures of square-well spheres.

There are other reasons for considering multistep potentials. We have an interest in "dissecting" the potential function to understand the role played by each segment of the potential. For example, we have found that adding steps at $\lambda > 1.7$ is necessary for representing the behavior of n-alkanes, but results in significant errors for liquid density when applied to a spherical molecule like methane. We would like to gain a similar understanding of impacts in the critical region, and of transport properties. Another motivation is the development of specialized approximations for different segments. A mean field theory may be applicable to segments with $\lambda > 1.3$, whereas integral equations may provide high accuracy for $\lambda < 1.3$.[13] Having a potential model that accurately matches experimental properties, but offers the ability to be "dissected" in this manner opens the door to many possible developments.

To illustrate briefly, we present here the development of a potential model for methane. The procedure for n-alkanes, benzene, or alcohols is similar and will be reported separately. The saturation pressures, liquid densities, and liquid internal energies were taken from Setzmann and Wagner[14] for T = 90.6K to 180K. Our procedure is to simulate the reference fluid with shells in r-space of r/σ = 1, 1.05, 1.1, 1.2, 1.3, ..., 1.9, 2.0. The average number of atoms in each shell around each atom is computed at simulated time intervals of 10 ps with the simulation proceeding to 35 ns with the temperature initially set at 300K. These data are archived during the simulation, not averaged and discarded. The archived data can be regrouped according to the step potential during post-simulation analysis.

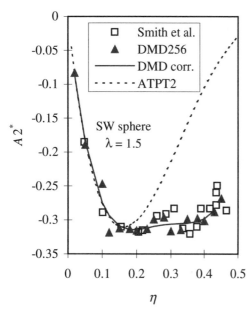

Figure 3. The second order perturbation term for square-well spheres.

For example, to construct the perturbation terms for a potential with λ = {1.2, 1.5}, the shells at 1.05, 1.1, 1.2 are combined at each time step, as are the shells at 1.3, 1.4, 1.5. Then $A1^*$ and $A2^*$ are computed from their definitions. Then the well depths are regressed from experimental data. All of this post-simulation analysis requires negligible computation time.

We proceeded with a stepwise forward regression using the square-well fluid with λ=1.5 as the starting point in all cases. Then we added steps to the potential and evaluated the optimal parameterization of each potential. We also computed the rms%err for pressure, density, and internal energy individually and summarized them as (errP, errD, errU) in the discussion below. The results are summarized in Table 1. The potential model with {λ} = {1.3, 1.5, 1.7, 2.0} has the lowest error (0.34, 0.74, 0.67), but the one with {λ} = {1.2, 1.5, 1.8, 2.0} is nearly as accurate (0.38, 0.78, 0.66). In both cases, the

errors in vapor pressure, density, and internal energy are less than 1%.

Conclusions

Three key points should be emphasized. First, molecular simulations at $\beta\varepsilon > 0$ and $\eta > 0.28$ may often be unnecessary because the physical properties representative of a given potential model are quantitatively determined by the DMD/TPT2 result. Second, a high degree of precision is required in applying TPT quantitatively, and ATPT2 does not appear to suffice. Third, generalized discontinuous potential models in the form of multi-step potentials provide the capacity for quantitative molecular modeling of experimental physical property data with at least 20 times greater computational efficiency than a simulation in which TPT2 is ignored. While DMD/TPT2 is not as efficient as ATPT2, it is directly applicable to non-spherical molecular models, as we will demonstrate in a separate communication.

Table 1. Stepwise Forward Regression of a Step Potential for Methane.

λ_1	λ_2	λ_3	λ_4	$\varepsilon_2/\varepsilon_1$	$\varepsilon_3/\varepsilon_1$	$\varepsilon_4/\varepsilon_1$	rmsErr
1.5							4.137
1.3	1.5			0.883			3.953
1.3	1.5	1.7		0.744	0.330		1.695
1.3	1.5	1.7	2.0	0.697	0.483	0.086	0.702
1.2	1.5			0.852			3.814
1.2	1.5	1.8		0.880	0.219		1.387
1.2	1.5	1.8	2.0	0.823	0.309	0.061	0.718

References

Barker, J. A. and D. Henderson (1967). *J. Chem. Phys.*, **47**, 2856.

Chang, J. and S. I. Sandler (1994). *Mol. Phys.*, **81**, 745.

Chapela, G. A., Martinez-Casas, S. E., and J. Alejandre (1984). *Mol. Phys.*, **53**, 139.

delRio, F. and L. Leonel (1987). *J. Chem. Phys.*, **87**, 7179.

Elliott Jr. and L. J. R. Hu (1999). *J. Chem. Phys.*, **110**, 3043.

Hu, L., Cui, J., Rangwalla, H., and J. R. Elliott Jr. (1999). *J. Chem. Phys.*, **111**, 1293.

Kofke, D. A. (1993). *J. Chem. Phys.*, **98**, 4149.

Liu, J.-X., Bowman, T. L., and J. R. Elliott Jr. (1994). *Ind. Eng. Chem. Res.*, **33**, 957.

Martin, M. G. and J. I. Siepmann (1999). *J. Phys. Chem. B.*, **103**, 4508.

Orkoulas, G. and A. Z. Panagiotopoulos (1999). *J. Chem. Phys.*, **110**, 1581.

Rapaport, D. C. (1980). *J. Comput. Phys.*, **34**, 184.

Setzmann, U. and W. Wagner (1991). *J. Phys. Chem. Ref. Data.*, **20**, 1061.

Smith, W. R., Henderson, D., and J. A. Barker (1970). *J. Chem. Phys.*. **53**, 508.

Vega, L., deMiguel, E., Rull, L. F., Jackson, G., and I. A. McLure (1992). *J. Chem. Phys.*, **96**, 2296.

LIQUID MIXTURE ACTIVITY COEFFICIENT PREDICTION VIA OSMOTIC MOLECULAR DYNAMICS

Paul S. Crozier and Richard L. Rowley
Department of Chemical Engineering
Brigham Young University
Provo, UT 84602

Abstract

The osmotic molecular dynamics (OMD) method was used to calculate activity coefficients in liquid mixtures for model non-polar and polar fluids. In so doing, the method was refined by implementing a rapid equilibration algorithm and by implementing a particle-particle/particle-mesh method (P^3M) to calculate long-range Coulombic interactions. OMD simulations were performed for model fluids representing two binary liquid mixtures: methanol/*n*-hexane, and *n*-pentane/*n*-hexane. The simulations are predictive in the sense that all model parameters were obtained from the literature and no adjustable parameters were used. Nevertheless, results agreed well with experimentally measured activity coefficient data for both binary mixtures. The OMD simulation method appears to be a viable method for the prediction of real liquid activity coefficients, limited primarily by the availability of computational resources and the accuracy of the intermolecular potential models.

Keywords

Methanol, *n*-hexane, *n*-pentane, Activity coefficients, Chemical potential, Phase equilibria, Osmosis, Molecular dynamics, Particle-particle/particle-mesh.

Introduction

Methods for indirect determination of activity coefficients have generally involved particle insertions. Activity coefficients at phase boundaries can be extracted in standard ways from Gibbs Ensemble Monte Carlo (Panagiotopoulos, 1987) simulations that rapidly produce equilibrium between two cells by particle transfers between them. Widom's particle insertion method (1963) can be used to obtain activity coefficients in single-phase systems. However, for complex, structured molecules at liquid densities, particle insertions become inefficient. A nice feature of osmotic molecular dynamics (OMD) simulations (Rowley, et al., 1994) is that activity coefficients can be obtained directly from the mechanical variables of the system as it equilibrates across a semi-permeable membrane. This allows activity coefficients to be determined at any composition, temperature, and

pressure (not limited to phase boundaries), but its reliance on the natural establishment of equilibrium through mass transfer makes its efficiency problematic. We describe here methods to lessen this efficiency problem, and we show how the new method is an effective way to accurately predict mixture activity coefficients in non-ideal mixtures.

Model Fluids

OMD simulations were performed on binary mixtures including molecular models for methanol, *n*-hexane, and *n*-pentane. The repulsion and dispersion potentials were represented in the models with united-atom (UA), pair-wise additive, site-site, LJ potentials located at all heavy-atom nuclei. Values for the LJ site parameters were obtained from the literature (Jorgensen, 1980; Jorgensen, et al., 1984). Lorenz-Berthelot combining rules were used to obtain values for all interactions between heterogeneous interaction sites. Additionally, point charges located at nuclear centers, with values obtained from the literature, were used to model the permanent dipole interactions for methanol (Jorgensen, 1980). Bond lengths and angles were fixed at their equilibrium positions.

The torsional potentials for *n*-hexane and *n*-pentane were modeled as a function of dihedral angle N with the Ryckaert-Bellemans cosine series (Ryckaert and Bellemans, 1975). All three dihedral angles in *n*-hexane, and both dihedral angles in *n*-pentane were assumed to be equivalent and given by the Ryckaert-Bellemans values.

OMD Simulations

OMD Thermodynamics

Rowley et al. (1995) showed that when osmotic equilibrium is established across a membrane that is perfectly permeable to solute molecules and perfectly impermeable to solvent molecules, the activity coefficient on the two sides of the membrane (*A* and *B*) are given by

$$\ln \gamma_A - \ln \gamma_B = \ln x_B - \ln x_A + \frac{1}{RT} \int_{P_A}^{P_B} \tilde{V} dP \qquad (1)$$

where x_A is the mole fraction of solvent on side *A* of the simulation cell, x_B is the mole fraction of solvent on side *B* of the simulation cell, P_A is the pressure on side *A*, P_B is the pressure on side *B*, and $\tilde{V}$ is the pure solvent molar volume as a function of pressure. All property averages are collected only after equilibrium has been established.

Fast Equilibration Method

Because mechanical equilibration is so much faster than mass transfer, we preemptively change the volume (and thus the pressure) on side *B* of the simulation cell once a net migration of solvent is detected. The net average velocity from side *A* to side *B* of all solvent molecules in the simulation is computed, and then net mass transfer is prohibited by rescaling *z* direction (perpendicular to the membrane) velocities of the solvent molecules. Thus, net migration of solvent across the membrane is artificially forced to zero. The natural chemical potential driving force producing the net migration is then eliminated by increasing or decreasing the side *B* volume in proportion to the observed mass transfer. The competing effects of osmosis and permeation are artificially forced to counterbalance while a natural balance is gradually achieved by changing the pressure difference across the membrane until the equilibrium osmotic pressure is obtained. In other words, chemical potential equilibrium across the membrane is achieved and net mass transfer is prohibited in the meanwhile.

In addition to dramatic reductions in equilibration time, this technique allows the user to hold compositions of constituent components (x_A and x_B) fairly constant throughout the duration of the simulation --- net flux of solvent through the membrane is forced to zero --- leaving the pressure difference across the membrane as the only variable affecting the prediction of the activity coefficient. It becomes clear that chemical potential equilibrium between sides *A* and *B* has been achieved once this pressure difference converges to a constant value (the osmotic pressure).

Adequate precision for the measured bulk properties required by OMD, including the flux measurements needed for the fast equilibration scheme, necessitates the simulation of a large number of molecules. For this work, we have used 100 parallel simulations of 1000 molecules each, where each simulation is a member of the same canonical ensemble (same NVT and geometry).

Force Calculations

As can be seen, OMD simulations require that a large amount of phase space be sampled. Even with optimization, parallelization, and reasonable levels of approximation, large amounts of CPU time are needed. Ideally, we would like to compute long-range coulombic interactions using a standard Ewald summation, but the computation cost is too high to be practical for this work. Instead, we use the particle-particle/particle-mesh (P^3M) algorithm that can be much faster than a standard Ewald summation, and still yield sufficient accuracy. The P^3M method is well described elsewhere; we have followed the formulas and recommendations of Deserno and Holm (1998a; 1998b).

Simulation Specifics

All simulations were started at low density, with individual molecules randomly oriented and distributed (at the desired mole fractions, x_A and x_B). The simulation cells were then gradually compressed to the desired liquid density according to a simple proportional control scheme. Once liquid density was achieved, the density control was turned off, and the pressure controller implemented. In both cases, the box length of the entire cell in the *x* and *y* directions, *L*, was varied to control the density, and then pressure of the bulk fluid on side *A*.

The *z* direction length of the simulation cell on side *A*, Lz_A, was fixed at 30 Å, while the *z* direction length of the simulation cell on side *B*, Lz_B, varied between 25 and 35 Å, and was the control variable that brought the chemical potential driving force of solvent molecules through the membrane to zero. Only the central one-third of side *A* and side *B* were used for property calculations. Outside this region, the membrane interactions distort the normal bulk fluid structure.

The cutoff distance for the real space force calculations was chosen to be 10 Å, with a mesh size of $16 \times 16 \times 8$ in the x, y, and z directions for the reciprocal space calculations.

A time step size of 2.0 fs was used throughout. Even with the fast equilibration method, 100,000 equilibration time steps were used to provide for local mass transfer, and to completely equilibrate the system. Subsequently, property averages were taken for at least 10,000 additional time steps.

Bond angles and bond lengths were constrained to their equilibrium values, using Gauss' principle of least constraint (differential feedback). Likewise, a molecular Gaussian thermostat was used to maintain isothermal conditions. Drift of bond lengths and angles due to numerical round off was corrected using a proportional feedback mechanism.

Activity Coefficient Parameter Regression

Once the OMD simulation was equilibrated, and the raw data subsequently collected, the activity coefficient of the solvent on side A ($\ln\gamma_A$) relative to the activity coefficient of the solvent on side B ($\ln\gamma_B$) was calculated using Eqn. (1). Each simulation yielded a difference between $\ln\gamma$ values at two different compositions. We performed ten OMD simulations for both binary mixtures of this study. The activity coefficients for both components along the entire composition range were then obtained from these differences by parameters in an activity coefficient model equation that best fit the OMD simulation data.

We chose to use the Wilson equation for this purpose because it is widely regarded as having a suitable functional form for activity coefficients of these types of mixtures. We used the method of Hirata et al. (1976) to regress the parameters by minimizing the objective function (OF):

$$OF = \sum_{i=1}^{n} (Q_{exp} - Q_{cal})_i^2 \qquad (2)$$

where

$$Q = x_1 \ln \gamma_1 + x_2 \ln \gamma_2 \qquad (3)$$

Reliability Tests

The magnitude of the OF is indicative of the error in the fit and the accuracy and thermodynamic consistency of the raw data obtained from the simulations. Also, each component's activity coefficient curve can be regressed independently in order to perform the integral test for thermodynamic consistency. Or, individual raw data points can be compared in pairs using the differential test for thermodynamic consistency.

Ideally, we would like to compare our results with other computer simulation results that use the same molecular models, but a different method of chemical potential determination. However, the scarcity of simulation-generated activity coefficient predictions for model binary mixtures of structured molecules in the literature makes such a comparison difficult, if not impossible. For the purposes of this work we simply compare our results with experimentally measured activity coefficients of the real chemicals.

Results and Conclusions

Figures 1 and 2 show OMD predicted and experimentally measured activity coefficients (Gmehling, et al., 1980; Gmehling and Onken, 1977) for the two systems investigated in this study. Component 2 is *n*-hexane in each case, and ln• is plotted versus liquid mole fraction of component 1 for each binary mixture. OMD predicted activity coefficients closely follow experimentally measured activity coefficients for both cases, and the prediction method clearly distinguishes between the nearly ideal *n*-pentane/*n*-hexane system and the highly non-ideal methanol/*n*-hexane system.

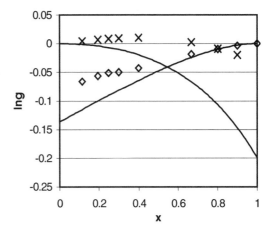

Figure 1. Experimental data (points) and OMD predicted curve (line) for n-pentane/n-hexane at 25° C.

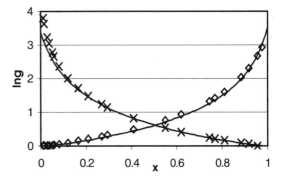

Figure 2. Experimental data (points) and OMD predicted curve (line) for methanol/ n-hexane at 35° C.

This is truly remarkable when one considers that no empirical data about the binary mixtures was used to perform this calculation; only the simple molecular models and the Lorenz-Berthelot combining rule.

Figures 3 and 4 show the OMD predicted and experimentally measured *x-y* diagrams for the *n*-pentane/*n*-hexane system and the methanol/*n*-hexane system respectively. Experimental vapor pressures were used along with the OMD predicted activity coefficients to produce the OMD *x-y* curves. Again, we see rather remarkable agreement between experiment and OMD prediction.

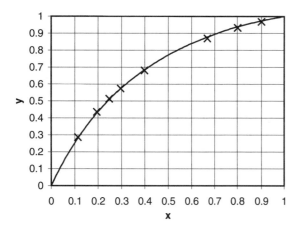

Figure 3. VLE experimental data (points) and OMD predicted curve (line) for n-pentane/n-hexane at 25° C.

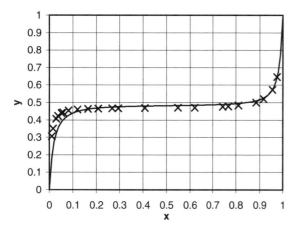

Figure 4. VLE experimental data (points) and OMD predicted curve (line) for methanol/n-hexane at 35° C.

Implementation of a fast equilibration method and a P³M algorithm enhances the efficiency of the OMD method, making it practical for calculation of activity coefficients in non-ideal mixtures. The improved OMD method was used here to predict activity coefficients in both a fairly ideal and a highly non-ideal system. Predicted results for both systems were excellent.

References

Deserno, M., and C. Holm (1998a). How to mesh up Ewald sums. I. A theoretical and numerical comparison of various particle mesh routines. *J. Chem. Phys.*, **109**, 7678-7693.

Deserno, M., and C. Holm (1998b). How to mesh up Ewald sums. II. An accurate error estimate for the particle-particle/particle-mesh algorithm. *J. Chem. Phys.*, **109**, 7694-7701.

Gmehling, J. and U. Onken (1977), Data Tables In., D. Behrens and R. Eckermann (Ed.), *Vapor-Liquid Equilibrium Data Collection, Organic Hydroxy Compounds: Alcohols*, Vol. 1, Part 2a. DECHEMA, 6000 Frankfurt/Main, W. Germany, 255.

Gmehling, J., U. Onken, and W. Arlt (1980), Data Tables In., D. Behrens and R. Eckermann (Ed.), *Vapor-Liquid Equilibrium Data Collection, Aliphatic Hydrocarbons, C_4-C_6*, Vol. 1, Part 6a. DECHEMA, 6000 Frankfurt/Main, Federal Republic of Germany, 122.

Hirata, M., S. Ohe, and K. Nagahama (1975). Equations for calculation of the vapor-liquid equilibrium In, *Computer-Aided Data Book of Vapor-Liquid Equilibria*. Kodansha Ltd., Tokyo, Japan, 5-8.

Jorgensen, W. L. (1980). Structure and properties of liquid methanol. *J. Am. Chem. Soc.*, **102**, 543-549.

Jorgensen, W. L., J. D. Madura, and C. J. Swenson (1984). Optimized intermolecular potential functions for liquid hydrocarbons. *J. Am. Chem. Soc.*, **106**, 6638-6653.

Panagiotopoulos, A. Z. (1987). Direct determination of phase coexistence properties of fluids by Monte Carlo simulation in a new ensemble. *Mol. Phys.*, **61**, 813-826.

Rowley, R. L., T. D. Shupe, and J. C. Perry (1995). A direct method for determination of chemical potential with molecular dynamics simulations. Part 2. Mixtures. *Mol. Phys.*, **86**, 125-137.

Rowley, R. L., T. D. Shupe, and M. W. Schuck (1994). A direct method for determination of chemical potential with molecular dynamics simulations. Part 1. Pure components. *Mol. Phys.*, **82**, 841-855.

Ryckaert, J. P., and A. Bellemans, (1975). Molecular Dynamics of liquid *n*-butane near its boiling point. *Chem. Phys. Lett.*, **30**, 123-129.

Widom, B. (1963). Some topics in the theory of fluids. *J. Chem. Phys.*, **39**, 2808-2812.

PREDICTING ACTIVITY COEFFICIENTS
THROUGH GLOBAL OPTIMIZATION

Aileen Cheung and Claire S. Adjiman
Centre for Process Systems Engineering,
Imperial College of Science, Technology and Medicine
London SW7 2BY, U.K.

Petr Kolar and Takeshi Ishikawa
Mitsubishi Chemical Corporation, Mizushima Plant
Kurashiki 712-8054, Okayama, Japan

Abstract

In the context of solvent selection for complex solutes such as pharmaceutical molecules, we describe a predictive approach which aims to identify the best solvent in terms of solubility. The method relies on the use of molecular mechanics to calculate interaction energies for clusters of solvent and solute molecules. A trade-off between computational cost and accuracy is sought, and different strategies to partially sample the conformational space are investigated through two case studies: *n*-butane/ ethylamine, a small but non-ideal system for which we predict vapor-liquid equilibrium, and the solid-liquid equilibrium of paracetamol, a medium-size pharmaceutical, in six different solvents. The interaction energies are used to compute interaction parameters for the UNIQUAC activity coefficient model. We find that the identification of all the minimum configurations of the solvent-solute complexes is of great importance in obtaining reliable rankings, motivating the development of more rigorous global optimization techniques.

Keywords

Solvent selection, Global optimization, Interaction energies, Solubility, Activity coefficients.

Introduction

The selection of materials such as solvents during process development significantly affects the overall economics, safety and environmental impact of the process. In order to *optimally* select these materials, the relevant physical properties of a large number of materials must be accessible at relatively low cost. Thus, in the context of solvent selection, the prediction of phase equilibria for complex molecules constitutes an important challenge in applied thermodynamics. Solutes typical of the agrochemical and pharmaceutical industries are flexible molecules with 20 to 70 atoms, often containing aromatic rings and heteroatoms (O, N, P, S, F,...). The combination of the electronegative heteroatoms and aromatic delocalized π-electrons may lead to specific interactions of these molecules with small polar and aprotic solvents.

As a result, the observed solubilities in such systems are often far greater than those predicted by regular solution theory. The identification of such favorably interacting solvents is of great practical interest and consequently the main focus of this paper.

UNIFAC is a low-cost prediction technique for phase equilibria which has been used for phase equilibrium calculations (e.g. vapor-liquid equilibria, VLE). However, its accuracy decreases as molecular size increases and many group parameters remain unavailable, making it unsuitable for complex systems. On the other hand, more rigorous methods based on molecular dynamics simulations (Gubbins, 1993) are very demanding computationally, especially for medium and large solutes. Given that specific interactions are largely responsible for

167

negative deviations from ideality (solubilities greater than ideal), we propose an intermediate approach which focuses on the energetics of the solvent-solute system as a means to determine the suitability of different solvents for a given pharmaceutical solute.

We first briefly review recent work on the prediction of phase equilibria through the energy minimization of molecular clusters. We then address key issues in such a procedure by applying it to the prediction of VLE for *n*-butane-ethylamine, a non-ideal system. Finally, we investigate its applicability to the selection of a solvent for paracetamol.

Literature Review

Approaches based on energy minimization (Meniai and Newsham, 1992; Jónsdóttir *et al.*, 1994; Sum and Sandler, 1999), usually involve two steps. First, interaction energies between the participating molecules are calculated. The interaction energy ΔU_{ij} between two molecules i and j is given by $\Delta U_{ij}=U_{ij}-(U_i+U_j)$, where U_{ij} is the total energy of the molecular pair, and U_i and U_j are the energies of molecules i and j respectively. These energies are then used to compute interaction parameters A_{ij} and A_{ji} for an activity coefficient model such as UNIQUAC (Abrams and Prausnitz, 1975), which forms the basis of property predictions.

Jónsdóttir *et al.* (1994, 1996 and 1997) calculated the interaction energies for alkane/alkane, alkane/ketone, alkane/amine systems and for molecules with –OH groups in aqueous solution using molecular mechanics. Their force field of choice was the Consistent Force Field (CFF). First, conformers for individual molecules were identified. The global energy minima U_{ij} of solvent/solvent, solute/ solute and solvent/solute pairs for different conformers were calculated using a stochastic optimization procedure, the Boltzmann Jump Procedure (QUANTA 4.0, Molecular Simulations, Cambridge, UK). The interaction energies for the three types of molecular pairs were obtained from a Boltzmann average of the interaction energies at the different minima. For the glycols, glycerols and glucose in aqueous solution, all minima were included in the calculation of a linearly-averaged interaction energy. This approach gave good results for simple molecules but deviated from experiments for more complex molecules.

Meniai and Newsham (1995) also developed a method involving molecular pairs, and used the CHEM-X molecular graphics system. They found good agreement with experimental values for the phenol/water system.

Ab initio calculations were used by Sum and Sandler (1999) to compute the interaction energies. Binary systems of water with alcohols, carboxylic acids, aldehydes and ketones were studied. The authors observed that a cluster of eight molecules (four of each type) gave a good compromise between robustness and efficiency.

The energies obtained between pairs within the cluster were linearly averaged to obtain the interaction parameters. Using these computed parameters in the UNIQUAC equation (Abrams and Prausnitz, 1975) resulted in predictions which compared well with experiments, while poor results were obtained when used in the Wilson equation. This led the authors to conclude that the interaction parameters in the UNIQUAC equation are more closely related to interaction energies.

The theoretical justification for establishing a correspondence between interaction energy and interaction parameters is indeed important if reliable predictions are to be obtained. However, Flemr (1976) showed that the assumption that interaction parameters are only dependent on interaction energies is inconsistent with the assumptions of the local composition model used to develop the UNIQUAC equation. On this basis, local composition equations should be regarded as semi-empirical techniques. Quantitatively valid predictions cannot generally be expected from the direct use of interaction energies in the calculation of interaction parameters.

Proposed Approach

The methodology we propose revolves around the calculation and use of the interaction energies of solvent-solute complexes. We aim to develop a consistent approach for the determination of the interaction energies through molecular mechanics. The relative importance of different energy minima and an appropriate representation of the solvation shell are two issues of particular interest.

Calculation of Interaction Energy

The main motivation for choosing Molecular Mechanics to calculate interaction energies is the relatively inexpensive nature of potential energy function evaluations. Initially, pairs of molecules are used to compute the interaction energy. Two strategies are tested:

- The global minimum energy for the pair is found. The corresponding interaction energy is taken as the interaction energy of the system.
- All minima for the pair are found. The interaction energy of the system ΔU_{ij}^{av} is calculated as a Boltzmann average, with each interaction energy ΔU_{ij}^{m} weighted by a probability p_m based on total energy:

$$p_m = \frac{e^{\frac{-U_{ij}^m}{kT}}}{\sum_n e^{\frac{-U_{ij}^n}{kT}}} \qquad (1)$$

where k is the Boltzmann constant. The Boltzmann averaged interaction energy is then given by:

$$\Delta U_{ij}^{av} = \sum_m p_m \Delta U_{ij}^m \qquad (2)$$

The use of Boltzmann averaging corresponds to a partial sampling of the conformational space of the molecular pair, and can be expected to give more reliable results as the flexibility of the molecules increases. A yet more accurate representation would account for the width of the energy wells, as is the case with Monte Carlo simulations for instance.

To further refine the strategy, the number of solvent molecules is increased to represent the first solvation shell. This is especially important when the solute is much larger than the solvent molecules, as often occurs with pharmaceutical solutes. We have used six solvent molecules which are randomly arranged around a solute molecule. One atom in each solvent molecule is fixed, but the solvent molecules are otherwise free to rotate during the minimization procedure. The interaction energy for a given minimum is calculated by linearly averaging the energies of all solvent/solute pairs.

Minimization of Potential Energy Function

Since we are particularly interested in solvent/solute systems with strong interactions, we allow full internal flexibility of all the molecules in the system. The COMPASS (Condensed-phase Optimized Molecular Potentials for Atomistic Simulation Studies) force field (Sun, 1998) is used. It was developed to model systems at finite temperatures and to give accurate results for most common organic and inorganic molecules under most feasible conditions. It is parameterized using condensed-phase properties in addition to *ab initio* and empirical data.

The minimization of the potential energy function is a non-convex multivariable optimization problem. For a molecule of N atoms, the number of local minima increases as $\exp(N^2)$ (Hoare, 1979). We use the Boltzmann Jump Procedure to identify energy minima. However, this stochastic optimization technique is not guaranteed to find the global minimum, nor all other minima.

Solvent Ranking

To find favorable solvents, interaction parameters A_{ij} and A_{ji} are obtained from the interaction energies:

$$A_{ji} = \frac{\Delta U_{ji} - \Delta U_{ii}}{R q_i} \qquad (3)$$

where R is the gas constant and q_i is the surface area fraction of molecule i. These parameters can then be used in the UNIQUAC equation.

Results and Discussion

VLE for n-Butane / Ethylamine System

The first system studied is a polar hydrogen bonding alkane/amine pair. This small system is used to illustrate the importance of global and local minima. Ethylamine and n-butane and have two conformers each. The vapor-liquid equilibrium of n-butane/ethylamine was reproduced using the UNIQUAC interaction parameters calculated from the interaction energies using only the global minimum energy of the system and the Boltzmann averaged interaction energy.

Fig. 1 compares predicted VLE data with experimental data (Wolff *et al.*, 1964). Both methods predict an azeotrope but two coexisting liquid phases are found with the global minimum, which is in contrast with experiments. The Boltzmann averaged interaction energies yield a better representation.

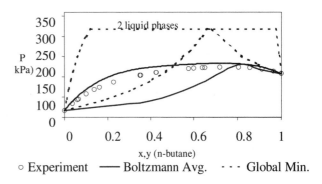

Figure 1. VLE data for n-butane/ethylamine at 293 K.

Solubility of Paracetamol in Several Solvents

Paracetamol (4-hydroxyacetanilide, $C_8H_9NO_2$), a medium-size molecule with an aromatic ring and a flexible tail, is an important analgesic and antipyretic agent. Its solubility in 26 solvents was reported by Granberg and Rasmuson (1999). It has a low solubility in nonpolar and chlorinated hydrocarbons. In solvents of medium polarity (e.g., ketones, esters, alcohols) the solubility is close to ideal (calculated from the melting point T_m and heat of fusion ΔH_{fus} of paracetamol). The solubility of paracetamol in n-alcohols and ketones decreases with increasing length of the carbon chain. In aprotic solvents (DMSO, DMF) and diethylamine (DEA), paracetamol shows strong negative deviations from ideality and excellent solubility.

Six solvents were chosen to test the methodology: diethylamine (DEA), methanol (MeOH), ethanol (EtOH), acetone, acetic acid (AcH) and ethyl acetate (AcOEt).

Interaction energies computed using the global minimum or the Boltzmann average are shown in Table 1. The calculated and experimental solubilities are shown in Table 1 and Fig. 2.

Diethylamine is identified as the best solvent for both the global optimum and Boltzmann average. Using the global optimum only, ethyl acetate was predicted to be the second best solvent while in reality, it is the worst solvent. Using the Boltzmann average however, it was predicted to be a poor solvent. Acetic acid was rightly predicted to be a poor solvent but the quantitative prediction deviates by a large factor. Acetone was predicted to be ranked above methanol and ethanol. During the minimization procedure for ΔU_{22}, however, the global minimum is found repeatedly, suggesting that it may be predominant. If the interaction energy corresponding to the global minimum energy of the pair is used instead, an improved solubility value of 0.162 is found. Other solubility predictions are qualitatively good. When the Boltzmann average is used, most values of the interaction energy decrease to varying degrees. In all instances, the predicted solubilities improve.

Table 1. Interaction Energies for Paracetamol (1) and Solvent (2) (kcal/mol), Solubilities.

Calculations based on global minimum				
Solvent	ΔU_{12}	ΔU_{22}	x_s^{calc}	x_s^{expt}
DEA	-10.21	-5.80	0.361	0.389
Methanol	-8.05	-6.60	0.242	0.073
Ethanol	-8.52	-7.12	0.198	0.066
Acetone	-8.03	-4.82	0.313	0.041
Acetic acid	-12.55	-13.50	1.1E-05	0.032
Ethyl Acetate	-8.27	-4.80	0.326	0.006
Calculations based on Boltzmann average				
Solvent	ΔU_{12}	ΔU_{22}	x_s^{calc}	x_s^{expt}
DEA	-7.97	-3.66	0.364	0.389
Methanol	-7.11	-6.28	0.171	0.073
Ethanol	-7.15	-6.13	0.098	0.066
Acetone	-6.29	-3.23	0.309	0.041
Acetic acid	-9.23	-11.0	2.3E-05	0.032
Ethyl Acetate	-6.10	-4.23	0.011	0.006

The interaction energies were also obtained for the diethylamine case using six solvent molecules surrounding one solute molecule. The average interaction energy for paracetamol/DEA was found to be −7.19 kcal/mol. The DEA/DEA interaction energy was found to be −1.27 kcal/mol. Consequently, the predicted solubility was found to increase to 0.387.

Conclusions

A methodology for selecting solvents for complex solutes through activity coefficient prediction has been proposed. It is based on the identification of the energy minima of molecular pairs or clusters using molecular mechanics. The method has been applied to paracetamol in six different solvents: correct rankings were obtained for four of the six solvents. The results have shown that it is important to take all energy minima into account when determining the interaction energies. Thus, future work will focus on the development of a more robust global optimization technique based on deterministic techniques which will help identify these minima.

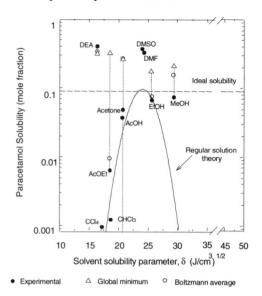

Figure 2. Solubility of paracetamol at 303 K.

Acknowlegments

Aileen Cheung, MCC Fellow, gratefully acknowledges Mitsubishi Chemical Corporation for financial support.

References

Abrams, D. S., and J. M. Prausnitz (1975). Statistical thermodynamics of liquid mixtures: A new expression for the excess Gibbs energy of partly and completely miscible systems. *AIChE Journal*, **21**, 116-128.

Burkert, U., and N. L. Allinger (1982). *Molecular Mechanics.* ACS Monograph 177. American Chemical Society, Washington D. C.

Flemr, V. (1976). A note on excess Gibbs Energy Equations based on local composition concept. *Coll. Czech. Chem. Comm.*, **41**, 3347-3349.

Granberg, R. A. and A. C. Rasmuson (1999). Solubility of paracetamol in pure solvents. *J. Chem. Eng. Data*, **44**, 1391-1395.

Gubbins, K. E. (1993). Molecular simulation of fluid phase equilibria. *Pure & Appl. Chem.*, **65**, 935-940.

Hoare, M. R. (1979). Structure and dynamics of simple microclusters. *Adv. Chem. Phys.*, **40**, 49-135.

Jónsdóttir, S. O., K. Rasmussen, and A. Fredenslund (1994). UNIQUAC parameters determined by Molecular Mechanics. *Fluid Phase Equil.*, **100**, 121-138.

Jónsdóttir, S. O., R. A. Klein, and K. Rasmussen (1996). UNIQUAC interaction parameters for alkane / amine systems determined by Molecular Mechanics. *Fluid Phase Equil.*, **115**, 59-72.

Jónsdóttir, S. O., and R. A. Klein (1997). UNIQUAC interaction parameters for molecules with –OH groups on adjacent carbon atoms in aqueous solution determined by molecular mechanics – glycols, glycerol and glucose. *Fluid Phase Equil.*, **132**, 117-137.

Meniai, A. H., and D. M. T. Newsham (1992). The selection of solvents for liquid-liquid extraction. *Trans. IchemE*, **70, Part A**, 78-87.

Meniai, A. H., and D. M. T. Newsham (1995). Computer-aided method for interaction parameter calculations. *Trans. IChemE*, **73, Part A**, 842-848.

Sum, A. K., and S. I. Sandler (1999). A novel approach to phase equilibria prediction using *ab initio* methods. *Ind. Eng. Chem. Res.*, **38**, 2849-2855.

Wolff, H., A. Hoepfner, and H. M. Hoepfner (1964). *Ber. Bunsenges. Phys. Chem.*, **68**, 410.

PARTIAL DERIVATIVES OF MOLECULAR DYNAMICS CALCULATIONS

Jelena Stefanovic, Bruno Fouchet and Constantinos C. Pantelides
Centre for Process Systems Engineering
Imperial College of Science, Technology and Medicine
London SW7 2BY, United Kingdom

Abstract

The molecular dynamics technique can be viewed as a deterministic mathematical mapping between, on one hand, the force field parameters that describe the potential energy interactions and input macroscopic conditions, and, on the other, the calculated corresponding macroscopic properties of the bulk molecular system.

The differentiability of such a mapping in conventional molecular dynamics calculations is affected by the discontinuities introduced through the use of the minimum image convention and other simplifications commonly employed in the calculation of interparticle potential and forces. A modified force function which is almost everywhere continuous and differentiable, and exhibits a natural periodicity has recently been proposed by Stefanovic and Pantelides (1999). These characteristics make it possible to apply standard methods for the computation of the partial derivatives of the molecular dynamics mapping based on the integration of either the adjoint equations, or the sensitivity equations of the classical Newtonian equations of motion. We present procedures for these computations in the standard microcanonical (N, V, E) ensemble, and compare the computational efficiency of the two approaches.

As an illustration, we apply these techniques to a system of flexible ethane molecules, computing the partial derivatives of temperature and pressure with respect to density, energy and all potential function parameters.

Keywords

Molecular dynamics, Partial derivatives, Parameter estimation.

Introduction

Performing a molecular dynamics simulation can be viewed as the computation of a function of the form:

$$p = \mathsf{P}(q, \vartheta) \qquad (1)$$

Here q denotes the vector of macroscopic input quantities, the values of which are specified (e.g. the internal energy E and the system density ρ), ϑ is a vector of interparticle potential parameters (e.g. the Lennard-Jones potential parameters ε and σ), and p denotes one or more macroscopic properties (e.g. temperature, pressure,

diffusion coefficients etc.) that are obtained by the computation. The above function is, in principle, well-defined, provided the same deterministic procedure for generating the initial condition of the system, and the same time horizon and equilibration period are always used.

From a practical point of view, the continuity and differentiability of function (1) would be highly desirable if, for instance, it is to be embedded within higher level computations (e.g. for designing chemical processes or new materials exhibiting certain desirable properties), or

172

if it is somehow to be inverted (e.g. in order to estimate values of the parameters ϑ from a set of experimental values $q^{[k]}$, $p^{[k]}$, $k = 1, 2,.., NE$). In most of these applications, it is important to be able to compute not only the values of p for given values of q and ϑ, but also the values of the partial derivatives $\partial p/\partial q$ and $\partial p/\partial \vartheta$ – assuming, of course, that these derivatives exist.

In conventional implementations of molecular dynamics, the differentiability of mapping (1) is affected by the occurrence of a large number of discontinuities that are caused by the use of the minimum image convention or other simplifications employed in the calculation of interparticle potential and force.

In an attempt to address these problems, Stefanovic and Pantelides (1999) recently proposed a novel framework for molecular dynamics that is based on only two premises:

- the physical system of interest exhibits a spatial periodicity expressed in terms of the infinite replication of a cube of given size L;
- the non-bonded interactions in the system of particles under consideration can be described in terms of a set of pairwise interactions, each characterized by a continuous and twice-differentiable potential function.

No other assumptions or simplifications are introduced. This leads to the following modified force function being used to describe non-bonded interactions between particle pairs (shown here in the x-direction):

$$F^x(X,Y,Z) \equiv -\frac{1}{L} \sum_{k=-\infty}^{+\infty} \sum_{k'=-\infty}^{+\infty} \sum_{k''=-\infty}^{+\infty} \frac{X-k}{R_{kk'k''}} \frac{\partial U^{NB}}{\partial R}\Bigg|_{R_{kk'k''}} \quad (2)$$

where X, Y, Z denote the normalised interparticle distance components (e.g. $X \equiv (x_i - x_j)/L$), and $R_{kk'k''}$ is the corresponding normalised interparticle distance ($R_{kk'k''} \equiv [(X-k)^2 + (Y-k')^2 + (Z-k'')^2]^{1/2}$). U^{NB} is any general non-bonded interparticle potential which is a continuous and twice-differentiable function of R. Similar expressions can be derived for F^y and F^z. The modified force function (2) is continuous and differentiable for all physically attainable values of X, Y, Z. Moreover, despite its apparent complexity, both the function and its partial derivatives may be computed to arbitrary accuracy with little cost beyond what is required for conventional force functions (Stefanovic, 2000).

The use of the force function (2) renders mapping (1) a continuous and differentiable function of its arguments. We present procedures for the computation of the partial derivatives $\partial p/\partial q$ and $\partial p/\partial \vartheta$ below.

Partial Derivatives of Molecular Dynamics Mappings

Molecular dynamics in the standard microcanonical (N, V, E) ensemble solves the Newtonian equations of motion for a system of N interacting particles given by:

$$\dot{\mathbf{r}}_i = \mathbf{v}_i ; \quad \dot{\mathbf{v}}_i = \frac{\mathbf{F}_i(\mathbf{r},\alpha)}{m_i} \quad \forall i = 1,..,N \quad (3)$$

subject to the initial conditions:

$$\mathbf{r}(0) = \mathbf{r}^0(\alpha); \quad \mathbf{v}(0) = \mathbf{v}^0(\alpha) \quad (4)$$

where vectors $\mathbf{r}$ and $\mathbf{v}$ denote the particle positions and velocities respectively, $\mathbf{F}_i$ is the total force exerted on particle i, m_i is the mass of particle i, and the set of parameters α is defined as $\alpha \equiv \{\vartheta, q\}$.

Most macroscopic properties p of interest determined by molecular dynamics involve time-integral expressions of the form:

$$p(\alpha) = \int_0^{t_f} \varphi(\mathbf{r}(t,\alpha), \mathbf{v}(t,\alpha), \alpha)\, dt \quad (5)$$

where t_f is some final time and φ is a continuous and differentiable function of the positions $\mathbf{r}$, the velocities $\mathbf{v}$ and the parameters α.

Equations (3)-(4) describe an initial value problem which, given values of the parameters α, can be solved to determine $\mathbf{r}(t)$ and $\mathbf{v}(t)$ for $t \in [0, t_f]$. This then allows the computation of the quantity p via Eqn. (5). Thus, p is solely a function of α, as indicated on the left-hand side of Eqn. (5). The partial derivatives $\partial p/\partial \alpha$ can be obtained by two alternative variational calculus techniques.

Partial Derivatives via the Adjoint System

One way of obtaining the partial derivatives $\partial p/\partial \alpha$ is via the solution of the adjoint system (Bryson and Ho, 1975), which, for the Newtonian equations of motion (3)-(4), can be formulated as follows:

$$\dot{\boldsymbol{\lambda}}_i^r = -\frac{1}{m_i} \sum_{j=1}^N \mathbf{M}^{(j,i)} \boldsymbol{\lambda}_j^v - \left(\frac{\partial \varphi}{\partial \mathbf{r}_i}\right)^T \quad \forall i = 1,..,N \quad (6)$$

$$\dot{\boldsymbol{\lambda}}_i^v = -\boldsymbol{\lambda}_i^r - \left(\frac{\partial \varphi}{\partial \mathbf{v}_i}\right)^T \quad \forall i = 1,..,N \quad (7)$$

$$\dot{\lambda}^\alpha = -\sum_{i=1}^N \frac{1}{m_i}\left(\frac{\partial \mathbf{F}_i}{\partial \alpha}\right)^T - \left(\frac{\partial \varphi}{\partial \alpha}\right)^T \quad (8)$$

subject to the final time conditions:

$$\boldsymbol{\lambda}^r(t_f) = 0; \quad \boldsymbol{\lambda}^v(t_f) = 0; \quad \lambda^\alpha(t_f) = 0 \quad (9)$$

Here λ^r, λ^v and λ^α are vectors of adjoint variables, and $\mathbf{M}^{(j,i)} \equiv \partial \mathbf{F}_i / \partial \mathbf{r}_j$ is a 3×3 matrix of the spatial partial derivatives of the force function. The above system is integrated backwards from $t = t_f$ to $t = 0$, and the required partial derivatives $\partial p / \partial \alpha$ can be obtained from:

$$\frac{\partial p}{\partial \alpha} = \left(\frac{\partial \mathbf{r}^0}{\partial \alpha}\right)^T \lambda^r(0) + \left(\frac{\partial \mathbf{v}^0}{\partial \alpha}\right)^T \lambda^v(0) + \lambda^\alpha(0) \quad (10)$$

Equations (8) and (10) have to be written separately for each parameter α; moreover, Eqns. (6)-(8) and (10) have to be written separately for each property p.

Partial Derivatives via the Sensitivity Equations

An alternative approach to the computation of the partial derivatives $\partial p / \partial \alpha$ is provided by the solution of the sensitivity equations derived by differentiating Eqns. (3)-(4) with respect to parameters α to obtain:

$$\dot{\mathbf{r}}_{i,\alpha} = \mathbf{v}_{i,\alpha} \qquad \forall i = 1,..,N \quad (11)$$

$$\dot{\mathbf{v}}_{i,\alpha} = \frac{1}{m_i}\left(\sum_{j=1}^{N} \mathbf{M}^{(j,i)} \mathbf{r}_{j,\alpha} + \frac{\partial \mathbf{F}_i}{\partial \alpha}\right) \qquad \forall i = 1,..,N \quad (12)$$

subject to the initial conditions:

$$\mathbf{r}_{i,\alpha}(0) = \frac{\partial \mathbf{r}_i^0}{\partial \alpha}; \quad \mathbf{v}_{i,\alpha}(0) = \frac{\partial \mathbf{v}_i^0}{\partial \alpha} \quad \forall i = 1,..,N \quad (13)$$

Here, we employ the notation $\mathbf{r}_{i,\alpha}(t) \equiv \partial \mathbf{r}_i(t)/\partial \alpha$ and $\mathbf{v}_{i,\alpha}(t) \equiv \partial \mathbf{v}_i(t)/\partial \alpha$. The sensitivity equations (11)-(12) can be integrated simultaneously with the original Eqn. (3) to obtain the sensitivities $\mathbf{r}_{i,\alpha}(t)$ and $\mathbf{v}_{i,\alpha}(t)$ for $t \in [0, t_f]$. By also differentiating Eqn. (5) with respect to α, we obtain an expression for the required partial derivatives:

$$\frac{\partial p}{\partial \alpha} = \int_0^{t_f}\left[\sum_{i=1}^{N}\left(\frac{\partial \varphi}{\partial \mathbf{r}_i}\mathbf{r}_{i,\alpha} + \frac{\partial \varphi}{\partial \mathbf{v}_i}\mathbf{v}_{i,\alpha}\right) + \frac{\partial \varphi}{\partial \alpha}\right] dt \quad (14)$$

Equations (11)-(14) have to be written separately for each property α; also Eqn. (14) has to be written for each property p.

Computational Considerations

Both the adjoint system (6)-(8), and the sensitivity equations (11), (12) and (14) allow the computation of the desired gradients $\partial p / \partial \alpha$. Here we examine the relative computational efficiency of the two approaches.

For a system involving N interacting particles, the number, NA, of adjoint equations (6)-(8) is given by:

$$NA \equiv (6N + N_\alpha) N_p \quad (15)$$

while the number NS of sensitivity equations (11), (12) and (14) is given by:

$$NS \equiv (6N + N_p) N_\alpha \quad (16)$$

where N_p and N_α denote the number of properties p and parameters α respectively. Thus, from Eqns. (15) and (16), it is clear that the sensitivity system is more compact than the adjoint one (i.e. $NS < NA$) if there are fewer parameters α than properties of interest, p.

We note that the integration of both adjoint and sensitivity equations requires knowledge of $\mathbf{r}(t)$ and $\mathbf{v}(t)$, $t \in [0, t_f]$, and involves the same functions of the vectors $\mathbf{r}$, $\mathbf{v}$ and α. Sensitivity equations involve more expensive matrix-vector operations than the adjoint ones, but they have the advantage of being integrated in the forward direction together with the original Newtonian equations (3). In contrast, the integration of the adjoint system is carried out backwards in time, and hence the original equations also have to be integrated backwards together with the adjoint equations (despite the fact that they have already been integrated in the forward direction). Overall, whether adjoints or sensitivities should be used for computation of partial derivatives will depend on the specific application under consideration and the number of parameters and properties involved.

Application: Dynamics of Systems of Ethane Molecules

As an illustration of the methodology presented in this paper, we now apply it to the system of flexible ethane molecules interacting via the NERD force field (Nath *et al.*, 1998).

We simulate a bulk ethane fluid in the modified force framework. The simulation involves 108 ethane molecules at a density of 360 kg/m³ and energy of 1000 J/mol. The results of the base case simulation are shown in Table 1. Here the simulation time horizon t_f is 22 ps; the time over which the averaging is performed is taken as $t/2$. An Adams-Bashforth integration method of order 3 with a fixed time step of 2 fs is used for the solution of the equations of motion.

Table 1.   Results of Base Case Simulation.

E (J/mol)	ρ (kg/m³)	T (K)	P (MPa)
1000	360	348.8	42.5

Table 2 presents the partial derivatives of temperature with respect to the input energy and density, and the potential parameters $\vartheta \equiv \{k_\theta, d_0, \varepsilon, \sigma\}$, and compares these with the values obtained using first-order finite difference approximations (FD). The latter are computed by applying *relative* perturbations of magnitude δ on the variable with respect to which the partial derivative is taken.

We observe that both the adjoint and the sensitivity formulations predict the same values of all partial derivatives within (at least) the four significant digits reported here. These values are generally similar to those obtained from finite difference approximations. Interestingly, the finite difference results indicate that

small perturbations applied to the values of α result in correspondingly small changes in the values of outputs, i.e. there is no evidence of chaotic behavior in the mapping $p(\alpha)$. However, as is usually the case with complex functions, the selection of an appropriate size for the finite difference perturbation is not straightforward, and varies from one partial derivative to another.

Table 2. Partial Derivatives of Temperature With Respect to Energy, Density and Potential Parameters.

	$\partial T/\partial E$ (K mol/J)	$\partial T/\partial \rho$ (K m³/mol)	$\partial T/\partial k$ (Å²)	$\partial T/\partial d$ (K/Å)	$\partial T/\partial \varepsilon$ (-)	$\partial T/\partial \sigma$ (K/Å)
F.D. $\delta = 1 \times 10^{-2}$	3.370×10^{2}	2.160×10^{2}	-2.135×10^{-3}	-2.216×10^{1}	3.152	2.051×10^{2}
F.D. $\delta = 1 \times 10^{-3}$	3.370×10^{2}	2.230×10^{2}	-2.073×10^{-3}	-2.195×10^{1}	3.152	2.168×10^{2}
F.D. $\delta = 1 \times 10^{-4}$	3.400×10^{2}	2.230×10^{2}	-2.073×10^{-3}	-2.208×10^{1}	3.151	2.180×10^{2}
F.D. $\delta = 1 \times 10^{-5}$	4.000×10^{2}	2.250×10^{2}	0.000	-1.948×10^{1}	3.181	2.196×10^{2}
Adjoints:	3.371×10^{2}	2.245×10^{2}	-2.130×10^{-3}	-2.196×10^{1}	3.151	2.175×10^{2}
Sensitivities:	3.371×10^{2}	2.245×10^{2}	-2.130×10^{-3}	-2.196×10^{1}	3.151	2.175×10^{2}

Concluding Remarks

This paper has shown that it is possible to establish rigorous procedures for the computation of all partial derivatives of all quantities computed by molecular dynamics simulations. Moreover, these derivatives are computed within the same degree of accuracy as the quantities themselves and are, therefore, numerically consistent with them.

Partial derivative information is likely to be valuable in a wide variety of applications, such as the use of sophisticated techniques for the estimation of values of the potential parameters. For example, suppose that we have available a number of experimentally determined points $(E^{[k]}, \rho^{[k]}, p^{[k]})$, $k = 1, 2, .., NE$. We could then use these to estimate the force field parameters ϑ by posing the following nonlinear least squares problem:

$$\Phi = \min_{\vartheta} \sum_{k=1}^{NE} (p(E^{[k]}, \rho^{[k]}, \vartheta) - p^{[k]})^2 \qquad (17)$$

where $p(E, \rho, \vartheta)$ denotes the value of the property p computed by the molecular dynamics mapping for given E, ρ, and ϑ. The use of sophisticated optimization algorithms to perform the above minimization requires the gradients of the objective function Φ with respect to the unknown parameters ϑ:

$$\frac{\partial \Phi}{\partial \vartheta} = 2 \sum_{k=1}^{NE} (p(E^{[k]}, \rho^{[k]}, \vartheta) - p^{[k]}) \frac{\partial p}{\partial \vartheta}\bigg|_{[k]} \qquad (18)$$

The above expression involves the partial derivatives $\partial p/\partial \vartheta$ which can be obtained accurately using the techniques presented here. On the other hand, the use of numerically inaccurate partial derivative values (such as those computed from finite difference approximations) may severely impair the convergence properties of the optimization algorithm.

Acknowledgment

The financial support of the Mitsubishi Chemical Corporation is gratefully acknowledged.

References

Bryson, A. E., and Y.-C. Ho (1975). *Applied Optimal Control* Wiley, New York.

Nath, S. K., F. A. Escobedo and J. J. de Pablo (1998). On the simulation of the vapor-liquid equilibria for alkanes. *J. Chem. Phys.*, **108**, 9905-9911.

Stefanovic, J. and C. C. Pantelides (1999). Towards tighter integration of molecular dynamics within process and product design computations. *Foundations of Computer Aided Process Design* (FOCAPD-99), Breckenridge, Colorado.

Stefanovic, J. (2000). *On the Mathematics of Molecular Dynamics*. PhD Thesis, University of London.

GENERATING MOLECULAR STRUCTURES HAVING SPECIFIC PROPERTIES WITHOUT SUFFERING FROM COMBINATORIAL EXPLOSION

Peter M. Harper and Rafiqul Gani
Computer Aided Process Engineering Center
Department of Chemical Engineering, Technical University of Denmark
Lyngby, Denmark

Abstract

An efficient multi-level algorithm for Computer Aided Molecular Design (CAMD) is presented. The algorithm is based on an enumeration approach and is capable of generating highly detailed molecular models from fragments without suffering from the so-called "combinatorial explosion". The starting point is a collection of building blocks (molecule fragments). The overall algorithm has 4 levels. Each level consists of a generation step and a screening step. In the generation step molecules are formed while in the screening step the properties are estimated and compared against the target properties. Molecules fulfilling the requirements are passed on to the next level. The first levels use a macroscopic representation of molecules while the later levels employ a detailed representation.

Keywords

Computer-aided molecular design, CAMD, Property prediction, Generation of structures, Generate and test, Combinatorial explosion.

Introduction

Computer Aided Molecular Design (CAMD) of substances having specific physical and chemical properties is a cost efficient alternative to extensive experimental efforts. By using a computer algorithm to identify compounds showing potential for being suitable for a given application the experimental resources can be concentrated on the most promising candidates resulting not only in a more thorough examination but also in a more effective use of expensive experimental facilities.

The goal of a CAMD algorithm is to "create" compounds possessing certain qualities by assembling molecular fragments into molecules. The creation of the molecules from fragments is performed using a generation algorithm while the qualities (properties) are evaluated using predictive techniques linking molecular structure to properties. The primary challenges in computer aided molecular design (CAMD) are the accurate prediction of physical and chemical properties

based on structural information and the efficient generation of structural alternatives.

Methods for the generation of molecules having specific physical and chemical properties have previously been reported by Duvedi and Achenie (1996), Venkatasubramanian et al. (1995), Raman and Maranas (1998), Vaidyanathan and El-Halwagi (1996) as well as others (Mavrovouniotis (1998) lists a series of examples with references in his review of CAMD) and have been applied in areas such as solvent design for liquid-liquid separation, entrainer, solvent replacement, refrigerant and polymer design. However, most of the proposed methods create compounds containing limited structural information thereby restricting the range of property prediction methods that can be applied as well as making it difficult to distinguish between structural isomers.

The computer aided design of molecules (and especially molecules possessing a high degree of molecular detail) is a complex process due to the size of the search space and the associated risk of encountering the so-called "combinatorial explosion" where the number of alternatives considered becomes so large that it is infeasible to solve the problem within a reasonable timeframe with the available computational resources. In the work presented here this has been achieved by using a multi-level approach where the molecular structures are refined as the generation progresses. The starting point is a collection of building blocks (molecule fragments). Each level consists of a generation step and a screening step. In the generation step molecules are formed while in the screening step the properties are estimated and compared to the target properties. Molecules fulfilling the requirements are passed on to the next level. The first levels use a macroscopic representation of molecules while the later levels employ a detailed representation.

Theoretical Background

Regardless of the nature of the design problem (i.e. what the compound is being designed for) a CAMD problem can be formulated in terms of a set of property constraints expressing the *Essential* and the *Desirable* properties a compound must have in order to be suitable for the specific application (Harper and Gani, 2000). Apart from the overall problem formulation in terms of what properties the designed compounds should have, it is also necessary to specify the types of compounds that the CAMD algorithm should generate. This is achieved by performing a preselection on the building blocks used to assemble the molecules. Furthermore it helps control the size of the problem by only allowing the examination of compound classes of interest.

A very important aspect of the application of CAMD methods lies in the availability of suitable property prediction methods. Since CAMD uses predictive techniques exclusively it is necessary that suitable property prediction methods for the evaluation of the essential and desirable properties are available. While many prediction methods based on group contribution exist for a wide range of properties it is not possible to estimate all properties of interest just by using a group based molecular description. Furthermore, the use of group based prediction methods rarely allows for the distinction between isomers. In order to break the limitations of group contribution it is necessary to use other prediction methods based on more structural information.

Algorithmic Framework

The CAMD algorithm has four levels. Each level has its own generate and test algorithms. Higher levels use additional molecular structural information compared to lower levels.

The fundamental basis for the developed algorithm is the continuous refinement of the results obtained from each level. The lower levels have a low computational complexity (i.e. it is possible to generate a large number of alternatives without excessive calculations) but do not in all cases generate all the information necessary to perform the estimation of the important properties. The higher levels are more complex and cannot handle a very large number of alternatives without application of a significant computational effort. As a result of this, the design strategy of the developed algorithm is a hybrid approach where the lower levels are used to "pick out" promising candidates from the search space while the higher levels use the output from the preceding level as input. After each level the applicable property prediction methods are applied and the results compared to design specifications. Molecules violating the specifications are discarded and are not passed on to the next level.

The net effect of this approach is that the results are refined from level to level without spending computational resources on candidates, which are unable to fulfil the requirements. In the following subsections the characteristics of the levels are given in outline form with emphasis on the generation of the molecules (fig. 2 illustrates the relationship and role of the different levels).

Level 1 - Generation of Group Vectors from UNIFAC Groups

Level 1 generates vectors of groups (fragments) by combining groups from the UNIFAC group-set. These sets are capable of forming at least one feasible molecular structure. Simultaneous calculation of related properties (that are dependent only on first order groups) and screening of the generated structures are performed to control the problem size and execution time. The algorithm is based on the classification work of Gani et al. (1991) but uses a different and more efficient method of group vector assembly. In short the algorithm is based on:

- Building blocks are classified according to type.
- Feasibility rules are based on the number of groups from a specific class a compound may contain.
- Valency rules are used to determine the number of groups with 1, 2, 3 & 4 connections to be used in molecule structure generation.

Level 2 - Generation of Structural Isomers from Group Vectors

This level generates molecular structures by combining elements of the individual fragment sets from level 1 to form molecular structures. First- and second-order groups (Constantinou and Gani, 1994) are considered in the calculation of properties in this level.

The goal of the generation in level 2 is to:

- Increase the dimensionality of the molecular model.
- Provide a foundation for improving the quality of the predicted properties as well as allowing estimation of properties that cannot be handled using only UNIFAC groups.

The generation is performed by combining the groups from each of the results from level 1 into connected graphs with groups as vertices and bonds as edges. When considering the generation of acyclic compounds the problem is that of generating all spanning trees in a base-graph with the added constraint of restrictions on the valence of each of the vertices. It is a requirement that compounds should be chemically feasible as well as adheres to the rules of application of UNIFAC groups. In order not to generate multiple identical compounds from different group vectors and in order to ensure that the "promotion" into chemical structures is "reversible". The requirement of reversible promotion is addressed by defining rules for how groups can be combined/connected. The rules imposed cause the base-graph to be incomplete in all but the simplest cases. Instead, the base-graph storing is a map of all compounds super-imposed onto each other creating a *compound-superstructure* representing all possible combinations as shown in Fig. 1.

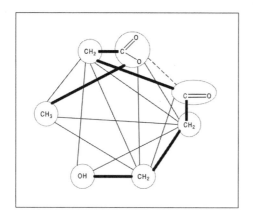

Figure 1. Example of compound super-structure with a spanning tree (thick lines).

An added requirement to the problem of identifying the spanning trees and later the rings in cyclic molecules is the necessity of generating unique structures only and avoiding graph isomorphism (the problems related to graph isomorphism have been described by Raman and Maranas (1998)).

Level 3 - Creation of Atomic Based Adjacency Descriptions

In level 3 the compound descriptions obtained from level 2 are subjected to further refinement and structural

variation. The goals of level 3 are to bring the compounds closer to a 3D structure and to enable the use of estimation methods not based on the original group-set or combinations hereof. Because atomic representations also define the connectivity of the molecules, property prediction methods based on connectivity indices can be employed to predict properties that could not be predicted earlier (due to unavailable group contributions).

The level 3 generation algorithm transforms the group based connectivity information (the adjacency matrix from level 2) into atom-based information. This is achieved by expanding each group into its corresponding atom-based adjacency matrix. When performing the expansion into an atomic representation it is possible to experience group-based descriptions yielding more than one atomic description. This is the case with compounds containing groups having a ring element with 1 or more free connections because of the ambiguously defined distance (in the ring) between the free bonds (as in ortho/meta/para) or between heteroatoms and bonds in aromatic rings (as in Pyridine derivates).

Level 4 - Generation of 3D Structures

In this level generation and testing enter an interactive mode. For any selected candidate from level 3 it is possible to use molecular modelling programs such as MOPAC or Chem3D (CambridgeSoft Inc., 1997). A three-dimensional graph (or molecular model) is created by applying a set of standard bond lengths and angles for the various types of connections. As a result the true molecular model of a compound which can be further analysed in terms of conformers, stability, properties, etc. is obtained.

The added dimensionality of a 3D representation cause the possibility of additional structural variations. The structural isomers possible to generate and distinguish in level 4 are the ones related to the relative steric placement of bonds and atoms. The isomer types theoretically possible to distinguish and generate are:

- Z/E isomers
- R/S isomers
- cis/trans isomers
- Boat/Chair isomers
- Anti/Gauche isomers

The latter two isomer types are conformational isomers while the rest are configurational isomers. Conformational isomers can be created by rotating single bonds and are controlled by the internal energies of the compound. The configurational isomers, however, cannot be transformed into each other by rotation around single bonds. The generation algorithm of level 4 considers only the distinction between configurational isomers and leaves the conformational isomer analysis and distinction to be handled as part of the analysis of the results. The reasons for this lie in the fact that the conformational

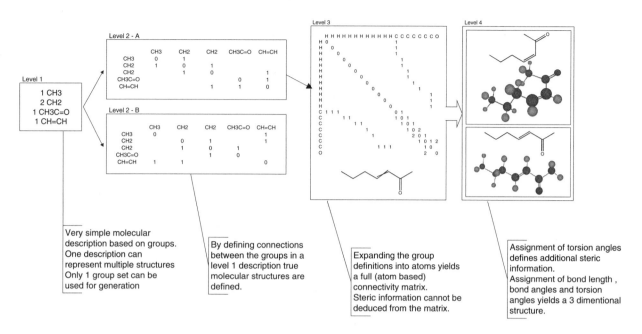

Figure 2.   Levels of the molecule generation algorithm.

isomer behaviour (or simply the conformation) of a compound is dependent on the state of a compound (temperature, pressure) as well as the presence of other compounds in the immediate environment and requires specialised tools in order to analyse the conformational space. Furthermore, in a bulk phase of a compound no single conformer will be the only one present but rather a distribution of the possible (stable) conformers (Jonsdottir, 1995).

Conclusions

An efficient CAMD algorithm for the design of compounds with specific properties has been developed. The algorithm is capable of designing compounds with a high level of molecular detail without suffering from combinatorial explosion. The capability to design highly detailed molecular models is achieved by subdividing the generation process into multiple levels as shown in Fig. 2 and only applying structural refinements to candidates showing the most potential. By being able to design compounds with a high degree of molecular detail it is possible to apply a more complex and diverse set of estimation techniques than traditionally applied in CAMD techniques used in engineering, bridging the gap in molecular design between the engineer (where the use of estimation techniques based on group contribution is prevalent) and the chemist (where the use of molecular modelling tools is commonplace).

In order to be applied in the design of compounds for use in chemical processes the generation algorithm must be applied as part of a general CAMD framework assisting in the formulation of the CAMD problem and the analysis of the results.

References

Duvedi, A. P. and L. E. K. Achenie (1996). Designing environmentally safe refrigerants using mathematical programming. *Chem. Eng. Sci.*, **51**(15), 3727.

CambridgeSoft Inc. (1997). *Chem3D Users Guide.* Cambridge, MA.

Constantinou L. and R. Gani (1994). New group-contribution method for the estimation of properties of pure compounds. *AIChE J.*, **10**, 1697-1710.

Gani, R. B. Nielsen and Aa. Fredenslund (1991). A group contribution approach to computer-aided molecular design. *AIChE J.*, **37**(9), 1318-1332.

Jonsdottir, S. O. (1995). *Theoretical Determination of UNIQUAC Interaction Parameters.* Ph.D. thesis, Technical University of Denmark, Department of Chemical Engineering.

Harper, P. M. and R. Gani (2000). A Multi-step and multi-level approach for computer aded molecular design. *Proceedings of PSE2000*, accepted.

Mavrovouniotis, M.L. (1998). Design of chemical compounds. *Comp. Chem. Eng.*, **22**(6), 713-715.

Raman, V. S. and C. D. Maranas (1998). Optimization in product design with properties correlated with topological indices. *Comp. Chem. Eng.*, **22**(6), 747.

Vaidyanathan R. and M. El-Halwagi (1996). Computer-aided synthesis of polymers and blends with target properties. *Ind. Eng. Chem. Res.*, **35**, 627.

Venkatasubramanian, V., K. Chan and J. M. Caruthers (1995). Genetic algorithmic approach for computer-aided molecular design. *ACS Symp. Series*, **589**, 396.

RELATIONSHIP BETWEEN TOXICITY
AND ELECTRONIC STRUCTURES OF DIOXINS:
DFT AND NEURAL-NETWORK APPROACHES

Satoshi Itoh[1] , Heihachiro Ueki[2], Masao Arai[3], Kazuaki Kobayashi[3],
and Umpei Nagashima[4]
[1]Toshiba Corporate R&D Center, Kawasaki, Kanagawa, 212-8582, JAPAN
[2]Toshiba CAE Systems, Kawasaki, Kanagawa, 212-8582, JAPAN
[3]National Institute for Research in Inorganic Materials, Namiki, Tsukuba, Ibaraki,
305-0044, Japan
[4]National Institute for Advanced Interdisciplinary Research, Azuma, Tsukuba,
Ibaraki, 305-8562, Japan

Abstract

We have firstly reported the electronic structures of poly-chlorinate dibenzo-p-dioxins (PCDDs) based on the density functional theory (DFT).   In order to clarify relationship between the electronic structures and its toxicity, an application of a neural network to chemo-metric analysis has been performed.   It suggests that the amount of charges on oxygen atoms plays an important role for its toxicity.

Keywords

Dioxins, Electronic structure, DFT, Toxicity, Neural network, Chemo-metric analysis.

Introduction

Over twenty years ago, the poly-chlorinates dibenzo-*p*-dioxins (PCDDs) have received increasing attention as toxic environment pollutants. The mechanism by which PCDDs act as poisons is not completely understood at present, and, therefore, there are many researches from the viewpoints of biological, physiological and chemical (Cheney and Tolly).   It is known that there are 75 different structural formula (isomer) for PCDDs. Among them, 2,3,7,8-tetrachlorodibenzo-p-dioxin (TCDD, see Fig.1.) shows the highest toxic ability, and the other PCDDs, e.g. OCDD is actually harmless.

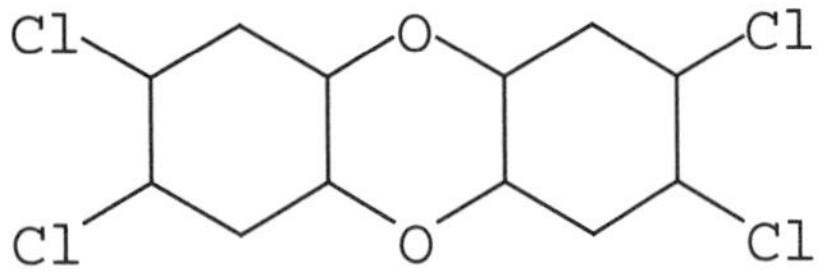

Figure 1. 2,3,7,8-TCDD.

From the viewpoint of quantum chemistry, it would be expected that the electronic structure of TCDD has an effect on the toxicity, so that, there are several reports concerning electronic structure calculations of PCDDs. All of previous calculations, as far as the authors know, have been carried out based on the semi-empirical model for a hamiltonian, such as CNDO, PPP, AM3, and so on. Several papers of these calculations show the fact that chemical hardness of electronic structure of molecules, which is mainly defined by energy difference between LUMO and HOMO levels, gives a good characteristic for toxicity of PCDDs (Kobayashi et al., 1991).   Although it seems that the charge distribution in a molecule play an important role for that toxicity, there is no report on precious numerical results of charge distribution at present. In this paper, the electronic structures of PCDDs based on *ab-initio* calculations will be presented, and we will point out the relation between the electronic structures and the toxicity of PCDDs based on neural network approaches.

Electronic Structures of Dioxins

The electronic structure calculations were carried out based on the density functional theory (DFT) in which Vosko, Wilk and Nusair exchange-correlation functional is used. This functional is so-called local approximation, and we checked the fact that there is no large difference in calculated results by using the gradient corrected exchange-correlation functional which was proposed by Perdew and Wang. Since atomic geometries of PCDDs are not determined experimentally (Kende et al.), the structural optimization was performed by estimating forces acting on each atom and total energy calculations. Optimized structures of 75 types of PCDDs have been determined by same numerical tolerance. The charge distribution was estimated by Mulliken's gross population analysis.

Fig. 2 shows LUMO-HOMO gap energies ($\bullet$E) of PCDDs.

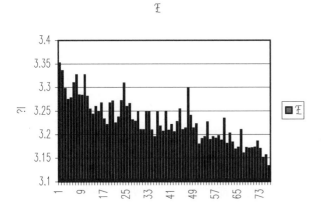

Figure 2.   LUMO-HOMO gap energy (eV).

In Fig.2, the x-axis means isomers of PCDDs , where the left hand of the x-axis indicates the isomers with a few *Cl* atoms contained and the right hand of the x-axis indicates the isomers with many *Cl* atom contained. The gap energy gradually decreases in accordance with the number of *Cl* atoms contained in the dioxin molecules. The dipole moment of molecules plays an important role for reactions between proteins and small molecules. Figure 3 shows the calculated dipole moments of PCDDs. The amount of dipole moment of molecules is strongly related to the symmetry of the molecules. The molecular structure of 2,3,7,8-TCDD has high symmetry, so that its dipole moment is nearly zero. Thus, the 2,3,7,8-TCDD weakly interacts with proteins and other biological molecules based on the higher order multipole interactions. Nevertheless, this TCDD shows the highest toxicity for living things, and it seems that the toxicity of PCDDs is related to microscopic charge distribution in the molecules.

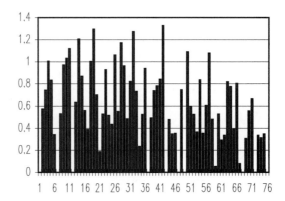

Figure 3.   Calculated dipole moment (a.u.).

Chemo-metric Analysis Based on Neural Networks

From 60's, a neural network (NN) has been extensively investigated from the viewpoint of artificial intelligence (AI). Recently, the NN is recognized as one of the methods for multi-dimensional data analysis or optimization methods in multi-dimensional space (Zupan and Gasteiger).

The NN consists of "neurons", and neurons connect each other (see Fig. 4). In this network, connecting weights are variable, and these weights are determined so as to obtain output signals which are desirable. This procedure is called "learning", and a teaching signal is used during the learning of the NN.

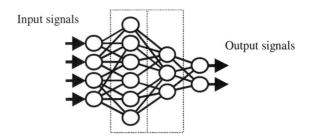

Figure 4.   Neural network with hidden layers.

There are several learning methods of the NN, and, in the present study, the "back propagation of errors" method has been used, where the network topology has hidden layers. Charge distributions in the PCDD molecules and toxicity are used as input signals and teaching signals, respectively. The index of the toxicity is given by the TEF (toxic equivalency factor) value determined by WHO (World Health Organization) in 1993. The charge distributions are strongly dependent on the number and positions of *Cl* atoms contained in PCDD molecules, so that it is not expected that all of the charge distributions relate with the toxicity for each isomer. Among the constituent atoms of PCDD molecules, oxygen atoms and its adjacent carbon atoms are not bonded to *Cl* atoms directly in all isomers, and, this hexagonal ring structures which appear in the center of PCDD molecules are proper structures of PCDDs independent from the number of *Cl* atoms contained in PCDDs. Then, the amount of charges on oxygen atoms and its adjacent carbon atoms is used as input signals. After successive learning cycles, we have obtained an optimized NN.

From the analysis of the learned NN, several characteristics for relationship between toxicity and charge distributions of PCDDs have been obtained. The amounts of charge on the two oxygen atoms are nearly equal for high TEF value PCDDs, and the small charge transfer between oxygen and the other atoms gives rather high TEF values for PCDDs.

The recommendation vales of TEF given by WHO are given for several isomers of PCDDs, and for the other of PCDD isomers, the TEF values of these isomers are to be zero in some reports. However, from the viewpoint of the present study, some of the remainders may be harmful for living things; e.g. from chemo-metric analysis based on the charge distribution and the NN, isomers of PCDDs containing a few *Cl* atoms like one or two *Cl* atoms are expected to be harmful. As for TCDDs, i.e. four *Cl* atoms are contained in the molecules, 1,2,3,7-TCDD, 1,2,3,8-TCDD, 1,3,7,8-TCDD, and 1,3,7,9-TCDD are predicted to show a high TEF value same as the 2,3,7,8-TCDD TEF value. Since there is no report concerning the toxicity of these molecules, it seems that a more extensive experimental study of toxicity for individual isomer of PCDDs is required.

Conclusions

This paper presented the relationship between toxicity and electronic structures of dioxin molecules. The electronic structures of PCDD isomers have been calculated non-empirically by using DFT We have performed chemo-metric analysis of calculated electronic structures, and have utilized NNs with hidden layers. As a result, it was pointed out that the charge distributions on oxygen atoms are related to its toxicity, which suggests a new way to a harmless process of PCDDs. The DFT and NN approaches are very effective for environment problems whose mechanisms are very complicated.

Acknowledgements

One of authors (S.I.) is grateful to Naoki Imazeki for helpful suggestions. A part of this work was funded by National Institute for Research in Inorganic Materials.

References

Cheney, B. V. and T. Tolly (1979). *Int. J. Quantum Chem.,* **16**, 87-110.
Distefano, G. et al. (1983). *J. Chem. Soc. Perkin Trans.,* **2**, 1109-1112.
Fronza, G. and E.Ragg (1982). *J. Chem. Soc. Perkin Trans.,* **2**, 291-293.
Isu, Y., Nagashima, U., Aoyama, T., and H. Hosoya (1996). *J. Chem. Info. Comp. Sci.,* **36**, 286-293.
Kende, A. S. et al. (1974). *J. Org. Chem.,* **9**, 931-937.
Kobayashi, S. et al. (1991). *Chem. Pharm. Bull,.* **39**, 2100-2105.
Kobayashi, S. et al. (1995). *Chem. Pharm. Bull.,* **43**, 1780-1790.
Poland and J. C. Knutson (1982). *Ann. Rev. Pharmacal. B. Toxicol.,* **22** 517-554.
Safe, B. H. (1986). *Ann. Rev. Pharmacal. Toxicol.,* **26**, 371-399.
Zupan, J. and J. Gasteiger (1993). Neural networks for chemists. *VCH Verlagsgesellschaft mbH.*

SELF-ASSEMBLY OF REVERSE MICELLES BY MOLECULAR DYNAMICS SIMULATION

H. D. Cochran, S. Salaniwal, S. T. Cui, and P. T. Cummings
Oak Ridge National Laboratory
Oak Ridge, TN 37831-6224
and
University of Tennessee
Knoxville, TN 37996-2200

Abstract

We have studied reverse micelles with aqueous cores in CO_2 by molecular dynamics simulation using a potential model for the dichain surfactant, $(C_7H_{15})(C_7F_{15})CHSO_4^-Na^+$, that we developed for this purpose. It was assembled from existing models for the sulfate head group, sodium ion, alkane tail and perfluoroalkane tail. We simulated 33 surfactant, 1,175 water, and 12,800 CO_2 molecules at T = 298 K and CO_2 density = 0.848 g/cm³, the subcritical state point of small angle neutron scattering experiments. The system rapidly self-assembled into aggregates with water in the core and surfactants at the interface. Numerous rapid changes in the number of aggregates by ±1 represent collisions of the surfactant tails, which did not result in coalescence. When collisions resulted in contact between aqueous cores, coalescence almost always occurred. The surfactant tails, therefore, provide steric stabilization of the aggregates. The rate of self-assembly was determined to be consistent with the theory of Smoluchowski for diffusion-limited coalescence. The strong coulombic forces and high diffusivity of molecules and aggregates, explains why the self-assembly was rapid in CO_2 compared to the rate in organic liquids. The aggregates formed had the appearance of reverse micelles and their structural characteristics were consistent with experimental results.

Keywords

Reverse micelle, Self-assembly, Molecular dynamics, Surfactant, Aggregation, Carbon dioxide.

Introduction

Advances in computational technology have enabled molecular simulation of surfactant self-assembly in aqueous systems (Desplat and Care, 1996; Esselink *et al.,* 1994; Karaborni *et al.,* 1994a,b). But very few simulation studies (Brown, 1991; Linse, 1989; Tobias and Klein, 1996) have addressed the phenomenon of surfactant aggregation in non-polar solvents which results in the formation of aggregates known as reverse micelles (RMs). RMs have hydrophilic (often aqueous) cores surrounded by the hydrophobic surfactant tails, the reverse of the familiar structure of oil-in-water micelles. RMs find numerous applications in fields such as lubrication, emulsion polymerization reactions, drug delivery, and enhanced oil recovery.

Recent experiments (e.g., Hoeffling et al., 1991; Newman et al. 1993) have investigated the use of carbon dioxide (CO_2) as a non-polar solvent medium for the formation of RMs. CO_2 is an inexpensive and essentially benign alternative to potentially hazardous industrial solvents. Formation of RMs in CO_2 may enable a wide range of industrial operations in a medium with minimum environmental impact. Consani and Smith (1990) showed that most industrially available surfactants are incapable of forming stable RMs in CO_2 because of their negligible solubility. Thus, a number of efforts (e.g., DeSimone, 1994; Fulton et al., 1995; Harrison et al., 1994; Johnston et al., 1996; McClain et al., 1996) have sought surfactants having favorable interactions with CO_2. These studies have identified a few CO_2-philic surfactant molecules, but little is known about the factors determining CO_2-philicity. Most of the successful CO_2-philic surfactants have contained perfluorinated chains; however, the reasons for CO_2-philicity of perfluoro moities is not well understood. Molecular-level knowledge could be helpful in guiding efforts to discover or design better CO_2-philic surfactants for industrial applications. Molecular simulation techniques can play an important role in developing this much-needed knowledge base.

We performed molecular dynamics simulations of a dichain surfactant + H_2O + CO_2 system. The dichain surfactant molecule $[(C_7F_{15})(C_7H_{15})CHSO_4^-Na^+]$ is one of the few CO_2-philic surfactant molecules that have been shown (Harrison et al., 1994) to form RMs with aqueous cores in CO_2. Recently, Eastoe et al. (1996) performed a small-angle neutron scattering (SANS) study on this system and determined the size and shape of the RMs. In this paper, we present simulations intended to reproduce the experimental system. We describe the detailed molecular model for the dichain surfactant molecule. We examine the mechanism of self-assembly of the RM-like aggregates and present quantitative comparisons of the simulation and experimental results with additional microscopic structural details.

Potential Models and Simulation Methodology

For quantitative comparison with the experimental work, our simulations employed detailed and realistic models for the surfactant, H_2O, and CO_2. Although this makes the system computationally demanding, it minimizes artificiality that might affect the dynamics of surfactant aggregation and/or the morphology of the RMs. The potential model for the hybrid surfactant molecule used in this simulation study was constructed like a mosaic by assembling existing models for the alkane tail, perfluoroalkane tail, sulfate head group, and the sodium counter-ion. For the alkane tail we used the united atom model of Siepmann et al. (1993) which accurately predicts phase envelopes of the n-alkanes. More importantly, as shown by Cui et al. (1999), it predicts the solubility of n-alkanes in CO_2 consistent with experiments. In this model, each CH_2 and CH_3 functional group is represented by a single spherical interaction site (united atom) centered on the carbon atom. Sites are connected via rigid bonds of length 1.54 Å, and those on different surfactant molecules and on the same molecule separated by more than three bonds interact via the Lennard-Jones (LJ) potential. The intra-molecular interactions include bond-angle bending and torsional potentials.

We used the perfluoroalkane model of Cui et al. (1998), which is quite similar to the alkane model. It was fitted to the phase envelope of n-perfluroralkanes and also predicts pefluoroalkane/CO_2 phase equilibrium (Cui et al., 1999). United atoms with LJ potentials are used to represent the CF_2 and CF_3 functional groups of the tail, and the bond length is fixed at 1.54 Å. The bond angle potential is similar to that of the alkane tail but has a slightly larger equilibrium angle (114.6°). The torsional potential has a twisted trans minimum at ~ 17° and gauche minima at ~ 125° and ~ 83°.

For the sulfate head group, we used the fully atomistic model of Cannon and co-workers (1994) with a rigid tetrahedral structure. To account for electrostatic interactions, the sulfur atom carries a charge of +2e while three oxygen atoms carry a charge of –1e. The fourth oxygen atom does not carry any negative charge as it is attached to the tails. A sum of coulombic and LJ terms describes the intermolecular interactions. A single LJ sphere with charge +1e represents each sodium ion.

We used the rigid CO_2 model of Harris and Yung (1995) which accurately reproduces the vapor-liquid co-existence curve and critical point. It consists of three LJ sites with a charge on each. Finally, for H_2O, we used the rigid SPC/E model of Berendsen et al. (1987); it has a single LJ site for oxygen with a charge of –0.8476e. The hydrogens are represented by +0.4238e point charges.

Potential parameters for all models and additional details are given by Salaniwal et al. (1999, 2000a, 2000b, 2000c). All interactions between unlike sites were modeled by the Lorenz-Berthelot combining rules because no basis is available for more suitable ones for, say, the fluorinated groups with which the conventional rules might be questioned

Molecular dynamics simulations of two different system sizes were performed. A small system consisting of 30 surfactant, 132 H_2O, and 2,452 CO_2 molecules was chosen for computational economy and to provide qualitative insight in exploratory calculations. The state conditions used with this system were 310 K and CO_2 density 0.482 g/cm³. The large system had 33 surfactant, 1,175 H_2O, and 12,800 CO_2 molecules at 298 K and CO_2 density 0.848 g/cm³ to mimic the concentrations, temperature, and density in the SANS study and corresponds to one average-sized RM.

The equations of motion for each site were solved with the Rattle algorithm (Anderson, 1983) to constrain bond lengths, using a time step of 1.483 fs. Simulations used both an aggregated and a scattered starting configuration. The aggregated starting configuration had all the H_2O molecules in a sphere. The surfactants with fully extended tails were on the surface of the sphere with head groups toward the center, resembling a RM. In the scattered starting configuration all the H_2O molecules were scattered within the simulation box. Of the infinite configurations that can be used as starting conditions, these were chosen for convenience.

Results and Discussion

We observed three stages in the self-assembly process: almost instantaneous ion pairing, followed by rapid hydration of the ion pairs, and finally slower aggregation of the hydrated ionic surfactants. This mechanism was confirmed by observing the distribution of energy during self-assembly, by observing the fraction of H_2O associated with surfactants (f_s, as shown in Fig. 1), and by observing visualizations of the process. The other results (omitted for brevity) were consistent with this mechanism for both large and small systems with both aggregated and dispersed starting configurations. The strong coulombic forces and high diffusivity of molecules and aggregates, explains why the self-assembly was rapid in CO_2 compared to the rate in organic liquids.

The number of surfactant aggregates present, $N(t)$, as seen in Fig. 2, was calculated by the algorithm of Sevick et al. (1988) in which two particles are in the same cluster if their centers lie within 3.3 Å. The rate of self-assembly obeys a power law with exponent -0.43, consistent with Smolukowski's theory of diffusion-limited aggregation (See Ziff et al., 1985).

Aggregate diffusivities determined from the simulations varied as the inverse of the radius of gyration, also consistent with the theory. The many rapid changes of $N(t)$ by ±1 are instances of tail-tail collisions; when collisions involved contact between aqueous cores, coalescence almost always occurred. Thus, the aggregates are kinetically stabilized by their tails.

The water to surfactant ratio, W_0, the radius of gyration, R_g, and the surface area per surfactant molecule, A_s, of the large RM-like aggregate shown in Fig. 3 are reasonably consistent with the SANS results as shown in Table 1. Consistent with the SANS results, a surprisingly large amount of the aqueous core is directly exposed to the CO_2 solvent.

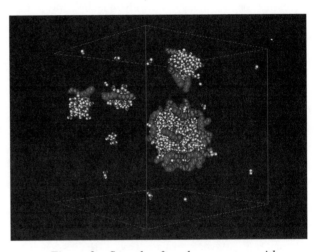

Figure 3. Snapshot from large system with aggregated starting configuration after 1 ns.

Table 1. Characteristics of Reverse Micelles.

	This work	SANS
W_0	35.2	35
R_g (Å)	15.4 ± 0.2	20.5 ± 1.0
A_s (Å^2)	115 ± 3	140

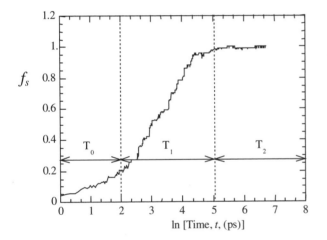

Figure 1. Fraction of H_2O associated with surfactant in the small system with dispersed starting configuration.

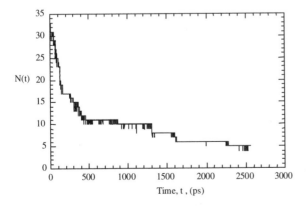

Figure 2. Kinetics self-assembly of large system with dispersed starting configuration.

The surfactant head groups and counter ions were located close to the CO_2-H_2O interface, and the H_2O in the first and second solvation shells of the ions was strongly oriented by the coulombic forces. However, the majority of the H_2O was bulk-like with density ~1.0 g/cm3 and substantial hydrogen bonding. This observation holds promise for the possibility of performing aqueous chemistry within RMs dispersed in environmentally friendly CO_2. Interestingly, the CO_2 concentration in the aqueous cores was observed to be essentially zero, probably as a result of the high ionic strength. Likewise, the H_2O concentration in the CO_2 phase was substantially less than the equilibrium solubility, again, probably as a result of the high ionic strength in the core.

Concluding Remarks

Molecular dynamics simulation of self-assembly of surfactant aggregates with aqueous cores in CO_2 has been performed using realistic molecular models. The resulting aggregates have the appearance of RMs, and their structural characteristics agree with prior experimental results. The self-assembly process was found to occur rapidly in three stages–ion pairing, hydration, and diffusion-limited aggregation. The aggregates were kinetically stabilized by the surfactant tails.

In preliminary work (Salaniwal et al., 2000c), we have begun examining the influence of the molecular architecture and chemical characteristics of the surfactant on self-assembly and aggregate stability. This work will be continued to provide guidance for the possible design of better surfactants. The industrial applications of surfactants for stabilizing RMs with aqueous cores in CO_2 may include enzymatic reactions, dispersion polymerizations, and many others. The potential of green chemistry for the future with such systems appears very promising.

Acknowledgments

This work was supported by the DOE and the NSF. Calculations were performed with generous allocations from NERSC, ORNL, and UT. ORNL is managed by UT-Battelle, LLC, under contract DE-AC05-00OR22725.

References

Andersen, H. C. (1983). Rattle: a velocity Verlet version of the shake algorithm for molecular dynamics calculations. *J. Comput. Phys., 52*, 24-34.

Berendsen, H. J. C., J. R. Grigera, and T. P. Straatsma (1987). The missing term in effective pair potentials. *J. Chem. Phys., 91*, 6269-6271.

Brown, D. and J. H. R. Clarke (1991). Molecular dynamics simulation of a model reverse micelle. *J. Phys. Chem., 92*, 2881-2888.

Cannon, W. R., B. M. Pettitt, and J. A. McCammon (1994). Sulfate anion in water: model structural, thermodynamic, and dynamic properties. *J. Phys. Chem., 98*, 6225-6230.

Consani, K. A. and R. D. Smith (1990). Observations on the solubility of surfactants and related molecules in carbon dioxide at 50C. *J. Supercrit. Fluids, 3*, 51-65.

Cui, S. T., *et al.* (1998). Intermolecular potentials and vapor-liquid phase equilibria of perfluorinated alkanes. *Fluid Phase Equil., 146*, 51-61.

Cui, S. T., H. D. Cochran, and P. T. Cummings (1999). Vapor-liquid phase coexistence of alkane carbon dioxide and perfluoralkane carbon dioxide mixtures. *J. Phys. Chem., 103*, 4485-4491.

DeSimone, J. M., *et al.* (1994). Dispersion polymerizations in supercritical carbon dioxide. *Science, 265*, 356.

Desplat, J. C. and C. M. Care (1996). A Monte Carlo simulation of the micellar phase of an amphiphile and solvent mixture. *Mol. Phys., 87*, 441-453.

Eastoe, J., et al. (1996). Droplet structure in a water-in-CO_2 microemulsion. *Langmuir, 12*, 1423-1424.

Esselink, K. *et al.* (1994). Molecular dynamics simulations of model oil/water/surfactant systems. *Colloids and Surfaces A, 91*, 155-167.

Fulton, J. L., *et al.* (1995). Aggregation of amphiphilic molecules in supercritical carbon dioxide: A small angle x-ray scattering study. *Langmuir, 11*, 4241-4249.

Harris, J. G. and K. H. Yung (1995). Carbon dioxide's liquid-vapor coexistence curve and critical properties as predicted by a simple molecular model. *J. Phys. Chem., 99*, 12021-12024.

Harrison, K. L., *et al.* (1994). Water-in-carbon dioxide microemulsions with a fluorocarbon-hydrocarbon hybrid surfactant. *Langmuir, 10*, 3536-3541.

Hoefling, T. A., R. M. Enick, and E. J. Beckman (1991). Microemulsions in near-critical and supercritical CO_2. *J. Phys. Chem., 95*, 7127-7129.

Johnston, K. P., *et al.* (1996). Water-in-carbon dioxide microemulsions: an environment for hydrophiles including proteins. *Science, 271*, 624-626.

Karaborni, S. *et al.* (1994a). Simulating the self-assembly of gemini (dimeric) surfactants. *Science, 266*, 254-256.

Karaborni, S. *et al.* (1994b). Simulating Surfactant self-assembly. *J. Phys. Cond. Matter, 6*, A351-A356.

Linse, P. (1989). Molecular dynamics study of the aqueous core of a reversed ionic micelle. *J. Chem. Phys., 90*, 4992-5004.

McClain, J. B., *et al.* (1996). Design of nonionic surfactants for supercritical carbon dioxide. *Science, 274*, 2049-2052.

Newman, D. A., *et al.* (1993). Phase behavior of fluoroether-functional amphiphiles in supercritical carbon dioxide. *J. Supercrit. Fluids, 6*, 205-210.

Salaniwal, S., *et al.* (1999). Self-assembly of reverse micelles with aqueous cores in supercritical carbon dioxide via molecular simulation. *Langmuir, 15*, 5188-5192.

Salaniwal, S., *et al.* (2000a). Self-assembly in a dichain surfactant/water/carbon dioxide system via molecular simulation. I. Structural properties of surfactant aggregates. *Langmuir,* submitted.

Salaniwal, S., *et al.* (2000b). Self-assembly in a dichain surfactant/water/carbon dioxide system via molecular simulation. II. Aggregation dynamics. *Langmuir,* submitted.

Salaniwal, S., *et al.* (2000c). Molecular Dynamics Simulation of Reverse Micelles in Supercritical Carbon Dioxide. *Ind. & Eng. Chem. Res.,* submitted.

Sevick, E. M., P. A. Monson, and J. M. Ottino (1988). Monte Carlo calculations of cluster statistics in continuum models of composite morphology. *J. Chem. Phys., 88*, 1198-1206.

Siepmann, J. I., S. Karaborni, and B. Smit (1993). Simulating the critical behavior of complex fluids. *Nature, 365*, 330-332.

Tobias, D. J. and M. L. Klein (1996). Molecular dynamics simulations of a calcium carbonate/calcium sulfonate reverse micelle. *J. Phys. Chem., 100*, 6637-6648.

Ziff, R. M., E. D. McGrady, and P. Meakin (1985). On the validity of Smoluchowski's equation for cluster-cluster aggregation kinetics. *J. Chem. Phys., 82*, 5269-5274.

PHASE EQUILIBRIA OF MULTICOMPONENT SYSTEMS USING PARALLEL MOLECULAR DYNAMICS ALGORITHMS

Lev D. Gelb
Florida State University
Tallahassee, FL 32306

María E. Suárez and Erich A. Müller
Universidad Simón Bolívar
Caracas, Venezuela

Abstract

Molecular modeling can be applied to model complex systems of industrial relevance, especially those where experimental data is difficult or expensive to obtain, or otherwise unavailable. In this work we propose algorithms to simulate vapor-liquid and liquid-liquid phase behavior of real systems with multipolar interactions. Simulations are done using molecular dynamics (MD) quench techniques, after which an interface-detection algorithm is run to discern between the two phases formed. Examples presented are the phase diagram of benzene and the liquid-liquid equilibria of the sulpholane/n-octane/benzene ternary mixture. We present the results of this work placing emphasis on the fact that parallel computers and algorithms have become cost-effective tools for solving industrially relevant problems.

Keywords

Molecular dynamics, Phase equilibria, Liquid-liquid, Liquid-vapor, Quench, Parallel computers, Benzene, Octane, Sulpholane, Keesom potentials, Multipolar.

Introduction

In this work we propose algorithms to simulate the vapor-liquid and liquid-liquid phase behavior using Molecular Dynamics (MD). An example application to the sulpholane/n-octane/benzene ternary mixture is given. Sulpholane is an oxygenated thiophene which, added to a hydrocarbon mixture, promotes a phase separation into an aromatic rich phase and alkane rich phase. The properties of sulpholane-containing mixtures are difficult to correlate using standard thermodynamic equations of state and activity models due to the large multipolar moments associated with the oxygenated solvent and the aromatic. Only sparse experimental data points are available (Lee and Kim, 1998), which are used to obtain limited range correlations based on liquid state activity models. Due to the complex electrostatic interactions present in this mixture, the current predictive methods (e.g. cubic equations of state) may be unreliable. We demonstrate that molecular simulation using parallel computers and algorithms has become cost effective tool for solving industrially relevant problems.

We report MD simulations using a parallel algorithm on a PC-based cluster. The system is modeled at constant pressure and overall composition and the temperature is allowed to drop (quenching) from a uniform one-phase region to produce a phase split. After localized equilibration, the compositions of the coexisting phases are calculated using local composition histograms. The quenching process allows the visualization of the phase split and the study of diffusion mechanisms in the mixture. The advantage over conventional Gibbs

Ensemble Monte Carlo simulations is that high densities may be conveniently explored and additionally, the dynamics of the phase separation may be monitored. We also note that molecular dynamics simulations of this type are feasible for very large and/or complex molecules, which are difficult to treat by Monte Carlo methods. The programs have been ported into BUHO, a 16-processor Beowulf type II machine, installed at the thermodynamics group at Simón Bolívar University, consisting of 8 dual-processor PCs connected via a fast Myrinet switch.

Intermolecular Potentials

As will be discussed in the next section, we have used a MD algorithm that solves the 3N (where N is the number of particles) simultaneous differential equations of motion. The inclusion of geometrical or energetic anisotropies would require the additional specification of at least a directional vector in each molecule (3N additional variables) and 3N corresponding differential equations to account for the angular momentum conservation.

In an effort to maintain simplicity without compromising the applicability of the methods we have chosen to use an intermolecular potential which takes into account repulsive and dispersive (attractive) forces, but also takes into account multipolar moments, specifically dipolar and/or quadrupolar moments. The molecules are all assumed spherical, thus geometrically isotropic.

The intermolecular potential (U) potential consists of two parts,

$$U = U^{LJ} + U^{multipolar} \qquad (1)$$

where the superscripts LJ and $multipolar$ denote the Lennard-Jones and multipolar contributions respectively. The usual potential is used for the normal attraction/dispersion, namely,

$$U^{LJ} = 4\varepsilon_{ij}\left[\left(\frac{\sigma_{ij}}{r_{ij}}\right)^{12} - \left(\frac{\sigma_{ij}}{r_{ij}}\right)^{6}\right] \qquad (2)$$

where r_{ij} is the center-center intermolecular distance and σ and ε are the size and energy parameters. The usual Lorentz-Berthelot combination rules (geometric mean for ε_{ij} and arithmetic mean for σ_{ij}) are used to relate the cross-interaction parameters to those of the pure substances. The multipolar contributions take into account permanent dipoles and/or quadrupoles. A rigorous formulation should take into account both the separations between molecules and the orientations (see for example Gray and Gubbins, 1984). However, if one takes an average over all orientations with each orientation weighted according to its Boltzmann factor, one arrives to an equation that represents the average multipolar attraction. The original suggestion to use such an approximation comes from Keesom (1922), thus the name "Keesom potentials". The result may be expressed as a sum of terms when expanded in terms of powers of $\beta = 1/kT$, where k is Boltzmann's constant and T is the temperature. The leading term of this series can be found in Reed and Gubbins (1973). The multipolar attraction is considered to be caused by either dipole-dipole, dipole-quadrupole and/or quadrupole-quadrupole interactions,

$$U^{multipolar} = -\frac{\beta}{3}\frac{(\mu_i\mu_j)^2}{r_{ij}^6} - \frac{\beta}{2}\frac{\left[(\mu_iQ_j)^2 - (\mu_jQ_i)^2\right]}{r_{ij}^8} - \frac{7\beta}{5}\frac{(Q_iQ_j)^2}{r_{ij}^{10}}, \qquad (3)$$

where μ_i and Q_i are the permanent dipole and quadrupole moments respectively. While these integrations are only valid for either weak multipolar moments and/or high temperatures, they represent a consistent way of adding to a LJ potential the effect of electrostatic interactions. The advantage of using these potentials over simple LJ potentials becomes evident when studying mixtures of polar and non-polar fluids. In these cases, the cross-interactions are well defined in our method and do not require the use of ad-hoc binary interaction parameters.

The potentials used cannot be applied to chain-type molecules nor to molecules which present strong directional attractions, such as hydrogen-bonding fluids. Ionic systems are also excluded from these studies.

Table 1. Potential Parameter Values.

Substance	ε/k (K)	σ (Å)	μ (Cm)	Q (Cm^2)
Benzene	358.74	5.236	0	$4.09 \cdot 10^{-39}$
Sulpholane	579.81	5.230	$4.57 \cdot 10^{-29}$	0
n-Octane	432.22	4.748	0	0

The VLE and LLE of generic dipolar/nonpolar mixtures described by Keesom potentials have been presented by Sadus (1996a,b), however we are unaware of any application to real systems. In this particular application we have calculated the phase diagram of benzene and the liquid-liquid equilibria of the benzene/sulpholane/n-octane mixture. The potential parameters ε and σ for these substances are found by comparison (regression) between published values for the critical properties of the multipolar model potentials (van Leeuwen et al., 1993; Stapleton et al., 1989) and experimental critical temperatures and densities. The critical properties of the real fluids are found from tabulations or in the case of sulpholane estimated from group contribution methods (Reid et al., 1987). Table 1 summarizes the values used throughout this study.

Program Details

Molecular dynamics were performed using a 5th-order Gear predictor-corrector algorithm (Allen and Tildesley, 1987). Constant temperature (NVT) simulations are performed with the Nosé-Hoover (Nosé, 1984, Hoover, 1985) thermostatting method, in which the total kinetic energy of the system is coupled to an external degree of freedom which acts as a piston, and ensures a canonical distribution of momenta at equilibrium conditions. In isothermal-isobaric (NPT) simulations, this approach is extended to also couple the volume of the simulation cell to a second external degree of freedom which drives the system towards an imposed constant pressure.

We have implemented parallel versions of the programs using a domain-decomposition approach, in which a rectilinear volume is assigned to each processor in a three-dimensional decomposition scheme. Communica-tions between processors are handled using Message Passing Interface (MPI) library. Each processor is responsible for the storage of data pertaining to particles within its assigned volume, for the integration of the equations of motion for these particles, for the computation of the forces these particles exert on each other, and for computation of one-half of the interactions (forces) between its own particles and those in neighboring domains. The algorithm used is well documented (Plimpton, 1995). At the beginning of each time-step, each processor updates its own particles' positions using the first half of the predictor-corrector integration algorithm. Particles that are within potential-function range of neighboring domains are then packed into arrays and sent to those processors; particles near to edges and corners of a domain will have to be sent to more than one other processor. Upon receipt of all positions, each processor then computes interactions between its own particles and those that it has just been sent. In the next phase, these 'partial force vectors' are sent back to the processors that own the relevant particles, where they are added together to obtain the complete force on every particle in the system. At this point, the corrector step of the integration algorithm is made, completing the time-step.

For the generation of distribution probabilities of density and mole fraction, the volume of the simulation cell is divided into sub-cells, in which the mole fractions of each species and the total density are calculated and accumulated into histograms. For relatively small simulation cells, and for systems in which phase separation occurs relatively slowly, these sub-cells may contain significant portions of two or more phases, which leads to unphysical broadening of the histogram peaks corresponding to each phase. We have, therefore, developed an algorithm for detecting when this situation occurs, thus only collecting histogram data in sub-cells that contain entirely one phase. In the case of interface detection based only on density, the algorithm works as follows. Particles that are in or very near to an interface between two phases have coordination numbers reflecting the interfacial region; e.g., they have fewer neighbors than particles in the liquid phase, and more than particles in the vapor phase. By counting the local coordination of every particle, these can be identified unambiguously. Sub-cells that contain more than 30% "interfacial" particles are then (arbitrarily) excluded from the histogram count. In the case of discrimination by mole-fraction, a similar algorithm can be applied to a single species; for every particle, one counts the total number of neighbors of a single species, and locates particles on the interface by values of this quantity that fall between those of the pure phases. In preliminary studies of vapor-liquid coexistence we have determined that this method is effective for interface location provided that sensible definitions of the coordination sphere are used and that rough estimates of the coexistence densities are available.

Simulation Details

The systems were initially equilibrated at a reasonably high temperature (typically $T^* = 5$) and quenched to the required temperature using an NVT ensemble. This quenching does not guarantee the required pressure specification, so the resulting configuration was used as the input of an NPT program. We do not use the NPT program to perform the quench since the density fluctuations are too large and the system may evolve into metastable vapor states by excessively increasing the available volume.

The programs were run on BUHO, a Beowulf type II cluster built in the thermodynamics group at Simón Bolívar University, as part of the High Performance Computing Laboratory. It consists of eight dual-processor Pentium III machines connected using a high speed (1 GB/s) Myrinet connection. Each node runs at 550 MHz and each pair shares 512 MB of memory. The machine itself is reasonably simple to build and maintain and offers an excellent performance/price ratio.

With the given configuration we can run systems of up to 5×10^5 molecules without problems. Here we have used 8000 and 64000 molecules for pure systems and mixtures respectively. Typical equilibration/production of $2 \times 10^5 + 4 \times 10^5$ time steps are used. On BUHO these runs amounted to approximately 24 CPUh on a single node.

Results

In Fig. 1 we show the temperature-density diagram of pure benzene (Perry et al., 1984) and the simulation data corresponding to the use of the proposed intermolecular potential. Some discrepancies are observed, noticeably in the critical region.

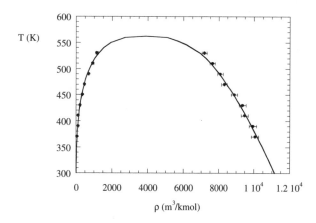

Figure 1. *Temperature(T)-density(ρ) diagram of benzene. Solid lines are experimental values, circles are MD-quench results using Eqn.(1).*

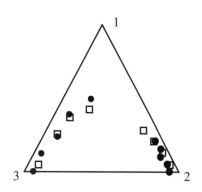

Figure 2. *Liquid-liquid phase diagram for the benzene (1) / sulpholane (2) / n-octane (3) system at 402.35K. Open squares are experimental results of Lee and Kim (1998); solid circles are MD-quench results using Eqn.(1).*

In Fig. 2 we show the results for the liquid-liquid equilibria of the ternary mixture. Tie-lines which connect corresponding equilibrium points between alkane-rich (at the left of the plot) and sulpholane-rich phases (at the right of the plot) are ommitted for clarity. In this case, the reasonably quantitative between theory and experiment except in the critical region. However, it is interesting to point out that even common engineering correlations, with a number of adjustable parameters give similar results. In this case we have no binary or ternary adjustable parameters, i.e. the simulations are actual predictions, and in this sense the results are very encouraging.

Conclusions

We have shown how the use of PC-cluster parallel computer running MD programs is a viable alternative to existing methods for calculating phase equilibria of real systems.

In spite of their underlying assumptions, the proposed potentials give reasonable qualitative results for vapor-liquid and liquid-liquid equilibria. It is expected that more detailed potentials, including shape effects and a more accurate description of the multipolar moments should improve the results. State-of-the-art parallel PC-clusters are more than sufficient for these type calculations.

Acknowledgments

E.A.M. wishes to acknowledge the financial support given by the Agenda Petroleo (project 97-003530) to this work.

References

Allen M.P. and D.J. Tildesley (1987). *Computer Simulation of Liquids*, Clarendon, Oxford.

Gray C.G., and K.E. Gubbins (1984). *Theory of Molecular Fluids,* Vol. 1, Clarendon, Oxford.

Hoover W.G. (1985) Canonical dynamics: Equilibrium phase-space distributions. *Phys. Rev. A*, **31**, 1695-1697

Keesom W.H. (1922). *Comm. Leiden, Supl.* 24a, 24b; *Phys.Z.* **22**, 129

Lee S. and H. Kim (1998). Liquid-liquid equilibria for the ternary systems sulpholane + octane + benzene, sulpholane + octane + toluene, and sulpholane + octane + p-xylene at elevated temperatures. *J. Chem. Eng. Data*, **43**, 358-361

Nosé S. (1984). A molecular dynamics method for simulations in the canonical ensemble. *Mol. Phys.*, **52**, 255-268.

Perry R.H., Green D.W. and J.O. Maloney, ed. (1984). *Perry's Chemical Engineer's Handbook.* 6th edition. McGraw-Hill, New York.

Plimpton S. (1995). Fast parallel algorithms for short-range Molecular Dynamics. *J. Comp. Phys.*, **117**, 1-19.

Reed T.M., and K.E. Gubbins (1973). *Applied Statistical Mechanics.* McGraw-Hill, New York. Appendix F.

Reid R.C., Prausnitz J.M. and B.E. Poling (1987). *The Properties of Gases and Liquids*, 4th ed. McGraw-Hill, New York.

Sadus R.J. (1996a). Molecular simulation of the vapour-liquid equilibria of pure fluids and binary mixtures containing dipolar components: The effect of Keesom interactions. *Mol. Phys.*, **87**, 979-990.

Sadus R.J. (1996b). Molecular Simulation of the liquid-liquid equilibria of binary mixtures containing dipolar and non-polar components interacting via the Keesom potential. *Mol. Phys.*, **89**, 1187-1194.

Stapleton M.R., Tildesley D.J., Panagiotopoulos A.Z., and N. Quirke (1989). Phase equilibria of quadrupolar fluids by simulation in the Gibbs Ensemble. *Mol. Simulation.*, **2**, 147-162.

van Leeuwen M.E., Smit B., and E.M. Hendriks (1993). Vapor-liquid equilibria of Stockmayer fluids: computer simulations and perturbation theory. *Mol. Phys.*, **78**, 271-283.

PREDICTION OF THERMOPHYSICAL PROPERTIES OF MIXTURES USING MOLECULAR SIMULATIONS BASED ON DENSITY FUNCTIONAL THEORY

Maurizio Fermeglia and Sabrina Pricl
Department of Chemical Engineering
University of Trieste
34127 Trieste, Italy

Andreas Klamt
COSMOlogic GmbH&Co.KG
51381 Leverkusen, Germany

Abstract

The present paper investigates the possibility of applying molecular simulations, based on DFT, to the prediction of thermophysical properties of simple fluids containing several types of atoms as well as of their binary mixtures. The main goal of the paper is to establish a simple and not expensive procedure for calculating, with *ab initio* techniques, the properties of interest, which will be fitted later to existing and reliable equations of state (*EOS*) for describing a spectrum of thermophysical properties in a wide range of process conditions.

Keywords

Phase equilibria, Equation of state, Perturbed-hard-sphere-chain theory, Binary mixtures, Density functional theory, Molecular simulations.

Introduction

Modern industry must face the actual challenge of designing, and hence producing, molecules (or molecular ensembles) having well defined chemico-physical characteristics (such as pharmaceutics, polymers and additives) by fast and economical processes. The recent progresses in computer architecture and the relevant molecular simulation software have transformed the theoretical models from topics of exquisite academic interest into practical tools for molecular and process design, thus becoming of industrial relevance. Notwithstanding the progressively increasing importance of molecular simulation in several industrial sectors, still a very few papers can be found in the current literature documenting the potentiality of such applications.

Recently, we (Fermeglia and Pricl, 1998, 1999a, 1999b) reported some examples on how molecular and process simulations can be successfully bridged via recourse to the equation of state (*EOS*) theory, both for simple and long-chains molecules. The basic machinery of the procedure consists in the application of molecular mechanics - molecular dynamics techniques for the determination of physically-based *EOS* parameters. Thus, we developed a strategy to calculate the perturbed hard sphere chain (*PHSCT*) *EOS* parameters A^*, V^* and E^* (Fermeglia *et al.*, 1997 and 1998) and the Lattice Fluid (*LF*) *EOS* parameters P^*, ρ^* and T^* (Sanchez and Lacombe, 1976). In the case of the *PHSCT EOS*, the proposed method essentially consists in the rough estimation of the molecular surface areas and volumes, related to the corresponding *EOS* parameter A^* and V^*, by a combination of molecular mechanics and graphical algorithms, whereas the third, energetic parameter E^* can be obtained from molecular dynamics simulations (*MD*) in the gas state. For the *LF EOS*, the procedure consists in determining *PVT* data sets via *MD* simulations at different decreasing temperatures. The two *EOS* parameters P^* and ρ^* can be directly obtained by data extrapolation to 0 K whereas the third parameter T^* can be evaluated by inserting P^*, ρ^* and a set of simulated *PVT* data into the *EOS* expression.

The aim of the present paper is to present an alternative procedure for the prediction of phase equilibria for pure components and their mixtures. To this purpose, we resorted once again to the *PHSCT EOS*, and in particular to its extended version for mixtures (Fermeglia *et al.*, 1998), and used a set of well-known compounds as standard testing materials. For the calculation of the *EOS* parameter sets, this time we propose a procedure based on the description of the solvation phenomena at a molecular level using the *COSMO* approach (Klamt and Schüürmann, 1993). The *COSMO* model has been extended beyond the dielectric approximation to real systems (*COSMO-RS*), covering neutral molecules containing *H, C, N, O, Cl* (Klamt *et al.*, 1998) and, more recently, *F* (Andzelm *et al.*, 1998). As we shall discuss later in more detail, the quality of the prediction, in conjunction with the relatively fast calculation time required for the simulation, can be considered extremely satisfactory. In our opinion, this procedure could be of great help in integrating molecular and process simulations, particularly in those cases where *EOS* parameters must be obtained for molecules with scarce if not absent experimental data sets available.

Theory and Simulation Details

The Conductor-Like Screening Model for Real Solvent (COSMO-RS)

The *COSMO-RS* theory takes the ideally screened molecules as starting points for the description of molecules in solution. The deviation from ideal screening, which unavoidably occur in any solvent, are described as pairwise misfit interactions of the ideal screening charges on contacting parts of the molecules in the fluid. Since *COSMO-RS* does not longer depend on experimental data or any parametrization for the solvent, it efficiently enables the calculation of the chemical potential of almost any solute in almost any solvent. Thus, it is capable of treating almost the entire equilibrium thermodynamics of fluid systems and, in our opinion, can constitute a powerful alternative to group/fragment based methods like *UNIFAC*. In this work, we resort to the *COSMO-RS* version as implemented in the *DMol*[3] version of *Cerius*[2] (v. 3.9 from Molecular Simulations Inc., USA). This choice relies on the fact that *DFT* calculations yield ground-state properties as reliable as Hartree-Fock calculations with higher order correlation corrections but at much lower costs. These, even at the default level, have sufficiently good tails to reliably reproduce quantities such as dipole

moments and polarizabilities, which are of crucial importance for any solvation calculation.

The gas-phase reference energies for all the structures of the data sets were obtained from non-local *DFT VWN-BP* gas-phase optimization applying the *DNP* basis (v.4.0.0) of *DMol*[3] and a real space cut-off = 5.50 Å. All structures were then re-optimized in a continuum conductor (i.e., with *COSMO/DMol*[3] and f(ε) = 1, using a number of segments = 92). Finally, the phase equilibrium properties of each binary mixture considered, necessary for the determination of the *EOS* parameters according to the maneuver reported below, were obtained from the *COSMO-RS* files using the new software *COSMOtherm* from *COSMOlogic* (Leverkusen, Germany).

The PHSCT EOS for Binary Mixtures

In the new, alternative procedure proposed in this work, the entire set of parameters of the *PHSCT EOS* version for mixtures can be obtained from *COSMO/DMol*[3] and *COSMOtherm* calculations in the following fashion. First, from the values of the area and volume of the cavity in which the solute molecule is embedded, the two geometrical molecular parameters A^* and V^* can be easily derived. The third, energetic parameter of the pure component set E^* can than be evaluated by inserting A^*, V^* and the experimental vapor pressure value into the *EOS* expression.

Beside the pure components parameter sets, the *PHSCT EOS* version for binary mixtures requires the knowledge of the so-called binary interaction parameter k_{ij}. For the estimation of this last parameter, the procedure we suggest consists in fitting the simulated activity coefficient data obtained as a result of *COSMOtherm* calculations, using the pure components parameter sets just obtained. For both E^* and k_{ij} determination, as well as for the subsequent *VLE* prediction, *ad-hoc* in-house *FORTRAN* codes have been developed, with particular attention given to reducing *CPU* time and to improve the reliability of the search procedure (Fermeglia *et al.*, 1998).

All molecular simulations were performed on a *Silicon Graphics Origin 200* (microprocessor *MIPS RISC* 10000, 64 bit *CPU*, 128 MB *RAM*), whereas the *COSMOtherm* program and the *FORTRAN* routines were run on *PC*s.

Results and Discussion

As a first result, Table 1 shows the values of the calculated *EOS* parameters for all the pure components considered. The calculated parameters reported in Table 1 were then inserted in the *PHSCT EOS* expression and the corresponding thermodynamical properties of the binary systems considered have been predicted and compared with the relevant experimental data.. Table 2 reports, together with the relevant k_{ij} values obtained from

COSMOtherm simulated data, the results of this comparison in terms of root-mean-square deviation (*RMSD*), defined as:

$$RMSD = 100 \sum_i \sqrt{\frac{(M_i^{exp} - M_i^{calc})^2}{N(M_i^{exp})^2}} \qquad (1)$$

where M_i^{exp} is the experimental value of a generic property M, M_i^{calc} is the corresponding calculated value and N is the total number of data points, whereas two graphical examples are given by Figure 1 (for the systems (1)/(3) (top)and (1)/(8) (bottom), respectively).

Table 1. PHSCT EOS Parameters for the Pure Components Considered.

Compound	A^* $(10^{-9}$ $cm^2/mol)$	V^* (cm^3/mol)	E^* $(bar$ $dm^3/mol)$
Methanol (1)	*4.102*	*25.76*	*58.02*
Ethanol (2)	*5.314*	*33.03*	*64.84*
1-Butanol (3)	*7.514*	*41.73*	*88.16*
Trichloromethane (4)	*7.043*	*39.44*	*68.42*
Cyclohexane (5)	*7.963*	*43.14*	*77.20*
Hexane (6)	*9.476*	*52.22*	*78.59*
Benzene (7)	*7.344*	*41.04*	*73.42*
Pyridine (8)	*7.091*	*39.75*	*81.17*

Table 2. Deviations Between Predicted and Experimental Data for a Selected Set of Binary Systems.

System	T (K)	k_{ij}	ΔP	Δy
(1)/(2)	*298*	*0.001*	*0.11*	*0.44*
(1)/(3)	*298*	*0.025*	*0.52*	*2.49*
(1)/(4)	*298*	*0.041*	*1.79*	*5.03*
(1)/(5)	*298*	*0.089*	*5.25*	*11.7*
(1)/(6)	*318*	*0.0869*	*11.5*	*10.5*
(1)/(7)	*298*	*0.061*	*3.46*	*7.92*
(1)/(8)	*373*	*0.021*	*11.0*	*0.628*
Average			*4.80*	*5.53*

If we take into account the low computational costs required to obtain one set of *PHSCT EOS* parameters for each molecule (from the molecule building to its *COSMOtherm* calculations) and that the knowledge of only one single experimental datum is required, the quality of the results reported in Table 2 are, in our opinion, to be considered more than satisfactory.

Further, we may consider the fact that the cost of collecting one vapor-liquid equilibria (*VLE*) data point (i.e., one temperature and composition for just one binary

mixture) has been estimated to be around \$2,600 and to take 2 days (Gubbins and Quirke, 1996). With these techniques, we can reasonably hope to minimize the experimental sessions necessary to characterize the *VLE* behavior of industrially important mixtures in a wide range of temperature and composition and, even more importantly, to create an on-line liaison between the molecular and process simulation worlds, being the time-scales of the two techniques now at least comparable, as illustrated in Figure 2.

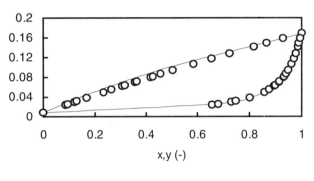

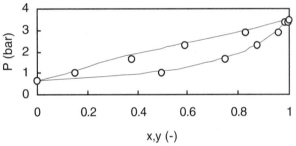

Figure 1. Comparison between experimental (symbols) and predicted VLE behavior (lines) for the binary system methanol/1-butanol (top) and methanol/pyridine (bottom).

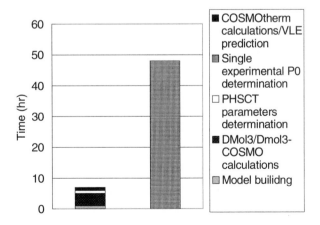

Figure 2. Comparison of the time scale for the prediction of the VLE behavior (left) and of a single experimental vapor pressure determination (right).

Conclusions

In this work we presented the results obtained with a new, alternative procedure for estimating *EOS* parameters from computer simulations. The issue of estimating reasonable parameters for *EOS* is topical in the analysis and synthesis of chemical processes and in the use of process simulators. In this last case, for instance, chemical engineers need to input *EOS* parameters for molecules that have not yet been synthesized, or for compounds for which experimental data cannot be easily accessed (due to the peculiar process or experimental conditions). The original method proposed in this work gives good results, is relatively computationally inexpensive, absolutely general and can be applied in principle to any *EOS*, provided the parameters have a well defined physical meaning.

References

Andzelm, J., K. Stark and A. Klamt (1998). Incorporation of solvent effects into density functional theory. *Proceedings of the AIChE Topical Conference Applying Molecular Modeling and Computational Chemistry*. AIChE Annual Meeting, Miami Beach (USA), -443.

Gubbins, K.E. and N. Quirke (1996). In K.E. Gubbins and N. Quirke (Eds.), *Molecular Simulations and Industrial Applications: Methods, Examples and Prospects*. Gordon & Breach, Amsterdam, 1-69.

Fermeglia, M., A. Bertucco and D. Patrizio (1997). Thermodynamic properties of pure hydrofluorocarbons by a perturbed hard-sphere-chain equation of state. *Chem. Eng. Sci.,* **52**, 1517-1527.

Fermeglia, M., A. Bertucco and S. Bruni (1998). A perturbed hard-sphere-chain equation of state for application to hydrofluorocarbons, hydrocarbons and their mixtures. *Chem. Eng. Sci.*, **53**, 3117-3128.

Fermeglia, M. and S. Pricl (1998). Molecular dynamics prediction of PVT behavior and determination of related thermophysical properties of pure polymers. *Proceedings of the AIChE Topical Conference Applying Molecular Modeling and Computational Chemistry*. AIChE Annual Meeting, Miami Beach (USA), pp.345-345.

Fermeglia, M. and S. Pricl (1999a). A novel approach to thermophysical properties prediction for chloro-fluoro-hydrocarbons. *Fluid Phase Equil.*, **166**, 21-37.

Fermeglia, M. and S. Pricl (1999b). Equation-of-state parameters for pure polymers by molecular dynamics simulations. *AIChE J.,* **45**, 2619-2627.

Klamt, A. and G. Schüürmann (1993). A new approach to dielectric screening in solvents with explicit expressions for the screening energy and its gradient. *J. Chem. Soc., Perkin Trans.*, **2**, 799-805.

Klamt, A., V. Jonas, T. Bürger and J.C.W. Lohrenz (1998). Refinement and parametrization of COSMO-RS. *J. Phys. Chem.*, **A**, **102**, 5074-5085.

Sanchez, I.C. and R.H. Lacombe (1976). An elementary molecular theory of classical fluids. Pure fluids. *J. Phys. Chem.*, **80**, 2352-2362.

COMPLETE PHASE DIAGRAMS FOR BINARY MIXTURES VIA GIBBS-DUHEM INTEGRATION

Monica R. Hitchcock and Carol K. Hall
North Carolina State University
Raleigh, NC 27695-7905

Abstract

Vapor-liquid, liquid-liquid, and solid-liquid coexistence curves are calculated for a binary mixture of Lennard-Jones spheres with the following parameters: $\sigma_{11} = \sigma_{22} = 1.0$, $\sigma_{12} = 0.85$, $\varepsilon_{11} = 1.0$, $\varepsilon_{12} = 0.82$, and $\varepsilon_{22} = 1.2$, using Monte Carlo simulation and the Gibbs-Duhem integration technique. These calculations allow us to construct a complete phase diagram, *i.e.,* showing equilibrium between vapor, liquid, and solid phases.

Keywords

Monte Carlo simulation, Phase equilibria, Vapor-liquid, Liquid-liquid, Solid-liquid, Lennard-Jones.

Introduction

Phase equilibria for binary mixtures of spherically symmetric molecules has been the subject of intensive investigation for decades. Van Konynenburg and Scott (1970, 1980) analyzed and classified the phase diagrams predicted by the van der Waals equation of state for binary mixtures. Remarkably, the simple van der Waals equation of state was found to exhibit five of the six types of fluid phase behavior observed experimentally. This landmark study has been followed by similar analyses for other equations of state, such as Redlich-Kwong (Deiters and Pegg, 1989) and Carnahan-Starling-Redlich-Kwong (Kraska and Deiters, 1992).

Most of the research directed at understanding how intermolecular interactions affect phase behavior has focused exclusively on fluid phase equilibria. However, in real systems solid phases form and often interrupt the complex fluid phase behavior (Schneider, 1978). In order to obtain a better overall picture of mixture phase behavior, all types of equilibrium between gas, liquid and solid phases must be considered. Valyashko (1986) classified complete phase diagrams for binary mixtures based on experimental results. Garcia and Luks (1999) recently examined complete phase diagrams for binary mixtures using the van der Waals equation of state and a simple solid state fugacity model.

Molecular simulation has become a popular way to study the phase behavior of model systems (Panagiotopoulos, 2000). Numerous simulations investigating binary mixture phase behavior for the Lennard-Jones intermolecular potential, the quintessential model of a spherically symmetric molecule, have been conducted for both fluid-fluid (Georgoulaki *et al.*, 1994; Guo *et al.*, 1994; Harismiadis *et al.*, 1991; van Leeuwen *et al.*, 1991) and solid-liquid (Hitchcock and Hall, 1999) phase equilibria. Vlot and coworkers (1997) calculated complete phase diagrams for symmetric (equal diameters, $\sigma_{11} = \sigma_{22}$; equal attractions, $\varepsilon_{11} = \varepsilon_{22}$) Lennard-Jones mixtures by using Monte Carlo simulation for selected state points to determine the excess free energy as a function of composition. This free energy versus composition data was fit with a two-parameter Redlich Kister polynomial and the convex envelope construction method was used to determine the phase diagram

In this paper we calculate a complete phase diagram for a binary Lennard-Jones mixture via Gibbs-Duhem integration and semigrand canonical Monte Carlo simulation. This is the first time that a complete phase diagram has been determined directly from molecular simulation.

The remainder of the paper is organized as follows. We first outline the Gibbs-Duhem integration method and describe how we applied the technique to the calculation of a complete phase diagram. We then present a complete phase diagram for a binary Lennard-Jones mixture.

Method

The coexistence lines were calculated using Gibbs-Duhem integration (Kofke, 1993a, 1993b). In this method, phase coexistence is determined by integrating the Clapeyron differential equation. The Clapyeron equation for equilibrium between two binary phases (α and γ) at constant pressure is:

$$\frac{d\beta}{d\xi_2} = \frac{(x_2^\alpha - x_2^\gamma)}{\xi_2(1-\xi_2)(h^\alpha - h^\gamma)}, \quad (1)$$

where β is the reciprocal temperature, $1/kT$, with k the Boltzmann constant and T the absolute temperature, ξ_2 is the fugacity fraction of species 2, $\xi_2 = f_2/\Sigma f_i$, with f_i the fugacity of species i in solution, x_2 is the mole fraction of species 2, and h is the molar enthalpy. The right hand side of Eqn. (1) can be integrated numerically to find an equation for β as a function of ξ_2 if we have an initial condition describing the temperature, fugacity fraction, enthalpies, and compositions at one coexistence point.

In our work, a convenient choice for the initial coexistence condition is the vapor-liquid or solid-liquid equilibrium condition for either of the pure components. The slope of the integrand in Eqn. (1) is undefined for pure components ($\xi_2 = 0$, $x_2 = 0$ or $\xi_2 = 1$, $x_2 = 1$) but it can be estimated using the limiting case of infinite dilution. This procedure is described elsewhere (Hitchcock and Hall, 1999; Mehta and Kofke, 1994) and will not be repeated here.

The enthalpies and mole fractions needed for the integration of Eqn. (1) were obtained by semigrand canonical Monte Carlo simulation (constant temperature, pressure, total number of molecules, and fugacity fraction) of the two phases. In this work, the simulations were run with a system size of 500 particles per phase. The pressure was held constant at reduced pressure $P^* = P\sigma_{11}^3/\varepsilon_{11} = 0.05$. The temperature and fugacity fraction were varied according to the values specified by the predictor-corrector algorithm used to numerically integrate Eqn. (1).

In this mixture, we encountered occurrences of three-phase lines, such as is found for a heteroazeotrope (liquid-liquid-vapor). In this case, two Gibbs-Duhem integrations were conducted, each starting from the vapor-liquid coexistence condition of the pure components ($\xi_2 = 0$ and $\xi_2 = 1$). At some ξ_2 (unknown at the commencement of the two integrations) the vapor phase coexistence lines will cross, thus, determining the temperature, fugacity fraction, and coexistence compositions of the three coexisting phases: liquid(1), liquid(2), and vapor. The liquid phase

mole fractions and enthalpies then become the initial condition for the liquid-liquid coexistence curve found below the heteroazeotrope temperature.

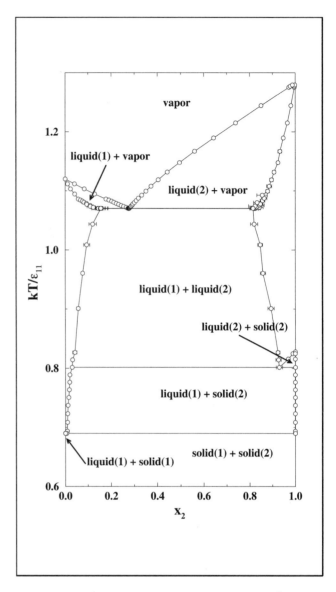

Figure 1. Complete T-x phase diagram at $P^* = 0.05$. The Lennard-Jones parameters are given in the text. The circles represent data from Gibbs-Duhem integration simulations. Error bars are shown when they are larger than the width of the symbol. Lines are drawn through the simulation points for clarity.

Results

In Fig. 1, we present a complete T-x phase diagram for a binary Lennard-Jones mixture with the following parameters: $\sigma_{11} = \sigma_{22} = 1.0$, $\sigma_{12} = 0.85$, $\varepsilon_{11} = 1.0$, $\varepsilon_{12} = 0.82$, and $\varepsilon_{22} = 1.2$, at $P^* = 0.05$. On this phase diagram, vapor-liquid coexistence curves originate from both $x_2 = 0$ ($\xi_2 = 0$) and $x_2 = 1$ ($\xi_2 = 1$). Both sets of curves decrease in temperature as fugacity fraction is varied until they meet and form a heteroazeotrope at reduced temperature, $T^* = kT/\varepsilon_{11} = 1.07$. At this temperature three phases coexist: a liquid mixture rich in component 1 (liquid(1)), a vapor mixture, and a liquid mixture rich in component 2 (liquid(2)). Below this temperature liquid(1) and liquid(2) are in equilibrium.

At $T^* = 0.828$, the solid and liquid phases of pure component 2 are in equilibrium. Solid-liquid coexistence curves originate from $x_2 = 1$ ($\xi_2 = 1$) and decrease in temperature as fugacity fraction is varied until they meet the liquid(1)-liquid(2) coexistence curves and form a monotectic at $T^* = 0.802$. At this temperature, three phases coexist: liquid(1), liquid(2), and a solid mixture rich in component 2 (solid(2)). Below this temperature, liquid(1) and solid(2) are in equilibrium.

At $T^* = 0.691$, the solid and liquid phases of pure component 1 are in equilibrium. Solid-liquid coexistence curves originate from $x_2 = 0$ ($\xi_2 = 0$) and decrease in temperature as fugacity fraction is varied until they meet the liquid(1)-solid(2) coexistence curves to form a eutectic at $T^* = 0.689$. At this temperature, three phases coexist: solid mixture rich in component 1 (solid(1)), liquid(1), and solid(2). Below this temperature, solid(1) and solid(2) are in equilibrium.

A fluid phase diagram for this mixture has been calculated by Canongia Lopes (1999) using Gibbs ensemble Monte Carlo (GEMC). We find that the heteroazeotrope temperature and compositions calculated with Gibbs-Duhem integration agree with the heteroazeotrope determined by GEMC. Below the heteroazeotrope, the two phase diagrams differ significantly. At $P^* = 0.05$, Canongia Lopes finds a lower critical solution temperature between $T^* = 0.8 - 0.9$, while our complete phase diagram shows the appearance of a monotectic near $T^* = 0.8$. We believe that the difference in the liquid-liquid equilibrium predictions over the range, $T^* = 0.8 - 0.9$, is due to the fact that inaccuracies arise when GEMC is used for high density phases, such as those encountered here ($\rho\sigma_{11}^3 > 0.9$). Achieving equilibration for two high-density phases is difficult with GEMC because of the required particle exchanges between the phases. The Gibbs-Duhem method, however, is well suited for determining phase equilibria of high-density systems because there are no particle exchanges involved.

Summary

The Gibbs-Duhem integration technique was combined with semigrand canonical Monte Carlo simulations to calculate a complete T-x phase diagram for a binary Lennard-Jones mixture. We have compared this complete phase diagram with the fluid phase equilibria simulation data of Canongia Lopes and found that our results differ below the heteroazeotrope temperature.

Acknowledgements

We gratefully acknowledge helpful discussions with David Kofke. This work was supported by the GAANN Computational Sciences Fellowship of the U.S. Department of Education, and the Office of Energy Research, Basic Sciences, Chemical Science Division of the U.S. Department of Energy under Contract No. DE-FG05-91ER14181. Acknowledgement is made to the Donors of the Petroleum Research Fund administered by the American Chemical Society for partial support of this work.

References

Canongia Lopes, J. N. (1999). Phase equilibria in binary Lennard-Jones mixtures: phase diagram simulation.. *Mol. Phys.,* **96**, 1649-1658.

Deiters, U. K. and I. L. Pegg (1989). Systematic investigation of the phase behavior in binary fluid mixtures. I. Calculations based on the Redlich-Kwong equation of state. *J. Chem. Phys.,* **90**, 6632-6641.

Garcia, D. C. and K. D. Luks (1999). Patterns of solid-fluid phase equilibria: new possibilities? *Fluid Phase Equil.,* **161**, 91-106.

Georgoulaki, A. M., I. V. Ntouros, D. P. Tassios, and A. Z. Panagiotopoulos (1994). Phase equilibria of binary Lennard-Jones mixtures: simulation and van der Waals 1-fluid theory. *Fluid Phase Equil.,* **100**, 153-170.

Guo, M., Y. Li, Z. Li, and J. Lu (1994). Molecular simulation of liquid-liquid equilibria for Lennard-Jones fluids. *Fluid Phase Equil.,* **98**, 129-139.

Harismiadis, V. I., N. K. Kourtras, D. P. Tassios, and A. Z. Panagiotopoulos (1991). How good is conformal solutions theory for phase equilibrium predictions? *Fluid Phase Equil.,* **65**, 1-18.

Hitchcock, M. R. and C. K. Hall (1999). Solid-liquid phase equilibrium for binary Lennard-Jones mixtures. *J. Chem. Phys.,* **110**, 11433-11444.

Kofke, D. A. (1993a). Gibbs-Duhem integration: a new method for direct evaluation of phase coexistence by molecular simulation. *Mol. Phys.,* **78**, 1331-1336.

Kofke, D. A. (1993b). Direct evaluation of phase coexistence by molecular simulation via integration along the saturation line. *J. Chem. Phys.,* **98**, 4149-4162.

Kraska, T. and U. K. Deiters (1992). Systematic investigation of the phase behavior in binary fluid mixtures. II. Calculations based on the Carnahan-Starling-Redlich-Kwong equation of state. *J. Chem. Phys.,* **96**, 539-547.

Mehta, M. and D. A. Kofke (1994). Coexistence diagrams of mixtures by molecular simulation. *Chem. Eng. Sci.,* **49**, 2633-2645.

Panagiotopoulos, A. Z. (2000). Monte Carlo methods for phase equilibria of fluids. *J. Phys.: Cond. Matter,* **12**, R25-R52.

Schneider, G. M. (1978). High-pressure phase diagrams and critical properties of fluid mixtures. In M. L. McGlashan (Ed.), *Chemical Thermodynamics, Vol 2, Specialist Periodical Report.* The Chemical Society, London, pp. 105-146.

Scott, R. L. and P. H. van Konynenburg (1970). Van der Waals and related models for hydrocarbon mixtures. *Disc. Faraday Soc.,* **49**, 87-97.

Valyashko, V. M. (1986). Complete phase diagrams of binary systems with different volatility components. *Z. Phys. Chemie, Leipzig,* **267, No. 3,** 481-493.

Van Konynenburg, P. H. and R. L. Scott (1980). Critical lines and phase equilibria in binary van der Waals mixtures. *Phil. Trans. Roy. Soc. London (A)*, **298**, 495-540.

Van Leeuwen, M. E., C. J. Peters, J. de Swaan Arons, and A. Z. Panagiotopoulos (1991). Investigation of the transition to liquid-liquid immiscibility for Lennard-Jones (12,6) systems, using Gibbs-ensemble molecular simulations. *Fluid Phase Equil.*, **66**, 57-75.

Vlot, M. J., J. C. van Miltenburg, H. A. J. Oonk, and J. P. van der Eerden (1997). Phase diagrams of scalemic mixtures: a Monte Carlo simulation study. *J. Chem. Phys.*, **107**, 10102-10111.

DETERMINING THE PRESSURE-VISCOSITY COEFFICIENT BY MOLECULAR SIMULATION

Clare McCabe[1,2*], Peter T. Cummings[2,3], Shengting Cui[1,2],
Peter A. Gordon[4], and Roland B. Saeger[4]

[1]Department of Chemical Engineering, University of Tennessee
Knoxville, TN 37996-2200.
[2]Departments of Chemical Engineering, Chemistry and Computer Science
University of Tennessee
Knoxville, TN 37996-2200.
[3]Chemical Technology Division, Oak Ridge National Laboratory
Oak Ridge, TN 37831-6181.
[4]Corporate Strategic Research Centre, ExxonMobil Research and Engineering
Annandale, NJ 08801.

Abstract

An understanding of the relationship between chemical structure and lubricant performance is highly desirable from both a fundamental and a practical perspective, such knowledge being vital to improve the performance of mineral oils and to guide the design of future synthetic lubricants. The rheological properties of alkanes of intermediate molecular size are of particular interest as they are important constituents of synthetic lubricant basestocks. In this work we determine the pressure-viscosity coefficient for 9-octylheptadecane by molecular simulation. Good agreement with experimental data is achieved, further illustrating the value of molecular simulation in predicting lubricant properties and its potential for providing guidance in the design of synthetic lubricants with desired properties.

Keywords

Simulation, Molecular dynamics, Viscosity, Pressure-viscosity coefficient.

Introduction

An understanding of the rheological properties of lubricant components is vital to the petrochemical industry. In particular, the properties of alkanes in the C_{20}-C_{40} mass range are of fundamental importance in industrial applications as they are important constituents of lubricant base-stocks.

Given the range of operating conditions lubricants are subjected to in practical applications, a consideration of not only the viscosity, but also the viscosity-temperature and viscosity-pressure behaviour is vital to determine if a fluid would make a good lubricant component. Although viscosity is a function of both temperature and pressure, the effect on the viscosity of a small change in pressure

from atmospheric is quite small compared with the effect of changes in temperature. For that reason much less time has been devoted to examining the influence of pressure, and experimentally its effect is usually ignored. However, the performance of machine elements such as gears and roller bearings depends on lubricants that routinely are subjected to pressures in the GPa range. Relative to atmospheric conditions, these pressures can induce viscosity increases of several orders of magnitude.

In the literature there is very little experimental data available on the high-pressure rheological properties of alkanes, especially in the mass range of interest. The most comprehensive experimental study of the viscosity of

---

lubricants to high pressures was undertaken at Harvard University (ASME 1953) in the late 40s. Viscosity and density data were reported for 40 lubricants, including pure hydrocarbons, mineral oils and synthetic lubricants, to pressures of 1GPa and temperatures up to 200°C. More recently Bair and co-workers have measured the pressure-viscosity behaviour of liquid lubricants to pressures in the GPa range (Bair 1982, Bair 1992, Bair 2000).

Although additional experimentation is needed in order to further investigate the influence of pressure and molecular architecture on viscous behavior, it is both very costly and time consuming. Computer simulation studies are proving to be an attractive and valuable means with which to "fill in the gaps" and obtain this important information. Furthermore, the conditions commonly met in practical applications such as automobile engines and machinery, (i.e. giga-pascal pressures and nanoscale gaps between surface asperities), are very difficult to achieve and study experimentally, but pose fewer difficulties in a computer "experiment".

Over the last decade there have been a number of simulation studies reported in the literature on the rheolgical properties of industrially important molecules (Cummings 1992). For the hydrocarbons early work was limited to short, mainly linear molecules (see for example references, Berker 1992, Cui 1996, Edberg 1987, Morris 1991, Mundy 1994). However, with the recent advances in computational power and the development of more realistic models and suitable algorithms it is now possible to perform simulations of more complex systems. Recently longer chains and branched molecules have been examined, (Cui 1996, Khare 1997, Mondello 1998, Moore 1997, Morris 1991, Mundy 1996, Travis 1996) giving us the opportunity to gain molecular level insight into questions of industrial relevance. Generally previous simulation studies either focused on a single state point, or examined a number of state points at differing temperatures in order to investigate the influence of temperature on the rheological behavior. To our knowledge the work of Mundy and Klein (Mundy 1996) is the only prior determination by molecular simulation for an alkane molecule of the influence of pressure on viscous behavior and the pressure-viscosity coefficient. They examined the shear viscosities of *n*-decane and 4-propylheptane over a range of pressures using equilibrium molecular dynamics simulation and a united atom model fitted to critical parameters. They determined the shear viscosities for both molecules and the pressure-viscosity coefficient for *n*-decane, where they achieved quantitative agreement with experimental data.

In this paper we present preliminary results from a detailed study of the rheology of 9-octylheptadecane by molecular simulation. In particular we focus on the high-pressure behaviour and the calculation of the pressure-viscosity coefficient.

Simulation Details

The model used to describe the alkane molecules is that of Siepmann *et al.* (Siepmann 1993, Smit 1995), which was later extended to branched alkanes by Mondello and Grest (Mondello 1995). As in earlier work (Cui, Cummings et al. 1996), we use the modification of Mundy and co-workers in which the fixed bond length is replaced by a stiff harmonic potential to generate a fully flexible model (Mundy 1994). We will briefly describe the model and the simulation methods, the reader is directed to the original papers for full details.

In this united atom description of the alkanes the methyl, methylene and methyne groups are recognized and individually treated as single spherical interaction sites with the interaction centre located at the centre of each carbon atom. A Lennard Jones potential describes the intermolecular interactions, and the intramolecular interactions between sites separated by three or more bonds. The size parameters for the CH_3, CH_2, and CH groups are 3.93Å, 3.93Å and 3.81Å and for the energy 114K, 47K and 40K respectively. A cut off distance of 9.85Å ($2.5\sigma_{CH2}$) was used. Simple Lorentz-Berthelot combining rules were used to determine the cross or unlike interactions.

Bond stretching is described by a harmonic potential with an equilibrium bond distance of $r_{eq} = 1.54$ Å and force constant $k_a/k_B = 4529000 K rad^{-2}$ (where k_B is Boltzmann's constant). The bond angle bending term is also described by a harmonic potential with an equilibrium angle of $\theta_{eq} = 114°$ and force constant $k_b/k_B = 62400 K rad^{-2}$. In the extension to branched alkanes by Mondello and Grest (Mondello and Grest 1995) an *ad hoc* harmonic potential similar to the bending term is used to prevent the unphysical umbrella inversion of the tertiary carbon atoms ($\theta = 27.25°$ and $k_c/k_B = 40258 K rad^{-2}$). Finally, torsional motion characterising the preferred orientational and rotational barriers around all non-terminal bonds is described through the potential of Jorgensen (Jorgensen 1984).

The alkanes were simulated under planar Couette flow using the SLLOD equations of motion with a Nosé thermostat. The multiple time step technique from the work of Tuckerman *et al.* (Tuckerman 1992) and Cui *et al.* (Cui 1996) was used to integrate the equations of motion with all the intramolecular interactions treated as fast motions and the intermolecular interactions as the slow motion. For the fast mode motion a time step of 0.236fs was used, and for the slow mode motion the time step was 2.36fs. To generate the starting configurations each atom was given a small LJ diameter and the molecules placed on a lattice in the all trans conformation with the centre of mass at the cubic lattice points. During this initial equilibration period the atoms are grown to full size and then allowed to equilibrate for a further 5000ps. This equilibrium configuration provides the starting point for both the non-equilibrium molecular dynamics (EMD) and high strain rate non-equilibrium molecular dynamics

(NEMD) simulations. At the lowest strain rates the final configuration of a higher strain rate was used as the starting configuration in order to achieve a steady state faster. The strain rate dependent properties were calculated once the system reached a steady state under the influence of shear flow. The time allowed to reach steady state varied from 2500ps for high strain rates to 5000ps for the lower strain rates.

Results

In an effort to determine the influence of pressure on the viscosity of 9-octylheptadecane we have examined a number of state points at varying pressures and a single common temperature, 372K.

Table 1. State Points Studied.

State Point	Temperature/ K	Pressure/ GPa	Density/ gcm^{-3}
A	372	0.958	0.980
B	372	0.829	0.964
C	372	0.682	0.943
D	372	0.000101	0.751

Details of the state points simulated are given in Table 1. At each state point we have determined the strain rate dependent viscosity for 9-octylheptadecane over a wide range of strain rates (10^7-10^{11}s^{-1}) using NEMD simulation. The results for state points A through C, which are at pressures in the GPa range, are presented in Figure 1. The horizontal lines correspond to the experimental zero shear viscosity for each state point studied (ASME 1953, API 1966), and the arrows from left to right indicate the inverse of the rotational relaxation time for state points A to C respectively. In all cases the viscosity shows shear-thinning behavior at high strain rates, and levels off to a Newtonian plateau at the lower strain rates. From the figure we note the transition to the Newtonian regime correlates well with the inverse of the rotational relaxation time, and occurs at a higher strain rate for the less dense system, moving to lower strain rates with increasing density. The viscosity is greatest at the highest density, though in the shear-thinning region the viscosity is virtually indistinguishable for the three state points.

The zero-shear viscosity was determined by averaging the values at strain rates that appeared to fall within the plateau region. We see from the figure that the united atom model under predicts the zero-shear viscosity, which was anticipated and is thought to be due to the smoothness of the united atom model compared to the real molecular architecture. The pressure-viscosity coefficient can easily be determined from the definition of Barus (Barus 1893),

$$\eta(T,P)=\eta(T,0)\bullet\exp\left(\alpha(T)\bullet P\right) \qquad (1)$$

where η is the viscosity and α is the pressure-viscosity coefficient. Using this relation, from a semi-log plot of the viscosity versus pressure at constant temperature (given in figure 2) we can determine α. Experimentally for 9-octylheptadecane we obtain a value for α of 5.88 GPa^{-1} and from the NEMD simulation results a value of 5.66($\pm$0.04) GPa^{-1}, which is in very good agreement.

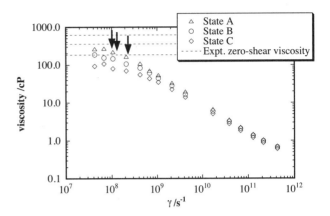

Figure 1. Shear-dependent viscosity of 9-octylheptadecane at state points A, B, and C from Table 1.

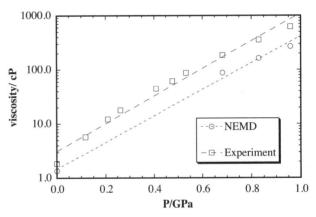

Figure 2. Semi-log plot of viscosity versus pressure for 9-Octylheptadecane at 372K.

Conclusions

We have presented the results of EMD and NEMD simulations for 9-octylheptadecane at 372K and pressures ranging from atmospheric to the GPa range. The rotational relaxation time was determined from EMD simulations for each state point and found to be in good agreement with the predicted transition from Newtonian to non-Newtonian behavior from the NEMD simulation results. The zero-shear viscosity was estimated from the NEMD simulations and seen to under predict the experimental value, which is consistent with previous studies. Additionally, from the zero-shear viscosity we determined the pressure-viscosity coefficient. Using the empirical relation defined by Barus to describe the pressure dependence of the viscosity we can easily calculate α. The predicted value is found to be in excellent agreement with the experimental result.

In earlier work we have extensively examined the influence of temperature on viscosity (Moore 2000, McCabe 2000, McCabe 2000), observing that while the united atom model under predicts the actual viscosity of these molecules, it does correctly describe the temperature dependence. By examining the influence of pressure on viscosity we provide further support for this observation, noting now that the model appears to capture relative changes in viscosity with pressure as well as temperature. With this in mind molecular simulation could prove an invaluable tool when used to probe the rheological properties of as yet unsynthesized molecules, such as synthetic lubricant candidates. As advances in machine design lead to more demanding operating conditions, the performance advantages of synthetic lubricants with tailored molecular structure will become increasingly attractive.

Acknowledgements

This work was supported in part by the Division of Materials Sciences of the U. S. Department of Energy. Oak Ridge National Laboratory is operated for the Department of Energy by Lockheed Martin Energy Research Corp. under contract number DE-AC05-96OR22464. Additional funding was provided by ExxonMobil Research and Engineering.

Nomenclature

γ = strain rate

ε = energy parameter in Lennard-Jones interaction

θ = angle between adjacent C-C bonds in an alkane chain

σ = distance parameter in Lennard-Jones interaction

τ = orientational relaxation time

a_i = parameters in torsional potential

P = pressure

T = temperature

k = spring constant in intramolecular interactions

k_B = Boltzmann's constant

α = pressure-viscosity coefficient

η = viscosity

References

ASME Research Committee on Lubrication (1953). Pressure Viscosity Report. Harvard University, New York.

API 42: *Properties of Hydrocarbons of High Molecular Weight* (1966). American Petroleum Institute, New York.

Bair, S. (2000). *Tribology Transactions*, **43**, 91-99.

Bair, S. and W. O. Winer (1982). *J. Lubrication Tech.*, **104**, 357-364.

Bair, S. and W. O. Winer (1992). *J. Tribology*, **114**, 1-13.

Barus, C. (1893). *Am. J. Sci.*, **45**, 87-96.

Berker, A., Chynoweth, S., Klomp, U. C. and Y. Michopoulos (1992). *J. Chem. Soc. Faraday Trans.*, **88**, 1719-1725.

Cui, S. T., Cummings, P. T. and H. D. Cochran (1996). *J. Chem. Phys.*, **104**, 255-262.

Cui, S. T., Cummings, P. T. and H. D. Cochran (1996). *J. Chem. Phys*, **104**, 255.

Cui, S. T., Gupta, S. A. and P. T. Cummings (1996). *J. Chem. Phys*, **105**, 1214-1220.

Cummings, P. T. and D. J. Evans (1992). *I&EC Res.*, **31**, 1237-1252.

Edberg, R., Morris, G. P. and D. J. Evans (1987). *J. Chem. Phys*, **86**, 4555.

Jorgensen, W. L., Madura, J. D. and C. J. Swenson (1984). *J. Am. Chem. Soc.*, **106**, 6638.

Khare, R., de Pablo, J. and A. Yethiraj (1997). *J. Chem. Phys.*, **107**, 6956-6964.

McCabe, C., Cui, S. T. and P. T. Cummings (2000). *Fluid Phase Equil.*, submitted.

McCabe, C., Cui, S. T., Cummings, P. T., Gordon, P. A. and R. B. Saeger (2000). *J. Chem. Phys.*, in preparation.

Mondello, M. and G. S. Grest (1995). *J. Chem. Phys.*, **103**, 7156.

Mondello, M., Grest, G. S., Webb, E. B. and P. Peczak (1998). *J. Chem. Phys.*, **109**, 798-805.

Moore, J. D., Cui, S. T., Cochran, H. D. and P. T. Cummings (2000). *J. Chem. Phys.*, submitted.

Moore, J. D., Cui, S. T., Cummings, P. T. and H. D. Cochran (1997). *AIChE J.*, **43**, 3260-3263.

Morris, G. P. (1991). *J. Chem. Phys.*, **94**, 7420-7433.

Mundy, C. J., Balasubramanian, S., Bagchi, K., Siepmann, J. I. and M. L. Klein (1996). *Faraday Trans.*, **104**, 17-36.

Mundy, C. J. and Klein, M. L. (1996) *J. Chem. Phys*, **100**, 16779-16781.

Mundy, C. J., Siepmann, J. I. and M. L. Klein (1994). *J. Chem. Phys.*, **102**, 3376-3380.

Siepmann, J. I., Karaborni, S. and B. Smit (1993). *Nature*, **365**, 330.

Smit, B., Karaborni, S. and S. J. I. (1995). *J. Chem. Phys.*, **102**, 2126.

Travis, K. P. and D. J. Evans (1996). *Mol. Sim.*, **17**, 157-164.

Tuckerman, M. E., Berne, B. J. and G. J. Martyna (1992). *J. Chem. Phys.*, **97**, 1990-2001.

STRUCTURE AND SLIDING FRICTION OF ADSORBED FRICTION-MODIFIER ADDITIVES

Michael L. Greenfield and Hiroko Ohtani
Chemistry Department
Ford Motor Company
Dearborn, MI 48121-2053

Abstract

Automatic transmission fluids contain surfactant-like "friction modifier" additives to provide precise control of the friction performance of wet clutches in the boundary lubrication regime. In this work we have used molecular dynamics simulations to address the structure and frictional properties of model friction modifier additives confined between two surfaces. At a high additive concentration, oscillations in normal pressure were found with increasing surface separation. Most additive molecules were adsorbed to the surface and some distinct layers of trapped fluid were observed between the adsorbed films. For mixtures of additive and solvent, significant desorption occurred and only patchy adsorbed layers remained. Preliminary results for sliding surfaces suggest the low-speed dynamic friction coefficient increases with sliding velocity.

Keywords

Molecular tribology, Tribology, Additive, Friction, Automatic transmission fluid, Lubricant, Molecular dynamics.

Introduction

Automatic transmission fluids (ATF's) are a complicated mixture of hydrocarbon base oil with numerous additives. The base oil and additives are selected in part to enable ATF to perform a number of different roles during transmission operation: dissipate heat away from frictional contacts, lubricate moving parts, move valves via hydraulic forces, inhibit metal corrosion (Papay, 1991). They are also selected based on less technical criteria, such as cost. Consequently there is motivation for both producers and consumers of ATF's and ATF additive packages to use an understanding of how additives function in order to select combinations and/or systems that will satisfy multiple requirements. Such an understanding would also help with designing more robust and durable systems and would facilitate developing steps to be taken should degradation of part of the additive package occur during use.

Of interest here is the surfactant-like "friction modifier" (FM) additive, which provides precise control of the friction performance of wet clutches in the boundary lubrication regime (Papay, 1983; Shirahama, 1994). Adding very small amounts of FM's to ATF's is known to decrease static and dynamic friction such that the low-speed dynamic friction increases with sliding velocity (Kugimiya, 1995; Shirahama, 1994, Slough, 1998). This smooths the "feel" of automatic transmission shifts by eliminating sharp friction increases as clutches engage into contact. This also eliminates undesirable friction-induced vibrations by providing a negative-feedback frictional system capable of damping out fluctuations in velocity (Kugimiya, 1995; Papay, 1991; Slough, 1995).

We are investigating the mechanisms of friction modification using two approaches rooted in so-called "molecular tribology" (Bhushan, 1995; Krim, 1996). The work described here uses molecular dynamics simulations of model lubricant and FM molecules confined between explicit atom surfaces, as in our previous work

(Greenfield, 1999). In conjunction with Steve Granick at the University of Illinois, we are also conducting surface forces apparatus (SFA) experiments on systems of similar molecules confined between mica surfaces (Ruths, 1999). In the molecular simulations, we have two specific objectives:

1. identifying the molecular-level structures of thin films formed by model FM's dissolved in hydrocarbon solvent and confined between two well-defined atomistic surfaces, and

2. predicting the velocity-dependent friction between such films under sliding conditions, at relevant sliding speeds.

Through elucidating the molecular details, we hope to learn which aspects of the friction modification process (detailed additive chemistry, surface concentration, chain length, etc.) are most crucial for achieving desirable and durable friction properties.

Method

Phase space under static and sliding conditions was explored using equilibrium and nonequilibrium molecular dynamics simulations. Velocity rescaling (60 ps) and microcanonical (120 ps) simulations were used to initialize the chain configurations and positions at the beginning of each run; a constant temperature was maintained using a Nose-Hoover thermostat (Allen, 1987) for production runs (200 ps or longer). Periodic boundary conditions were used in 2 of the 3 dimensions.

The model system was chosen to capture the key qualitative aspects of the model system used in the SFA experiments (Ruths, 1999). A 9-bead united-atom model was used for each solvent chain. Bead–bead interactions were calculated using a Lennard-Jones 12-6 potential, with parameters taken from our previous work (Greenfield, 1999). Four model FM's were used, each containing 9 united-atom beads. In FM's 1, 3, and 4, the Lennard-Jones ε_{FM-S} parameter of the first bead on each chain was increased to provide more favorable bead–bead and bead–wall interactions. In FM 2, the ε_{FM-S} parameter was increased in the same manner for the second bead of each FM chain. The corresponding parameters are listed in Table 1. The bond angle and torsion angle potential we used previously (Greenfield, 1999) was supplemented for FM 3 and 4 and in all sliding system calculations with a harmonic bond bending potential,

$$V = k(r - r_0)^2 \qquad (1)$$

with $k = 100$ kcal mol^{-1}Å^{-2} and $r_0 = 1.53$ Å.

The overall forces in the shear and normal directions were calculated by summing over all forces on each wall. Under static and sliding conditions, the average normal forces were equal in magnitude and opposite in sign to within 0.4%. Under sliding conditions, the calculated shear forces on each wall were similar to each other. Sliding was imposed on the upper wall at a fixed velocity in the $+x$ direction, with the lower wall held fixed. Changes in free energy with separation were extracted via a procedure described previously (Greenfield, 1999).

Table 1. Lennard-Jones Potential Energy Parameters for the End Group(s) of Each Friction Modifier Molecule Type.

FM molecule	# of end groups	ε_{FM-S} (kcal/mol)	σ (Å)
1	1	0.5	3.93
2	2	0.5	3.93
3	1	1.0	3.93
4	1	1.414	3.93

An in-house molecular dynamics code was used for FM 1 and 2; the LAMMPS molecular dynamics code (Plimpton, 1997) was used for FM 3 and 4 and in all sliding simulations. Under static conditions, there were 98 9-unit FM and solvent molecules in total and 600 atoms per surface (the surface atoms were not allowed to move). Under sliding conditions, the system size was increased by 9x (replicated 3x each in the x- and y-directions, followed by reinitialization) for a better signal-to-noise ratio. Calculations were performed on a Digital Alphastation 500/500, an 8-node PC Beowulf cluster (one Pentium II 400 MHz processor per node), and a multi-node SGI Origin (typically 8 R10000 nodes per run). A single node on the Origin was comparable in speed to the single-CPU Alphastation.

Results

The pressures found for different wall separations are shown in Fig. 1 for five cases: either pure FM 1 or 50 chains of FM 1, 2, 3, or 4 mixed with 48 solvent chains. For pure FM 1, oscillations in normal pressure and free energy were found with increasing surface separation, and three stable points can be seen where $P \approx 0$ and normal pressure decreases with increasing separation. At the smallest stable separation, all additive molecules were adsorbed to the surface. Based on changes in the total free energy (not shown), the outermost of these three separations is the most stable (Greenfield, 1999). This state corresponds physically to a monolayer film adsorbed on each wall (FM molecules oriented perpendicular to the surface) with two distinct layers of FM molecules trapped between the films and parallel to the surface. This configuration was not employed as an original input to the simulation; instead, some chains desorbed during the course of the MD simulation and localized near the center of the simulation cell.

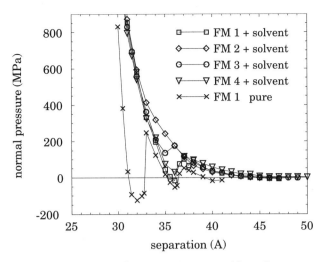

Figure 1. Change in pressure with wall separation for pure FM 1 and for mixtures of FM 1, 2, 3, or 4 with solvent.

For the mixtures of FM with solvent, the situation is less clear. The starting configurations for these cases had been at the smallest separation, with the FM and solvent molecules packed perpendicular to the wall in all-*trans* conformations. At the smallest separations, the chains remained adsorbed and ordered. For FM 1, 3, or 4 with solvent, this remained the case until a minimum in pressure was reached. As seen in Fig. 1, for FM 1 with solvent the system passed through a $P \approx 0$ state before this minimum was reached; for FM 3 or 4 with solvent this minimum was reached at some $P > 0$. With increasing simulation time and increasing wall separation, these regular structures then broke down and significant desorption occurred, leaving very disordered, liquid-like configurations. Most solvent and additives chains were more parallel than perpendicular to the surface in this case. There were mostly FM and few solvent chains directly adsorbed to the surface, also lying in parallel configurations. This orientation maximizes the magnitude of the favorable interactions between the surface and each single adsorbed molecule. Under these desorbed and disordered circumstances, the pressure decreased continuously with increasing wall separation. There were few differences among the pressures required to contain these three FM–solvent systems to the same volume.

The pressure decreases with increasing wall separation were somewhat different for FM 2 with solvent. No local minimum was found for the pressure with increasing separation, and visualization (see Greenfield (1999)) indicated that most chains had lost their perpendicular orientation relative to the surface and that FM and solvent molecules were somewhat segregated. Flat, parallel configurations were found instead, which allowed the second attractive group on FM 2 to locate closer to the surface. However, a few FM

chains did remain in perpendicular conformations as an isolated patch.

To investigate sliding friction, we have been conducting nonequilibrium molecular dynamics simulations in which the upper wall is slid at constant speed and the lower wall is held fixed. The ratio of tangential and normal forces on each wall then yields the friction coefficient. Preliminary results suggest that the friction coefficient increases with sliding speed, in agreement with experiment for actual friction modifiers. In ongoing work we are investigating how this sliding friction coefficient changes with the type and surface concentration of friction modifier.

Discussion

The relationships among normal load, fluid layer thickness, and wall separation found in the simulations dependencies observed agree favorably with those measured experimentally using a surface forces apparatus for model FM's in solvent (Ruths, 1999). In particular, a "hard wall" is seen at separations close to a separation equal to twice the fully extended length of the adsorbed FM additive. The simulations indicate that this corresponds to one perpendicular additive monolayer per wall. The series of stable states can also explain the hysteresis — purely repulsive forces when surfaces approach but attractive forces when they recede — observed in the SFA experiment (Ruths, 1999). According to Fig. 1 or to the free energy profile (Greenfield, 1999), layers can be squeezed out from between the adsorbed films as the separation is decreased; this leads to normal forces that appear to increase continuously. Upon pulling the surfaces apart, the system is pulled into a local free energy minimum, and attractive forces must be overcome in order to escape this minimum and reach larger separations.

For all three stable states of the pure FM 1 system, the additive molecules that remained adsorbed to the surface formed a structure similar to a self-assembled monolayer or a Langmuir-Blodgett film. Consequently the general shape and magnitudes of the free energy barriers between local minima can be compared with those measured experimentally for molecules confined between self-assembled monolayers. Our results may also be compared with those found by others in simulations of similar systems. In addition to the direct experimental comparisons for FM molecules (see above), we have found (Greenfield, 1999) that the barriers found via simulation are similar to those measured by SFA (Peanasky, 1995) for undecane confined between alkyltriethoxysilane self-assembled monolayers on mica (barriers of order 10 mN/m and distances of 3-4 Angstroms between local minima). Simulations in this area were reviewed in our previous work (Greenfield, 1999) and by Siepmann (1995). Siepmann also presented "hard wall" profiles for self-assembled monolayer films similar to those shown here; the changes in free energy with separation for pure FM 1 found here were comparable both to those and to those calculated by Tupper and Brenner (1994) for alkylthiols on gold. Increases in friction with sliding speed (observed via simulation) have been found by Glosli and McClelland (1993) for ordered-alkane chains and by Kong, Tildesley, and Alejandre (1997) for ionic surfactant monolayers. Self-assembled monolayers have also been used as models for dithiophosphate additives in engine oils (Jiang, 1997).

The FM systems diluted with solvent showed one significant departure from the experimentally observed behavior: significant desorption occurred. In the SFA experiments, the "hard wall" was observed in systems in which the model FM additive was present at 0.1 wt% concentration in solvent (Ruths, 1999). In the simulations, desorption had already occurred by ca. 50 wt%. We expect that this discrepancy is due to insufficient strength in the wall–end group interaction potential. Resolving this discrepancy and achieving a better match between the energetics of the model system chosen for the experiments and the interaction potentials chosen for the simulations will be one subject of future work.

Conclusions

Molecular dynamics simulations have been conducted of model "friction modifier" additives confined between atomically detailed surfaces, both pure and diluted to ca. 50 wt% with solvent. For the pure FM system, adsorbed monolayers comparable to Langmuir-Blodgett films or to self-assembled monolayers were found, and the system was mechanically stable ($P \approx 0$) at multiple separations. The lowest free energy state contained distinct layers of fluid trapped between the adsorbed films. For the system of FM additives diluted with solvent, the potentials employed were not sufficiently strong to prevent significant desorption from occurring, and the change in pressure with separation under such circumstances was not sensitive to the details of the end group–surface interaction strength. Using a model additive with two attractive end groups induced some changes in the pressure–separation profile and led to a state in which a small set of additives remained adsorbed in a perpendicular orientation as a "patchy" film.

Simulations of sliding conditions have not yet led to distinctive conclusions. For pure FM systems, first indications are that the low-speed dynamic friction coefficient will increase with sliding velocity, as observed experimentally. Studies of this type will be the focus of future work.

References

Allen, M. P. and D. J. Tildesley (1987). *Computer Simulation of Liquids.* Oxford University Press, New York.

Bhushan, B., J. N. Israelachvili, and U. Landman (1995). Nanotribology: Friction, wear and lubrication at the atomic scale. *Nature*, **374**, 607–616.

Glosli, J. N. and G. M. McClelland (1993). Molecular dynamics study of sliding friction of ordered organic monolayers. *Phys. Rev. Lett.*, **70**, 1960–1963.

Greenfield, M. L. and H. Ohtani (1999). Molecular dynamics simulation study of model friction modifier additives confined between two surfaces. *Tribo. Lett.*, **7**, 137–145.

Jiang, S., R. Frazier, E. S. Yamaguchi, M. Blanco, S. Dasgupta, Y. Zhou, T. Cagin, Y. Tang, and W. A. Goddard III (1997). The SAM model for wear inhibitor performance of dithiophosphates on iron oxide. *J. Phys. Chem. B*, **101**, 7702–7709.

Kong, Y. C., D. J. Tildesley, and J. Alejandre (1997). The molecular dynamics simulation of boundary-layer lubrication. *Mol. Phys.*, **92**, 7–18.

Krim, J. (1996). Friction at the atomic scale. *Sci. Am.*, **275**, 74–77.

Kugimiya, T., J. Mitsui, N. Yoshimura, H. Kaneko, H. Akamatsu, F. Ueda, T. Nakada, and S. Akiyama (1995). Development of automatic transmission fluid for slip-controlled lock-up clutch systems. SAE paper 952348.

Papay, A. G. (1983). Oil-soluble friction reducers – theory and application. *Lubr. Engr.*, **39**, 419–426.

Papay, A. G. (1991). Formulating automatic transmission fluids. *Lubr. Engr.*, **47**, 271–275.

Peanasky, J., H. M. Schneider, S. Granick, and C. R. Kessel (1995). Self-assembled monolayers on mica for experiments utilizing the surface forces apparatus. *Langmuir*, **11**, 953–962.

Plimpton, S. J., R. Pollock, and M. Stevens (1997). Particle-mesh Ewald and rRESPA for parallel molecular dynamics simulations. In *Proc. of the Eighth SIAM Conference on Parallel Processing for Scientific Computing*, Minneapolis, MN.

Ruths, M., H. Ohtani, M. L. Greenfield, and S. Granick (1999). Exploring the "friction modifier" phenomenon: Nanorheology of alkane chains with polar terminus dissolved in n-alkane solvent. *Tribo. Lett.*, **6**, 207–214.

Shirahama, S. (1994). Adsorption of additives on wet friction pairs and their frictional characteristics. *Japanese J. Tribo.*, **39**, 1479–1486.

Siepmann, J. I. (1995). Monte Carlo calculations for the mechanical relaxation of a self-assembled monolayer and for the structures of alkane/metal interfaces. *Tribo. Lett.*, **1**, 191–199.

Slough, C. G., H. Ohtani, M. P. Everson, and D. J. Melotik (1998). The effect of friction modifiers on the low-speed friction characteristics of automatic transmission fluids observed with scanning force microscopy. SAE paper 981099.

Tupper, K. J. and D. W. Brenner (1994). Compression-induced structural transition in a self-assembled monolayer. *Langmuir*, **10**, 2335–2338.

Yoshizawa, K., T. Akashi, and T. Yoshioka (1990). Proposal of new criteria and test methods for the dynamic performance of ATF. SAE paper 900810.

ATOMIC SCALE FRICTION OF
SELF-ASSEMBLED MONOLAYERS
BY HYBRID MOLECULAR SIMULATIONS

Shaoyi Jiang and Yongsheng Leng
Department of Chemical Engineering
University of Washington
Seattle, WA 98195

Abstract

We developed a hybrid molecular simulation method to extend simulation time scale to that in atomic force microscopy (AFM) experiments. The method combines a 'dynamic element' model for the tip-cantilever system in AFM and a molecular dynamics relaxation approach for the sample. We applied the hybrid simulation method to investigate atomic-scale frictional and mechanical properties of C_8 and C_{15} self-assembled monolayers (SAMs) as a function of chain length. A quite different behavior of atomic-scale friction was found from that by conventional molecular dynamics simulation. Stick-slip phenomena were observed when a soft spring was used. The coefficient of friction calculated is chain-length dependent, which is consistent with what was observed in AFM experiments. Our studies further show that elastic modulus is chain-length *independent*, whereas effective shear modulus is chain-length *dependent*. Thus, effective shear modulus is responsible for the chain-length dependence of friction.

Keywords

Molecular dynamics, Friction, Self-assembled monolayer, Atomic force microscopy.

Introduction

Understanding of atomic-scale frictional properties of self-assembled monolayers (SAMs) is critical in microelectromechanical systems (MEMS) and surface engineering (Maboudian, 1998). In recent years, extensive experiments were carried out to probe sliding friction of SAMs (Kiely, 1999; McDermott, 1997; Lio, 1997; Li, 1999). In the sliding friction experiments of SAMs by atomic force microscopy (AFM), n-alkanethiols on Au (111) showed a chain-length dependenct behavior of friction: the coefficient of friction decreased as the number of methylene groups in the alkane chain increased (McDermott, 1997; Lio, 1997; Li, 1999). The low friction of -CH_3 terminated SAMs largely comes from low surface energy due to their hydrophobic tail groups and high load bearing capacity due to their compact structures. In parallel, there have been a few papers concerning molecular dynamics (MD) simulations of friction of SAMs (Glosli, 1993; Tupper 1994a; Bonner, 1997; Fujihira, 1999; Tutein, 1999). However, the key problem encountered in MD is the short time scale (or fast sliding velocity). A typical sliding velocity used in MD simulations is 1~100m/s, many orders of magnitude higher than that in AFM experiments. With such a high sliding velocity, the physical picture obtained is questionable when compared with AFM experimental results. This time-scale problem has recently been solved (Leng, 2000a; Leng, 2000b) through the development of a novel 'hybrid' molecular simulation method, which combines a 'dynamic element' model (Johnson, 1998) for the tip-cantilever assembly (TCS) in AFM and a molecular dynamics relaxation (MDR) approach for the molecular sample. The proposed method is suitable for probing the dynamic processes in which the relaxation of sample molecules is quite fast. We applied the hybrid simulation method to examine atomic-scale frictional and mechanical properties of C_8 and C_{15} SAMs as a function of chain length.

Models and Simulation Methodology

In our tip-based simulations SAMs are modeled by 90 alkyl thiolate chains [$-S(CH_2)_{n-1}CH_3$] (C_n, n=8 or 15) chemisorbed onto a static gold (111) surface. The system is extended with periodic boundary conditions parallel to the substrate. The tip is represented by a rigid pyramidal one exposing (001) facet and contains 50 Au interacting atoms. The bottom layer of the pyramidal tip is composed of 9 atoms, corresponding to an effective contact area of $0.75nm^2$. For SAMs we adopt the united atom model developed by Hautman (1989). The harmonic bond vibrations of SAMs and the interactions between SAM molecules and Au atoms come from Tupper (1994b). At low temperature, C_8 and C_{15} SAM molecules exhibit an all-trans, well-ordered herribone structure with a tilt angle of 36°, and a lattice constant of 0.5nm, corresponding to a $\sqrt{3} \times \sqrt{3}$ sulphur adlattice.

At low temperature (e.g., at 0.1K), upon any small perturbation to the system caused by a tiny scanning displacement of an AFM tip at slow motion (e.g., 4000Å /s), we found that the relaxation of crystalline SAM films was quite fast. Fig. 1 shows how the kinetic temperature, and the bond stretching, bending, and torsion potential energies of SAMs vary with time during relaxation after the tip scans over SAMs by 100 TCS time steps (equivalent to 0.2Å). The relaxation of SAMs is very fast, roughly less than 0.2ps at 0.1K. This enables us to consider the evolution process of SAM molecules as quasi-static. In other words, when we deal with the sliding dynamics of the AFM tip at time t, we are only concerned about the equilibrium configuration of SAMs at this time instant, rather than their detailed dynamic relaxation processes. Therefore, we can use a larger time step to integrate the tip equations of motion, followed by a fast molecular dynamics relaxation (MDR) of SAMs with a smaller time step. This algorithm maintains the continuity of the tip motion with discrete SAM molecular configurations (Leng 2000a). For simulations of friction, TCS is driven by a support (x_M, y_M) where $x_M = vt$ moving at a constant scanning velocity v under a constant-force mode. This is described by the following differential equations of motion

$$M \ddot{x}_t = k_x(vt - x_t) + W_x(x_t, y_t, z_t; X(t)) \quad (1)$$

$$M \ddot{y}_t = k_y(y_M - y_t) + W_Y(x_t, y_t, z_t; X(t)) \quad (2)$$

$$M \ddot{z}_t = W_z(x_t, y_t, z_t; X(t)) - W_{ext} \quad (3)$$

where the coordinates (x_t, y_t, z_t) denote the center of mass of CTS, k_x and k_y are the effective spring constants of the cantilever along x and y directions, W_{ext} is the constant normal load, and $X(t)$ corresponds to the SAM molecular configuration at time t. The effective inertial mass of TCS is measured as 10^{-11}kg through our AFM calibrations (Li, 1999). In equations (1)-(3), the surface forces exerted on the tip by SAMs, W_x, W_y and W_z, were directly calculated from the summation of the interactions between tip atoms and SAM chains at the molecular configuration $X(t)$. The friction forces in scanning and raster directions were calculated from spring elongations, i.e., $F_x = k_x(x_M - x_t)$, and $F_y = k_y(y_M - y_t)$ (Holscher, 1997). For simulations of indentation and separation processes, we set v to zero and varied the external force W_{ext} in equation (3). The time steps for the tip and SAM equations of motion are $0.5\mu s$ and $1fs$, respectively (Leng, 2000a). The MD simulations of the SAMs were performed at constant temperature. For simulations of friction, the scanning velocity v of the support was set to 4000Å/s, similar to that used in our AFM experiments. To simplify analysis, we assume that the spring constant in y-direction is sufficiently large (k_y=100N/m) to prevent two dimensional scanning.

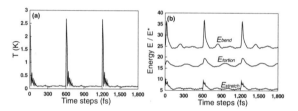

Figure 1. Relaxation of SAMs as described by (a) kinetic temperature, and (b) energies of bond stretching, bending and torsion. The 'jump-up' peaks are induced by tip sliding increments. The energy unit E^* is 88.1K.

Results and Discussion

To compare the results from the hybrid and the conventional MD methods, Fig. 2 shows the stick-slip behavior derived from both methods at a normal load of 0.5 nN with a compliant spring (k_x=1N/m) of TCS. The sliding velocity was set to 100m/s for MD and 4000A/s for hybrid simulations. The inertial mass of TCS was ~10^{-24}kg in MD and ~10^{-11}kg in hybrid simulations measured through our AFM calibration. In many tip-based MD simulations, no elastic response of the cantilever was included. This is equivalent to an infinite spring constant. The so-called 'stick-slip' seen by conventional MD at extremely high sliding velocity simply reflects the atomic feature of the surface. Even when the elastic response of the cantilever with very small inertial mass (Bonner, 1997) is taken into account, the MD method still does not correctly describe the real picture of the stick-slip phenomena in AFM experiments. As shown in Fig. 2, for a compliant spring, curve *a* shows a double-slip of the MD tip followed by an irregular motion. To prevent this double-slip, one has to add a damping γ artificially (Bonner, 1997) to get a regular motion (curve *c*) which shows a stick-slip pattern. However, the penalty of doing this is that frictional force is artificially increased, in contrast to our results from the hybrid method (curve *b*).

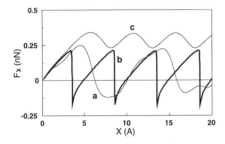

Figure 2. Stick-slips derived from the hybrid method (curve b) and the conventional MD (curves a and c). Curve a represents a 'double-slip' of the MD tip, while curve c depicts the regular pattern of the so-called 'stick-slip' with an artificial damping.

The hybrid method was applied to examine the stick-slip phenomena and the chain-length dependence of friction of C_8 and C_{15} SAMs scanned with an AFM tip. The stick-slip motion in AFM is the unstable mechanical behavior of a friction measurement system (Johnson, 1998; Holscher, 1997; Tomanek, 1991), and is related to the properties of both substrate and the stiffness of TCS (Robbins, 1996). For C_8 SAMs, at a contact load of 0.5nN corresponding to a contact pressure of 0.67 GPa, the critical spring constant k_c was 2.5N/m (Leng, 2000a; Leng, 2000b). When k_x was greater than k_c (e.g., k_x=132N/m), motions of the Au tip were stable, and the instantaneous frictional force F_x varied smoothly with the scanning distance x. However, when k_x was below k_c (e.g., k_x=1N/m), stick-slips occurred, resulting in higher energy dissipation. Our earlier results (Leng, 2000a) showed that the mode of friction energy dissipation was controlled by the spring constant k_x. Fig. 3 shows variations of the instantaneous friction force F_x and the work of friction (dissipation energy) W_f versus scanning distance for C_8 SAMs. The coefficients of friction for the 'hard' and 'soft' springs are calculated to be 0.0886 and 0.12, respectively, which are comparable to our AFM experiments (Li, 1999). Our detailed investigations (Leng, 2000a; Leng, 2000b) revealed that the sliding of the AFM tip on SAM surfaces led to excitations of SAM molecules mainly in bond stretching, bending and torsion vibration modes, which in turn supplied friction dissipation channels. These investigations showed that the stiffness of the measurement system played an important role in AFM, as demonstrated in the quasi-static analysis

by Tomanek (1991). In the case of C_{15} SAMs, for smooth sliding (hard spring) and for stick-slips (soft spring), the magnitude of the friction work dramatically decreases as compared with that of C_8, implying that the mean friction force $<F_x>$ is significantly lower. The coefficients of friction are 0.04 and 0.045 for the 'hard' and 'soft' springs, respectively. This strongly suggests that frictional properties of SAMs are chain-length *dependent* (McDermott, 1997; Lio, 1997; Li, 1999).

Figure 4 shows how the mean friction force $<F_x>$ varies with the normal load F_n for C_8 and C_{15} SAMs. The chain-length dependence of friction is obvious. For hard spring scanning (k_x=132N/m), the extent to which frictional forces increase with load is comparably smaller than that for soft spring scanning (k_x=1N/m). Our detailed investigations (Leng, 2000b) revealed that for soft spring scanning, the unstable stick-slips made the tip to penetrate SAM film to a larger depth than for hard spring scanning, activating higher excitations in SAM molecules. This explains why frictional force is comparably higher in stick-slips under higher loads (see Fig. 4). As also shown in Fig. 4, C_{15} can support larger loads up to 1.3 nN (equivalent to a contact pressure of 1.7GPa) than C_8 does. Beyond these critical loads, SAM films will be penetrated by the tip, resulting in higher friction forces mainly from plowing components. In AFM experiments, whether or not the stick-slip occurs depends on the magnitude of contact

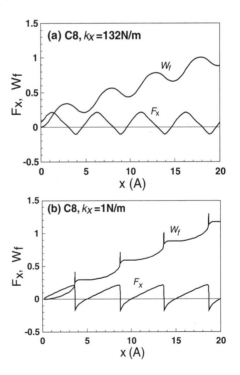

Figure 3. Variations of the instantaneous friction force F_x and the work of friction W_f versus tip scanning distance for C_8 SAMs. Units of force and friction work are in nN and nN·Å, respectively.

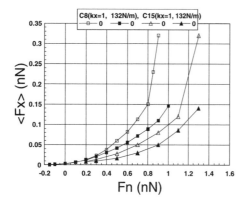

Figure 4. Variations of mean friction force versus contact load with different cantilever spring constants. The chain-length dependence of friction is obvious.

load and spring constant. If the load increases, the interfacial potential will increase (Fujisawa 1998), and the critical spring constant as the second derivative of this potential will also increase (Tomanek, 1991). In that case, stick-slips will be more likely to occur, as verified by Fujisawa (1998). The nonlinear increase of frictional force with load comes from the significant increase of the contact pressure due to the geometry of the tip used, leading to non-constant coefficient of friction. This is consistent with the earlier *ab initio* calculations by Zhong (1990) for friction between atomically flat interfaces. Since the AFM tip has a larger tip radius of curvature, the dependence of contact pressure on load will be significantly decreased (Zhong, 1990). Therefore, we expect that in AFM experiments of SAMs, the mean contact pressures should correspond to the load range of F_n =0.2~0.8 nN in Fig. 4.

The hybrid method was also applied to examine mechanical properties of C_8 and C_{15} SAMs. For indentation, it is interesting to find that surface deformations are almost the same for both C_8 and C_{15} and can be well described by the Hertz theory, except for the small load region where surface adhesion is prevailing. This suggests that the elastic modulus of -CH_3 terminated SAM films is chain-length *independent*, which is consistent with the prediction through molecular mechanics calculations by Henda (1998) Therefore, it seems that the continuum Hertzian contact mechanics is still valid at the atomic level. The Young's modulus of both SAMs at low temperature was estimated to be E=20±10 Gpa in this work. For sliding, we examined the molecular trajectories of C_8 and C_{15} just beneath the tip in the x-z plane. It is interesting to observe that the lateral displacements of atoms for the C_8 molecule along the sliding direction are much larger than those for the C_{15} molecule, suggesting that C_{15} is stiffer than C_8. The effective shear moduli of SAMs were estimated to be 3.0 and 5.3Gpa for C_8 and C_{15}, respectively in this work. Thus, effective shear modulus is chain-length *dependent*.

Energetic considerations (McDermott, 1997; Lio, 1997) suggest that with increases in alkyl chain lengths in SAMs, friction coefficient decreases because cohesive energy between alkyl chains increases. Our analysis above shows that longer chains with higher cohesive energy will have larger lateral stiffness, which can significantly reduce the magnitude of UA displacements, thus the excitations of SAM molecules.

Conclusions

We have developed a hybrid molecular simulation method combining a 'dynamic element' model for the TCS in AFM and an MDR approach for the sample. The proposed method extends simulation time scale to that in AFM experiments. The method is suitable to probing the dynamic processes in which the relaxation of sample molecules (e.g., SAMs) is quite fast. Conventional MD simulations of AFM employed an unphysical high sliding velocity, and failed to yield a correct description of atomic-scale friction. With the hybrid method, we observed the stick-slip phenomena and the chain-length dependence of frictional properties of SAMs as in AFM experiments. Our further studies with the hybrid method show that the mechanical properties of SAMs are strongly anisotropic - elastic modulus is chain-length *independent*, whereas effective shear modulus is chain-length *dependent*. Thus, longer chains have higher lateral stiffness than shorter chains, leading to lower coefficients of friction.

References

Bonner, T. and A. Baratoff (1997). *Surf. Sci.*, 377-379, 1082.

Fujihira, M. and T. Ohzono (1999). *Jpn. J. Appl. Phys.*, **38**, 3918.

Fujisawa, S., Yokoyama, K., Sugawara, Y., and S. Morita (1998). *Phys. Rev. B*, **58**, 4909.

Glosli, J. N. and G. M. McClelland (1993). *Phys. Rev. Lett.*, **70**, 1960.

Hautman, J. and M. L. Klein (1989). *J. Chem. Phys.* **91**, 4994.

Henda, R., Grunze, M., and A. J. Pertsin (1996). *Trib. Lett.*, **5**, 191.

Holscher, H., Schwarz, U. D., and R. Wiesendanger (1997). *Surf. Sci.*, **375**, 395.

Johnson, K. L., and J. Woodhouse (1998). *Trib. Lett.*, **5**, 155.

Kiely, J. D. and J. E. Houston (1999). *Langmuir*, **15**, 4513.

Leng, Y. S. and S. Jiang (2000). Extending time scale in molecular simulations of atomic force microscopy.. resubmitted to *Phys. Rev. Lett.* 2000.

Leng, Y. S. and S. Jiang (2000). Atomic Indentation and Friction of Self-Assembled Monolayers by Hybrid Molecular Simulations. submitted to *J. Chem. Phys.*

Li, L., Yu, Q., and S. Jiang (1999). *J. Phys. Chem., B*, **103**, 8290.

Lio, A., Charych, D. H., and M. Salmeron (1997). *J. Phys. Chem. B*, **101**, 3800.

Maboudian, R. (1998). *Surf. Sci. Reports*, **30**, 207.

McDermott, M. T., Green, J. D., and M. D. Porter (1997). *Langmuir*, **13**, 2504.

Robbins, M. O., and E. D. Smith (1996). *Langmuir*, **12**, 4543.

Tomanek, D., Zhong, W., and H. Thomas (1991). *Europhys. Lett.,* **15**, 887.

Tupper, K. J., and D. W. Brenner (1994). *Thin Solid Films,* **253**, 185.

Tupper, K. J.; and D. W. Brenner (1994). *Langmuir*, **10**, 2335.

Tutein, A. B., Stuart, S. J., and J. A. Harrison (1999), *Langmuir,* **16**, 291.

Zhong, W., and D. Tomanek (1990). *Phys. Rev. Lett.,* **64**, 3054.

DISAPPEARING MINIMA OF ENERGY LANDSCAPES UNDER STRESS

Daniel J. Lacks
Department of Chemical Engineering
Tulane University
New Orleans, LA 70118

Abstract

Stress is shown to lead to major changes in energy landscapes, such as the disappearance of energy minima. These changes in the energy landscape alter the system dynamics, which in turn alters the system properties. In particular, these landscape changes are shown to lead to amorphous-amorphous phase transitions, enhanced diffusion in flowing liquids, plastic deformation in glassy materials, and a critical polydispersity for colloidal crystals.

Keywords

Energy landscapes, Phase transitions, Diffusion, Colloidal crystals.

Introduction

The effects of stress on disordered systems are still not fully understood, despite their importance. Two obstacles to this understanding are the difficulties of (i) describing the dynamics of disordered systems in terms of a simple molecular-level picture (in contrast, the dynamics of crystals can largely be described in terms of vibrational motion about well-defined structures and the motion of defects such as dislocations), and (ii) addressing long-time phenomena in molecular-level simulations. Although molecular dynamics simulations are widely used to study disordered systems, it is often difficult to extract physical understandings of the phenomena solely from such simulation results, and to address phenomena occurring with time scales longer than several nanoseconds.

Recently, the energy landscape or "inherent structure" formalism has emerged as a framework that allows the dynamics of disordered systems to be physically described in a simple way (Stillinger and Weber, 1984). The system dynamics can be considered the sum of vibrational motion within local minima and transitions between local minima. The vibrational motion is similar to that in a crystal, while the transitions between minima manifest in fluid-like properties such as diffusion and flow. For crystal phases, the system remains in the energy minimum corresponding to that crystal structure, and the system dynamics consists of vibrations within this energy minimum (for perfect crystals). For liquid and glassy phases, the system moves between local energy minima corresponding to disordered structures – the transitions between local minima are relatively frequent in liquids, whereas the transitions are relatively infrequent in glasses.

The present paper will describe how stress changes the potential energy landscape, and how these landscape changes are shown to lead to amorphous-amorphous phase transitions, enhanced diffusion in flowing liquids, plastic deformation in glassy materials, and a critical polydispersity for colloidal crystals.

Computational Methods

The present simulations are carried out for systems of 100-2000 particles interacting through empirical force fields, with periodic boundary conditions. The strain-induced changes in the potential energy landscape are determined as follows: Local potential energy minima corresponding to disordered structures are first obtained by quenching liquid systems to zero temperature using a variable-metric minimization algorithm. Strains are then imposed in extremely small increments (on the order of 0.0001-0.001). After each strain increment, the potential energy is minimized with respect to the atomic coordinates (at constant strain). System properties are calculated at these minimum energy configurations, to see how the local minima change with strain. Saddle points between local minima are obtained with a conjugate gradient algorithm.

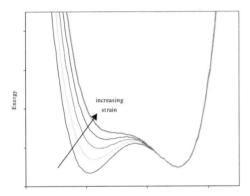

Figure 1. Schematic of the strain-induced disappearance of an energy minimum.

Strain-Induced Disappearances of Energy Minima

The strains resulting from applied stresses lead to disappearances of energy minima and barriers in disordered systems, as shown schematically in Fig. 1 (Malandro and Lacks, 1997). The evidence for these disappearances, as shown in Fig. 2 (for a monatomic glass), is that at certain strains: (1) The curvature of the local minimum decreases to zero in one direction (equivalently, a normal mode frequency decreases to zero). (2) The height of an energy barrier that separates this local minimum from another decreases to zero. (3) The curvature of the barrier decreases to zero. (4) The distance between this barrier and the local minimum decreases to zero. These changes in the potential energy landscape establish that a local energy minimum disappears when the system is strained.

These disappearances of local energy minima give rise to locally discontinuous and irreversible changes in the structure and properties of the system, as the system suddenly moves to another energy minimum.

The strain-induced disappearance of energy minima is found to be a localized phenomenon; this localization contrasts to mechanical instabilities in crystals, which are delocalized (e.g., McGann and Lacks, 1999). Based on the localization of the mechanical instabilities to finite-sized groups of atoms, as the system-size increases: (1) the disappearance of energy minima occur more frequently because the number of groups of atoms which

undergo instabilities is an extensive quantity, and (2) the magnitudes of the discontinuous property changes will decrease because a decreasing fraction of the system is involved in the instability. For these two reasons, the properties will appear continuous for macroscopic systems.

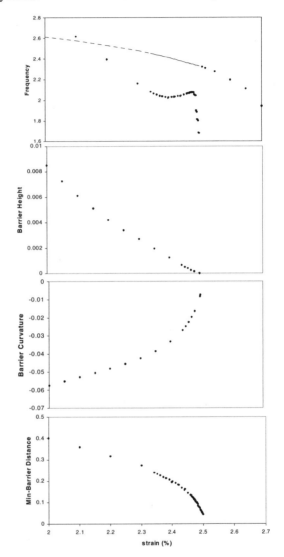

Figure 2. Evidence for the disappearance of a local energy minimum, which occurs here at a strain of 2.5%: A normal mode frequency at the energy minimum, the height of an energy barrier, the minimum-barrier distance, and the curvature of the barrier decrease to zero.

Plastic Deformation in Glassy Materials

As a glassy material is strained, the stress usually increases continuously and nearly linearly with strain, but discontinuous stress drops punctuate these increases. These stress drops correspond to the disappearance of energy minima, as described above, and are irreversible. In this way, the strain-induced disappearances of energy minima give rise to yielding and plastic flow in glassy materials (Malandro and Lacks, 1999).

Enhanced Diffusion in Flowing Liquids

Shear stresses and strains are generated in liquids by flow (due to viscous effects at the boundaries). These flow-induced strains lead to disappearances of energy minima (Malandro and Lacks, 1998). After an energy minimum disappears, the system relaxes towards another energy minimum, with an associated change in atomic positions.

The atomic displacements associated with the disappearances of energy minima correspond to steps in a random walk (in the directions perpendicular to the shear); the sum of these steps gives rise to self-diffusion that is strain-activated rather than thermally activated. These strain-activated contributions to diffusion augment the usual thermally activated contributions.

To demonstrate the validity of the proposed mechanism for enhanced self-diffusion in liquids, NEMD simulations are carried out to determine the self-diffusion constants as a function of shear rate and temperature, using the sllod equations of motion with the temperature maintained by a Gaussian thermostat. The NEMD results, shown in Fig. 3, demonstrate that the dependence of the diffusion constants on shear rate extrapolate in the low temperature limit towards the result based only on the disappearance of energy minima (Malandro and Lacks, 1998).

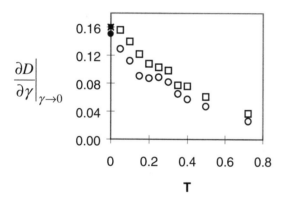

Figure 3. Dependence of diffusion constant D on shear rate γ, as a function of temperature. Open symbols are the result of NEMD simulations, and filled symbols (at T=0) are results arising solely from strain-induced changes of the potential energy minima. Squares and circles represent diffusion constants in different directions.

Amorphous-Amorphous Phase Transitions

The results for the pressure as the glass is compressed and then decompressed (at zero temperature) are shown

in Fig. 4. The path followed by the system is not reversed on decompression due to the strain-induced disappearance of local minima of the potential energy surface (Lacks, 1998) (these disappearances give rise to the discontinuities in the pressure-volume curve evident in Fig. 4). The low density (high volume) branch of this pressure-volume curve can be associated with a low-density amorphous (lda) phase, and the high density (low volume) branch can be associated with a high-density amorphous (hda) phase.

A comparison of the Gibbs free energies G of the lda and hda phases gives insight in regard to the long time dynamics of the system, which cannot be addressed by molecular dynamics (Lacks, 2000). These free energies are evaluated at the lda and hda energy minima described above, and are calculated in the limit of zero temperature such that G = E + PV. As shown in Figure 2, the free energy curves of the lda and hda phases cross at P=3 GPa, indicating that a first-order transition occurs between the lda and hda phases at P=3 GPa.

The first-order transition from lda to hda requires thermal activation, because a barrier exists between these energy minima. The first-order phase transition will therefore be kinetically hindered at some "low temperatures". If the first-order transformation is kinetically hindered, the lda phase will persist until the pressure exceeds its spinodal (i.e., metastability limits). The transformation from lda to hda at zero temperature, as shown in Fig. 4, occurs by what can be considered spinodal decomposition – i.e., the energy minima corresponding to the lda phase disappear above certain pressures. This spinodal decomposition takes place over a broad pressure range due to the variety of local environments in an amorphous system (i.e., different local atomic arrangements become unstable at different pressures), and is thus manifested as a continuous change in structure.

In summary, we predict that there is a pressure-induced first-order phase transitions from lda to hda in silica, characterized by a discontinuous change in volume. The first-order phase transition can be kinetically hindered at low temperatures, in which case the lda phase persists until it undergoes a gradual spinodal decomposition at a higher pressure.

Order-Disorder Transitions in Polydisperse Colloids

We have also shown for colloidal systems that increases in polydispersity lead to the disappearance of energy minima corresponding to ordered structures, analogously to the effects of stress (Lacks and Wienhoff, 1999). The disappearances are detected by the decrease to zero of the curvature of the energy minimum and the ensuing discontinuous decrease in the extent of order. These results demonstrate that there is a terminal polydispersity above which the crystal phase cannot possibly exist, even as a metastable state.

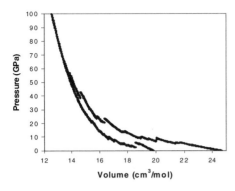

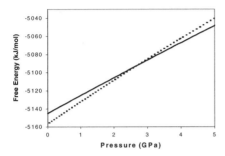

Figure 4. (top) Pressure of system as a function of volume. The low density (high volume) branch corresponds to the compression of the initial glass structure, and the high density (low volume) branch corresponds to the subsequent decompression from 100 GPa. (bottom) Gibbs free energies of the lda and hda structures at zero temperature. The points represent the lda free energies, and the solid line represents the hda free energies.

Conclusions

Stress leads to important changes in energy landscapes, which alter the system dynamics and ultimately the system properties. For example, stress can lead to the disappearance of local minima of the potential energy landscape. These landscape changes are shown to lead to amorphous-amorphous phase transitions, enhanced diffusion in flowing liquids, plastic deformation in glassy materials, and a critical polydispersity for colloidal crystals.

Acknowledgments

Funding for this project was provided by the National Science Foundation (grant number DMR-9624808).

References

Lacks, D. J. (1998). Localized mechanical instabilities and structural transformations in silica glass under high pressure. *Phys. Rev. Lett.,* **80**, 5385-5388.

Lacks, D. J. (2000). First-order amorphous-amorphous transformation in silica. *Phys. Rev. Lett.,* **84**, 4629-4632.

Lacks, D. J. and J. R. Wienhoff (1999). Disappearances of energy minima and loss of order in polydisperse colloidal systems. *J. Chem. Phys.* **111**, 398-401.

Malandro, D. L. and D. J. Lacks (1997). Volume dependence of the potential energy landscape in glasses. *J. Chem. Phys.,* **107**, 5804-5810.

Malandro, D. L. and D. J. Lacks (1998). Molecular-level mechanical instabilities and enhanced self-diffusion in flowing liquids. *Phys. Rev. Lett.,* **81**, 5576-5579.

Malandro, D. L. and D. J. Lacks (1999). Relationships of shear-induced changes in the potential energy landscape to the mechanical properties of ductile glasses. *J. Chem. Phys.,* **110**, 4593-4601.

McGann, M. R. and D. J. Lacks (1999). Entropically-induced Euler buckling instabilities in polymer crystals. *Phys. Rev. Lett.,* **82**, 952-955.

Stillinger, F. H. and T. A. Weber (1984). Packing structures and transitions in liquids and solids. *Science,* **225**, 983-989.

A NOVEL PROTOCOL FOR SIMULATION OF HIGHLY VISCOUS MOLECULAR SYSTEMS

Jerry W. Jenkins and Peter J. Ludovice
School of Chemical Engineering
Georgia Institute of Technology
Atlanta, GA 30332-0100

Abstract

Computer simulation of viscous materials, such as bulk polymers or self-assembled two-dimensional films, are often required to produce meaningful structure-property relationships for such systems. Current constraints on computational resources restrict typical Molecular Dynamics (MD) simulations to time scales much shorter than the relevant relaxation times for these materials. We have characterized the performance of a new algorithm, called Protracted Colored-Noise Dynamics, to efficiently sample phase space for such viscous systems. The algorithm significantly reduced the Mean First Passage Time for a one-dimensional double-well oscillator. This algorithm appears to work because it locally increases the entropy of the conformational transition state.

Keywords

Simulation, Colored noise, Molecular dynamics, Glass, Viscous, Convergence.

Introduction

For reasonably large molecular systems, comprised of many thousands of atoms, the typical time scale for Molecular Dynamics (MD) simulations rarely exceeds several nanoseconds. Such simulations are used to relax molecular systems as viscous as amorphous glasses. Given the viscosity of such systems, such a short MD simulation will provide only a minimal amount of relaxation that is insufficient to relax a glass from an arbitrary starting conformation to its relaxed state. Such approaches only work when the initial conformation of the molecular system is quite close to the relaxed glassy state as seen in simulations of atactic polypropylene (Theodorou, 1985) and Bisphenol-A polycarbonate (Hutnick, 1991). These systems used a fairly random initial conformation, but because they are amorphous, this represented a reasonable starting conformation. Polymer glasses that contain some level of order are more difficult (Ludovice, 1999), but can be modeled if the initial conformation is properly biased to include this order (Ahmed, 1999; 2000). However, this requires *a priori* knowledge of the polymer structure.

The proposed algorithm was originally developed to sample phase space in two-dimensional self-assembled phospholipid monolayers. We found that the structure of the alkane tails in these phospholipid films relaxed on a short time scale (Mauk, 1998a), while the area of films did not (Mauk, 1998b). The proposed algorithm uses large scale stochastic fluctuations to traverse a large region of conformation space in a relatively small amount of simulation time to effectively relax these viscous systems that include self assembled films and polymer glasses.

Simulation Algorithm

Phase space sampling algorithms range from MD, which is completely deterministic, to Monte Carlo (MC) methods, which are completely stochastic as seen in Fig. 1. Unlike the MD approach which solves deterministic equations of motion, the MC approach applies random conformational perturbations and selects an ensemble of these perturbations so as to reproduce the appropriate statistical mechanical distribution (Metropolis MC method). This approach can be more efficient than the MD approach provided the random perturbations are energetically reasonable. For complex molecules in condensed phases (i.e. polymer glasses or self-assembled films) random perturbations are not very efficient given the high degree of interaction or entanglement among such molecules. The efficiency of MC methods for these viscous systems can be improved by utilizing a more realistic conformational perturbation to increase the acceptance ratio of these MC moves (Smart MC). Various polymer moves involving more local changes in the chain conformation (Dodd, 1999), or changes in response to the surrounding polymer matrix (de Pablo 1992, 1993) have been used to improve the efficiency of MC methods for polymer chains. However, these Smart MC approaches require some *a priori* knowledge of the molecular dynamics.

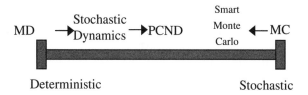

Figure 1. Contiuum of phase-space sampling techniques.

Rather than try to deduce these efficient moves we propose a more general approach, which adds stochastic fluctuations to the deterministic method of MD simulation. Similar approaches have been used to describe the Brownian motion of a surrounding solvent on molecules. These approaches are based on various forms of the Langevin equation and are referred to as Langevin, Brownian or Stochastic Dynamics. We propose to increase the stochastic fluctuations of such an approach to manipulate the efficiency of phase space sampling, as opposed to mimicking the surrounding solvent. Eqn. (1) is the simplified Langevin equation where the mass times the second derivative of the position (left side) is equal to the negative of the potential energy gradient plus a stochastic force term (R_i), less a damping term.

$$m\ddot{x}_i = -\nabla_i U + R_i - \gamma p_i \qquad (1)$$

The damping term is a drag coefficient (γ) times the momentum of the current atom or particle (p_i). The stochastic fluctuations are Gaussian distributed with a zero first moment and a second moment that is a function of the damping coefficient. These fluctuations are also uncorrelated in time as seen by Eqn. 2 below.

$$\langle R_i(t) \cdot R_i(t') \rangle = 2mkT\gamma\delta(t - t') \qquad (2)$$

The coupling of the damping coefficient and the stochastic fluctuation term seen in Eqn. 2 is a consequence of the Fluctuation Dissipation Theorem (FDT) and limits the stochastic fluctuations of the system. We propose two modifications to the Langevin equation to broadly sample phase space. First, the Dirac delta function in Eqn. 2 is replaced to facilitate the correlation of the stochastic forces in time. This replaces the random, or white, noise of the Langevin equation with colored noise, which is correlated in time. Although any function can be chosen, we will choose the exponential decay of the Orstein-Uhlenbeck process for simplicity (Uhlenbeck, 1930). The use of colored noise does allow systems to more broadly sample phase space (Hängii, 1989). To further increase the stochastic fluctuations, we will decouple the damping and stochastic terms in Eqn. 1. This does represent a violation of the FDT, and consequently an accurate dynamic trajectory is not guaranteed for the proposed method. However, we intend to use this method only to traverse significant paths in phase space to move a highly viscous system across prohibitively large energetic barriers to a more reasonable region of phase space. Such movement through phase space would not be possible with conventional MD simulation, yet this method does not require any knowledge of efficient MC perturbations. Once such a relaxed region of phase space is found, conventional MD simulation can be used to obtain an accurate trajectory. Simulations of Lennard-Jones glasses, using this method, relax to the crystal state approximately 10^4 times faster than conventional MD methods.

The fundamental equation of motion for this method is still given by Eqn. 1, but Eqns. 3 and 4 are the constraining equations for the damping coefficient and the stochastic force.

$$\dot{\gamma} = \frac{1}{Q}\left(\frac{T}{T_{Bath}} - 1\right) \qquad (3)$$

$$\{\langle R_i(t) \cdot R_i(t') \rangle\} = \frac{D}{\tau} e^{-|t-t'|/\tau} \qquad (4)$$

The damping coefficient is determined by an integral controller that constrains the average system temperature (T) to be equal to that of a desired external bath temperature (T_{Bath}). This is essentially the controller developed by Nosé and Hoover (1985), where the controller relaxation time is $Q^{1/2}$. The Dirac delta function of equation 2 is replaced by an exponential function in time that is governed by a relaxation time τ. The parameter D/τ is the overall strength of the stochastic fluctuations and is not coupled to the damping coefficient. The random Gaussian forces are generated according to the above constraints using a variation of the

Box-Meuller algorithm (Knuth, 1962) developed by Fox (1988, 1991). Because this algorithm utilizes colored noise and extends or protracts the effect of the stochastic force, we refer to it as Protracted Colored Noise Dynamics (PCND). Its increased used of stochastic fluctuations makes the algorithm more stochastic in nature as pictured in Fig. 1. The braces outside the brackets in Eqn. 4 denote an average over initial conformations which Fox (1991) determined was needed to get good statistical convergence.

Results and Discussion

As a preliminary test we applied the PCND algorithm to a one-dimensional, two-site harmonic oscillator of mass m, with an energy barrier of ΔE between the two sites separated by distance σ. The Mean First Passage Time (MFPT) between the two wells was recorded for a number of simulations. The simulations were carried out for sufficient time such that the 95% confidence interval about the mean of the MFPT was less than 5% of the mean. Typically this required from 80 to 100 million dimensionless time steps. Fig. 2 is a graph of the MFPT as a function of both the fluctuation strength D/τ and the relaxation time τ. Dimensionless parameters are used to describe time (t), temperature (T) and the strength of the fluctuations (D) in terms of their dimensional counterparts (t^*, T^*, D^*) as see in Eqn. 5.

$$ t = \frac{t^*}{\sqrt{\frac{m\sigma^2}{\Delta E}}} \qquad T = \frac{kT^*}{\Delta E} \qquad D = \frac{D^*\sigma}{m^{1/2}(\Delta E)^{3/2}} \qquad (5) $$

A constant value of 0.1, in dimensionless time units squared, was used for the thermostat parameter Q. As one would expect, the MFPT decreases initially as D/τ increases due to the increased fluctuations. The MFPT is less sensitive to the colored-noise relaxation time τ, which appears to have an optimal value as seen in the shallow local minima in MFPT in Fig. 2.

Exactly why the algorithm reduces the MFPT appears related to the entropy of the transition-state. By considering 1/MFPT as an effective rate constant, some insight into the algorithm can be made. An Arrhenius analysis can be made with the results of simulations carried out at different temperatures. Fig. 3 contains the results of these simulations for different values of τ and D/τ as a function of temperature. Regardless of the values of these parameters, the slopes of the logarithm of 1/MFPT vs the inverse of the dimensionless temperature are essentially the same. This suggests the algorithm does not change the effective height of the barrier between the two states. There is some minor deviation from linearity at low temperatures. We suspect that this occurs because the fluctuations dominate in this regime. While the slope is fairly constant, the intercept changes drastically with changes in τ and D/τ. As stochastic fluctuations increase, the intercepts of the lines in Fig. 3 also increase. The intercept of these lines is proportional to the entropy change of the transition-state. Therefore, it appears that increasing the strength of the protracted colored-noise, or the relaxation time τ increases the entropy of the transition-state thereby decreasing the free energy of the transition state. This increase in the entropy of the transition-state is what facilitates efficient sampling of phase space.

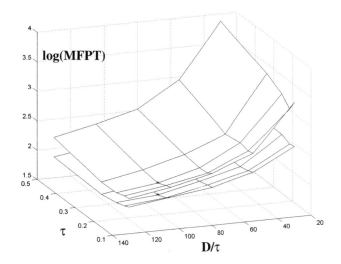

Figure 2. Base 10 logarithm of MFPT as a function of model parameters (T*=0.3 upper curve, T*=0.5 lower curve).

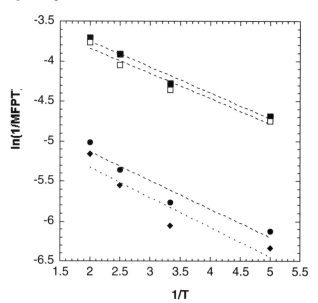

Figure 3. Arrhenius analysis for various parameter values (■ τ=10 D/τ=75, ● τ=2 D/τ=375, ❑ τ=10 D/τ=60, ◆ τ=2 D/τ=300, dotted lines are linear fits).

An additional effect of τ can be seen by analyzing the simulation results in the context of the generalized Langevin Eqn. 6.

$$m\ddot{x}_i = -\nabla_i U + R_i - \int_0^t K(t-s)\dot{x}(s)ds \qquad (6)$$

Eqn. 6 is the generalized version of Eqn. 1, where K(t) is the friction kernel that describes the hydrodynamic memory of the system. Eqn. 1 simply uses a dirac Delta function times a constant for the friction kernel to obtain a constant damping parameter. Using the procedure developed by McCammon (Berkowitz, 1981) we can extract the friction kernel from our PCND simulations. Regardless of the dynamics of the system, the friction kernel should eventually go to zero at equilibrium. The friction kernel for the simulations above is plotted as a function of dimensionless time in Fig. 4. If τ is too small (τ=1), the simulation does not appear to equilibrate. While the other values of τ equilibrate, this parameter can clearly affect the hydrodynamic behavior of the system, which is the object of current investigations.

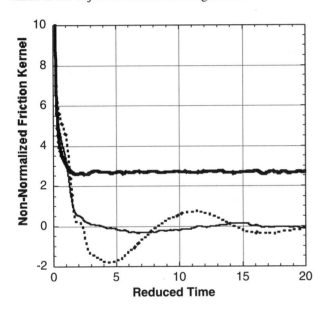

Figure 4. Friction kernel as a function of dimensionless time with $D/\tau=100$ and $Q=0.1$ (bold line $\tau=1$, solid line $\tau=4$, dotted line $\tau=10$).

Conclusions

Protracted Colored-Noise Dynamics, a new simulation algorithm reduced the MFPT of a double-well oscillator. The algorithm dramatically increases the stochastic fluctuations of MD algorithms by using colored-noise and decoupling the noise from the damping parameter. The effectiveness of the algorithm is derived from an increase of the local entropy of conformational transition-states.

Acknowledgements

We greatly acknowledge the financial support of the Whitaker Foundation under Biomedical Research Grant # RG-95-0334. We also acknowledge financial support from the ONR sponsored Molecular Design Institute and the Polymer Education Research Center at Georgia Tech, as well as the Emory-Georgia Tech Joint Biomedical Research Center. Helpful discussions with Prof. Ron Fox of the Georgia Tech School of Physics were greatly appreciated.

References

Ahmed, S., Bidstrup, S.A., Kohl, P., and P. Ludovice (1998). Prediction of stereoregular poly(norbornene) structure using a long-range RIS model. *Macromol. Symp.*, **133**, 1-10.

Ahmed, S., Kohl, P., and P. Ludovice, (2000). Microstructure of 2.3-erythro di-isotactic polynorbornene from atomistic simulation. *J. Comp. Theor. Polym. Sci.*, **10**, 221-233.

Berkowitz, M., Morgan, J.D., Kouri, D.J. and J.A. McCammon (1981). Memory kernels from molecular dynamics. *J. Chem. Phys.*, **75**, 2462-2463.

de Pablo, J.J., Laso, M. and U.W. Suter (1992). Estimation of the chemical potential of chain molecules by simulation. *J. Chem. Phys.*, **96**, 6157.

de Pablo, J.J., Laso, M. and U.W. Suter (1993). Simulation of polyethylene above and below the melting point. *J. Chem. Phys.*, **96**, 2395.

Dodd, L.R. and D.N. Theodorou (1993). A concerted rotation algorithm for atomistic monte carlo simulation of polymer melts and glasses. *Molec. Phys.*, **78**, 961-996.

Fox, R.F., Gatland, I.R., Rajarshi, R. and G. Vemuri (1988). Fast accurate algorithm for numerical simulation of exponentially correlated noise. *Phys. Rev.*, **A38**, 5938-5940.

Fox, R. (1991). Second-order algorithm for the numerical integration of colored noise problems. *Phys. Rev.*, **A43**, 2649-2654.

Hänggi, P., Jung, P. and F. Marchesoni (1989). Escape driven by strongly correlated noise. *J. Stat. Phys.*, **54**, 1367-1380.

Hoover, W.G. (1985). Canonical dynamics: equilibrium phase space distributions. *Phys. Rev.*, **A31**, 1695-1697.

Hutnick, M., Gentile, F.T., Ludovice, P.J., Suter, U.W. and A.S. Argon (1991). An atomistic model of the amorphous glassy polycarbonate of 4,4'-Isopropylidene. *Macromolecules*, **24**, 5962-5969.

Jenkins, J.W. and P.J. Ludovice (2000). Equilibration of glasses with a novel colored-noise stochastic simulation algorithm, in preparation.

Knuth, D.E. (1962). *The Art of Computer Programming*, Vol II. Addison Welsley, Reading, MA.

Ludovice, P., Ahmed, S., Van Order, J. and J. Jenkins (1999). Simulation of intermediate order in polymer glasses. *Macromol. Symp.*, **146**, 235-244.

Mauk, A., Chaikof, E.L. and P.J. Ludovice, P.J. (1998). Structural characterization of self assembled lipid monolayers by computer simulation. *Langmuir*, **14**, 5255-5266.

Mauk, A. (1998). *Determination of the Structure of the Magainin II Transmembrane Channel*. Ph.D. Thesis, Georgia Institute of Technology.

Theodorou, D. N. and U.W. Suter (1985). Detailed molecular structure of a vinyl polymer glass. *Macromolecules,* **18**, 1206.

Uhlenbeck, G.E. and L.S. Ornstein (1930). On the theory of the brownian motion. *Phys. Rev.,* **36**, 823-841.

THERMAL CONDUCTIVITY SIMULATION
USING HEAT RESERVOIRS

Jurivan Ratanapisit and James F. Ely
Chemical Engineering and Petroleum Refining Department
Colorado School of Mines
Golden, CO 80401

Dennis J. Isbister
School of Physics, University College
University of New South Wales, ADFA
Canberra ACT 2600, Australia

András Baranyai
Department of Theoretical Chemistry
Eötvös University
1518 Budapest 112, Hungary

Abstract

In this study we investigate the use of a modified form of Baranyai's original method to evaluate the thermal conductivity of a Lennard-Jones fluid. In this method the heat flux is generated by a linear temperature gradient driven by thermostatted heat baths (regions) within a rectangular simulation cell. We show that by using Nosé-Hoover thermostats and a nearly symplectic integration technique, larger simulation time-steps may be used and that at conditions removed from the critical point, a single simulation (rather than a series of simulations at varying thermostat widths) may be used to obtain the thermal conductivity.

Keywords

Thermal conductivity, Lennard-Jones fluid, Molecular dynamics, Symplectic integrator.

Introduction

Over the years there has been interest in developing non-equilibrium molecular dynamics (NEMD) simulation methods for the thermal conductivity that overcome the long time equilibrium molecular dynamics (EMD) simulation problems and the fact that boundary driven heat flow violates periodic boundary conditions (Evans and Morriss, 1984). Examples of NEMD methods that have been developed are synthetic field methods (Evans, 1986; Hoover, *et al.*, 1985; Massobrio and Ciccotti, 1984), the fluid wall method (Ashurst, 1974), and the momentum exchange method of Müller-Plathe (Müller-Plathe, 1997).

A recent modification of the fluid wall method was proposed by Baranyai (1996). In his approach, alternating hot and cold thermostatted fluid reservoirs are simulated and regions in between the reservoirs ("Newtonian" regions) are treated by normal NVE-EMD. Periodic boundary conditions are satisfied in all regions and the equations of motion were solved using a fifth order Gear predictor-corrector method. In Baranyai's study, the reservoir temperatures were maintained using Gaussian thermostats and smooth weighting functions were introduced in the reservoir regions to prevent momentum discontinuities at the boundaries. Our interest in this method arises from the fact that in Baranyai's original report, he was only able to use the method with a small time step—attempts to use larger time steps resulted in instabilities in the thermostatted regions.

In this study, we have investigated and simplified Baranyai's algorithm as follows. First, we investigated the role of the weight function by examining the number of particles in each region, temperature stability in the reservoirs, and density variations in the Newtonian regions, both with and without weight functions. We found that there is no substantial difference between the results obtained with and without the weight function. Secondly, we studied the usefulness of a Nosé-Hoover thermostat as opposed to the Gaussian thermostat (Hoover, 1991). Finally, we investigated the stability of the simulation algorithm using a symplectic integrator (SI) (McLachlan and Atela, 1992; Yoshida, 1990; Yoshida, 1993) as opposed to the Gear predictor-corrector. The SI applied in this work is the position Verlet method as proposed by Zhang (1997).

Simulation Details

Baranyai's system is composed of series of cubic cells extending in the y-direction. The system contains two heat reservoirs separated by a Newtonian (NVE) region. The thermal reservoirs are thermostatted at different temp-eratures to generate a heat flow through the intervening Newtonian region. The original equations of motion used temperature and momentum constraints, *viz.*

$$\dot{\mathbf{q}}_i = \mathbf{p}_i / m_i \quad \text{and} \quad \dot{\mathbf{p}}_i = \mathbf{F}_i - f_\beta\left(y_i\right)\left[\kappa_\beta \mathbf{j} + \alpha_\beta \mathbf{p}_i\right] \quad (1)$$

where $\mathbf{p}_i$ is the momentum, $\mathbf{q}_i$ is the position, $\mathbf{j}$ is the unit vector in the y-direction which is parallel with the dir-ection of the temperature gradients, α is the thermo-statting multiplier, and κ is the momentum drift multi-plier for zeroing any momentum drift in the y-direction. The subscript $\beta = \pm 1$ refers to the hot and cold reservoir, respectively. The function $f_\beta(y_i)$ is a weight function introduced into the system to smooth the boundaries of reservoirs. These weight functions originally had identical shapes (cubic prisms) and varied smoothly between zero and one. It was unity for particles completely inside in a reservoir, zero for particles completely outside reservoir and intermediate for particles in the transitional wall region separating the reservoir from the Newtonian region

Given the temperature gradient and heat flux, the thermal conductivity is calculated *via* Fourier's law. The temperature gradient can be calculated from reservoir temperatures (which are calculated directly from kinetic energy of the particles in a given reservoir) and the simulation cell geometry. There are two methods that can be used to calculate the heat flux—the Irving-Kirkwood heat flux for particles in the Newtonian regions

$$\mathbf{J}_Q V = \sum_i \frac{\mathbf{p}_i}{2m_i}\left(\frac{p_i^2}{m_i} + \sum_i \sum_{j=i+1} u_{ij}\right) - \sum_i \sum_{j=i+1} \frac{\mathbf{r}_{ij}\mathbf{F}_{ij} \cdot \mathbf{p}_i}{2m_i} \quad (2)$$

or by the energy dissipation from the reservoirs

$$\mathbf{J}_Q \cdot \hat{\mathbf{e}}_y = \frac{dH}{dt} = \frac{\alpha_\beta}{2} \sum_{i=1}^N p_i^2 \quad (3)$$

The thermal conductivity can be calculated from the heat flux by

$$\lambda = \frac{\mathbf{J}_Q \cdot \hat{\mathbf{e}}_y}{L_x L_z\left(\partial T / \partial y\right)} \quad (4)$$

where $\partial T/\partial y$ is the temperature gradient and $L_x L_z$ is the area perpendicular to the heat flux.

Simulation Model

The system size was 2048 particles (8 cubic cells with 256 particles per cell) with an overall reduced density of 0.8. The simulations were started from a fcc lattice structure with a Maxwell-Boltzman velocity distribution and a neighbor list was used to update particles' positions and momenta if particles exceeded a maximum specified distance. The simulation time was fixed at 1200 reduced time units. Temperature gradients, density variations and heat flux variation along the system in y-direction were investigated using the method of planes (Todd, Davis and Evans, 1995). In the reservoirs the instantaneous values of the y-momentum, temp-erature, and energy output were monitored. Steady state was assumed if the rates of energy output of the reservoirs were approximately equal (to within 10%) and opposite in sign. We used the fourth-order Gear non-symplectic algorithm (G4) and a second order near-symplectic algorithm, the position Verlet (pV2) extended to include a thermostatting term as needed. In addition we invest-igated the removal of the continuous $f(y)$ functions and replacing them with their limiting step function forms. The relaxation of the smoothing function to this dis-continuous form did not influence the results.

Results

The results of study are separated into two parts. For the original Baranyai approach, results are shown in Table 1 using the G4 method for solving the equations of motion. The perpendicular reduced area for heat flow is 46.784 and the uncertainties reported in the table were calculated from the rate of heat transfer. The thermal conductivity was calculated at three different temperature gradients (0.4, 0.3 and 0.2), each of which was simulated at three reduced reservoir widths (7, 3.5, and 1.75). The estimated thermal conductivity, λ_{ext} is obtained by extra-polating the apparent thermal conductivities obtained at varying reservoir widths to zero width.

Conclusions from these results are, that to within statistical uncertainties, the thermal conductivity is independent of the temperature gradient and that given an appropriate reservoir width, it may be possible to obtain the thermal conductivity from a single simulation. Fig. 1 illustrates this conclusion for the case where $\Delta T^* = 0.3$. Comparisons with other NEMD methods show good agreement. For example, Evans' synthetic field method for a system of 500 LJ particles at a reduced temperature 0.9 and reduced density 0.8 is gives a thermal conductivity of 5.92 as compared to our average of 6.046.

The results in Table 1 were obtained using a small time step, e.g., $\Delta t^* = 0.002$. As part of our investigation of symplectic algorithms, we have explored the possibility of increasing the size of the time step in the thermal cond-uctivity algorithm. The results of that study using the Gaussian + PV2 method are shown in Fig. 2. As can be seen from this figure, use of the PV2 method provides accurate results even at large time steps.

Increasing the size of the time step in the Gaussian +G4 thermostatted regions lead to large discrepancies in the energy dissipation rates in the hot and cold reservoirs—up to a 50% difference. Replacement of G4 with PV2 method provided significantly better results. For example, the difference in energy dissipation rates were less than 10%, even in the limit of a large time step ($\Delta t^* = 0.02$). We also studied some alternatives to the Gaussian thermostat algorithm, namely the Nosé-Hoover and the *ad hoc* velocity rescaling algorithms.

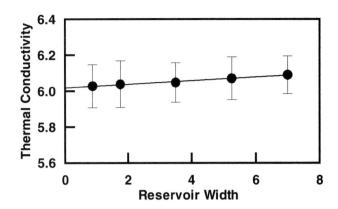

Figure 1. Simulated thermal conductivity as a function of reservoir width.

Results for the thermal conductivity coefficient based on the Irving-Kirkwood with various integration methods are reported in Table 2. We should also point out that even though G4 does not provide reasonable results for the energy dissipation, results for static properties such as temperature gradient and density in the y-direction in the system are excellent.

Results obtained using the Nosé-Hoover thermostat are excellent without generating any non-canonical disturbances into the system. Also, pure ad hoc velocity rescaling provides excellent (perfect) temperature control, but it isn't clear if one is really solving the equations of motion for the particles. The results obtained while using

Table 1. Thermal Conductivity Calculated Using Baranyai's Method and the Gaussian+G4 Algorithm.

$T_{res,hot}$	1.1			1.05			1.0		
$T_{res,cold}$	0.7			0.75			0.8		
Res. Width	7.0	3.5	1.75	7.0	3.5	1.75	7.0	3.5	1.75
$(dH/dt)_{hot}$	9.701	9.096	8.501	7.3317	6.750	6.203	5.146	4.456	4.334
$(dH/dt)_{cold}$	9.617	9.040	8.381	7.327	6.698	6.208	4.978	4.346	4.283
$(dH/dt)_{ave}$	9.659	9.068	8.441	7.329	6.724	6.205	5.048	4.401	4.309
λ	6.455	6.168	6.122	6.091	6.048	6.038	7.411	6.591	6.554
λ_{ext}	5.978 ± 0.259			6.016 ± 0.386			6.144 ± 0.443		

Table 2. Results of Heat Flux Calculated from G4 and PV2 Algorithms Using Different Thermostatting Methods.[1]

Width	NH+G4	G*+G4	G+G4	*ad hoc* +G4	NH+PV2	G+PV2	*ad hoc*+PV2
7.0	9.0341	9.2703	9.5064	9.6246	9.0931	9.2112	9.3293
5.0	8.3256	8.5506	8.6069	8.2694	8.1006	8.6069	8.6631
3.5	7.7429	7.7978	7.9626	7.9076	7.3585	8.0175	7.9626
2.5	7.0479	7.2106	7.5359	7.6443	6.8310	7.3732	7.6443
Extrap.	6.1166	6.1601	6.2526	6.4326	5.6909	6.1148	6.4666

[1]NH = Nosé-Hoover thermostat, G* = Gaussian thermostat with feedback, G = Gaussian thermostat, *ad hoc* = velocity rescaling

the Gaussian thermostat with feedback are good, but the magnitude of the constraint multiplier and the energy dissipation are different than those observed with the original Gaussian and Nosé-Hoover thermostats. This is not surprising since this method is used to suppress the numerical drift without concern to the canonical nature of the flow. To obtain good results using the Gaussian with feedback method, a trial and error process must be used to find the optimal feedback proportionality constant. Otherwise, this method can do more harm than good. Heat fluxes obtained from *ad hoc* rescaling seem to be larger than those obtained with the other thermostats. The Gaussian + feedback and Nosé-Hoover algorithms are very competitive. Results using the Nosé-Hoover thermostat with G4 and PV are also competitive. It is difficult to make clear comparisons of the results for PV2 and G4 since the G4 simulations were rescaled every 100 steps. It is fair to say, however, that the PV2 method seems to track the true physical trajectory better than G4. Fig. 2 shows the advantage of the PV2+Gaussian thermostat algorithm for large time steps. The uncertainties increase with increasing time step size, but the results do not show any large discrepancies in the average values. These results support results from other studies on the superiority of SI methods for molecular dynamics simulations (Tuckerman, Berne and Martyna, 1991).

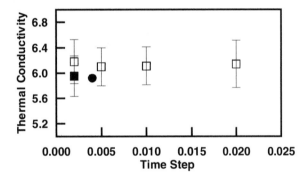

Figure 2. Simulated thermal conductivity as a function of time step size using Gaussian+PV2. phone - this work, mailbox - Evans, letter - Baranyai.

In addition to the near triple point studies described above, we have also compared the Baranyai method results to those obtained from the synthetic field NEMD method at other conditions. In particular, we performed one series of simulations at a reduced temperature of 1.35 for reduced densities of 0.4, 0.6, and 0.8, and second series at a reduced temperature of 2.0 for reduced densities of 0.8 and 0.6. We found that the synthetic field and fluid wall methods agree to within 10% which is within the combined uncertainties of the two methods. The state where the reduced temperature and density are

1.35 and 0.4 is close to critical point and both methods show a strong non-linear dependence on the external field or heat reservoir size. We attribute this to the longer range dynamic correlations found in the critical region.

Summary and Conclusions

The fluid wall algorithm for thermal conductivity based on Baranyai's approach has been reinvestigated. In particular, it has been revised for simplicity and modified to use different thermostat and integration algorithms. We did not find a significant difference between Gaussian and Nosé-Hoover thermostats provided that for the Gaussian thermostat, the velocity is rescaled at appropriate intervals or a proportional feed-back is applied. The *ad hoc* velocity rescaling seems to give a higher thermal conductivity but it is still in agreement with other results to within their combined uncertainties. At conditions removed from the critical point it is possible to perform a single simulation with a relatively large time-step using the Gaussian+PV2 method. This method offers the possibility of decreasing the time required for the simulation by a factor of two or more as compared to conventional NEMD.

Acknowledgements

JR acknowledges a scholarship from the Royal Thai government. This work was supported by the U. S. Department of Energy, Office of Basic Energy Science, grant No. DE-FG02-95ER14568

References

Ashurst, W. T. (1974). Determination of thermal conductivity coefficient via nonequilibrium molecular dynamics. Paper presented at the 12th International Conference on Thermal Conductivity.

Baranyai, A. (1996). Heat flow studies for large temperature gradient by molecular dynamics simulations. *Phys.Rev.A*, **54**(6), 6911-6917.

Evans, D. J. (1986). Thermal conductivity of the Lennard-Jones fluid. *Phys.Rev.A.*, **34**(2), 1449-1453.

Evans, D. J., and G. P. Morriss (1984). Nonequilibrium molecular dynamics. *Comp. Phys. Rep.*, **1**, 299-343.

Hoover, W. G. (1991). *Computational Statistical Mechanics*. Elsevier Science B. V., Amsterdam.

Hoover, W. G., G. Ciccotti, G. Paolini, and C. Massobrio (1985). Lennard-Jones triple-point conductivity via weak external fields: Additional calculations. *Phys. Rev. A*, **32**(6), 3765-7.

Massobrio, C., and G. Ciccotti (1984). Lennard-Jones triple point conductivity *via* weak external fields. *Phys. Rev. A*, **30**(6), 3191-3197.

McLachlan, R. I., and P. Atela (1992). The accuracy of symplectic integrators. *Nonlinearity*, **5**, 541-562.

Müller-Plathe, F. (1997). A simple nonequilibrium molecular dynamics method for calculating the thermal conductivity. *J. Chem. Phys.*, **106**(14), 6082-6085.

Todd, B. D., Davis, P. J. and Evans, D. J. (1995). *Phys. Rev. E: Stat. Phys., Plasmas, Fluids, Relat. Interdiscip. Top.*, **51,** 4362-8.

Tuckerman, M. E., B. J. Berne, and G. J. Martyna (1991). Molecular dynamics algorithm for multiple time scales: System with long range forces. *J. Chem. Phys.*, **94**(10), 6811-6815.

Yoshida, H. (1990). Construction of higher order symplectic integrators. *Phys. Let. A.*, **150**, 262-268.

Yoshida, H. (1993). Recent progress in the theory and application of symplectic integrators. *Celest. Mech. Dynam. Astron.*, **56**, 27-43.

Zhang, F. (1997). Operator-splitting integrators for molecular dynamics. *J. Chem. Phys.*, **106**(14), 6102-6106.

TRANSPORT AND EQUILIBRIUM PROPERTIES
OF LARGE GLOBULAR MOLECULES

Lydia Zarkova and Peter Pirgov
Institute of Electronics BAS
1784 Sofia, Bulgaria

Uwe Hohm
Institut für Physikalische und Theoretische Chemie der Technischen Universität
D-38106 Braunschweig, Germany

Abstract

The goal of this work is to prove the applicability of the isotropic temperature-dependent potential (ITDP) to some large globular molecules and to obtain reliable data for their transport and equilibrium properties. The four parameters of this effective Lennard-Jones type potential are determined by minimization of the sum of squared deviations between experimental and calculated viscosity and second virial coefficient data normalized to their relative experimental errors. First results are obtained for the molecules tetramethylmethane, $C(CH_3)_4$, and tetramethylsilane, $Si(CH_3)_4$. The intermolecular potentials obtained are used to calculate reliable data for viscosity, self-diffusion and second virial coefficients in the temperature range between 250 and 800 K. These data might be useful for modeling mass transport relevant in chemical-vapor deposition (CVD). Consequences for the intermolecular interaction potentials of the higher homologues $Ge(CH_3)_4$, $Sn(CH_3)_4$, and $Pb(CH_3)_4$ are discussed.

Keywords

Intermolecular potential, Globular molecules, Viscosity, Second virial coefficient, Self-diffusion, Tetramethylmethane, Tetramethylsilane.

Introduction

The large globular molecules of the type $X(CH_3)_4$, (X =C, Si, Ge, Sn or Pb), recently have attracted much interest because all of them might have an important impact on the quality and/or the price of the life. Some of them like $Ge(CH_3)_4$ (Venkatasubramanian et al., 1989) and $Si(CH_3)_4$ (Lee and Sanchez, 1997; Bauhofer and Ruter, 1996) find applications in semiconductor industries and chemical vapor deposition (CVD) technologies. Others like $Pb(CH_3)_4$ have been important additives in fuels (Fielden and Greenway, 1989) and are therefore relevant to flame-combustion processes. Moreover, especially $Pb(CH_3)_4$ contributes to the increase of heavy metals and harmful radicals in the atmosphere and the environment. Understandably there is a need for

having reliable data on thermophysical properties of these substances. Additionally, knowledge of intermolecular interaction potentials of these gases allows for a modeling of their thermophysical behavior in wide ranges of temperature and pressure.

We will concentrate on only two gases, tetramethylmethane (TMM), $C(CH_3)_4$, and tetramethylsilane (TMS), $Si(CH_3)_4$. There is an appreciable amount of thermophysical data available for TMM. These include measurements on the gas-phase viscosity η, and on the second pVT (B)- and acoustic (β) virial coefficients. The situation is much worse for TMS, where only few experimental data on η and B are given in the literature. For both substances the intermolecular interaction potentials obtained with different methods and different input data are generally not able to reproduce the measured values. It is worth mentioning that in the case of the other tetramethyl compounds no experimentally determined gas-phase thermophysical data are available. Of course, these data can be estimated from suitable intermolecular interaction potentials. At the moment, however, reliable potentials do not exist. The Lennard-Jones (12-6) potential parameters ε and σ have been reported by Rummens, Raynes and Bernstein (1968) for all of the above mentioned tetramethyl compounds $X(CH_3)_4$. They were obtained from critical data and by using scaling laws. However, such intermolecular potentials can hardly be used to calculate thermophysical properties. Recently, Baonza, Alonso, and Delgado (1994) have applied a generalized van der Waals model to TMS and four different (ε - σ) pairs have been obtained for LJ(12-6), LJ(18-6), and LJ(24-6) potentials. Their interesting approach was developed on the background of $B(T)$ data and also failed to predict transport properties.

The goal of our study is to fit simultaneously all available experimental thermophysical data for a given compound by means of an isotropic temperature-dependent potential (ITDP). The potential parameters achieved can be used for e.g. prediction of temperature-dependent equilibrium and transport properties in a temperature range that exceeds the range of the existing experimental data.

Procedure of Defining the Intermolecular Potential and Computational Code

The isotropic temperature-dependent potential (ITDP) $U(r,T)$ is a Lennard-Jones (n-6) potential which can be written in the form

$$U(r,T) = \frac{\varepsilon(T)}{n-6}\left\{6\left[\frac{r_m(T)}{r}\right]^n - n\left[\frac{r_m(T)}{r}\right]^6\right\} \qquad (1)$$

where $r_m(T)$ is a temperature-dependent intermolecular distance at which $U=\varepsilon(T)$. It is defined via

$$r_m(T) = r_m(T=0) + \delta(T) \qquad (2)$$

where $\delta(T)$ is an effective enlargement of the molecule which can be calculated with the aid of the vibrational partition function. The effective temperature-dependent well-depth $\varepsilon(T)$ is subsequently calculated via

$$\varepsilon(T) = \varepsilon(T=0)\left[\frac{r_m(T=0)}{r_m(T)}\right]^6 \qquad (3)$$

Therefore, the ITDP is an effective potential with slightly temperature dependent parameters r_m and ε reflecting the vibrational excitation of the molecule. In brief, we calculate this dependence via the vibrational partition function. In general the four parameters n (repulsive parameter), $r_m(T=0)$ (equilibrium distance at $T=0$ K), $\varepsilon(T=0)$ (potential well-depth at $T=0$ K), and δ (the enlargement of the molecule when excited from the ground state to the first vibrationally excited level) are obtained by minimizing the sum of squared deviations F between the experimental and calculated thermophysical (second virial coefficients, viscosity) input data.

Following the model described above we have built up an efficient computational code. This code enables us to make different kind of calculations, such as scanning the potential "surface" in order to localise the minimum of F, solving the ill-posed problem, defining the potential parameters, producing tables of thermophysical properties, covariance-variance matrix, etc. The only input data needed are the molecular mass, the normal vibrational frequencies, and the experimental data with their relative experimental errors.

The universality and the reliability of the model and the computational procedure have been proved in our recently published results for a number of globular molecules (see, for example, Zarkova (1996) for CF_4 and SF_6, Zarkova and Pirgov (1995) for UF_6, Zarkova and Pirgov (1996) for WF_6 and MoF_6).

Results

Normal Vibrational Frequencies

The problem which appears when applying the ITDP to complex molecules consisting of many atoms is given by the relatively large number of normal vibrational frequencies (45 for the molecules of type $X(CH_3)_4$). All of them have to be taken into account in the calculation of the vibrational partition function $Z(T)$. The relative populations of the vibrational states at different temperatures are subsequently obtained by means of $Z(T)$. If some of the vibrational frequencies are not known as it is in our case additional measurements or calculations of these frequencies have to be performed.

The potential parameters at $T=0K$ (ε in K, r_m in 10^8 cm and δ in 10^{10} cm) obtained in this work for TMM and TMS are given in Table 1. The number of the measured values of B, η or β which were taken into consideration is denoted by m_B, m_η, m_β, respectively. rms is the root mean square deviation in units of the accepted experimental error.

Table 1. ITDP Potential Parameters.

Gas	ε	r_m	n	δ	m_B	m_η	m_β	rms
TMM	586.3	5.779	28.02	1.41	74	29	8	1.20
TMS	674.8	5.905	20.79	1.88	38	17	-	0.74

In Fig. 1 an example of the ITDP is given for $Si(CH_3)_4$ - $Si(CH_3)_4$ interaction calculated at 0 K, 300 K and 600 K.

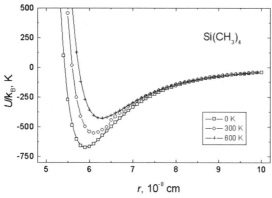

Figure 1. Intermolecular interaction potentials for $Si(CH_3)_4$ calculated at different temperatures.

Tetramethylmethane

The thermophysical properties of TMM have been relatively well studied by experiment. Since 1952 measured thermophysical data have been reported in 18 papers in the temperature range between 250 and 550 K: 13 for B, 4 for η, and 1 for β. However, we have taken into consideration only 16 data sets which seem to be reliable and self-consisted.

In Figures 2, 3, and 4 our calculations are given by full lines, whereas the experimental input data are denoted by circles. Figure 2 shows results on the second pVT- virial coefficient. The gas-phase viscosity is presented in Fig. 3, and the second acoustic virial coefficient in Fig. 4 .

To the best of our knowledge, this is the first example that speed-of-sound data are used in a simultaneous fit of thermophysical properties of vapors in order to determine the intermolecular interaction potential. The method for obtaining $\beta(T)$ (Ewing et al., 1987) is quite different from those used for $B(T)$. Additionally, $\beta(T)$ can be obtained with an exceptional high accuracy of much less than 1%. Including such data makes the solution much more reliable. There are two main reasons why this valuable source of information has not been considered in the determination of intermolecular potentials. Firstly, this is due to a lack of such experimental data (β and C_p). Secondly, the procedures of minimization require a great number of repeated calculations which in the case of β are rather time-consuming (Hurley, 2000). Our computational code is, however, fast and flexible enough to obviate the second problem.

Tetramethylsilane

Compared to TMM much less measurements of the thermophysical properties of gas phase TMS have been

carried out in contrast to its much wider application in the microelectronic technologies. Only four papers deal with the measurement of $B(T)$, one of which could not be used because of its systematically wrong data. Experimentally obtained gas-phase viscosity data have been reported in one paper. Altogether we have taken into consideration 55 experimental points in the range between 263 and 572 K.

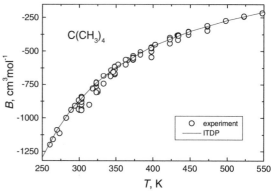

Figure 2. Second virial coefficient of tetramethylmethane.

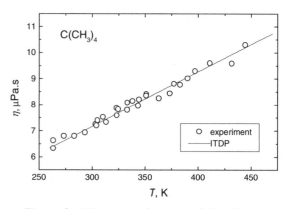

Figure 3. Viscosity of tetramethylmethane.

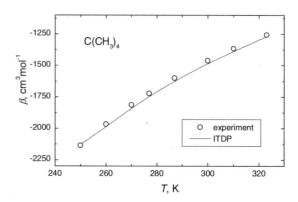

Figure 4. Second acoustic virial coefficient of tetramethylmethane.

We considered these data to be reliable and self-consistent in order to obtain the intermolecular potential.

The results for Si(CH₃)₄ are shown in Fig. 5 ($B(T)$) and Fig. 6 ($\eta(T)$).

Unfortunately, there are no experimentally determined speed-of-sound data of TMS available to compare with. Calculations of $\beta(T)$ in this case are hardly possible because of the lack of experimental C_p/R data.

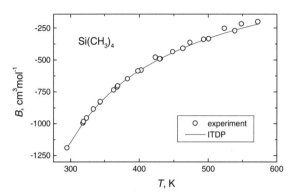

Figure 5. Second virial coefficient of tetramethylsilane.

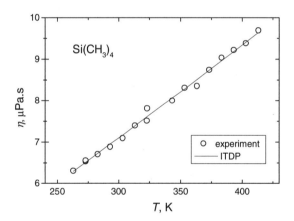

Figure 6. Viscosity of tetramethylsilane.

Discussion

In general, all experimental data for both substances are fitted with good accuracy with respect to the experimental error. This observation allows for a reliable prediction of recommended thermophysical properties B, η, and ρD (ρ = density, D = diffusion coefficient) in an extended temperature range between 250 and 800 K. Tabulated temperature dependencies can be re-approximated by convenient algorithms and used for easy estimations. One can make calculations by means of ITDP at fixed temperatures as easy as with a LJ (n-6) potential using the repulsive parameter n (Table 1) and tabulated temperature dependencies of the effective potential parameters $r_m(T)$ and $\varepsilon(T)$ (more detailed numerical information will be published soon).

Conclusion

The observed convergence of the simultaneous fit of different available transport and equilibrium data proved the validity of the ITDP approach to the gases TMM and TMS. The so defined intermolecular interaction potentials can be used for calculations of different potential-dependent properties. Tabulated $B(T)$, $\eta(T)$, $\rho D(T)$ are convenient for interpolation and may be useful for estimations at temperatures which are out of the experimental temperature range.

Our future projects are concerned with the possibility of finding a way to determine the intermolecular interaction potentials of the tetramethyl compounds of germanium, tin, and lead, for which gas-phase thermophysical properties are not yet investigated.

References

Baonza V. G., M. C. Alonso, and J. N. Delgado (1994). Generalized van der Waals model applied to tetramethylsilane. *Ber. Bunsenges. Phys. Chem.* **98**, 53-58.

Bauhofer, W., and D. Ruter (1996). Luminescent waveguide structures deposited from liquid organosilane vapor sources. *Appl. Surf. Sci.,* **102**, 319-322.

Ewing, M. B., Godwin A. R., M. L. McGlashan, and J. P. M. Trusler (1987). Thermophysical properties of alkanes from speeds of sound determined using a spherical resonator I. Apparatus, acoustic model, and results for dimethylpropane. *J. Chem. Thermo.,* **19**, 721–739.

Fleiden, P. R., and G. M. Greenway (1989). Diffusion apparatus for trace level vapor generation of tetramethyllead. *Anal Chem.,* **61**, 1993-1996.

Hurly, J. J. (2000). Thermophysical properties of gaseous tungsten hexafluoride from speed-of-sound measurements. *Int. J. Thermophys.,* **21**, 185-205.

Lee, Y. L., and J. M. Sanchez (1997) Simulation of chemical vapor-deposited silicon carbide for a cold vertical reactor. *J. Cryst. Growth,* **178**, 505-512.

Rummens, F. H. A., W. T. Raynes, and H. J. Bernstein (1968). Medium effects in nuclear magnetic resonance. V. liquids consisting of nonpolar magnetically isotropic molecules. Medium effects in nuclear magnetic resonance. *J. Phys. Chem.,* **72**, 2111-2119.

Venkatasubramanian, R., R. t. Pikett, and M. L. Timmons (1989). Epitaxy of germanium using germane in the presence of tetramethylgermanium. *J. Appl. Phys.,* **66**, 5662-5664.

Zarkova, L. (1996). An isotropic intermolecular potential with temperature-dependent parameters for heavy globular gases. *Mol. Phys.,* **88**, 489-495.

Zarkova, L., and P. Pirgov (1995). Transport and equilibrium properties of UF₆ gas fitted by an effective isotropic potential with temperature-dependent parameters. *J. Phys. B,* **28**, 4261-4281.

Zarkova, L., and P. Pirgov (1996). The isotropic temperature-dependent potential desciding the binary interactions in gaseous WF₆ and MoF₆. *J. Phys. B,* **29**, 4411-4442.

COMPUTER SIMULATION OF
ORGANOSILANE-BASED COATING AGENTS

Robert A. Hayes, Graeme W. Watson, and David J. Willock
Department of Chemistry
University of Wales, Cardiff
Cardiff, CF10 3TB

Hywel Edwards
Welsh Technology Centre
Corus
Port Talbot, West Glamorgan, SA13 2NG

Abstract

Organosilane coating agents are an important class of compounds used in corrosion protection of metals. In this work they have been modelled using both density functional theory and molecular dynamics simulations. The preferential binding site for hydrogen, water and a simple silane on Fe{111} have been determined using DFT methods. The packing energies of 3-Glycidoxypropyltrimethoxysilane (3-GPMS) have been calculated on the three lowest index surfaces of iron: Fe{100}, Fe{110} and Fe{111}, with a preferential unit cell shape for packing being identified. The moment of interia and the inclination angle for 3-GPMS averaged over the film have also been determined for severals surface coverages on Fe{111}.

Keywords

Density functional theory, Molecular dynamics simulations, Organosilanes, Corrosion protection, Packing energy, Moment of inertia, Inclination angle.

Introduction

Organofunctional silanes have been used for many years as coupling agents to improve paint adhesion to metals[1]. More recently organosilanes have been used as an independent corrosion inhibition barrier without the use of an organic based topcoat[2]. They can be applied to a variety of metal substrates including galvanized and cold-rolled steel, aluminium and all of its alloys. Organosilane pretreatments also offer an environmentally friendly replacement for currently used pretreatment technologies that include both chromate and phosphate.

Organofunctional silanes have the general formula $X_3Si(R)_nY$, where X is a hydrolysable group (*e.g.* methoxy, ethoxy or chloro), R is an alkyl-chain of length 3 – 20 carbons, and Y is an organofunctional group (*e.g.* amine, epoxide or carboxylic acid). The hydrolysable groups, X, can react with water, either present in the pretreatment solution or adsorbed onto the surface of the substrate, to form silanol groups (Si–OH). Several theories have been proposed[3, 4, 5, 6] for the mechanism by which these silanol groups react with surface hydroxyls to create the surface coating. The covalent bond theory suggests that a strong oxane bond (*e.g.* Fe-O-Si) is formed between the silanol groups of the hydrolyzed coupling agent and the OH groups from the surface of the substrate, thus leaving the functional organic group at the other end of the molecule free to react with an organic polymer (*e.g.* paint, adhesive or coating).

Plueddemann and Stark[7] proposed an interpenetrating polymer network theory which supports the formation of an oxane bonds at the substrate interface, and suggests that the secondary organic polymer coating forms a polymer network with the silanes at the interface between coupling agent and organic coating.

In this paper we report the use of two complementary computer simulation techniques to study the chemical bonding of organosilanes to iron surfaces and the intermolecular interactions that occur between the organosilane side-chains.

Methodology

Density Functional Theory (DFT) Calculations

To study the Fe-O-Si bond we have used a DFT based model. For these calculations a ten-atom cluster model of the Fe{111} surface containing two Fe layers containing 7 and 3 atoms, which we denote Fe(7,3), is employed.

Calculations are carried out using the DMOL code obtained from MSI[8] for all DFT cluster calculations. The DNP (double numeric with polarization functions) basis set is used together with restricted spin. The local spin density functional proposed by Vosko, Wilk and Nusair (VWN)[9] with gradient corrections for exchange from Becke (B88)[10] and for correlation from Perdew and Wang (PW91)[11] is used throughout. To improve SCF convergence an Xfine numerical integration grid was used and electron smearing was set to 0.05 Hartrees. We calculated the optimised structure and the total energy for each adsorbate keeping all the Fe atoms fixed.

Initial calculations have focused on the chemisorption of small molecules i.e. hydrogen and water onto the iron cluster. For top site adsorption the adsorbate was placed above the central Fe of the 7 atom layer and for hollow site adsorption we used the centre of the 3 atom layer (cluster inverted), this avoids interaction of the adsorbate with low co-ordination Fe atoms. Validity of the cluster model has been checked by comparison to fully periodic DFT calculations using VASP[12, 13]. In a cluster model artificially high adsorption energies can result from interactions with low co-ordination edge atoms, in the periodic model all atoms are fully co-ordinated. We find that the cluster calculations for the adsorbates agree to within at least 18 % of the more computationally intensive periodic method and always show the same trend in relative adsorption energies. Following the validation of the cluster-DFT approach for the adsorption of small molecules onto iron surfaces, the calculations have been extended to study the chemisorption of low molecular weight silanes onto a hydroxylated Fe{111} surface.

Molecular Dynamics (MD) Simulations

To study the packing energy of 3-GPMS molecules on Fe, we used potential-based MD simulations. Simulations were carried out with the CCP5 code DLPOLY[14]. Potentials for bond-stretching, bond-bending, and van der Waals (VDW) interactions were obtained from the cvff force-field[15]. The organosilane coating agent used in this

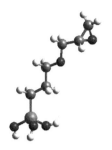

Figure 1.   Ball and stick representation of 3-GPMS. Atoms shaded according to: O = dark grey, Si = light grey, large, C = light grey, small, H = white.

study was 3-GPMS (figure 1). The starting structure was generated from a single 3-GPMS molecule bound to an Fe atom in a surface unit cell chosen to give the required surface coverage. A fixed O-Fe bond was used to represent tethering to the Fe surface.

This cell was duplicated to form a simulation cell with the appropriate number of independent 3-GPMS molecules. Usually the simulation cell contained 25 3-GPMS molecules in a 5x5 array of single molecule cells. This was not possible for the cases of half coverage (1 silane per 2 surface Fe atoms) and 5/16ths coverage (5 silanes per 16 surface Fe atoms) when supercells containing 32 molecules and 20 molecules respectively were used instead.

The metal surface was represented by a fixed layer of Fe atoms. These atoms were simply used as a restraining surface for the film. To achieve this the Fe atoms interact with the silane layer through a Lennard-Jones (LJ) potential derived in the following manner. The repulsive part was defined so that the parameters for Fe-Fe interaction gave a minimal energy at the Fe-Fe nearest neighbour separation in the lattice and the attractive part was set to zero. The standard cvff combining rules were then used to generate interaction potentials between Fe and atoms in the silane molecules. This prevents unrealistic configurations being generated at low coverage, but has only a small influence on the packing energy calculated.

The packing energy is defined by plotting the average MD energy as a function of surface coverage, this tends to be a constant value at high dilutions, which was used as our reference energy. Hence the packing energy is the energy gained on forming a film from infintely separated, surface bound silane molecules. The packing energy has been calculated for 3-GPMS on the three lowest index surfaces of iron: Fe{100}, Fe{110} and Fe{111}.

The method of generating the supercell results in a film containing a regular array of silane molecules. In our initial studies we were concerned that the MD simulation should be capable of generating a randomized set of silane conformations as would be expected in the real film. We found this was particularly difficult to achieve for the densest packing unless a high temperature "annealing" step was introduced. After several trials the following protocol was developed and is applied for all MD simulations presented here.

Simulations were initially run for 100 picoseconds (ps) to equilibrateat 300 K, before a further 100 ps annealing run at 1700 K is used to completely randomize the starting conformations. The final run of 300 ps is made up of 100 ps equilibration and 200 ps production all at a more realistic simulation temperature of 300 K. All data for the MD simulations is collected in this final 200 ps run. All simulations were performed with the periodic boundary conditions of the supercell applied, and a time step of 1fs.

Results and Discussion

Density Functional Theory (DFT) Calculations

The most stable adsorption site for both surface hydrogen and hydroxyl groups was found to be the hollow-site above a second layer Fe atom. The DFT-cluster calculations show that hydrogen adsorption is 92.5 kJmol⁻¹ more stable in the hollow-site than in the top site (table 1). The same trend is also followed for the DFT-periodic calculations, where hollow-site adsorption is 79.4 kJmol⁻¹ more favorable.

Table 1. Overall Reaction Enthalpies for the Adsorption of Hydrogen and Water (H and OH species) on the Fe{111} Surface.

	Cluster		Periodic	
	Top Site (kJmol⁻¹)	Hollow Site (kJmol⁻¹)	Top Site (kJmol⁻¹)	Hollow Site (kJmol⁻¹)
Hydrogen	-24.2	-116.7	-18.5	-97.9
Water				
(H Top)	-79.9	-89.3	-89.4	-103.8
H(Hollow)	-126.1	-135.5	-129.1	-143.5

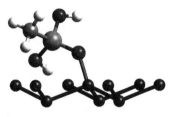

Figure 2. Minimised structure of methylsilane adsorbed onto the top-site of the Fe(7,3) cluster. Atoms shaded according to: Fe = dark grey, O = middle grey, Si = light grey, large, C = light grey, small, H = white.

Water adsorption onto Fe, where both the hydrogen and hydroxyl species are in the hollow-site, is the most

favorable conformation by 9.4 kJmol⁻¹ for the cluster calculations, and 14.4 kJmol⁻¹ for the periodic case.

Methylsilane (CH₃Si(OH)₃) chemisorption onto the same Fe(7,3) cluster (figure 2) shows a much greater affinity for the top-site, which is 22.5 kJmol⁻¹ more stable than the hollow site (table 2). This is probably due to the steric hindrance from surface Fe atoms around the adsorption site.

Table 2. Reaction Enthalpy for the Adsorption of Methylsilane on the Fe(7,3) Cluster.

	Cluster	
	Top Site (kJmol⁻¹)	Hollow Site (kJmol⁻¹)
CH₃Si(OH)₃	-92.9	-70.4

Molecular Dynamics (MD) Simulations

The packing energy of 3-GPMS on the 3 surfaces of Fe studied follows a similar trend. At high coverages the interaction energy increases rapidly with coverage, as the coverage is reduced a definite minima is observed before the packing energy tends to zero in a monotonic fashion.

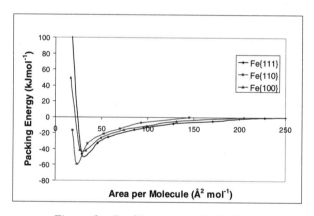

Figure 3. Packing energy for 3-GPMS molecules onFe{100}, Fe{110} and Fe{111}.

For Fe{111} full coverage, which equates to all surface sites being occupied, a very unfavorable packing energy is found due to the repulsive intermolecular interactions particularly between the organosilane head groups (figure 3). The lowest packing energy is observed when half the surface sites are occupied. For Fe{100} and Fe{110} the most favorable surface coverage is when only 1 in 4 surface sites are occupied. However, the lowest packing energy in Fe{110} occurs at a slightly higher surface coverage of 3-GPMS molecules than for the other two surfaces. This indicates that there is a preferred two-dimensional unit cell shape in which to pack, with the density of the film at optimum packing energy following the order

Fe{110} > Fe{100} > Fe{111}

The moments of inertia (MoI) have been used to carry out a detailed geometry analysis of the 3-GPMS side chains on Fe{111}. The smallest MoI (sMoI) can be used to define the molecular axis of the silanes. This element can highlight two very important pieces of information about the 3-GPMS molecules: how elongated the molecules are in the film, and the average inclination angle of a molecule with respect to the Fe substrate.

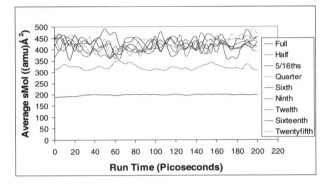

Figure 4. Smallest moments of inertia for all coverages of 3-GPMS on Fe{111}.

From figure 4 the average sMoI for the densest packing on Fe{111} is seen to be 197.9 ± 2.9 (amu) $Å^2$, indicating that all the organosilanes are elongated, and are packed close together. At optimum coverage (half) the average MoI is 321.59 ± 11.3 (amu) $Å^2$, showing that there is far more room for movement and the silanes are not "artificially" elongated. For all other coverages between 5/16ths and the reference (one twenty fifth) coverage, the average sMoI's are very similar.

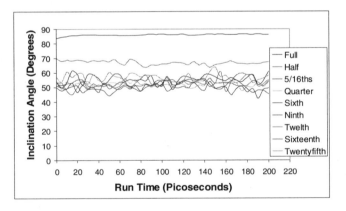

Figure 5. Inclination angle for all coverages of 3-GPMS on Fe{111}. Angle defined from surface, so that a 90° chain is perpendicular to surface.

The average inclination angle (*figure 5*) for full coverage confirms that the 3-GPMS molecules are standing in an upright fashion i.e. almost perpendicular to the Fe substrate ($85.9° \pm 0.5°$). Optimum coverage has an average inclination angle of $66.9° \pm 1.5°$, and as with the

sMoI data the inclination angles for all other coverages are overlayed at around 55°.

Conclusions

From the DFT calculations on smaller adsorbates, it was noted that smaller molecules bind preferentially in the hollow-site. Larger molecules e.g. silanes prefer to adsorb onto the top site.

From the MD results it is clear that there is a preferred unit cell shape in which to pack 3-GPMS molecules. Also it is apparent that the 3-GPMS molecules are elongated compared to the lowest densities considered, and that there is an average inclination angle of about 70° on the Fe{111} surface.

These calculations indicate that surface bound silanes will tend to form dense films and we are currently investigating the effect of chain length on the optimum packing densities.

Acknowledgments

The authors wish to thank Dr. B. J. Hewitt, Programmes Corus R, D & T, and Mr. A. C. Lewis, Manager, Light Construction, Welsh Technology Centre, for the permission to publish this paper. This work was carried out with a financial grant from the European Coal and Steel Community.

The calculations presented in this paper were carried out using "Glyndwr", a Silicon Graphics multiprocessor Origin 2000 machine at the department of Chemistry, University of Wales, Cardiff. This facility was purchased with support from the EPSRC, ICI Katalco and Open Computers and Finance (OCF).

References

Becke, A. D. (1988). *J. Chem. Phys.*, **88**, 2547.

Broutman, L. J. (1970). *Adhesion*, **2**, 147.

Child, T. and W. J. van Ooij (1998). *Chemtech*, **26**.

DMOL code, MSI, San Diego, CA 92121.

Fleischmann, R. B., Johannson, O. K., Stack, F. O., and G. E. Vogel (1967). *J. Comp. Mat.*, **1**, 278.

Forester, T. R. and W. Smith. Daresbury Laboratory, Warrington, England, WA4 4AD.

Furthmüller, J. and G. Kresse (1996a). *Phys. Rev. B.*, **54**, 11169.

Furthmüller, J. and G. Kresse (1996b). *Comp. Mat. Sci.*, **6**, 15.

Hagler, A. T., Huler, E., and S. Lifson (1974). *J. Am. Chem. Soc.*, **96**, 5319.

Hobbs, P. M. and A. Kinloch (1997). *J. Adhesion.*, **00**, 1.

Kwei, T. K. (1965). *J. Poly. Sci.*, **3**, 3229.

Nusair, M., Vosko, S., and L. Wilk (1980). *Can. J. Phys.*, **58**, 1200.

Perdew, J. P. and Y. Wang (1992). *Phys. Rev. B.*, **45**, 13244.

Plueddemann, E. P. (1974). *Interface in Polymer Matrix Composites.* Academic Press, New York.

Plueddemann, E. P. and G. L. Stark (1980). *Proceedings of the 35th Ann. Tech. Conf., Reinforced Plastics/ Composites.* Inst. Soc. Plast. Ind., Section 20-B.

SIMULATION STUDY OF CRYSTALLIZATION AND MECHANICAL STABILITY OF PARTICLE GELS

Danial Irfachsyad
Department of Chemistry
University of Southampton
Southampton, UK SO17 1BJ

Dominic Tildesley
Unilever Research Port Sunlight Laboratory
Wirral, UK CH63 3JW

Abstract

We report the Brownian Dynamics simulation of colloidal particles immersed in an ionic solution, varying the ionic concentration to alter the inverse Debye length. We show that the inverse Debye length determines the aggregation and the onset of crystallization. The mechanical stability is studied by applying a step-strain into the structure at different stages to study the effect of aging. The step strain is varied to investigate the change in yield stress as the gel ages.

Keywords

Brownian dynamics, Crystallization, Flocculation, DLVO, Gels, Step-strain.

Introduction

In an aqueous colloidal suspension, the addition of electrolyte can progressively screen the repulsive Coloumbic potential. Under certain condition, the van der Waals attractive interaction dominates and induces aggregation between colloidal particles. This aggregate can form a network spanning the whole system which is commonly referred to as a gel. In this study, we investigate the effect of salt concentration on the resulting aggregation and obtain a good understanding of the mechanical stability of colloidal structure, which will enable us to predict the rigidity of the aggregate structures.

The relationship between the addition of electrolyte and the resulting effective interaction can be described by the standard DLVO theory. In this approach, the potential energy, $U_{DLVO,}$ is the sum of the electrostatic repulsion, U_{el}, and the van der Waals attraction, U_{vdw}. An additional soft-core repulsion prevents particle overlap. The electrostatic interaction, which represents the

repulsion by the double layer formed by the counter-ions surrounding the particles, is written as,

$$U_{el}(r_{ij}) = \pi \, \varepsilon_r \, \varepsilon_o \, \sigma \, \psi_o^2 \ln \left[\, 1 + \exp(\, -\kappa \, (r_{ij} - \sigma)) \, \right] \quad (1)$$

where ψ_o is the surface potential of the particle, σ is its diameter, ε_r is the relative dielectric constant, ε_o is the permittivity of free space and κ is the inverse Debye length. The van der Waals attraction energy between spherical particles is,

$$U_{vdw}(r_{ij}) = -\frac{A_H}{12} \left(\frac{\sigma^2}{r_{ij}^2 - \sigma^2} + \left(\frac{\sigma}{r_{ij}} \right)^2 + 2\ln \left[1 - \left(\frac{\sigma}{r_{ij}} \right)^2 \right] \right) \quad (2)$$

where A_H is the Hammaker constant.

Two important parameters in our study are the surface potential, ψ_o, and the inverse Debye length, κ. A high surface potential will produce a high barrier, known as the primary maximum, preventing two particles from an irreversible coagulation. As we are mainly interested in the flocculation problem where all the aggregations occur in the secondary minimum, we set the value of surface potential to give a high primary maximum.

We vary the inverse Debye screening length, κ, to alter the shape of the potential profiles through the value of salt concentration, C, which is directly related to κ,

$$\kappa = \sqrt{\frac{2e_o^2 z^2 C}{\varepsilon_o \varepsilon_r k_B T}} \, , \qquad (3)$$

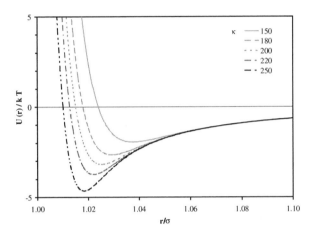

Figure 1. The profiles of potential energy at various inverse Debye length, κ.

where e_o is the elementary charge and z is the valency. Figure 1 shows the potential profiles for different κ.

The Brownian dynamics method is used to simulate the system. In this method, each particle moves under the action of forces according to the Langevin equation,

$$m \ddot{\mathbf{r}}_i = \sum_j \mathbf{F}_{ij}(r_{ij}) + \mathbf{R}_i - \zeta \dot{\mathbf{r}}_i \, . \qquad (4)$$

m is the mass and $\mathbf{r}_i$ is the vector position of i. The dot represents a time derivative. $\mathbf{F}_{ij}$ is the force acting on particle i due to its neighbours j. $\mathbf{R}_i$ is the Brownian force which models the diffusion of i due to random collisions with solvent particles. The Brownian force has a Gaussian probability distribution whose average is zero and variance $6k_B T \Delta t/\zeta$, where Δt is the timestep and ζ is the friction coefficient, related to the viscosity of the solvent, η, through $\zeta = k_B T/3\pi\eta\sigma$. We neglect the many body hydrodynamic interactions. In this study, we use the following parameters suitable for colloidal particles ($\sigma = 1$ μm, $A_H = 0.1$ eV, $\psi_o = 17.7$ mV) immersed in water ($\eta = 8.61 \times 10^{-4}$ Pa.s) at room temperature in the presence of a monovalent salt. $\kappa\sigma$ is varied between 150 and 250. All the variables are in reduced units and the particle diameter, σ, the viscosity, η, and the elementary charge, e_o, are taken as unity. Time is in units of a diffusive

particle relaxation time, τ_R, which is the time for a particle to diffuse a distance on the order of one diameter in the absence of other forces.

Results and Discussion

For the system under consideration, we compute the phase diagram by employing hard sphere perturbation theory (Victor, 1985), (Kaldasch, 1996). The results are shown in Fig. 2. The chosen potential is of sufficiently short range that the liquid-vapour critical point falls below the fluid-

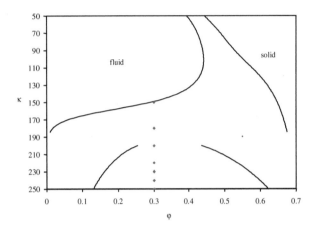

Figure 2. The phase diagram for DLVO potential for $\Psi_o = 17.7$ mV. The dots show the states at which the BD simulations were performed.

solid binodal line (Gast, 1983). We also compute the spinodal lines to show the unstable region where the structure will decompose into a co-exisiting flocculated solid and vapour.

We use the computed phase diagram to determine the range of κ that we will simulate. In this study, we choose a fixed volume fraction, $\phi = 0.3$ and vary the salt concentration. For a system with κ less than 220, the equilibrium state is reached almost immediately. The resulting structure for these systems is a homogeneous fluid with no aggregation. As we increase κ to 220, we observe the formation of a percolating network which fills the entire simulation box. At this point the plot of the configurational energy as a function of time displays an inflection point (see Fig. 3(a)). The decrease in energy then proceeds slowly as the system relaxes to equilibrium. We also note that this transition occurs at earlier time as κ increases. This transition is also observed by Soga (1999) using a depletion potential and as we shall show, it relates to the onset of crystallization in the gel structure. The crystallization at $\kappa = 250$ occurs very early in the simulation so that no transition is visible and the potential energy decreases continuously immediately following the quench.

We also see a clear change in the mean square displacement profiles, $W(t)$, for different κ as shown in Fig 3(b). The mean square displacement at time t is defined as

$$W(t) = \frac{1}{6} \left\langle \left[\mathbf{r}(t) - \mathbf{r}(0) \right]^2 \right\rangle. \qquad (5)$$

For a system with κ less than 220, $W(t)$ increases monotonically with time, showing typical fluid behaviour. The plot for $\kappa = 220$ shows a clear decrease in slope of $W(t)$ at the onset of crystallization. For κ above 220, $W(t)$ increases slowly towards a plateau as a function of time indicating a gelled structure.

We study the structure of the system by counting the number of particles which have coordination number 12, the maximum at close packing. We specify that a particle is a nearest-neighbour if its separation is closer than a cut-off distance, defined as the distance at which the potentials – k_BT. Although this distance varies slightly with κ, the difference is only about 0.01% between the largest and the smallest κ values.

Figure 4 shows how crystallization develops as a function of time. If we compare the plot with the potential energy profile, we see that the crystallization appears at the same time as the change in the curvature in the energy profiles shown in Fig. 3(a). We observe that there are fewer 12-coordinated particles as the potential becomes deeper. This is probably due to easier structural rearrangement in systems with a shallower primary minimum. Figure 5 shows a snapshot from the simulation with $\kappa = 250$ after it forms a gel network.

We also consider the mechanical stability of the gels in our simulation. We apply a step-strain method to measure the strength of the gel network. For a dispersed system, we expect that the structure will not be able to sustain any stress. If a dispersed system is subjected to a stress by applying a step-strain, the stress will quickly dissipate and the system will completely relax after a short time. The system is said to exhibit viscoelastic liquid behavior. A gel on the other hand will be able to sustain some of the stress within its structure.

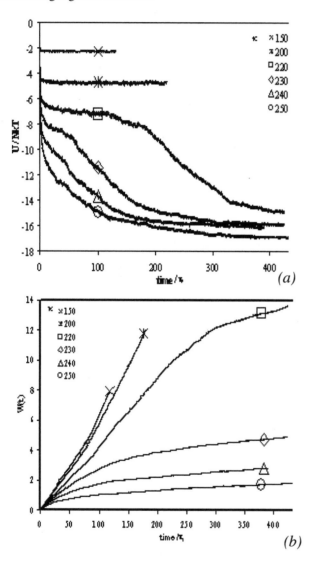

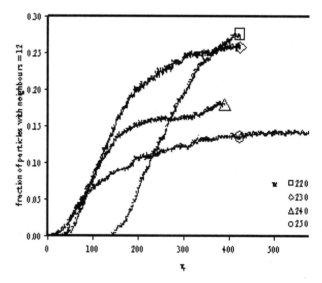

Figure 4.   Fraction of particles with 12-fold coordination as a function of time at different κ values.

Figure 3.   (a) The configurational energy and (b) the mean square displacement as a function of time at different inverse Debye Length, κ.

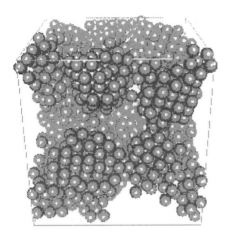

Figure 5. A picture of the aggregated structure at $\kappa = 250$ and $\tau_R = 235.6$. The stronger shade is in the fore-ground.

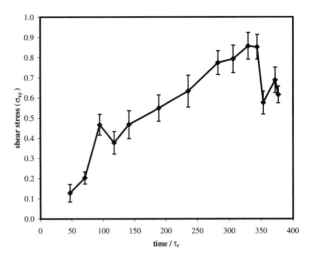

Figure 6. Shear stress at particular times after gelation.

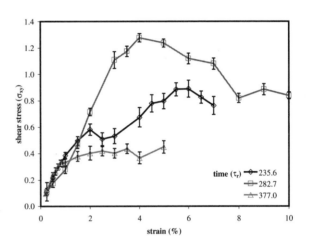

Figure 7. Shear stress as a function of strain at different stages of gelation starting from time of $\tau_R = 235.6$, 282.7 and 377.0.

We study the changes in the mechanical stability as the gels evolve with time. We start with configurations from the Brownian dynamics simulation at several points along the trajectory. This provides gels of a different age. Another series of simulations is performed using these as the initial configurations. A step strain of 3% of simulation box size is applied at the first step and then the simulation proceeds for $1\tau_R$ (around 200,000 steps). The stress resulting from the strain is quickly dissipated and it reaches a more or less a steady state after $\frac{1}{2}\tau_R$. We average the strain-stress calculation over 10 different runs from the same starting configuration. We choose the gel formed at $\kappa = 250$ as we expect this should be the strongest due to the depth of its potential and should easily sustain a 3% step strain. We show the stiffness of gel generally increases with its age in Fig. 6. At the later stages of gel formation, i.e. greater than 300 τ_R, the sustained stress drops. We could not see any significant structural change in the gel at this age compared to the earlier gels.

We have tried to understand this behaviour by running a series of simulations at different strains. The results are shown in Fig 7. For a small step-strain, the shear stress increases linearly with the amount of strain applied and the gel will go back to its initial configuration (i.e. deform elastically) as it relaxes. The behaviour persists until the strain is large enough to force the particles within the gel structure to deform plastically. This is associated with the levelling out of the stress-strain curve. Figure 7 shows that the plastic deformation occurs at lower strain for older gels. It is likely that an older, denser gel structure finds it more difficult to undergo elastic deformation than a more open one when the same magnitude of step-strain is applied and this leads to plastic deformation. The dip in the shear stress for old gels shown in Fig. 6 may be related to the observation that for a dense structure, plastic deformation may have already occurred at a step-strain of 3 %.

A deviation from the expected linear behaviour at small step-strain in Fig. 7 implies that the stress is sustained by different chains within the network. As the strain increases some of the chain might have reached their yield stresses and deformed plastically while others still show elastic behaviour. At large enough strain, all the chains reach their yield stresses and undergo the plastic deformation.

Conclusion

We show that the presence of electrolytes in solution can induce gel formation. We observe that the speed of onset of nucleation depends on the amount of electrolyte present, through the strength of the attractive interaction. We measure the mechanical stability of the gel structures and show how the stiffness of the gel varies with age. The strain needed to induce plastic deformation is also shown to decrease with the gelling time.

Acknowledgments

A studentship from Unilever plc at the University of Southampton is gratefully acknowledged by DI.

References

Gast, A.P., Hall, C.K. and W.B. Russel (1983). Polymer-induced phase separations in nonaqueous colloidal suspensions. *J. Colloid. Interf. Sci.*, **96**, 251-267.

Kaldasch, J., J. Laven and H. N. Stein (1996). Equilibrium phase diagram of suspensions of electrically stabilized colloidal particles. *Langmuir.*, **12**, 6197-6201.

Soga, K. G., J. M. Melrose and R. C. Ball (1999). Metastable states and the kinetics of colloid phase separation. *J. Chem. Phys.*, **110**, 2280-2288.

Victor, J. M. and J.-P. Hansen (1985). Spinodal decomposition and the liquid-vapour equilibrium in charge colloidal dispersion. *J. Chem. Soc., Faraday Trans.*, **81,** 43–61.

MOLECULAR MODELING STUDIES ON THE CONFORMATIONAL PROPERTIES OF POLYCARBONATES

M. S. Sulatha[†], R. Vetrivel[‡] and S. Sivaram[†]
[†]Polymer Chemistry Division, [‡]Catalysis Division
National Chemical Laboratory
Pune - 411 008, India

Abstract

This paper describes the studies on the conformations and conformational properties of bisphenol A polycarbonate (BPAPC) and structurally modified polycarbonates using Molecular Mechanics and RIS Metropolis Monte Carlo method (RMMC). Conformational analysis about the C_α atom of various polycarbonates indicated that the freedom of rotation of the backbone phenyl rings is restricted by structural modification. Conformational properties like root mean square end to end distance, root mean square radius of gyration, persistence length and characteristic ratio were derived using RMMC method. The low values of the computed characteristic ratios (2.7-3.0) indicate compact average conformations for the substituted polycarbonate chains. RMMC simulations for BPAPC were carried out in the temperature range of 300-500 K and the temperature coefficient of the mean square end to end distance was found to be $- 0.415 \times 10^{-3}$ deg^{-1} which agrees with the experimental value.

Keywords

Conformational analysis, Rotational Metropolis Monte Carlo method, Polycarbonates, Substituted polycarbonates, Conformational properties.

Introduction

Polycarbonates are an extremely useful and industrially important class of engineering thermoplastics because of many desirable properties like high Tg, heat resistance, dimensional stability, toughness and transparency. The interplay between the physical properties and the molecular structure for these polymers has been the subject of several theoretical and experimental studies (Hutnik *etal*, 1991).

The chain dimensions as given by root mean square end to end distance and radius of gyration represent the most important conformational properties of a polymer chain. Experimentally, the chain dimensions are determined by dilute solution measurements involving light scattering and viscometry. Theoretically, RIS theory (Flory, 1969) has been applied widely to study the single chain conformational statistics in the unperturbed state. RIS approach involves the derivation of statistical weights for discrete rotational states for different bonds in the polymer chain. RMMC method (Honeycutt, 1998) enables the calculation of unperturbed chain dimensions of polymer chains in the melt or in theta state. It calculates the conformational properties of polymer chains from atomistic simulations.

Molecular modeling studies on the conformation of substituted polycarbonates have received little attention when compared to that of BPAPC (Sundararajan, 1989). This work is part of our efforts to elucidate the effect of structural modification on the conformational characteristics of BPAPC. The substituent effect on the conformational flexibility of the backbone phenyl rings in substituted polycarbonates was studied by Molecular Mechanics and conformational properties of the polycarbonate single chains were derived using RMMC method.

Methodology

Conformational Analysis

Figure 1 depicts the repeat unit of BPAPC and the structure of bisphenols used for the present study. Conformational analysis by force field energy minimization technique was carried out for the various polycarbonate segments.

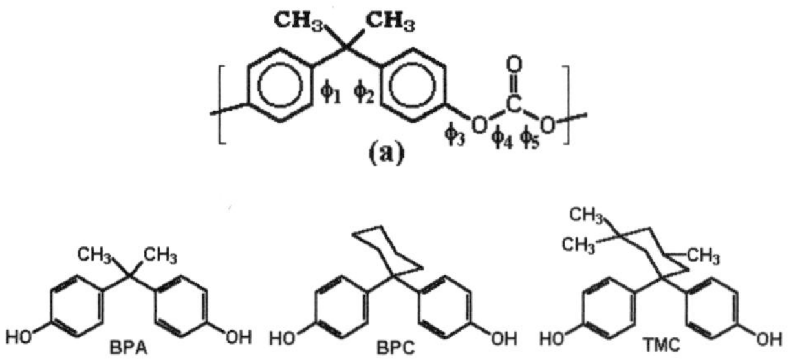

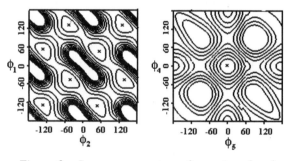

Figure 1.   (a) Repeat unit of BPAPC (b) BPA and substituted bisphenols.

PCFF parameters were used to calculate the potential energy surface. Potential energy was calculated as the sum of bonded and non-bonded interactions. It included a quadratic polynomial for bond stretching and angle bending and a three term Fourier expansion series for torsional motions. Coulombic terms and a 9-6 VDW potential was used to describe the non-bonded interactions.

The potential energy surface was scanned from -180 to 180° in 5° increments. The dihedrals ϕ_1 and ϕ_2 were constrained at discrete values by applying a force constant of 1000 kcal/mol rad^2 and the rest of the molecule was allowed to relax till the gradient of energy was less than 0.001 kcal/mol A°. For BPAPC, conformer search was also done about ϕ_4 and ϕ_5. The energy barriers were then compared with the results of *ab initio* computations reported on model compounds of BPAPC.

Rotational Metropolis Monte Carlo Method

In an RMMC calculation, the first step involves the geometry optimization of the single chain allowing the variation of bond lengths, bond angles and torsional angles. The second step involves the Monte Carlo simulation of the minimized chains, where only the torsional angles are allowed to vary. Bond lengths and bond angles are held constant. Torsional angles are varied continuously. To determine the interaction range of the potential either a bond-based cutoff or a distance-based cutoff is used. The Monte Carlo simulation is performed in two substages. First, the starting conformation is equilibrated and then the production stage is carried out with a large number of Monte Carlo steps. The polymer chain properties are averaged during the production stage specifying a fixed interval. A description of the method in detail can be found elsewhere (Honeycutt, 1998).

Monte Carlo simulations were done for preminimized single chains of BPAPC and substituted polycarbonates with 21 repeat units at 300K. A bond-based cutoff was used to determine the interaction range of the potential. The values assigned for the present simulations were 3 and 6 respectively for n_{min} and n_{max} and 2 for the dielectric constant. The number of equilibration and simulation steps was 500,000 and 2,000,000 steps, respectively. Data collection during simulation was carried out after every 200 steps. For BPAPC, calculations were also carried out at 400 and 500K to determine the temperature coefficient of the mean square end to end distance. There was no evidence of any long-term drift of energy during the simulation, which indicated that the chains were fully equilibrated.

The calculations were performed using the InsightII Polymer 4.00 molecular modeling package from MSI and a Silicon Graphics O2 workstation.

Results and Discussion

Conformational Studies

Iso-energy contours about ϕ_1 vs ϕ_2 and ϕ_4 vs ϕ_5 for BPAPC are given in Fig. 2. The iso-energy contours are drawn at intervals of 1 kcal/mol from the global minima. In all the iso-energy contour maps, the conformations having energy $\geq$ 10 kcal/mol were ignored. It is assumed that all such conformations are not statistically feasible under equilibrium conditions.

Figure 2.   Iso-energy contours for various bond pairs in BPAPC. The minima are marked as 'x'.

8 equivalent minimum energy conformations are possible due to rotation about ϕ_1 and ϕ_2 for BPAPC, where the phenyl rings are twisted out of the plane by 50°. The 4 local minima 0,±90 and ±90,0 (Morino conformation) are at a saddle point between equivalent global minima and the energy barrier for the transformation is 1.7 kcal/mol. The interconversion is also possible *via* the intermediate ±90, ±90 (Butterfly conformation) states and the energy barrier is 3.5 kcal/mol. As shown in Table 1, the energy barriers match well with the results of quantum chemical calculations reported on model compounds of BPAPC (Sun *etal*, 1995; Laskowski *etal*, 1988).

Table 1. Comparison of Conformational Energies (in Kcal/mol).

	HF-6-31G*	STO-3G	PCFF
(ϕ_1, ϕ_2)			
Optimized (50, 50)	0.0	0.0	0.0
Morino (0, 90)	2.17	1.9	1.7
Butterfly (90, 90)	3.93	-	3.5
(0, 0)	17.21	-	19.8
(ϕ_4, ϕ_5)			
trans, trans (0, 0)	-	0.0	0.0
trans, cis (0,180)	-	1.68	1.61

Iso-energy contours evaluated about ϕ_4 and ϕ_5 for BPAPC clearly indicates the preference for the *trans, trans* conformation of the carbonate group (0,0) over the *trans, cis* (0,180) conformation by 1.61 kcal/mol of energy. This agrees well with the value 1.68 kcal/mol obtained from *ab initio* computations on DPC (Laskowski *etal*, 1988). Thus it can be concluded that minimum energy conformations and torsional energy barriers obtained from PCFF are in agreeement with the quantum chemical calculations and hence this force field is well suited for the simulation for polycarbonates.

Substituted Polycarbonates

Figure 3 shows the iso-energy contours about ϕ_1 and ϕ_2 for BPCPC and TMCPC. The global minima correspond to the conformations in which the phenyl rings are twisted out of the plane by 55-70 degrees. The butterfly ($\pm90,\pm90$) and the two of the morino ($\pm90,0$) conformations represent the local minima which are at 1.7 and 0.8 kcal/mol above the global minima. The $0,\pm90$ conformers were of higher energy about 4.6 kcal/mol above the global minima. This difference in energy of the otherwise equivalent 'morino' conformations was studied and it was found that the iso-energy contours depended on the disposition of the cyclohexyl group in the starting model compound.

The cyclohexyl group can have two configurations as shown in Fig. 3c, which are equivalent in energy (global minimum). But for BPC1, $0,\pm90$ are preferred over $\pm90,0$ conformation and for BPC2 the reverse holds true. In each case, the preference for one conformation over the other was found to be due to the VDW interaction energy. In the model compound studied the cyclohexyl group was in BPC2 configuration.

For TMCPC, rotation about ϕ_1 and ϕ_2 leads to 4 minima at -70,-60; -70,120; 110,-60 and 110,120. Local minima are at ±90, 0 and ±90, ±90 conformations, which are 1.6 and 2.1 kcal/mol respectively above the global minima.

For both BPCPC and TMCPC, the iso-energy contours shows that a series of continuous lower energy

conformations are accessible provided one of the phenyl rings is constrained at a dihedral angle of 90°. But in both cases the mobility of the backbone phenyl rings is restricted by substitution of the cyclohexyl group at the C_α atom.

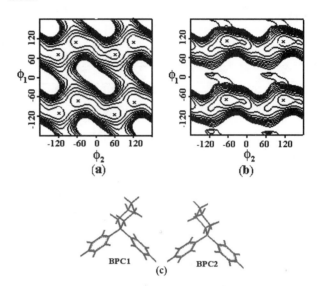

Figure 3. Iso-energy contours for (a) BPCPC and (b) TMCPC for the bond pair ϕ_1 and ϕ_2 (c) Two different configurations of the cyclohexyl group.

RMMC Simulations

The probability distribution of the end to end vector obtained from the single chain simulation of various polycarbonates was found to obey the gaussian chain statistics, typical of polymer chains in the θ state. Distributions of the various dihedrals as obtained from RMMC simulations of single chain polycarbonates are given in Fig. 4. For BPAPC, the distribution of ϕ_1 and ϕ_2 shows two states at -55 and 145 degrees for each. The two angles ϕ_1 and ϕ_2 are strongly coupled, which leads to the possible conformations as –55,-55; -55,145; 145,-55 and 145,145. Distribution about ϕ_3 shows 4 minima at -130, -50, 50 and 130 degrees, where as ϕ_4 shows single state corresponding to the *trans, trans* conformation of the carbonate group in the polycarbonate chain. These dihedral distributions are consistent with the results obtained from conformational analysis on BPAPC segment.

For BPCPC and TMCPC, ϕ_1 and ϕ_2 can take values of -70 and 110 degrees, the dihedrals being coupled similar to BPAPC. The various conformations possible are -70,-70; -70,110; 110,-70 and 110,110. The probability distribution about ϕ_3 shows 4 minima at -130, -50, 50 and 130. Distribution about ϕ_4 confirms *trans, trans* as the preferred conformation about the carbonate group in the substituted polycarbonate single chains also. The interaction between the substituents on the C_α atom and the carbonate group is expected to be negligible due to the large distance of separation between them. Hence the conformation about Ph-O bond and the carbonate group remains the same as in BPAPC.

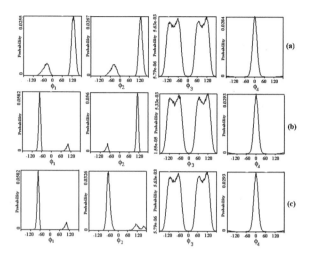

Figure 4.   Probability distributions of the various dihedrals in (a)BPAPC, (b)BPCPC and (c)TMCPC from RMMC simulations.

The various conformational properties calculated for polycarbonates are given in Table 2. For BPAPC, reasonable agreement is found for the calculated and the experimental values (Brandrup, 1975; Anwer *etal*, 1991). C_n was calculated from the mean square end to end distance of the single chains (Flory, 1969). For the substituted polycarbonates, no experimental values are available for comparison. To the best of our knowledge these properties are reported for the first time from our studies on substituted polycarbonates.

The simulations done for BPAPC alone at 400 and 500K gave interesting insights into the effect of temperature on conformation. It was observed that the unperturbed dimensions decreased with increase in temperature. No significant changes in the distribution of the dihedrals ϕ_1, ϕ_2 and ϕ_3 were noted at 400 and 500K. But, the distribution about ϕ_4 showed that the population of the *trans*, *cis* conformations increases with temperature. This can be explained on the basis that at lower temperatures, chains prefer the more expanded *trans,trans* conformations of the carbonate group. Increasing the temperature facilitates the polymer to have conformations of higher energy namely the contracted *trans,cis* states. This results in dimensional contraction of the chains in response to the increased temperature, and hence the reduction in the unperturbed dimensions with increase in temperature. A slightly negative temperature coefficient was obtained for the mean square end to end distance, the value being -0.415×10^{-3} deg^{-1}. Experimentally, this was found to be between zero and -0.6×10^{-3} deg^{-1} (Reddy *etal*, 1988). Thus, the predicted

value comes well within the experimentally reported value.

Table 2.   Conformational Properties of Polycarbonates from RMMC Simulations at 300K.

Polymer	$(<r^2>/M)^{1/2}$ $(A^{o2}mol/g)^{1/2}$	$(<s^2>/M)^{1/2}$ $(A^{o2}mol/g)^{1/2}$	C_n	A (A^o)
BPAPC	0.983	0.392	2.7	12.7
	(0.87-1.28)	(0.45)	(2.4)	(11.3)
BPCPC	0.925	0.371	2.9	12.9
TMCPC	0.884	0.353	3.0	13.3

Conclusions

Conformational analysis about the C_α atom for various polycarbonates were carried out and it was observed that substitution by the cyclohexyl group reduces the conformational flexibility of the backbone phenyl rings. The single chains were simulated using RMMC method to derive various conformational properties. The unperturbed dimensions are not drastically effected by substitution. The computed values of characteristic ratios are in the range of 2.7-3.0. The low values of the unperturbed dimensions and characteristic ratios could be an indication of compact average conformations of the substituted polycarbonate chains.

Acknowledgement

MSS thanks CSIR, New Delhi for a Senior Research Fellowship.

References

Anwer, A., R. Lovell, and A. H. Windle (1991). Conformational analysis of diphenyl linkages. In R. J. Roe (Ed.,) *Computer Simulation of Polymers*, Prentice Hall, New Jersey.

Brandrup, J., and E. H. Emmergut. (1975). *Polymer Handbook*, 2nd Edn. Wiley, New York.

Flory, P. J. (1969). *Statistical Mechanics of Chain Molecules*. Interscience, New York.

Honeycutt, J. D. (1998). A general simulation method for computing conformational properties of single polymer chains. *Comp. Theor. Polym. Sci.*, **8**, 1-8.

Hutnik, M., A. S. Argon, and U. W. Suter (1991). Conformational characteristics of the polycarbonate of 4,4'-Isopropylidenediphenol. *Macromolecules*, **24**, 5956-5961.

Laskowski, B. C., D. Y. Yoon, D. Mc Lean, and R. L. Jaffe (1988). Chain conformations of polycarbonates from *ab initio* calculations. *Macromolecules*, **21**, 1629-1633.

Reddy, G. V., M. Bohdanecky, D. Staszewska, and L. Huppenthal. (1988). Temperature and solvent effects on the unperturbed dimensions of bisphenol A polycarbonate. *Polymer*, **29**, 1894-1897.

Sun, H., S. J. Mumby, J. R. Maple, and A. T. Hagler. (1995). *Ab initio* calculations on small molecule analogues of polycarbonates. *J. Phys. Chem.* **99**, 5873-5882.

Sundararajan, P. R. (1989). The effect of substituents on the conformational features of polycarbonates. *Macromolecules*, **22**, 2149-2154.

FIRST PRINCIPLES INVESTIGATIONS OF MECHANICAL AND THERMAL PROPERTIES OF SILICON OXIDE - SILICON ION IMPLANTED TITANIUM INTERFACES

F. Tsobnang[1], I. Lado[1], A. Mavromaras[1], A. Moormann[2], and L. Wehnert[2]

[1]Institut Supérieur des Matériaux du Mans
7200, Le Mans, France
[2]Berlin Institute of Research and Development
14052 Berlin, Germany

Abstract

This work is a contribution towards the atomic scale understanding and optimization of the thermal and mechanical behavior of dental prosthodontics composed of porcelain and silicon ions implanted titanium. The specific goal is the study of the bond strength increase observed experimentally in these systems compared to those made up from porcelain / pure titanium interfaces. This is achieved by using Density Functional Theory (DFT) energy calculations performed on SiO_2/Ti and $SiO_2/$ Si-ion implanted titanium supercell models. Its is found that silicon impurities in titanium surfaces cause a reduction of bond length and an increase of the adhesive energy between the Ti and the SiO_2. These results are presented and discussed with respect to experiment and computational approximations.

Keywords

Silicon ion implantation, Porcelain-titanium interfaces, Density functional theory, Adhesion work, Bond strength, Bond length.

Introduction

After a certain amount of wear, the bond strength of titanium porcelain interfaces utilized in dental prosthodontic decreases (Walter et al., 1994; Eichner, 1968; Könönen and Kivilathi, 1994; Kimura et al., 1990; Moorman et al., 1998). As a result, the patient has to replace his apparel very frequently in order to avoid risk of infection and maintain a comfortable utilization. Recently, new apparels with extended life cycle, based on the implantation of silicon ions in titanium, have been developed [Eichner, 1968].

According to previous thermodynamic calculations and electron microscopy studies (Moorman et al., 1998), the different behavior exhibited by porcelain/pure titanium systems on one hand, and porcelain/silicon ions implanted titanium systems on the other hand, is related to the nature and structure of interfacial compounds. In porcelain/pure titanium systems, the high reactivity of Ti, particularly with O, gives rise to the formation of scaly crystalline structures in the Ti-porcelain contact zone. These structures mainly consist of stoichiometric and non-stoichiometric titanium oxides (Eichner, 1968) and are responsible of a continuing embrittlement of these interfaces (Walter et al., 1994; Könönen and Kivilathi, 1994; Kimura et al., 1990; Moorman et al., 1998; Brice, 1970). In porcelain/silicon ions implanted titanium systems, silicide compounds formed after the implantation process show a very high thermodynamic stability (Brice, 1970; Dearnaley, 1969; Kinchen et al., 1955), which prevents the formation of the scaly structures observed in conventional Ti/porcelain systems, which in turn gives rise to an increased bond strength between the Ti and the porcelain.

First principles investigations can provide both a quantitative fundamental understanding of the bond strength increase in porcelain/silicon ions implanted titanium systems and rational approaches to design new products with improved characteristics. However due to their complexity (existence of several phases, different kind of bonds, defects, dislocations, ...) these interfaces have been investigated insufficiently or not at all on the atomic scale.

This work is a contribution towards the atomic scale understanding and optimization of the thermal and mechanical behavior of theses materials. Below, we present the computational methodology, the preliminary results, the discussion of our work and our plans for future calculations.

Computational Methodology

Two-dimensional periodic slabs in the form of (1x1) supercells were used to model the interfaces (Fig. 1). The α-SiO$_2$ (0001) surface can be built up from neutral O-Si-O-O-Si-O-O-Si-O units. In this work we only consider the neutral, non-polar termination for this surface. Misfit dislocations are a characteristic feature of ceramic/metal interfaces and result from the mismatch between the ceramic and metal lattice parameters. In our models, misfit is avoided by a slight modification of the Ti lattice constants.

The bond strength was investigated by means of adhesive energy calculations on model systems obtained by varying the distance between the surfaces from 1Å up to 4Å. Adhesive energy is defined as the difference between the energy of the interface model minus the sum of the energies of the isolated free surfaces that compose the interface. The minimum of the curve of the adhesive energy vs separation distance gives an estimation of the adhesion work and the bonding length. In this preliminary study, energies were calculated without prior structure optimization.

Calculations were performed within the framework of density functional theory (Hohenberg and Kohn, 1964) in the local density approximation for electronic exchange and correlation, the exchange and correlation energy as a function of electron density being that of Vosko, Wilk and Nusair, as implemented in the Fast Structure code (Biosym Technologies). The crystal orbitals were expanded in a basis of the atomic orbitals occupied in the neutral atom for the core and valence electrons and an additional basis function for each valence electron (double numerical basis function). The core electrons were frozen (1s for the Si atom and 1s, 2s, 2p for the Ti atom).

This approach is similar to that employed by (Hong et al., 1993) to study adhesion at Mo/MoSi$_2$ interfaces with and without interface impurities.

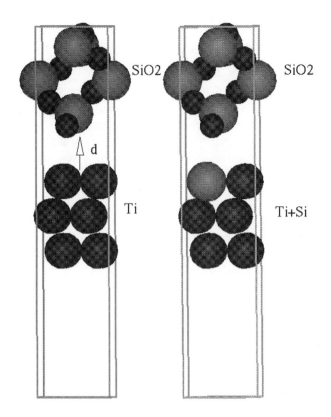

Figure 1.   Models of Ti/ SiO$_2$ and Ti-Si/ SiO$_2$ interfaces used in cohesion energy calculations.

Results and Discussion

The results obtained for the studied interfaces are shown in Figure 2. The Ti-Si/SiO$_2$ system is characterized by an adhesion work and equilibrium distance estimated to be -1.16 J.m^{-2} and 1.75 Å respectively. In the case of the Ti/SiO$_2$ system these quantities are -0.75 J.m^{-2} and 2.00 Å. These values are estimated within ± 0.25 J.m^{-2} and ± 0.06 Å for energy and distance respectively. These errors are essentially related to the interpolation procedure based on the small number of points available in this preliminary study.

Compared to results obtained by other authors in similar metal/ceramics systems (Smith et al., 1996) the above results are reliable. They indicate that Si impurities can cause changes in the adhesion work and bonding length. The absolute values of the magnitude of these changes are 0.41 J.m^{-2} and 0.25 Å. One could argue that these energy and distance differences are not large enough –at least for energies- to allow any conclusion. We believe that increasing the number of single point calculations around the equilibrium point will improve the resolution. Thus, the relative magnitude of the changes, which can be as important as ~35% and ~12% for energy and equilibrium separation distance respectively, can be considered as significant. Even so, it is too early to compare rigorously the results obtained to the experimental situation, which is too complex. To this end, complementary calculations on more complex models with improved computational approaches are needed.

More complex models include other surfaces with different compositions and terminations, different slab thickness or bigger supercell sizes. Improved computational approaches include the utilization of several k points, geometry optimizations or other functional. However, considering the accuracy of the DFT approach for relative energies calculations, one can say that these first calculations tend to match the experimental results. Once validated models and computational approaches are developed, thermo-mechanical properties including elastic constants and thermal expansion coefficients will be calculated. The influence of the formation of silicides on oxygen diffusion from SiO_2 into titanium will be also investigated.

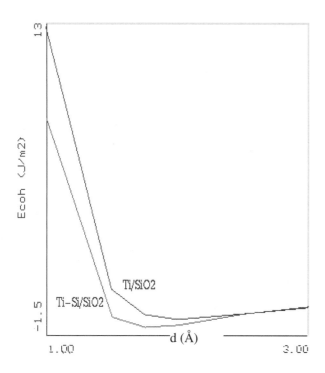

Figure 2.   Adhesive energy versus distance for Ti/ SiO_2 and Ti-Si/ SiO_2 interfaces.

Conclusion

Using Density Functional Theory (DFT) energy calculations performed on supercell models of SiO_2/Ti and SiO_2/ Si+Ti, we have found that silicon impurities can affect substantially adhesive energies and interface equilibrium distances. The observed trend agrees with the experimental fact that silicon ions implantated in titanium lead to the strengthening of the resulting titanium /ceramics based prosthodontics. However improved structural models and computational approaches are

necessary before strong correlations between experiment and modeling can be established. To this end, some calculations are in progress.

References

Biosym Technologies. *Fast_Structure*, San Diego, California, U.S.A.

Brice, D. K. (1970). Ion implantation depth distributions: energy deposition into atomic processes and ion location. *Appl. Phys. Lett.*, **16**, 103.

Dearnaley, G. (1969). Ion Bombardement+Implantation. *Reports on Progress in Physics*, **XXXII**, Part II.

Eichner, K. (1968). Über die Bindung von keramishen Massen und Edelmetall-Legierungen-Theirien und optishe sowie elektronenmikroskopishe Untersuchungen. *Dtsch. Zahnärztl Z*, **23**, 373.

Hohenberg, P.0. and W. Kohn (1964). *Phys. Rev. B*, **136**, 864.

Hong, T., Smith, J.R. and D. J. Srolovitz (1993). *Phys. Rev. B*, **47**, 13615.

Kimura, H., Hrong, C. J., Okazaki, and M., J. Takahashi (1990). Oxidations effects on porcelain-titanium interface reactions and bond strength. *Dent. Mat. J.*, **9**, 91.

Kinchin, G. H. and R. S. Pease (1955). The displacement of atoms in solids by radiation. *Reports Prog. Phys.*, **2**, 1.

Könönen, M. and J. Kivilathi (1994). Bonding of low fusing dental porcelains to commercially pure titanium. *J. Biomed. Mat. Res.*, **28**, 1027.

Moorman, A., Wehnert, L., and W. B. Freesmeyer (1998). New findings on metal-porcelain bonding. *Proceedings of the 22nd Annual Conference European Prosphodontic Association (EPA)*, Turku, Finland.

Smith, J.R., Raynolds, J.E., Roddick, E. R.and D. J. Srolovitz (1996). *J. Comp.-Aided Mat. Design*, **47**, 169-172.

Walter, M., Böning, K., and P. D. Reppel (1994). Clinical performance of machined titanium restorations. *J. Dent.*, **22**, 346.

THE MONTE CARLO AND MOLECULAR DYNAMICS STUDY FOR THE COLOR REWRITABLE FILMS

Hajime Okajima, Tatsuya Kawamura, Chikara Kongo, and Yasuaki Hiwatari
Department of Computational Science
Kanazawa University
Kakuma, Kanazawa 920-1192, Japan

Naohito Urakami
Department of Physics, Biology, and Informatics
Yamaguchi University
Yamaguchi, 753-8512, Japan

Ryoko Hayashi
School of Information Science
Japan Advanced Institute of Science and Technology
Nomi, Ishikawa 923-1292, Japan

Koichi Kato
Chemical Products R&D Center
RICOH Company Ltd.
Numazu-city, Shizuoka, 410-0004, Japan

Abstract

The structure of the developers plays an important role in the coloring/decoloring process for the color rewritable films. We study the structure of the developer through both the Monte Carlo and molecular dynamics simulations. In these simulations, the temperature dependence of the developer structure and the role of the potential parameter of the developers are investigated to study how the lamellar structure is constructed in two and three dimensions. We found the transition of the states from a disordered state to an aggregated state at a temperature (282K). We obtained the lamellar structure in a preferable condition of the potential parameters, while in either smaller or larger interactions, it is rather difficult for the developer chains to form lamellar structures.

Keywords

Color rewritable film, Leuco dye, Developer, Lamellar structure, Monte Carlo simulation, Molecular dynamics simulation.

Introduction

It is expected that the development of digital informatizing will contribute to paperless. However, the consumption of paper is still increasing. One reads articles easier in sheets of paper rather than in display. In view of the environmental problem, the development of rewritable thermal recording paper is absolutely necessary. The color rewritable (CR) film is one of the thermal rewritable recording papers, which is composed of the leuco dyes and developers.

The coloring/decoloring process of the CR film originates from the structure of the leuco dyes and the developers (Tsutsui, 1994, 1995, Furuya, 1999). In the decoloring state, the developers are crystallized and not directly contacted with the leuco dyes. When the CR films are heated from the decoloring states, the crystallized developers are melted and colored because the leuco dyes can contact with the developers and then induce changes of the electronic states. By slow cooling of this coloring state, the developers are crystallized and separated from leuco dyes and then the system returns to the decoloring state. By rapid quenching, however, the CR films are solidified in a coloring metastable state. For a system of pure developers without leuco dyes, the developers can take two different structures by slow cooling and raid quenching. These structures correspond to the coloring and decoloring structures of the CR films (Tsutsui, 1994, 1995). Thus different structures obtained by slow cooling and rapid quenching rely critically on the developers.

In this paper, we investigate the structure of the developers by using both Monte Carlo and molecular dynamics simulations. In Monte Carlo simulation, the temperature dependence of the structure of developers is studied at low concentrations. In molecular dynamics simulation, the interaction between head molecules of each developer is varied to study the change of the developer structure.

Model and Simulation Method

CR films are composed of the leuco dyes and developers. For the coloring/decoloring process of the CR films, we performed our simulations for a pure developer system as a first step. Typical developers have a long alkyl chain, as shown in Fig 1(a). The developer molecules are composed of two parts, the head monomer and the tail monomers of an alkyl chain. It is known that the coloring/decoloring process strongly depends on the kind of the head monomers (Furuya, 1999). We simplified the developer molecules as shown in Fig. 1(b).

(a)

$$CH_3(CH_2)_n\text{-}R$$

(b)

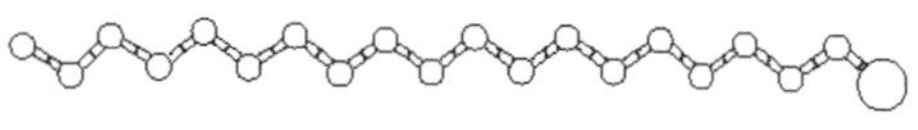

Figure 1.   (a) The molecular structure of a developer. (b) The model of the developer chain. The large circle is the head monomer and the small circles are the other tail monomers.

Monte Carlo Simulation

Each developer is composed of 10 monomers and each bond length is 2.71 angstrom. The bond angle is fixed at 90 degrees. The head monomer on each developer has an attractive interaction, while the other tail monomers a repulsive interaction.

We used the Lennard-Jones potential between the head monomers alone.

$$E_{attraction} = 4\varepsilon\left\{\left(\frac{\sigma}{r}\right)^{12} - \left(\frac{\sigma}{r}\right)^6\right\}, \qquad (1)$$

where r denoted the distance between a pair of head monomers. In our simulations, $\varepsilon=9\varepsilon_0$, where ε_0 was 0.1984Kcal/mol, and σ was 3.239 angstrom.

The repulsive interaction between tail monomers and between heads and tail monomers was assumed by

$$E_{repulsion} = \begin{cases} 4\varepsilon\left\{\left(\frac{\sigma}{r}\right)^{12} - \left(\frac{\sigma}{r}\right)^6 + \frac{1}{4}\right\} & r \leq 2^{\frac{1}{6}}\sigma \\ 0 & r > 2^{\frac{1}{6}}\sigma \end{cases}, \quad (2)$$

where $\varepsilon=\varepsilon_0$.

The torsional potential in the intrachain monomers was defined as

$$E_{torsion} = 10\varepsilon_0\left(1 - \cos(\theta - \theta_0)\right), \qquad (3)$$

where θ denoted the dihedral angle of a chain and θ_0 was 180 degrees.

For the intrachain interaction between ith and jth monomers, where $j>i+3$, we also used the potential energy given by Eqn. (2).

For the motion of the chain, translations and rotations were permitted. The rotations of the torsional angles were only permitted in each intrachain.

Monte Carlo simulations were performed in three dimensional space.

Molecular Dynamics Simulation

Each developer is composed of 20 monomers. In molecular dynamics simulations, interactions between all monomers were used the Lennard-Jones potentials, given by Eqn (1). The interaction between head monomers is assumed to have a different value of ε from the other interaction between monomers. For the interaction between the head monomers, we studied three different cases, i.e., $\varepsilon=3\varepsilon_0$, $6\varepsilon_0$, and $9\varepsilon_0$. For the interaction between heads and tail monomers and between tail monomers, we assumed $\varepsilon=\varepsilon_0$.

The angle bend potential energy was assumed to have the following form.

$$E_{\text{angle bend}} = \frac{1}{2} k_{ijk} \left(\theta_{ijk} - \theta_0 \right)^2 , \qquad (4)$$

where the subscript ijk denoted subsequent ith, jth, and kth monomers, θ_{ijk} was the bond angle between ij bond and jk bond, θ_0 was 108 degrees, and k_{ijk} was 500Kcal/mol/degree^2.

The bond length was fixed at 1.53 angstrom by using the non-interactive matrix method (Yoneya, 1994, 1995). We performed molecular dynamics simulations for the present model in two dimensions aiming at a more effective calculation in a limited short time than three dimensions as well as its simplicity. To save the computational time, we used the cell index method (Allen, 1987). The temperature was kept constant by the constraint method.

Results

Monte Carlo Simulation

We performed Monte Carlo simulations to investigate the temperature dependence of the developer structures in three dimensions. In our simulations, the number of the chain was 63 and the simulation box length was 58 angstrom. The concentration of the developer chains was thus very low. We changed the simulation temperature from 423 to 141K.

Figure 2 shows a typical configuration of developers. At high temperature T=423K, the developer chains dispersed almost uniformly in the simulation box. At low temperature T=141K, the developer chains aggregate and make some domains. The head monomers of the developers aggregate to the center of each domain because of the attractive energy of the head monomers. From the temperature dependence of the total energy, the developer configurations change rapidly from disordered states to the aggregated states around at the temperature 282K. In real CR films, it is predicted that the changes from the coloring state to the decoloring state also take place drastically.

Molecular Dynamics Simulation

We performed two dimensional molecular dynamics simulations. The number of the developer chains was 120 and the simulation box size was 160 angstrom. In these simulations, we investigated the configurations of the developers for three different head potential interactions. We used the Lennard-Jones potential, Eqn (1), between monomers. The head potential interactions were taken with $\varepsilon = 3\varepsilon_0$, $6\varepsilon_0$, or $9\varepsilon_0$. The other potential interactions were taken with $\varepsilon = \varepsilon_0$.

(a)

(b)

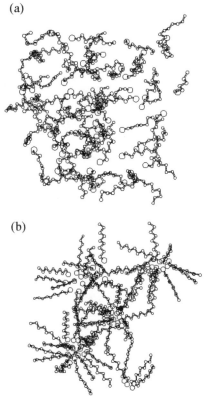

Figure 2.   The typical configurations of the developers obtained by the Monte Carlo simulation. The temperatures are (a) 423K and (b) 141K.

(a)

(b)

(c)

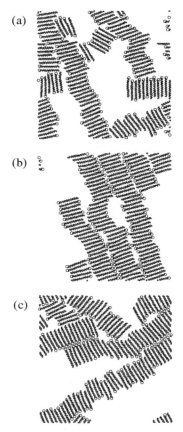

Figure 3.   The typical configurations of the developers obtained by molecular dynamics simulations. The Lennard-Jones potential parameters between head monomers are (a) $3\varepsilon_0$, (b) $6\varepsilon_0$, and (c) $9\varepsilon_0$, respectively.

In annealing, the temperature was decreased form 1100K to 400K during 8ns.

Figure 3 shows the typical configurations of the developer chains obtained after 4ns at a constant temperature, 400K. For the head potential parameter $\varepsilon=3\varepsilon_0$, the developer chains make some ordered (crystallized) domains. These domains disperse in parts, because the attractive energy of the head monomer is not enough large. For the potential parameter $9\varepsilon_0$, the developer chains also exhibit some crystallized domains. These domains aggregate each other closely but irregularly because the head potential energy is too large. For the head potential parameter $6\varepsilon_0$, the developer chains show a clear lamellar structure. From our analysis of the pair distribution function, the peak corresponding to the length of the developer chains is significantly higher for $\varepsilon=\varepsilon_0$ than that for other potential parameters.

It is suggested that the developer chains hardly form the lamellar structures, when the head potential energy is either too small or too large. Thus we conclude that the head potential energy of the developer chain plays an importance role to form the lamellar structures.

Conclusion

From Monte Carlo simulations for three dimensions, it turns out that the developer chains aggregate drastically at the temperature 282K. The developer chains exhibit some aggregated domains. In these simulations, any clear lamellar structure was not obtained because the concentration of the developer chains was very low. It is naturally predicted that if the concentration of the developer chains is increased, the domains of the developer chains aggregate each other and the developer chains will form lamellar structures at high concentrations. The transition from the dispersed state to the aggregated state corresponds to the coloring/decoloring process.

From the molecular dynamics simulations for two dimensions, we obtained a clear lamellar structure of the developer chains for the case that the interaction between the head monomers is $6\varepsilon_0$. The lamellar structure well corresponds to the coloring/decoloring structure of the developers. The head potential energy of the developer chains also plays a key role for obtaining the lamellar structure. If the attractive interaction between the head monomers is either too small or too large, it is rather difficult to form the lamellar structure.

The coloring/decoloring process depends on the cooling rate. We will further investigate the effect of cooling rate into the developer structure. These studies will be of a great importance for a new development of the color rewritable films, which is currently undertaken.

Acknowledgements

This research is supported by "Research and Development Applying Computational Science and Technology" of Japan Science and Technology Corporation.

References

Allen, M. P. and D. J. Tidesley (1987). Some tricks o the trade. In *Computer Simulation of Liquids*. Oxford University Press, Oxford. pp. 140-181.

Furuya, H, M. Torii, T. Tatewaki, H. Matsui, M. Shimada, F. Kawamura, and K. Tsutsui (1999). Improvement of coloring/decoloring characteristics for thermal rewritable recording Media using leuco dye, *Ricoh Technical Report*, No. 25, 6-14.

Tsutsui, K., T. Yamaguchi, and K. Sato (1994). Thermochromic properties of mixture systems of octadecylphosphonic acid and fluoran dye. *Japanese J. Appl. Phys.*, **33**, 5925-5928.

Tsutusi, K., T. Yamaguchi, and K. Sato (1995). Effects of alkyl-chain length of alkylphosphonic acids on fluoran dye. *J. Chem. Soc. Japan*, **1**, 68-73.

Yoneya, M., H. J. C. Berendsen, and K. Hirasawa (1994). Non-interactive matrix method for constraint molecular dynamics. *Mol. Simulation*, **13**, 395-405.

Yoneya, M. and T. Ouchi (1995). Molecular dynamics simulations on shared-memory multiple processor computers. *Mol. Simulation*, **15**, 273.

MOLECULAR DYNAMICS SIMULATION OF
LAYERED DOUBLE HYDROXIDES

Andrey G. Kalinichev, Jianwei Wang and R. James Kirkpatrick
Department of Geology
University of Illinois
Urbana, IL 61801

Randall T. Cygan
Geochemistry Department
Sandia National Laboratories
Albuquerque, NM 87185-0750

Abstract

The interlayer structure and the dynamics of Cl$^-$ ions and H$_2$O molecules in the interlayer space of two typical LDH phases were investigated by molecular dynamics computer simulations. The simulations of hydrocalumite, [Ca$_2$Al(OH)$_6$]Cl·2H$_2$O, reveal significant dynamic disorder in the orientation of interlayer water molecules. The hydration energy of hydrotalcite, [Mg$_2$Al(OH)$_6$]Cl·nH$_2$O, is found to have a minimum at approximately $n = 2$, in good agreement with experiment. The calculated diffusion coefficient of Cl$^-$ as an outer-sphere surface complex is almost three times that of inner-sphere Cl$^-$, but is still about an order of magnitude less than that of Cl$^-$ in bulk solution. The simulations demonstrate the unique capabilities of combined NMR and MD studies to understand the structure and dynamics of surface and interlayer species in mineral/water systems.

Keywords

Molecular dynamics, Layered double hydroxides, Interlayer structure, Hydrocalumite, Hydrotalcite.

Introduction

Layered double hydroxides (LDHs), also called "anionic clays", are among the few oxide-based materials with permanent anion exchange capacity. They occur in natural environments, are readily synthesized, and are receiving rapidly increasing attention from a wide variety of applications as materials for catalysis, environmental remediation, and medicine (Cavani et al., 1991; Ulibarri et al., 1995). LDHs have a layered structure based on that of Mg(OH)$_2$ or Ca(OH)$_2$. Isomorphous substitution of +3 cations (often Al) for +2 cations (often Mg or Ca) in the main hydroxide layer results in a permanent positive charge which is compensated by interlayer anions, often associated with water molecules.

For most LDHs, the arrangement of interlayer species is not understood well and is difficult to probe experimentally. Anions and water molecules strongly associate with particle surfaces, and some adsorbed H$_2$O is indistinguishable from interlayer water in thermal analysis. Molecular dynamics (MD) computer simulations can give significant new insight into the local structure and dynamical behavior of this class of compounds. We performed MD simulations of two typical LDHs – hydrocalumite, [Ca$_2$Al(OH)$_6$]Cl·2H$_2$O, and hydrotalcite, [Mg$_2$Al(OH)$_6$]Cl·nH$_2$O, with major emphasis on the interlayer structure and the dynamical behavior of the interlayer and surface species for the time scale from 10^{-15} to 10^{-10} sec. Hydrocalumite is well ordered and is the only LDH for which a single crystal structure refinement is available (Terzis, et al., 1987). Thus, it is currently the best model compound for understanding the structure and dynamical behavior of surface and interlayer species in LDHs. Such understanding is essential for full exploitation of the unique anion exchange capabilities of this important class of compounds.

Molecular Dynamics Simulations

In many computer simulations of layered materials, the atoms of the main oxide layers are often treated as fixed in a rigid lattice, except for the degrees of freedom associated with swelling and lateral displacements of the lattice as a whole. This approach is computationally efficient at providing useful structural information, but it has inherent limitations for dynamic modeling of surface and interlayer species. Due to the immobility of the lattice atoms, there is no exchange of momentum and energy between the atoms of the main layers and the interlayer/surface species. Thus, in these models the imposed momentum and energy conservation laws *a priori* prevent accurate representation of the dynamics of such processes as hydrogen bonding, adsorption, and surface complexation. Surface diffusion rates can be overestimated, and the structure of the water layers at the interface can be distorted.

In our simulations (Kalinichev et al., 2000; Wang et al., 2000), all atoms were treated as completely movable. Except for the 3-dimensional periodic boundary conditions (e.g., Allen and Tildesley, 1987) imposed on the simulation supercell, there were no additional symmetry constraints. All structures were treated as triclinic, and all cell parameters, a, b, c, α, β, γ, were considered independent variables during the isothermal-isobaric MD simulations.

The initial hydrocalumite structure was based on the single crystal refinement results (Terzis et al., 1987), and the simulation supercell contained 2×2×1 crystallographic unit cells in the a-, b- and c- directions, respectively. Simulations were performed for bulk crystals at temperatures between −100 and 500°C, for dehydrated crystals at 100°C, and for the interface of hydrocalumite with bulk liquid water at 25°C. In the latter case the model crystal was initially cleaved in the interlayer along the (001) crystallographic plane; half of the Cl$^-$ anions were left on each of the created surfaces, and the simulated periodic system consisted of 4 hydroxide layers (infinite in the a and b directions) interspersed in the c direction with a layer of water approximately 20 Å thick (Fig.1).

The structural models of Mg/Al hydrotalcite were based on a crystal structure obtained by a refinement of powder X-ray diffraction data using Rietveld methods (Bellotto, 1996). The simulation supercells contained 6×3×1 crystallographic unit cells in the a-, b-, and c-directions, respectively. To study the swelling behavior of hydrotalcite upon hydration, each supercell contained three interlayers, consisting of 6 Cl$^-$ ions and from 0 to 40 water molecules (Wang et al., 2000).

Each atom in the system had an assigned partial charge, and the total potential energy of the simulated system consisted of a Coulombic term representing the sum of all electrostatic interactions between partial atomic charges and a Lennard-Jones (12-6) term modeling the short-range van-der-Waals dispersive interactions. The force field used in the simulations was modified from the augmented ionic consistent valence (CVFF_aug) force field (Molecular Simulations Inc., 1999), and methods for choosing the potential energy expressions and their specific parameterization are described elsewhere in detail (Cygan et al., 2000; Kalinichev et al., 2000). For water, the flexible version of the simple point charge (SPC) interaction potential was used (Teleman et al., 1987).

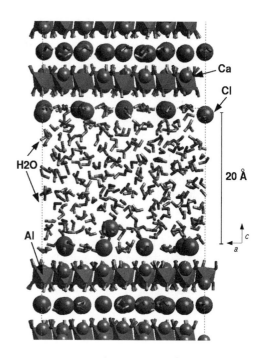

Figure 1. A snapshot of the simulation of hydrocalumite/water interface at 25°C.

Results and Discussion

Although all atoms were treated as movable and the size and shape of the simulation cell were not constrained, the structural results demonstrate a remarkable ability of the modified force field to reproduce and preserve the structure and density of the simulated materials. For hydrocalumite, the monoclinic angle β and the density are about 2% smaller than in the X-ray results, but all other parameters at room temperature are well within 0.5% of their measured values (Terzis et al., 1987). Angles α and γ remain within 0.2° of their nominal value of 90°.

For hydrotalcite, [Mg$_2$Al(OH)$_6$]Cl·nH$_2$O, the computed crystallographic parameters and configurations of the hydroxide layers do not vary significantly with H$_2$O/Cl ratio, n, in the interlayer. The average a- and b-axis lengths remain around 3.20 ±0.01 Å, the γ angle is

120±0.5°, and the α and β angles fluctuate around 90°. Only the computed c-axis dimension increases with increasing n (see Fig. 3a below). The calculated crystallographic parameters compare well with the available experimental data: $a=b=3.046$ Å, $\alpha=\beta=90°$, and $\gamma=120°$ (Bellotto, 1996).

According to the X-ray data for hydrocalumite, the interlayer of composition [Cl·2H$_2$O]$^-$ consists of a primitive hexagonal lattice, with Cl$^-$ ions forming almost regular triangles of this lattice and water molecules located at the center of each triangle. The orientation of these water molecules is such that each Cl$^-$ is coordinated by four H atoms forming ordered chains of hydrogen bonds along the a crystallographic direction. Each Cl$^-$ anion is additionally coordinated by six OH groups (three from two adjacent hydroxide layers). In our simulations these hydrogen bonds prevent chloride diffusion even at temperatures as high as 300°C.

The 10-coordinate arrangement of H-bonds to Cl$^-$ is well reproduced in our MD simulations. However, in contrast to the highly ordered orientations of the interlayer H$_2$O molecules interpreted from the X-ray diffraction measurements, the simulations reveal significant dynamic disorder in water orientations (Kalinichev et al., 2000). At all simulated temperatures, the interlayer water molecules undergo librations (hindered hopping rotations) around an axis essentially perpendicular to the layers. This results in breaking and reformation of hydrogen bonds with the neighboring Cl$^-$ anions. At any instant each water molecule is hydrogen bonded to 2 chlorides, but because of the hopping among the three possible pairs of chlorides, it is, on average, hydrogen bonded to each one about 2/3 of the time. From the perspective of the Cl$^-$ ion, this results in a time-averaged nearly uniaxial symmetry. Although at any instant there are four H-bonds to the neighboring water molecules, averaged over time there are six H-bonds with 2/3 occupancy.

The variable temperature ^{35}Cl NMR experiments for hydrocalumite (Kirkpatrick et al., 1999) indicate the existence of a dynamical order-disorder phase transition near ~6°C. Below the transition temperature, the Cl$^-$ ion is rigidly held in a triaxial environment, whereas above this temperature, the Cl$^-$ is in a dynamically averaged uniaxial environment. The frequency of the atomic motion causing the dynamical averaging which results in the phase transition, must be greater than ~10^5 Hz. Thus, the librational frequencies for water of about 10^{13} Hz determined from the present MD simulations are much more than sufficient to cause the dynamical averaging observed by NMR.

Power spectra of atomic motions in the translational, librational, and vibrational frequency ranges, calculated as Fourier transforms of the atomic velocity autocorrelation functions, allowed comprehensive and detailed analysis of the dynamics in this material. The spectral density of the low frequency vibrational motions of Cl$^-$ ions in the interlayer consists of two distinct frequency bands centered at approximately 50 and 150 cm^{-1} (Fig. 2). Decomposition of chloride velocities into components parallel and perpendicular to the interlayer plane indicates that the 50 cm^{-1} band is associated with the "in-plane" motions of the anions, and the 150 cm^{-1} band is associated with the low frequency vibrations perpendicular to this plane. Because each chloride ion is H-bonded to neighboring H$_2$O molecules and OH-groups, these two O···Cl···O modes of molecular motions are clearly analogous to the intermolecular O···O···O bending and stretching motions of water molecules in the H-bonded network, respectively (e.g., Eisenberg and Kauzmann, 1969). Two weaker spectral peaks at ~110 and ~190 cm^{-1} distinctly observable at lower temperatures (Fig. 2c) are merging with the 150 cm^{-1} peak into one broad spectral band at higher temperatures (Fig. 2a-b). The intensity of the stretching peak is greatly reduced for Cl$^-$ ions on the surface of hydrocalumite, where the shape of the Cl$^-$ low frequency vibrational spectrum more closely resembles the spectrum in a bulk aqueous solution (Fig. 2b). In the collapsed, dehydrated hydrocalumite structure at high temperature, low frequency vibrations in the plane of the interlayer are the only possible chloride motions, as shown by the very intense peak at ~50 cm^{-1} in Fig. 2a.

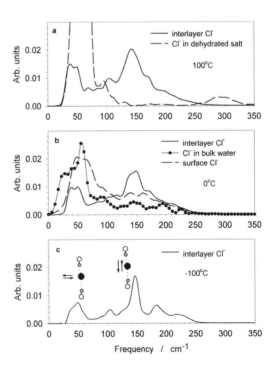

Figure 2. Spectral density of the interlayer Cl$^-$ low frequency vibrational motions.

In the simulations of Mg/Al hydrotalcite, the computed c-axis dimension increases with increasing n (Fig. 3a), but there are three important ranges where it is almost independent of the degree of hydration. For $n < 1$, the c-axis lengths are about 21.7 Å. Near $n = 2$ the Cl$^-$ and H$_2$O form a complete single layer in each interlayer region (similar to the interlayer of Ca/Al hydrocalumite, but significantly more disordered) and the computed *c*-axis dimension is 23.9 Å. This is in excellent agreement with the value of 23.6 Å obtained from X-ray diffraction for hydrated Mg/Al Cl^--hydrotalcite under ambient conditions (Boclair et al., 1999). Near $n = 5$ a second layer of water molecules is formed in each interlayer space, and the simulations give a *c*-axis dimension of 31.8 Å. In reality, however, Mg/Al Cl^--hydrotalcite does not expand beyond 23.9 Å at any relative humidity and atmospheric pressure, and this hydration state is not observed experimentally.

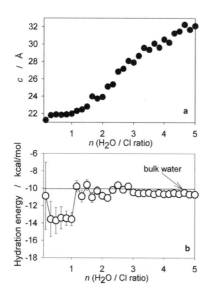

Figure 3. Hydrotalcite $[Mg_2Al(OH)_6]Cl \cdot nH_2O$ swelling upon hydration. (a) c-axis dimension of hydrotalcite. (b) Hydration energy.

The energy of hydrotalcite hydration (Fig. 3b) was calculated as the average potential energy of the system per mole of water with dry hydrotalcite taken as a reference state. For $n < 1$, the hydration energy is quite negative, indicating a strong tendency for water molecules to enter the dry structure. At $n \approx 2$, there is a clear minimum of the hydration energy that corresponds to the region of *n*-independent *c*-axis dimensions (Fig. 3a). The computed hydration energies suggest that almost dry Mg/Al hydrotalcite with just a few interlayer water molecules and hydrotalcite with two water molecules per formula unit should be stable, and indeed the second composition, $Mg_2Al(OH)_6]Cl \cdot 2H_2O$, is readily observed in experiments (Boclair et al., 1999). Large negative values of the hydration energy at low *n* suggest a strong tendency of dry hydrotalcite to sorb water even at low relative humidities.

Diffusion coefficients of interlayer and surface species were calculated for hydrocalumite. On the space and time scale of the present simulations no statistically appreciable Cl^- diffusion is observed in the interlayer, which is experimentally known to be about 10^{-8} cm^2/s (Buenfeld

and Zhang, 1998). However, the simulations show two distinct types of surface Cl^- with diffusion rates differing by a factor of ~3. The inner-sphere complex ($D_{is} = 8.1 \times 10^{-7}$ cm^2/s) is strongly bonded directly to the OH groups of the hydroxide layer and is most common. The other type ($D_{os} = 2.6 \times 10^{-7}$ cm^2/s) has one molecular layer of water between the anion and surface hydroxides (outer-sphere surface complexing), making them much more mobile (see Fig. 1). These results are in good qualitative agreement with the $^{35}Cl^-$ NMR data for hydrocalumite under controlled relative humidity conditions (Kirkpatrick et al., 1999). However, even the diffusion rates of the outer-sphere Cl^- are almost an order of magnitude smaller than those of Cl^- in bulk aqueous solution. The exchange of Cl^- between inner-sphere and outer-sphere environments is clearly observable on the 100 ps time scale of our simulations.

Acknowledgments

The study was supported by NSF (grants EAR 95-26317 and EAR 97-05746) and NCSA (grant EAR 990003N) and by the U.S. Department of Energy, Office of Basic Energy Sciences, Geosciences Research Program, under contract DE-AC04-94AL85000 with Sandia National Laboratories.. The computations were performed using the Cerius^2-4.0 software package from Molecular Simulations Inc.

References

Allen, M. P., and D. J. Tildesley (1987) *Computer Simulation of Liquids.* Oxford University Press, New York.

Bellotto M., B. Rebours, O. Clause, J. Lynch, D. Bazin and E. Elkaiem (1996). A reexamination of hydrotalcite crystal chemistry. *J. Phys. Chem.*, **100**, 8527-8534.

Boclair, J. W., P. S. Braterman, B. D. Brister and F. Yarberry (1999). Layer-anion interactions in magnesium aluminum layered double hydroxides. *Chem. Mater.* **11**, 2199-2204.

Buenfeld, N.R. and Zhang, J.-Z. (1998) Chloride diffusion through surface-treated mortar specimens. *Cement and Concrete Research*, **28**, 665-674.

Cavani, F., F. Trifiro and A. Vaccari (1991). Hydrotalcite-type anionic clays. Preparation, properties and applications. *Catalysis Today*, **11**, 173-301.

Cygan, R. T., J. Liang and A. G. Kalinichev (2000). Molecular models of hydroxide, oxyhydroxide, and clay phases and the development of a general forcefield. *J. Phys. Chem. B*, submitted.

Eisenberg, D. and W. Kautzmann, (1969) *The Structure and Properties of Water.* Oxford University Press, Oxford.

Kalinichev, A. G., R. J. Kirkpatrick and R. T. Cygan (2000). Molecular modeling of the structure and dynamics of the interlayer and surface species of mixed-metal layered hydroxides. *Am. Mineralogist*, **85**, in press.

Kirkpatrick, R. J., P.Yu, X. Hou and Y. Kim (1999). Interlayer structure, anion dynamics, and phase transitions in mixed-metal layered hydroxides. *Am. Mineralogist*, **84**, 1186-1190.

Molecular Simulations Inc. (1999). *Cerius^2-4.0 User Guide. Forcefield-Based Simulations*. MSI, San Diego.

Teleman, O., B. Jönsson and S. Engström (1987). An MD simulation of a water model with intramolecular degrees of freedom. *Mol. Phys.*, **60**, 193-203.

Terzis, A., S. Filippakis, H.-J. Kuzel and H. Burzlaff (1987). The crystal structure of $Ca_2Al(OH)_6Cl \cdot 2H_2O$. *Z. Kristallographie*, **181**, 29-34.

Ulibarri, M.A., I. Pavlovic, M. C. Hermosin and J. Cornejo (1995). Hydrotalcite-like compounds as potential sorbents of phenols from water. *Appl. Clay Sci.*, **10**, 131-145.

Wang, J., A. G. Kalinichev, R. J. Kirkpatrick and X. Hou (2000). Molecular modeling of the structure and energetics of hydrotalcite hydration. *Chem. Materials*, submitted.

A MECHANISTIC STUDY OF
LOW-DOSAGE INHIBITORS OF
CLATHRATE HYDRATE FORMATION

M. T. Storr and P. M. Rodger
Department of Chemistry
University of Warwick
Coventry CV4 7AL, UK

J.-P. Montfort and L. Jussaume
Laboratoire de Génie Chimique
UMR CNRS 5503, ENSIGC-INPT UPS
F-31078 Toulouse Cedex 4, France

Abstract

In this paper we present results from a combined theoretical and experimental study of low dosage inhibitors (LDIs) for hydrate formation. Molecular dynamics and Monte Carlo simulations have been used to develop a detailed analysis of the structural changes that occur on the molecular level when inhibitors are added to a hydrate interface. The results are compared with experimental measurements of the LDI activity, performed as part of this study. The combination of molecular modelling and experimental probes is shown to give valuable insight into the mechanisms of inhibition.

Keywords

Low dosage inhibitors, Clathrate hydrates, Gas hydrates, Kinetic inhibitors, Anti-freeze polymers.

Introduction

Clathrate hydrates are inclusion compounds formed when small, typically non-polar, guest molecules are incorporated into the interstices in a host water lattice. The result is a crystalline solid that is stable under conditions that typify oil and gas trasnportation pipelines. Thus hydrate formation represents a major flow-assurance problem for the oil and gas industry.

Most current methods of inhibition are based on the use of compounds, such as methanol, that shift the boundaries on the hydrate phase diagram to more extreme conditions. However, these have to be used at large concentrations (up to 50% by weight of water). A more promising approach was provided by the discovery in the late 1980s that polyvinylpyrrolidone (PVP) could induce substantial sub-cooling of gas/water mixtures at less than 1 %. This has spurned considerable interest in developing more effective "Low Dosage Inhibitors" (LDIs). Some of the monomers and polymers currently in use are listed in Table 1.

While a great deal of effort has gone into experimental screening programmes for better LDIs, there is still only a limited understanding of the molecular mechanisms by which these compounds inhibit hydrate formation or growth. Activity is believed to arise in a manner similar to that found for anti-freeze proteins (Lal, 1993): selective adsorption of the inhibitor onto specific crystal growth surfaces disrupts further growth of the hydrate. Indeed, a number of simulation studies have invoked this working hypothesis to try to develop a computational screen for new LDIs (Freer, 2000), but in general the methods have ignored too much of the inherent complexity of the hydrate interface (Carver *et al.*, 1995, 1996) to be reliable.

Table 1.   Types of Monomers and Names of Commercial Polymers

Monomer	Alias	Commercial Polymeric LDI
1-vinyl-2 pyrrolidone (1)	VP	VC713-THID78 (1+2+3)
6-caprolactam (2)	VCap	PVP (1)
1-formylpyrrolidine (3)	FP	PVCap (2)
2-dimethylaminoethyl Methacrylate (4)	DMM	FXAP1000 (3)
Proline (5)	Pr	
Acrylamide (6)	Acryl	
1-vinyl-2 valerolactam (7)	VV	

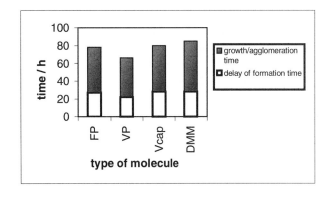

Figure 1.   Inhibition by monomer molecules in THF hydrate test.

The purpose of this paper is twofold. In the first place we show how the full complexity of the hydrate interface can be incorporated into a simple efficiency parameter based on the differential attraction of water and the LDI for the growing crystal. In the second part we show how molecular dynamics simulations (MD) can be used to examine the full complexity of an inhibited hydrate interface, and thereby to begin to establish a comprehensive mechanism for LDI activity that incorporates factors such as the partition of the LDI between the various phases of an oil/gas/water mixture, solvent effects on the LDI conformation, and structuring effects on both liquid and hydrate water.

An Efficiency Parameter for Predicting LDI Activity

Experimental Tests

A series of monomers were tested for the delay of hydrate formation in both THF and ethane hydrate. The THF tests used the immobilisation of ball bearings in a rocking cell to determine hydrate formation, while the ethane hydrate studies were performed in a specially designed isothermal semi-batch pressure crystalliser. Both systems have been described in detail elsewhere (Jussaume, 1999).

As an example, the results for four different monomers in the THF cells are presented in Fig. 1. Both the time to initial hydrate formation, and the time for hydrate formation to block the cell, have been measured. It can be seen that there is some differentiation between the different monomers, with DMM giving the greatest inhibition, and VP the least.

In Figure 2 we present results obtained with some patented polymers currently used as low dosage inhibitors for ethane hydrate. The measured efficiency is based on the cumulative effects of the induction time plus the time required for a one bar decrease in pressure in the crystallizer; the latter is a measure of global crystal growth and agglomeration effects. There is a correlation with Fig. 1 in that the most active polymers contain the most active monomers. However, the polymeric systems

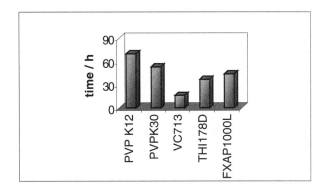

Figure 2.   Efficiency of commercial polymers (cf. Table 1). PVP K12 and K30 differ only in molecular weight.

show greater differentiation and polymer size effects; which suggests some synergism between the active units in the polymeric inhibitors.

Calculated Efficiency Parameter

In defining a theoretical efficiency parameter, we have followed most previous workers in assuming surface adsorption is the main mechanism underlying LDI activity. This is clearly an approximation, but does provide a useful starting point. Unlike previous work, we have taken considerable care to incorporate much of the underlying complexity of the hydrate surface.

The theoretical efficiency parameter was calculated as follows. A series of Metropolis Monte Carlo simulations were performed to examine the behaviour of a single monomer on a hydrate surface. From these simulations we claculated an average free energy for adsorption, A_{ads}, which constitutes a measure of both the entropic and enthalpic driving force for adsorption. Calculations were performed for each of the monomers in Table 1 and for water, and on each of the five main surface topologies that characterise the growth planes for ethane hydrate (Carver,

1996). The efficiency parameter is then based on the notion that inhibition results from a competition between water and inhibitor for the available adsorption sites, *i.e.* growth is delayed when the inhibitor adsorbs more strongly than further water molecules. This is incorporated into the efficiency parameter as:

$$\Delta A_{ads} = A_{ads}(water) - A_{ads}(monomer)$$

$$F(monomer) = \sum_{layers} \Delta A_{ads} \times H(\Delta A_{ads})$$

where $H(x)$ is zero for $x<0$, and 1 otherwise; a negative value for ΔA_{ads} would indicate that water adsorbs preferentially to the surface (*i.e.* sustained growth).

Calculations were implemented using the SORPTION module of Cerius^2 (version 3.5), and so the surface was immobilised for these calculations. The SPC potential was used for water, and the Dreiding force field for other compounds. A hydrate surface of at least 40×40 Å was used (with 2-D periodic boundaries) to allow an adequate sampling of the hydrogen disorder in the water lattice, and long simulations (10^6 configurations) were used to ensure that the inhibitor sampled the whole surface. All simulations were performed on a silicon graphics O² work station. A full description of the method has been published elsewhere (Jussaume, 1998).

Calculated efficiency parameters appear in Fig. 3 and compare favourably with the experimental results for THF hydrate (Figure 1). In particular, VCap and DMM return high efficiency scores, and VP a low score. It is also worth mentioning that a clear correlation exists between the efficiency parameters for THF and propane hydrate; this indicates that strong similarities exist between the two hydrates and that the THF test can be used as a preliminary screen for natural gas hydrate systems.

The SORPTION calculations were also used to examine the low-energy adsorption geometries. It was found that strong binding usually coincided with hydrogen-bond formation between water and either the N or O of the amide group in the LDI.

Molecular Simulations of the Hydrate / Hydrocarbon Interface

To obtain more detailed information about the mechanism of LDI activity we have carried out molecular dynamics (MD) simulations on a thin hydrate film under a hydrocarbon atmosphere, with and without inhibitor present. For this study we have used a small zwitterionic inhibitor akin to those used alongside polymeric inhibitors in commercial hydrate inhibitor blends: N,N,N-tributyl butylsulphonate amine (TBABS). Experimental studies on this inhibitor are currently in progress.

MD simulations were performed on a system of about 400 water molecules 250 methane molecules and 150 propane molecules. The water molecules were initially arranged to expose {111} surfaces of a type II hydrate;

given the complexity of the hydrate crystal structure, the upper and lower surfaces of the film had different topologies. The large cavities in the hydrate region were occupied by propane, and the small cavities by methane

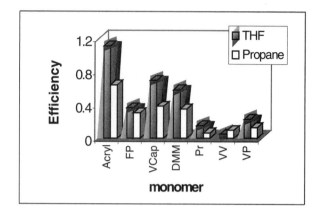

Figure 3. Scale of efficiency of monomers for inhibiting THF and propane hydrate.

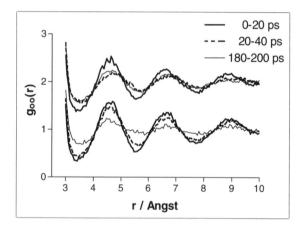

Figure 4. O–O radial distribution functions for water in the hydrate surface layer with (lower) and without (upper) TBABS.

molecules, while the remainder of the hydrocarbons provided the gas phase. This provided a 30Å thick film of hydrate. Calculations were performed in the $[N,V,T]$ ensemble using the Nosé-Hoover thermostat. SPC water was used, together with the CHARMm force. Simulations were conducted at 277 K and a pressure (4 kbar) for which the inner 20 Å of the hydrate film was stable. Trajectories were followed for 200 ps. Full details of the methodology are given elsewhere (Storr, 2000).

Results and Discussion

For the purposes of analysis, the hydrate region was divided into six slabs of 5.0 Å thickness; the repeat distance for the {111} surface of a type II hydrate is 10.0 Å. In this paper we focus on the surface region associated with the sixth slab, since this was the surface to which the inhibitor was added. Initially this surface corresponded to the {111;-0.001} surface, which was previously identified as the most significant for inhibitor activity (Carver, 1996).

Radial distribution functions calculated for the water oxygen atoms in layer 6 are presented in Fig. 4. The functions have been calculated at different times during the simulations. For the clean surface (*i.e.* the system without TBABS present) there is clear evidence of surface melting occurring immediately. The addition of one TBABS molecule leads to complex changes in the behaviour of the system. Initially it appears to stabilise the hydrate surface (Fig. 4), but once surface melting commences, the TBABS leads to a more complete loss of order than is apparent for the clean surface. Close examination of the simulation indicated that conformational changes occurred in the TBABS during the first 50 ps of the simulation. These resulted in shorter distances between the N^+ and SO_3^- groups. It is not yet possible to say whether these are responsible for the aparently contrary influences on the surface stability (initially stabilising, then destabilising), and more extensive calculations are needed to confirm this.

A series of order parameters—specifically designed to study ice and clathrate systems (Rodger, 1996)—have also been used to elucidate the influence of the TBABS on water structure (Table 5). These parameters take distinctly different values in hydrate and liquid water environments, and so are good measure of local water structure. It is apparent that the sulphonate group provides a very strong structural influence, perturbing the water structure away from clathrate and towards liquid-like order. Solvation of the terminal methyl groups in TBABS is more clathrate-like, but the effect is not strong. No difference was observed between the methane/propane solvation structure for the systems with and without TBABS.

The simulations were also examined to determine the disposition of the inhibitor at the interface. At 277 K the inhibitor appeared to be confined to the interfacial region, with typically two of the butyl chains lying at the interface and one pointing out into the hydrocarbon phase; the sulphonate head group was restricted to within a few Angtsroms of the interface. This is in contrast with simulations at 300 K (Storr, 2000), where considerable migration of the whole molecule into the aqueous phase was found to accompany surface melting.

Conclusions

We have presented some results from a combined experimental / theoretical investigation of clathrate hydrate inhibition. We have shown that it is possible to develop a theoretical efficiency scale that is based on the working hypothesis that surface adsorption of low dosage inhibitors disrupts further growth of the hydrate crystals.

Table 2. Order Parameters for Water Solvating Different Solute Atoms; Values for Bulk Hydrate and Liquid Water are Included for Comparison.

Environment	F_3	$F_{4\varphi}$	F_{4t}
Liquid	0.101	-0.01	0.26
Hydrate	0.014	0.73	0.47
S	0.081	0.11	0.28
O	0.082	0.11	0.27
N	0.070	0.13	0.29
CH3	0.065	0.19	0.30
CH4 / C3H8	0.035	0.51	0.40

Indeed, the scale is shown to produce the same order of inhibition activity as is found in experimental studies. None-the-less, more detailed molecular dynamics simulations of a thin hydrate film have confirmed that the surface adsorption model is a simplification, and that a number of other competing effects, such as the distribution of inhibitor across the interface and the inhibitor conformation, must also contribute to the effectiveness of low dosage inhibitors.

References

Carver, T. J., M. G. B. Drew and P. M. Rodger (1995). Inhibition of crystal growth in methane hydrate. *J. Chem. Soc., Farad. Trans.*, **91**, 3449.

Carver, T. J., M. G. B. Drew, and P. M. Rodger (1996). Characterisation of the {111} growth planes of a type II gas hydrate and study of the mechanism of kinetic inhibition by poly(vinylpyrrolidone). *J. Chem. Soc., Faraday Trans.*, **92**, 5029–5034.

Freer, E.M. and E.D. Sloan (2000). An engineering approach to kinetic inhibitor design using molecular dynamics simulations. *Ann. N.Y. Acad. Sci.*, **912**, 651.

Jussaume, L., J.P. Canselier, J. P Monfort, and P.M. Rodger (1998). Molecular modeling of the interaction of additives on type II gas hydrate surfaces: development of an efficiency parameter. In *Applying Molecular Modeling and Computational Chemistry*. Proceedings AIChE meeting, Nov. 15, Miami-Beach, 200.

Jussaume, L. (1999). *Etude d'Inhibiteurs Cinétiques d'Hydrates de Gaz : Méthodes Expérimentales et Modélisations Numériques*. PhD thesis, INP Toulouse.

Lal M., A. H. Clark, A. Lips, J. N. Ruddock, and D. N. J. White (1993). Inhibition of ice crystal-growth by preferential peptide adsorption – a molecular modeling study. *Faraday Discussions*, **95**, 299.

Rodger, P. M., T. R. Forester, and W. Smith (1996). A simulation study of the methane hydrate / methane gas interface. *Fl. Phase. Eq.*, **116**, 326.

Storr M. T. and P. M. Rodger (2000). A molecular dynamics study of the mechanism of kinetic inhibition. *Ann. N.Y. Acad. Sci.*, **912**, 669.

MOLECULAR-LEVEL INSIGHTS INTO CHEMICAL REACTIONS IN HIGH-TEMPERATURE WATER

Naoko Akiya and Phillip Savage
Department of Chemical Engineering
University of Michigan
Ann Arbor, MI 48109-2136

Abstract

We conducted molecular modeling and simulation studies of several model systems to elucidate the role of water in elementary reactions in high-temperature water (HTW). These computational studies, which were motivated by intriguing experimental observations, revealed that water could influence the reaction kinetics and equilibrium in different ways. Water can catalyze a reaction by stabilizing its transition state and facilitating the formation and cleavage of bonds. Water can affect the reaction kinetics and equilibrium through differential solvation of reactants, products, and transition states. Besides these molecular-level interactions, the bulk properties of water also play a role in the density dependence of reaction rate.

Keywords

Reaction kinetics, Solvent effects, Water, High temperature, SCWO, Hydrogen peroxide, Formic acid.

Introduction

High-temperature water (HTW), which includes both liquid water (T > 200 °C) and supercritical water (T > 374 °C, P > 218 atm), has been receiving increasing attention as an environmentally benign reaction medium (Savage, 1995; Savage, 1999). To elucidate the pathways, kinetics, and mechanisms for reactions in HTW, we have employed a combination of experimental determination of reaction kinetics, mechanism-based modeling, and molecular modeling and simulation of model reaction systems. In particular, molecular modeling studies provided insights into the solvent effects on reactions in HTW and a better understanding of the experimentally observed phenomena. The present manuscript presents a brief summary of these studies. Computational details and relevant references are provided in the works cited herein.

Role of Water in Supercritical Water Oxidation

Supercritical water oxidation (SCWO) is complete oxidation of organic compounds in SCW, and it is a highly effective waste treatment technology. Experiments show that changes in water density influence the oxidation rates differently for different reactants. For some reactants the rate increases with increasing water density, for some it decreases, and for others it remains unchanged. In addition, there is a lack of quantitative agreement between experiments and predictions by kinetics models based on gas-phase reaction mechanisms regarding density effects in SCWO. These observations suggest that water may play a role in the SCWO kinetics through solvation effects.

Solvation of Free Radicals in SCW

For mechanism-based kinetics models for SCWO, the fugacity coefficients for all species are needed to calculate the equilibrium constants for the elementary reactions. It is a customary practice in modeling SCWO kinetics to take these fugacity coefficients to be unity, *i.e.* the reaction mixture behaves as an ideal gas. We examined the validity of this assumption by calculating the fugacity coefficient of a hydroperoxyl radical (HO_2), a dominant reactive intermediate species in SCWO, in SCW using molecular dynamics (MD) simulations (Mizan et al., 1997).

We calculated the fugacity coefficients of HO_2 in SCW from the changes in the Gibbs free energy of solvation, which were determined by the generalized alteration of structure and parameters method (Severance et al., 1995). The flexible simple point charge model, or the "TJE model," was used to simulate water (Teleman et al., 1987). The equilibrium geometry, intramolecular potential, and Lennard-Jones parameters for HO_2 were taken from the literature. Partial charges were determined by fitting the quantum chemical electrostatic potential, calculated with UHF/6-311+G(d,p) model chemistry.

The simulation results (Fig. 1) clearly indicate that the ideal gas assumption is not valid for HO_2 in SCW at the state points examined. The fugacity coefficients predicted by MD simulations are inconsistent with the values calculated by CHEMKIN Real Gas, a widely used software package for mechanism-based kinetics modeling, which is designed to account for thermodynamic non-idealities. Interestingly, the agreement is much better between the simulation results and the fugacity coefficients for pure SCW generated by the Peng-Robinson equation of state. The data suggest that solvation of free radical intermediates could modify the equilibrium constants for elementary reaction steps in SCWO.

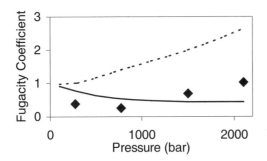

Figure 1. Fugacity coefficient of HO_2 in SCW at 773K. Simulation (◆), CHEMKIN Real Gas (dotted line), pure SCW (solid line).

Water Density Effects on Kinetics and Equilibrium

A recent experimental study showed that hydrogen peroxide dissociation into highly reactive hydroxyl radicals ($H_2O_2 = 2OH$) is 3 times faster in SCW than in the gas phase at the same temperature (Croiset et al., 1997). The study also showed that the rate of dissociation varies with water density. To account for these experimental observations, we examined the solvent effects on the kinetics and equilibrium of H_2O_2 dissociation in SCW using MD simulations (Akiya and Savage, 2000a;b). H_2O_2 dissociation is one of key elementary reaction steps of SCWO, and the global SCWO rate is highly sensitive to the rate constant for this

reaction (Brock et al., 1998; Phenix et al., 1998). Insights into the role of water in H_2O_2 dissociation can lead to a better understanding of the observed water density effects on SCWO kinetics.

Solvent effects on reaction kinetics and equilibria can be characterized within the framework of transition state theory. We quantified the density effects on the rate constant (k) and equilibrium constant (K_c) in terms of the activation volume ($\Delta v^{\ddagger}$) and reaction volume (Δv^{rxn}), respectively (Savage et al., 1995):

$$\left(\frac{\partial \ln k}{\partial \rho}\right)_T = \frac{-\Delta v^{\ddagger}}{\rho \kappa_T RT} \quad \left(\frac{\partial \ln K_c}{\partial \rho}\right)_T = \frac{-\Delta v^{rxn}}{\rho \kappa_T RT} + \frac{1}{\rho} \quad (1)$$

The activation volumes and reaction volumes are calculated from the partial molar volumes of the reactant, transition state, and product. We determined the partial molar volumes from Kirkwood-Buff theory (Kirkwood and Buff, 1951), using the solute-solvent and solvent-solvent pair correlation functions. Considering only the equilibrium solvation effects, one can express the rate constant in SCW in terms of the corresponding gas-phase rate constant and the change in free energy of solvation across the activation barrier, as $k_{SCW} = k_{gas}\exp(-\Delta A_{solv}^{\ddagger}/RT)$ (Billing and Mikkelsen, 1996). We obtained pair correlation functions and changes in free energy directly from MD simulations.

First, we obtained the quantum chemical gas-phase reaction energy profile of H_2O_2 dissociation through a series of partial geometry optimizations with the OO distance, which we took to be the reaction coordinate, fixed. We used the model chemistry B3LYP/6-311+G($3df,2p$) for all calculations. As the potential energy surface of this reaction lacks a saddle point, we postulated the structure of the transition state (TS) from the shape of this reaction energy profile. The calculations also provided changes in the reactant geometry and partial charges as functions of the OO distance.

MD simulations were performed for H_2O_2, OH, and TS in TJE water to calculate the partial molar volumes of these solutes. The intramolecular potentials for H_2O_2 and OH were taken from the literature. The geometry of TS was constrained. Free energy calculations were performed with statistical perturbation method (Zwanzig, 1954). Starting with the equilibrium geometry of H_2O_2, each perturbation corresponded to an incremental increase in the OO distance, and the reactant geometry and partial charges were updated according to the relationships obtained from the gas-phase calculations.

We were unable to find a literature model for the solute-water intermolecular potential, so we developed one for this study. We calculated quantum chemical interaction energies of H_2O_2-water and OH-water dimers with the "energy-distributed" dimer geometries generated using Jorgensen's approach (1979). We then fit a model comprising the Lennard-Jones and electrostatic interaction terms to the resulting potential energy surface. The partial charges for water and the solute molecule (H_2O_2, OH) were taken from the TJE model and from fitting the quantum chemical electrostatic potential for the solute molecule, respectively. Only the Lennard-Jones parameters were treated as adjustable parameters when fitting the model to the dimer potential energy surface.

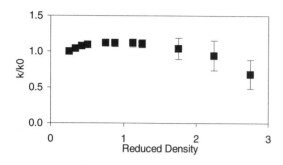

Figure 2. Rate constant for $H_2O_2 = 2OH$ in SCW at $T_r = 1.15$.

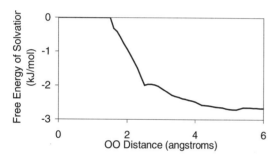

Figure 3. Change in free energy of solvation for $H_2O_2 = 2OH$ in SCW at $T_r = 1.15$.

Figure 2 shows the rate constant for H_2O_2 dissociation at $T_r = 1.15$. The first data point was arbitrarily chosen as a reference point for performing the numerical integration of Eqn. (1). The uncertainties were determined from the standard deviations of the partial molar volumes and were magnified as a result of numerical integration and error propagation. The rate constant goes through a maximum near $\rho_r = 1$. The equilibrium constant, on the other hand, is insensitive to density changes at low densities ($\rho_r < 1.25$) but decreases by an order of magnitude as the density is increased to $\rho_r = 2.75$ (Akiya and Savage, 2000a).

The free energy of solvation decreases as the reaction progresses (Fig. 3), which indicates that the solute-water interactions become more attractive as the reaction progresses. This preferential solvation of the TS relative to H_2O_2 is absent in the gas phase and could account for the observed increase in the H_2O_2 dissociation rate in SCW. The enhancement in the solute-water interactions is consistent with the modest increase in the rate constant with increasing water density at low densities (Fig. 2). The decrease in rate constant and in equilibrium constant

at high densities, on the other hand, could be attributed to the decreasing isothermal compressibility of water with increasing density. The fact that these solvent effects are not included in mechanism-based kinetics models could account for the lack of agreement between the model prediction and experimental data in some cases. Thus, to better understand the water density effects on SCWO kinetics, one must consider both the changes in solute-solvent interactions and bulk solvent properties with changes in water density.

Role of Water in Formic Acid Decomposition

Formic acid decomposes via two pathways: (1) dehydration, which produces carbon monoxide and water, and (2) decarboxylation, which produces carbon dioxide and hydrogen. Experimental studies show that CO/CO_2 ratio in the gas phase is about 10 (Blake et al., 1971) but about 1/100 in HTW (Yu and Savage, 1998). These data indicate that dehydration is the preferred pathway in the gas phase but decarboxylation is preferred in HTW. The reaction rate in HTW is also higher than that in the gas phase by about six orders of magnitude.

To elucidate the role of water suggested by these experimental observations, we obtained the reaction energy profiles and activation barriers for formic acid decomposition using computational quantum chemistry (Akiya and Savage, 1998). We chose as model systems *cis* and *trans* formic acid and transition states (TS) for dehydration, decarboxylation, and isomerization, all of which in the presence of 0, 1, or 2 water molecules. Calculations were performed using the model chemistry B3LYP/6-311+G($2d,p$)//HF/3-21G(d,p).

In the gas phase, the dehydration TS is lower in energy than the decarboxylation TS, which is consistent with the observation that dehydration is the preferred pathway in the gas phase. In the presence of water, the activation barriers for both pathways are reduced from the gas-phase values, which is consistent with the observation that the reaction rate is higher in water than in the gas phase. The decarboxylation TS is now lower in energy than the dehydration TS, making decarboxylation the preferred pathway in water. The same trend is observed in the presence of two water molecules. The optimized structures of the transition states show that water molecules actively participate in the bond-breaking and bond-forming processes in dehydration and decarboxylation and thereby facilitate the reaction. Isomerization, on the other hand, occurs independently of water.

Using the results of the quantum chemical calculations and transition-state theory (Laidler, 1987), we determined the rate constant corrected for tunneling effects (Johnston, 1966), for formic acid decomposition in the gas phase and in water at 700 K. We then estimated the CO/CO_2 ratio for the reaction in the gas phase and in the presence of water. The results are summarized in

Table 1. The orders of magnitude of the rate constants and CO/CO_2 ratios are in excellent agreement with the experimental data, indicating that the calculations were able to capture successfully capture the role of water in formic acid decomposition.

Table 1. Order of Magnitude Estimates at 700K.

Reaction	k (1/sec)		CO/CO_2	
	Calc.	Exp.	Calc.	Exp.
Gas phase	10^{-7}	10^{-7}	10^1	10^1
Aqueous phase	10^{-3}-10^0	10^{-1}	10^2	10^2

Conclusions

We applied molecular modeling and simulation tools to study the solvation of hydroperoxyl radical in SCW, the density effects on the kinetics and equilibrium of hydrogen peroxide dissociation in SCW, and the role of water in the kinetics of formic acid decomposition in HTW. These computational studies were motivated by experimental observations, which suggested unique roles of water in these systems. The studies revealed that water could influence the reaction kinetics on a molecular level, via differential solvation or formation of a stabilizing complex with the transition state. These molecular-level interactions are not easily studied in a laboratory. Thus, molecular modeling and simulation studies coupled with experimental determination of kinetics data provide a greater understanding of reactions in HTW.

Acknowledgments

Dr. Tahmid Mizan conducted the MD simulations of HO_2 in SCW. We thank Prof. Robert Ziff and Dr. Jurgen Schnitker for helpful discussions. We gratefully acknowledge the financial support by the National Science Foundation (DMR-9122341, CTS-9521698, CTS-9903373, and Graduate Research Fellowship), the Exxon Education Foundation, and the University of Michigan.

References

Akiya, N. and P. E. Savage (2000a). Effect of water density on hydrogen peroxide dissociation in supercritical water. 1. Reaction equilibrium. *J. Phys. Chem. A*, **104**, 4433-4440.

Akiya, N. and P. E. Savage (2000b). Effect of water density on hydrogen peroxide dissociation in supercritical water. 2. Reaction kinetics. *J. Phys. Chem. A*, **104**, 4441-4448.

Akiya, N. and P. E. Savage (1998). Role of water in formic acid decomposition. *AIChE J.*, **44**, 405-415.

Billing, G. D., and K. V. Mikkelsen (1996). *Molecular Dynamics and Chemical Kinetics*. John Wiley and Sons, New York.

Blake, P. G., H. H. Davies, and G. E. Jackson (1971). Dehydration mechanisms in the thermal decomposition of gaseous formic acid. *J. Chem. Soc. (B)*, **10**, 1923-1925.

Brock, E. E., P. E. Savage, and J. R. Barker (1998). A reduced mechanism for methanol oxidation in supercritical water. *Chem. Eng. Sci.*, **53**, 857-867.

Croiset, E., S. F. Rice, and R. G. Hanush (1997). Hydrogen peroxide decomposition in supercritical water. *AIChE J.*, **43**, 2343-2352.

Johnston, H. S. (1966). *Gas Phase Reaction Rate Theory*. Ronald Press, New York.

Jorgensen, W. L. (1979). Deriving intermolecular potential functions for the water dimer from *ab initio* calculations. *J. Am. Chem. Soc.*, **101**, 2011-2016.

Kirkwood, J. G. and F. P. Buff (1951). The statistical mechanical theory of solution. I. *J. Chem. Phys.*, **19**, 774-777.

Mizan, T. I., P. E. Savage, and R. M. Ziff (1997). Fugacity coefficients for free radicals in dense fluid: HO_2 in supercritical water. *AIChE J.*, **43**, 1287-1299.

Phenix, B. D., J. L. Dinaro, M. A. Tatang, J. F. Tester, J. B. Howard, and G. J. McRae (1998). Incorporation of parametric uncertainty into complex kinetic mechanisms: Application to hydrogen oxidation in supercritical water. *Combust. & Flame*, **112**, 132-146.

Savage, P. E. (1999). Organic chemical reactions in supercritical water. *Chem. Rev.*, **99**, 603-621.

Savage, P. E., S. Gopalan, T. I. Mizan, C. J. Martino, and E. E. Brock (1995). Reactions at supercritical conditions: Applications and fundamentals. *AIChE J.*, **41**, 135-144.

Severance, D. L., J. W. Essex, and W. L. Jorgensen (1995). Generalized alteration of structure and parameters: A new method for free-energy perturbations in systems containing flexible degrees of freedom. *J. Comp. Chem.*, **16**, 311-327.

Teleman, O., B. Jonsson, and S. Engström (1987). A molecular dynamics simulation of a water model with intramolecular degrees of freedom. *Mol. Phys.*, **60**, 193-203.

Yu, J., and P. E. Savage (1998). Decomposition of formic acid under hydrothermal conditions. *Ind. Eng. Chem. Res.*, **37**, 2-10.

Zwanzig, R. W. (1954). High-temperature equation of state by a perturbation method. I. Nonpolar gases. *J. Chem. Phys.*, **22**, 1420-1426.

WATER IN POROUS CARBONS:
A SIMULATION STUDY

John K. Brennan, Kendall T. Thomson, and Keith E. Gubbins
North Carolina State University
Raleigh, NC 27695

Abstract

We present a simulation study of the adsorption mechanism of water in a realistic carbon structure at 300 K. Water molecules are modeled using a recently developed fixed-point charge water model optimized to the vapor-liquid coexistence properties (Errington and Panagiotopoulos, 1998). Reverse Monte Carlo techniques (Thomson and Gubbins, 2000) are used to generate a realistic model of carbon pore morphologies composed of rigid, carbon basal plates. Arrangements of the carbon plates are driven by a systematic refinement of simulated C-C radial distribution functions to match experiment. The resulting pore morphology is generally non-slitlike and highly connected. The adsorption of water on activated (surface oxygen sites, =O) and non-activated (graphitic) carbon is investigated using the grand canonical Monte Carlo simulation method. The adsorption behavior is found to be strongly dependent on the presence of active sites with no appreciable adsorption in the graphitic carbon until the pressure is above the bulk saturation point. Additionally, water adsorption is found to increase as the active site density increases.

Keywords

Adsorption, Molecular simulation, Activated carbon, Microporous carbon, Water.

Introduction

The molecular behavior of water in graphitic and activated carbons is of significant industrial importance. For example, quite low humidities in gases entering industrial adsorbers are known to have a large effect on the capacity and selectivity for the removal of organic and inorganic contaminants. Similarly, small amounts of water often reduce the diffusion rates of alkanes in microporous media by an order of magnitude. The mechanism producing these effects is poorly understood. The behavior of water in porous carbons is significantly different than that of simple non-associating adsorbates such as nitrogen and hydrocarbons (Gregg and Sing, 1982). The main differences stem from strong water-water interactions, weak-water carbon interactions, and the formation of H-bonds with oxygenated groups on the surface.

Progress in understanding the behavior of water in activated carbons at the molecular level has been slow.

Activated carbons are derived from a wide variety of precursors, and their amorphous pore structure has proved to be among the most difficult to characterize. Experimental characterization of these materials faces a number of challenges, including determining (a) the distribution of carbon microcrystal sizes; (b) the densities and species of surface groups; (c) the topological nature of the connected pore structure; and (d) pore size distributions. The details of pore filling are sensitive to the size and arrangement of graphite microcrystals, pore size distribution, and the species and arrangement of the surface groups. Little is known of the effect of these variables on phase changes for confined water, although these transitions are known to be strongly dependent on the species and density of the surface sites.

There have been numerous experimental studies of water in activated carbons (Gregg and Sing, 1982; Rouquerol *et al*, 1999). Unfortunately, these measurements were made on poorly characterized carbons that make direct comparisons with theory or simulation results difficult. Previous molecular simulations of water on activated carbons (Muller *et al*, 1996; McCallum *et al*, 1999) have found the adsorption behavior to depend strongly on the density and geometric arrangement of the active sites on the carbon surface. The chief deficiency of these simulations, however, was an over-simplified treatment of the carbon structure that was modeled as unconnected slit pores. This work is an initial attempt to address this deficiency by considering a more realistic carbon model.

Microporous Carbons

The production of activated carbons involves two processes: (1) "carbonization", i.e. thermal decomposition of raw organic material (e.g. wood, peat, coal, lignite, synthetic polymers, coconut shell) in an inert atmosphere; and (2) "activation" to enhance porosity and surface area, usually by heating in the presence of an oxidizing gas stream such as air, steam or CO_2 (Gregg and Sing, 1982; Rouquerol *et al*, 1999). Such carbons are composed of graphite microcrystals having a range of sizes, some of which are connected by covalent bonds; inorganic atoms (typically O, H, N, S) are present at the edges of some of the microcrystals in the form of chemical groups.

Figure 1. Structural representation of microporous activated carbon generated by RMC. Black and gray spheres represent carbon atoms and surface sites, respectively.

Since the precursors and techniques used to prepare carbons vary widely and the reactions involved are unknown, it is difficult to develop simulation protocols that mimic the manufacturing process, such as that which has been done recently for porous glasses (Gelb and Gubbins, 1998). Alternatively, the Reverse Monte Carlo

(RMC) technique has been applied to this problem (Thomson and Gubbins, 2000). The starting configuration is a system containing graphite basal planes having a range of sizes (the mean size is typically about 2 nm) that is based on experimental evidence. The graphite microcrystals are treated in atomic detail, using carbon and graphite parameters and intermolecular parameters developed by Steele (1974). The microcrystals are placed with random positions and orientations in a periodic simulation cell, with a carbon density chosen to match the experimental material, and Monte Carlo moves of the position and orientation of the planes are attempted. It is possible to include both changes in the size of the basal planes, and cross-linking (bonding) between basal planes during this procedure. The acceptance criteria for the moves in the RMC method are designed to adjust the structure until the carbon-carbon structure factor or the radial distribution function matches a target experimental structure factor obtained from SANS or SAXS. The RMC method has already been demonstrated to converge to the target radial distribution function for a variety of carbons; an example is shown in Fig. 1. These model structures have been shown to have carbon densities, porosities, surface areas, and surface energy distributions that closely match those of the targeted materials (Thomson and Gubbins, 2000).

Having prepared model carbons with an appropriate topology, oxygenated groups need to be placed on the microcrystals; these are located at random edge sites on the microcrystals. The density and species of such groups will be guided by experimental characterization data. Details of the surface oxygenated groups are given below.

Water

Despite many attempts, an intermolecular potential model for water that is valid over a broad range of state conditions and gives satisfactory results for a variety of properties remains elusive. Most of the existing empirical potentials can be divided into: (a) fixed-point charge models, and (b) models that account for polarizability. Since the bulk and pore phases are at significantly different densities and reduced temperatures, it is important that the model describes the coexisting densities, vapor pressures and structure over a significant range of temperature. For these properties the more computationally-intensive polarizable models have so far not shown a significant improvement for pure water. The model used in this study is a recently proposed fixed-point charge model (Errington and Panagiotopoulos, 1998) that has been optimized to these properties. The model gives a good description of the vapor-liquid equilibria over the whole temperature range from sub-ambient to the critical point, in contrast to previous models. It is also superior to other fixed-point charge models in its description of the vapor pressure, second virial coefficient and critical parameters. The principle defects of the model are:

failure to describe the second peak in $g_{OO}(r)$; the predicted dielectric constant is too low; and the polarizability effects are only included through an effective dipole moment. In addition, like all fixed charge models, effects of bond flexibility, many-body forces and quantum effects are neglected. The most serious defect of the model for application in porous carbons may be the crude treatment of polarizability effects. Although the inclusion of many-body polarization does not appear to offer a significant improvement in the description of the thermodynamics and structure for bulk water, it may be significant for water near a carbon surface.

Water-Substrate

Water-substrate interactions are of two kinds, those with the carbon surfaces, and those with oxygenated groups. Typically, only dispersion and repulsion water-carbon interactions are included while water-oxygenated group interactions are modeled with a standard potential such as an OPLS potential (see e.g., Jorgensen and Tirado-Rives, 1988). In this work, the carbon-water interactions are taken from McCallum et al (1999) where dispersion interaction parameters were adjusted to fit very low-pressure adsorption data (in the pressure region where fluid-fluid interactions are negligible). We model the surface oxygen site (=O) as a single Lennard-Jones (LJ) sphere with a point charge fixed at its center. The LJ energy and size parameters are taken from the OPLS format (see e.g., Jorgensen and Swenson, 1985) to be $\varepsilon/k=105.76$ and $\sigma=0.296$ nm, respectively. The C=O bond length was taken as 0.1214 nm. Finally, the point charge is q=-0.37e, which was derived from the quantum chemical electrostatic potential (Sigfridsson and Ryde, 1998).

The dispersive interactions in the water and the surface oxygen site are accounted for by a Buckingham Exp-6 potential and Lennard-Jones potential, respectively. Therefore, estimates of the unlike pair interactions cannot be made directly. Recently (Panagiotopoulos, 2000) has shown that for simple fluids the parameters of an Exp-6 potential are transferable to a LJ potential by a judicious choice of the repulsion parameter, α, in Eqn. (1). Following this procedure, we estimated the LJ parameters of the water model to be $\varepsilon/k=159.78$, $\sigma=0.3226$ nm and $\alpha=10.8$. Afterwards, the Lorentz-Berthelot combining rules were used for ε_{wo} and σ_{wo}, while $\alpha_{wo}=(\alpha_w\alpha_o)^{1/2}$, where w=water and o=surface oxygen.

$$U_{LJ}(r) = \varepsilon\left[\frac{6}{\alpha-6}\left(\frac{r_m}{r}\right)^\alpha - \frac{\alpha}{\alpha-6}\left(\frac{r_m}{r}\right)^6\right] \quad (1)$$

Method and Simulation Details

Grand canonical ensemble Monte Carlo simulations were carried out at 300 K (Frenkel and Smit, 1996). Biased insertion/deletion techniques were used to facilitate convergence to equilibrated states. Electrostatic interactions were estimated using Ewald summation. A spherical cutoff of $3.9\sigma_w$ was applied without long range corrections. A virial equation of state was used to convert activity values used in the simulation into relative pressure (P/P_o) values (Muller et al, 1996). Cubic simulation cells with lengths of either 5.0 nm (porosity=0.485) or 10.0 nm (porosity =0.579) were used.

Results

Adsorption in Graphitic Carbon

The adsorption isotherms at 300 K for graphitic carbon at different porosities are shown in Fig 2. The dashed line is the corresponding bulk water saturation point. The isotherms are of IUPAC classification type V (S-shape). In the absence of surface sites, the carbon plates are hydrophobic and no appreciable adsorption occurs until relatively high pressures. Qualitatively this agrees with experimental data for graphitic carbons (Gregg and Sing, 1982; Rouquerol et al, 1999) although experimentally the onset of significant water uptake and capillary condensation may occur at lower pressures. This difference can be attributed to structural heterogeneity (e.g., non-aromatic rings of carbon atoms or heteroatoms that cause cross-linking of carbon plates) in actual carbons not present in our simplified representation of aromatic carbon plates. Such heterogeneity provides additional adsorption sites for the water molecules. The effect of porosity on the adsorption is most evident at high pressures in Fig 2. The higher porosity carbon contains more accessible volume for an increased amount of water to occupy. Desorption simulations for these systems are underway.

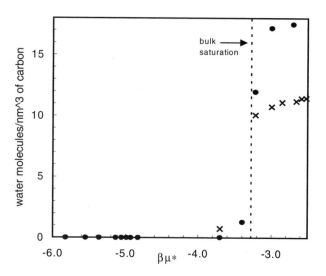

Figure 2. Adsorption isotherms for water at 300 K in graphitic carbon at porosities, 0.579(circles) and 0.485(crosses). μ^* is the reduced chemical potential and $\beta=1/kT$.

Adsorption in Activated Carbon

The adsorption behavior of water adsorbed in our activated carbon model is shown in Fig 3 for activated site densities, 0.25 and 0.65 sites/nm^2. Although all the isotherms in Figs 2 and 3 are of IUPAC classification Type V, the adsorption mechanism between the graphitic and activated carbons has changed significantly.

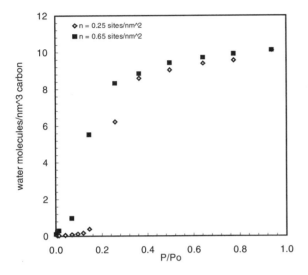

Figure 3. *Adsorption isotherms for water at 300 K at two different site densities for porosity 0.485.*

Figure 4. *Snapshots of water clusters adsorbed in carbon (n=0.65 sites/nm^2) at P/P$_o$=0.03.*

At low pressures, uptake of water occurs via the bonding of isolated water molecules to the oxygenated surface sites. This stage is followed by a buildup of water clusters around those sites. For the n=0.25 sites/nm^2 isotherm, when all sites are occupied by water clusters little further uptake occurs as the pressure increases. This

is attributed to the sites (and their resident water clusters) being located far from each other so that the bridging of proximal water clusters is unlikely. At the higher site density, continuous uptake occurs as the pressure increases. Water molecules search for locations where they can either bond to several pre-adsorbed water molecules or to a water molecule and a surface site simultaneously – a mechanism recently termed cooperative bonding by Muller et al (1996). For P/P$_o$>0.4, the adsorption behavior is dominated by water-water interactions and growing water clusters. Experimentally the onset of "capillary condensation" occurs at P/P$_o$~0.5. Fig 3 indicates "capillary condensation" occurs much lower than this. This discrepancy may be due to micropore filling in the model since the simulation cell (5x5x5 nm^3) is too small to contain mesopores. A snapshot of water clusters adsorbed in activated carbon with n=0.65 sites/nm^2 is shown in Fig 4. It should be noted that simulations for different random placement of the active sites found no significant changes in the adsorption behavior. Also, simulations of the desorption isotherms for these carbons are currently in progress.

Acknowledgments

This work was funded by the Division of Basic Sciences, Department of Energy (Grant No. DE-FA02-99 ER14847) and by the National Partnership for Advanced Computational Infrastructure (NPACI) through the San Diego Supercomputing Center. JKB and KTT thank NSF for support through a NSF-CISE Postdoctoral Fellowship (ACI-9896106).

References

Errington, J.R., and A.Z. Panagiotopoulos (1998). A fixed-point charge model for water optimized to the vapor-liquid coexistence properties. *J. Phys. Chem. B*, **102**, 7470-7475.

Frenkel, D. and B. Smit (1996). *Understanding Molecular Simulation.* Academic Press, New York.

Gelb, L. D. and K. E. Gubbins (1998). Characterization of porous glasses: Simulation models, adsorption isotherms, and the Brunauer-Emmett-Teller analysis method. *Langmuir*, **14**, 2097-2111.

Gregg, S.J. and K.S. Sing (1982). *Adsorption, Surface Area and Porosity, 2nd ed.* Academic Press, New York.

Jorgensen, W.L. and J. Swenson (1985). Optimized intermolecular potential functions for amides and peptides structure and properties of liquid amides. *J. Amer. Chem. Soc.*, **107**, 569-578.

Jorgensen, W.L. and J. Tirado-Rives (1988). The OPLS potential functions for proteins. Energy minimizations for crystals of cyclic peptides and crambin. *J. Amer. Chem. Soc.*, **110**, 1657-1666.

McCallum, C.L., T.J. Bandosz, S.C. McGrother, E.A. Muller and K.E. Gubbins (1999). A molecular model for adsorption of water on activated carbon: Comparison of simulation and experiment. *Langmuir*, **15**, 533-544.

Muller, E.A., L.F. Rull, L.F. Vega and K.E. Gubbins (1996). Adsorption of water on activated carbons: A molecular simulation study. *J. Phys. Chem.*, **100**, 1189-1196.

Panagiotopoulos, A.Z. (1999) On the equivalence of continuous-space and lattice models for fluids. *J. Phys. Chem. B* (accepted).

Rouquerol, F., J. Rouquerol and K.S. Sing (1999). *Adsorption by Powders and Porous Solids*. Academic Press, London.

Sigfridsson, E and U. Ryde (1998). Comparison of methods for deriving atomic charges from the electrostatic potential and moments. *J. Comp. Chem.*, **19**, 377-395.

Steele, W.A. (1974). *The Interaction of Gases with Solid Surfaces*. Pergamon Press, Oxford

Thomson, K.T. and K.E. Gubbins (2000), Modeling structural morphology of microporous carbons by reverse Monte Carlo. *Langmuir* (accepted).

GIBBS ENSEMBLE SIMULATION OF WATER IN SPHERICAL CAVITIES

I. Brovchenko and A. Geiger
Physikalishe Chemie
Universitat Dortmund
44221 Dortmund, Germany

D. Paschek
Department of Chemical Engineering
University of Amsterdam
1018 WV Amsterdam, The Netherlands

Abstract

Water in spherical cavities was simulated in equilibrium with a bulk reservoir. Cavities with radii from 6 to 20 A and various strengths of the water-surface interaction were analyzed. The bulk liquid water was simulated at T = 300 K and P = 1 bar. Chemical equilibration between the bulk and confined water was provided by molecular transfers in the Gibbs ensemble.

A liquid-vapour phase transition in the pore was observed at some threshold value of the water-surface interaction strength: water exists as a vapour in the hydrophobic pores and as a liquid in the hydrophilic pores.

The analysis of the structural properties shows that liquid water may exist in the pore only by forming two strongly orientationally ordered layers near the surface (with a thickness of about 7 A). The structure of the first outer layer is distorted from the tetrahedral structure towards a quasi-planar square-like or hexagonal-like structure. Due to the formation of the water layers the diffusivity of the water in the pore decreases with respect to the bulk.

The presented results indicate the importance of an equilibration with the bulk reservoir for a correct reproduction of the water properties in confining geometries.

Keywords

Confined water, Computer simulation, Chemical equilibria, Gibbs ensemble, Layering.

Introduction

Understanding the effect of confinement on the properties of fluids is of great importance in view of industrial applications of porous materials (Gelb, 1999). A correct equilibration of the confined water with the bulk reservoir is the main problem in computer simulations of water in pores. The neglect of this equilibration and a rather arbitrary choice of the water density in the pore leads to contradictory results on the degree of water layering near the surface, water diffusivity, etc. (Lee, 1994, Harting, 1998).

In the present paper the equilibration of the confined water with the bulk reservoir was done in the Gibbs ensemble (GE). The dependence of the water properties on the water-surface interaction and pore size is analyzed.

Gibbs Ensemble Simulations

512 TIP4P water molecules in a cubic box with periodic boundary conditions were used to reproduce bulk water at P = 1 bar and T = 300 K respectively 350 K. Intermolecular interactions were calculated in the bulk up to a cutoff distance of 12 A and long-range corrections were included for the Lennard-Jones potentials only. Spherical cavities with radii R_C from 6 to 20 A were considered. The water-water interactions in the cavity were calculated without cutoff. The water-substrate interaction was simulated as Lennard-Jones (9-3) potential depending on the distance between the oxygen atom and the substrate. The size parameter σ in this potential was fixed to 2.5 A, the parameter ε varied to change the well-depth U from -0.46 to -5.77 kcal/mol. This range of strength of the water-substrate interaction approximately corresponds to a variation from a hydrophobic to a hydrophilic pore.

The chemical equilibration of the water in the cavity with bulk water was provided by simulations in the Gibbs ensemble (Panagiotopoulos , 1992). New techniques for the molecular transfer were used in order to improve its efficiency. The insertion of molecules was done as follows: a) a position and orientation for insertion was chosen randomly; b) the distances between the atoms of the inserted molecule and the atoms of its neighbors were calculated; c) if at least one of these distances is shorter than some "critical" value, the new configuration is rejected immediately, otherwise energetical calculations follow. The choice of the "critical" interatomic distances was done on the basis of the different pair distribution functions, calculated for water both in the bulk and in the cavity. They correspond to the largest interatomic distances, which are never observed in long-term canonical MC simulations. The efficiency of the molecule deletion was improved by choosing loosely bound water molecules for the transfer. The probability of acceptance of the new configuration was corrected by multiplying it with some factor P_C, which is equal to the probability to find water molecules with energies higher than some chosen "cutoff" value U_C.

One GE simulation step consists of: 1) thermal equilibration with 1000 MC moves (translational and rotational) in both systems and 2 attempts to vary the volume of the bulk phase in order to impose constant pressure; with acceptance probabilities between 40 and 50% in all cases; 2) chemical equilibration of the two systems by the transfer of randomly chosen molecules: a series of $0.5*10^5$ to $1.5*10^6$ attempts yields a probability of a few to 50 percents to get one successful transfer, depending on the state of the systems. If the transfer is accepted, the number of water molecules in the cavity changes accordingly. The transfer from and to the bulk actually is ignored, because it is simulated at fixed P and T.

The dependence of the number of water molecules in the cavity on the well-depth U of the water-substrate interaction for cavities with R_C = 12 A is presented in Fig. 1. The presented data evidence the existence of liquid-vapour phase transition at some critical value of the water-substrate interaction parameter U. Liquid water exists in spherical cavities with R_C = 12 A only if U < -1.5 kcal/mol.

The average density of the water in the cavities depends essentially on the parameter U. It increases by 11%, 12%, 18% and 34% for cavities with R_C = 20 A, R_C = 15 A, R_C = 12 A and R_C = 9 A, respectively, when U changes from -1.93 to -4.62 kcal/mol. A temperature increase from 300 K to 350 K causes a slight decrease of the number of water molecules in the cavity by a few %.

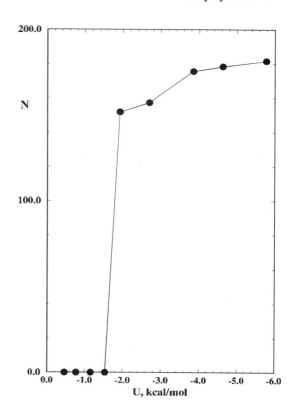

Figure 1.   Dependence of the equilibrium number of water molecules in a spherical cavity with R_C = 12 A on the water-substrate interaction well-depth U (T = 300 K).

Structural and Dynamical Properties of Water in Spherical Cavities

The integer numbers closest to the average values <N> of water molecules were used for the simulation and analysis of the structural and dynamical properties of water. MC simulations were done in the canonical NVT ensemble, structural analysis was performed every 1000 MC steps during runs of $2*10^5$ configurations. In Fig. 2 water density profiles along the cavity radius are presented. The water density was calculated, based on the positions of the O and H atoms. Strong layering of water exists in all systems. Even for the weakest water-substrate interaction for which water still exists in the cavity (U = -1.93 kcal/mol), the water density profile exceeds 2.5 g/cm³ in the first outer layer. This means that water can

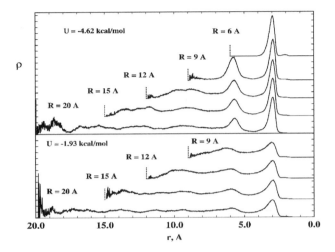

Figure 2. Water density along the cavity radius. One tick on the Y axis corresponds to 1.0 g/cm³ (the curves are shifted vertically). Vertical dashed lines indicate the centers of the cavities.

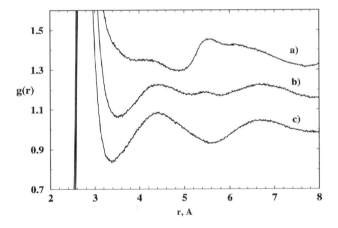

Figure 3. O-O radial distribution functions of water (T = 300 K): a) first, outer layer in the cavity with R_C = 15 A, U = -4.62 kcal/mol; b) first, outer layer in the cavity with R_C = 15 A, U = -1.93 kcal/mol; c) bulk water.

exist in cavities (in equilibrium with the bulk water at the given P and T) only, when prominent layers are formed. For small cavities (R_C = 6 A) only one water layer remains. The analysis of the structural properties of the cavity water was done separately for the different water layers. The layer thicknesses were determined by the distances between the minima in the water density profile.

In Fig. 3 radial distribution functions $g_{OO}(r)$ are presented for the first, outer layers of water in cavities with R_C = 15 A and for the bulk water. The tetrahedral water structure is distorted in the first layer: the shift of the second maximum of $g_{OO}(r)$ by -0.15 A corresponds to a change of the OOO-angle from 107° to 102°. The shift of the third maximum by -0.4 A reflects still stronger distortions of the long-range tetrahedral structure of water in the cavities. The position of the new maximum of $g_{OO}(r)$, appearing at 5.45 A, corresponds to a doubling of the first maximum distance of 2.75 A. All above mentioned features of the $g_{OO}(r)$ show that the tetrahedral structure of water diminishes in cavities and some features of a quasi-planar square or hexagonal lattice structure in the outermost water layer appear. The distortion of the tetrahedral structure increases with strengthening of the water-substrate interaction and decreasing cavity size.

To obtain dynamical properties, additional NVT Molecular Dyanmics simulations with a Berendsen thermostat were made. The isotropic short-time self-diffusion coefficients D were determined from the intermediate linear parts of the mean-square displacements of oxygen atoms as a function of time. The obtained average values are: D = 1.85*10⁻³ m²/s (R_C = 12 A, U = -4.62 kcal/mol) and D = 2.93*10⁻³ m²/s (R_C = 12 A, U = -1.93 kcal/mol). In both cases the self-diffusion coefficient is lower than the bulk value D = 3.61*10⁻³ m²/s, obtained at 300 K with the same interaction potential and at bulk density. The general decrease of water diffusivity in cavities may be noted.

Discussion

Equilibration with a bulk reservoir is the crucial problem in the simulation of water in confined geometries, since the structural and dynamical properties of water are highly sensitive to its density. In this paper we discuss the application of GE simulations for the equilibration of water in spherical cavities with bulk water. The main problem in GE simulations is connected with the necessity to provide the deletion of a molecule in one box and its insertion in another box in one move. For dense and highly associated systems like liquid water, efficiencies of both insertion and deletion are very low. We have improved the efficiency of the insertion by early rejection of the new configuration, based on the analysis of the shortest interatomic distances, and efficiency of the deletion by choosing loosely bound molecules.

This successful application of GE for the equilibration of two dense water systems shows perspectives for its use in simulations of complex aqueous systems. But even for simple aqueous solutions MC simulations in the usual canonical ensemble or MD simulations may not provide accurate equilibration. The system may be easily trapped in some metastable state and, for example, a possible aggregation of the solute molecules will never be observed due to the finite time of calculations. It may also be important in simulations of water in biological membranes, since many possible water distributions are unachievable in the course of usual MC or MD simulations. Water molecule transfers between different parts of the simulated system will provide the necessary chemical equilibration. The Gibbs ensemble may also be extended to simulations of liquid-liquid equilibria of aqueous solutions and, eventually, even of pure water at low temperatures near the conjectured second critical point in supercooled water (Stanley, 1998).

References

Gelb, L.D., K.E. Gubbins, R. Radahakrishnan, and M. Sliwinska-Bartkowiak (1999). Phase separation in confined systems. *Rep. Progr. Phys.*, **62** 1573-1659.

Lee, S.H., and P.J. Rossky (1994). A comparison of the structure and dynamics of liquid water at hydrophobic and hydrophilic surfaces - a molecular dynamics study. *J. Chem. Phys.*, **100(4)**, 3334 -3345.

Harting, C., W. Witshcel, and H. Spohr (1998). Molecular dynamics study of water in cylindrical pores. *J. Phys. Chem. B*, **102**, 1241-1249.

Panagiotopoulos, A.Z. (1992). Direct determination of fluid phase equilibria by simulation in the Gibbs ensemble: a review. *Mol. Simul.*, **9**, 1-23.

Stanley, H.E. et al. (1998). The puzzling statistical physics of liquid water. *Physica. A*, **257**, 213-232.

POSITIONAL AND ORIENTATIONAL CONFIGURATIONAL TEMPERATURES FROM COMPUTER SIMULATION

A. A. Chialvo[1,2], J.M. Simonson[1], P. G. Kusalik[5], and P. T. Cummings[2,3,4,6]

[1]Chemical and Analytical Sciences Division, Oak Ridge National Laboratory, Oak Ridge, TN 37831-6110
[2]Departments of Chemical Engineering, [3]Chemistry, and [4]Computer Science,
University of Tennessee, Knoxville, TN 37996-2200
[5]Department of Chemistry, Dalhousie University, Halifax, Nova Scotia, B3H 4J3, Canada
[6]Chemical Technology Division, Oak Ridge National Laboratory, Oak Ridge, TN 37831-6181

Abstract

We define two configurational temperatures associated with the positional r^N and orientational ω^N portions of the configurational phase space of the molecules in a system according to Hirschfelder's hypervirial theorem, and implement their calculation by Monte Carlo and molecular dynamics simulations. We use a simple quadrupolar Lennard-Jones fluid at normal conditions and the TIP4P model at low temperature to illustrate the use of these quantities to test the correctness of Monte Carlo simulation protocols, and as a new tool in the study of the poorly understood behavior of supercooled water.

Keywords

Configurational temperatures, Monte Carlo simulation, Molecular dynamics.

Introduction

Constants of motion are suitable targets to check the validity and accuracy of the integration procedure in the simulation of dynamical systems. Unfortunately, while the conservation of these quantities is a necessary requirement, it is usually not a sufficient condition of correctness of the algorithm, as many practitioners have found out the hard way. For non-dynamical simulation approaches, such as Monte Carlo techniques, there is a need for additional checks of validity not only for the random translational (Butler et al., 1998), but also for orientational moves (Chialvo et al., 2000). A subtler situation rises when dealing with systems at very low temperature such as water at sub-ambient conditions, for which the configurational/orientational relaxation times are rather long. While the kinetic temperature (the measure of the time-average kinetic energy) might be at the set point, the corresponding average configurational (positional and orientational) temperature might not. Yet,

there are other compelling reasons to be able to assess the temperature of a system by analyzing its average configuration. For instance, in the validation of microstructural information from x-ray, electron and neutron scattering spectra by means of reverse Monte Carlo techniques we require ways to diagnose the correctness of the procedure used in the processing and to test the internal consistency of the raw data (Tóth and Baranyai, 1999). Moreover, many researchers would like to be able to estimate the temperature of a system based on configurational information provided by those spectra (Heinze et al., 1997).

As a first step toward the achievement of those goals, in this paper we revisit an earlier theoretical development, the so-called *hypervirial theorem* developed by Hirschfelder (Hirschfelder, 1960), as a well-suited tool to make possible the molecular simulation route to *positional* and *orientational* configurational-temperatures. After developing the corresponding algorithms, we analyze specific molecular dynamics and Monte Carlo illustrative simulation cases where the two configurational quantities can be used to check either the validity of the simulation methodology, the proper configurational equilibration of the system, and/or the correctness of the simulation codes.

Configurational Temperatures

Starting from the hypervirial expressions (Hirschfelder, 1960),

$$\left\langle \left(\nabla_r \phi\right)^2 \right\rangle = kT \left\langle \nabla_r^2 \phi \right\rangle \tag{1}$$

$$\left\langle \left(\nabla_\omega \phi\right)^2 \right\rangle = kT \left\langle \nabla_\omega^2 \phi \right\rangle \tag{2}$$

where $\phi(r, \omega)$ is the potential depending on distance (r) and orientation (ω), and $\nabla_\omega(...) \equiv l \times \nabla_r(...)$ is the angular gradient operator, such that the torque τ becomes

$$\begin{aligned} \nabla_\omega \phi &\equiv l \times \nabla_r \phi \\ &= -\tau \end{aligned} \tag{3}$$

In equation (3) the operator "$\times$" denotes the cross-product of two vectors, and $l = l_x \hat{i} + l_y \hat{j} + l_z \hat{k}$ is the vector position of molecular sites with respect to the corresponding center of mass. Thus, the divergence of the torque becomes

$$\begin{aligned} \nabla_\omega \bullet \tau &= -\nabla_\omega \bullet \nabla_\omega \phi \\ &= -\nabla_\omega^2 \phi \end{aligned} \tag{4}$$

where the operator "$\bullet$" denotes a scalar product between two vector. Consequently, the configurational temperature associated with the orientational degrees of freedom (ω^N) reads,

$$\begin{aligned} kT &= \left\langle \tau^2 \right\rangle / \left\langle \nabla_\omega^2 \phi \right\rangle \\ &= -\left\langle \tau^2 \right\rangle / \left\langle \nabla_\omega \bullet \tau \right\rangle \\ &= -\left\langle \tau^2 \right\rangle / \left\langle \left(l \times \nabla_r\right) \bullet \left(l \times \nabla_r \phi\right) \right\rangle \end{aligned} \tag{5}$$

Note that the expression for the configurational temperature associated with the translational degrees of freedom (r^N), Eqn. (1), is similar to that derived through rather different approaches by Rugh (Rugh, 1998, Butler et al., 1998), and Baranyai (Baranyai, 2000).

Simulation Results and Discussion

We have performed NVT-MC simulations of a simple quadrupolar Lennard-Jones fluid (Reed and Gubbins,

1991), and NVT-MD simulations of supercooled TIP4P water (Jorgensen, 1981). The detailed account of the methodology and findings of this ongoing investigation will be presented elsewhere (Chialvo et al., 2000).

For illustration purposes, we present in Figures 1-2 the comparison between the kinetic and the two configurational temperatures resulting from the NVT-MD simulation of supercooled and ambient water. This comparison highlights an interesting (and, so far not considered) simulation scenario, i.e., one in which the average system configuration does not agree with the expected value from the kinetic temperature. In other words, even after several nanoseconds of simulation trajectory, the energy equipartition is not satisfied. Note that as the system (kinetic) temperature is increased, the two configurational temperatures approach the corresponding kinetic counterpart.

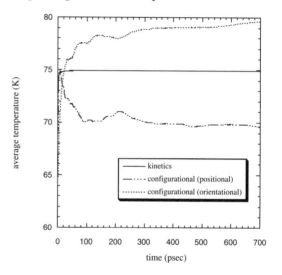

Figure 1. Configurational versus kinetic temperatures for supercooled TIP4P water at T=75K and ρ=1.0g/cc.

The TIP4P water sample at 75 K (Fig. 1) is in a highly metastable (glassy) state. Thermodynamic (energy) and kinetic (diffusion constant) data suggest that it is well below its glass transistion temperature, esitmated to be ~150 K. Structural analysis of the system reveals no indications of crystalline order.

Likewise, in Figure 3 we display the translational and orientational configurational temperatures from NVT-MC simulations of a quadrupolar Lennard-Jones fluid, and their response to an instantaneous change in the set-point simulation temperature. The corresponding responses for the pressure and configurational energies are given in Figure 4.

These pictures clearly indicate the difference between the structural relaxation associated with the translational and orientational degrees of freedom. While the translational configurational temperature adjusts almost instantaneously to the step function change, the orientational counterpart exhibits a rather long asymptotic relaxation toward the set point temperature. This behavior, in turn, affects the corresponding responses of the pressure and configurational energy as shown in Figure 4. These illustrations suggest the need for further investigation in the behavior of the configurational temperatures and their role in the thermophysical behavior of fluids at low temperature and high density.

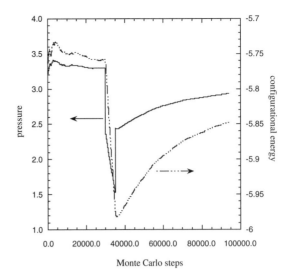

Figure 4. Response of the configurational energy and the pressure (in units of ε and σ) to a step function change (from 1.227 to 0.2 and back to 1.227) on the set point temperature for the same fluid as in Figure 3.

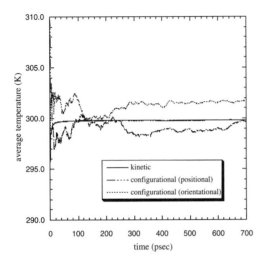

Figure 2. Configurational versus kinetic temperatures for ambient TIP4P water at T=300K and ρ=1.0g/cc.

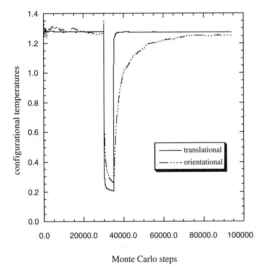

Figure 3. Response of the configurational temperatures (in units of ε / k) to a step function change (from 1.227 to 0.2 and back to 1.227) on the set point temperature for a Lennard-Jones quadrupolar fluid with Q*=0.5 at T*=1.227 and ρ*=0.85.

Acknowledgements

This research was sponsored by the Division of Chemical Sciences, Geosciences, and Biosciences, Office of Basic Energy Sciences, U.S. Department of Energy, under contract DE-AC05-00OR22725 with Oak Ridge National Laboratory, managed and operated by UT-Battelle, LLC. PTC was supported by the Division of Chemical Sciences, Geosciences, and Biosciences, Office of Basic Energy Sciences, U.S. Department of Energy. PGK acknowledges the financial support of the Natural Sciences and Engineering Research Council of Canada. AAC acknowledges fruitful discussions with Dr. Andras Baranyai during his visit to UTK (July 1999).

References

Baranyai, A. (2000). *J. Chem. Phys.,* **112**, 3964-3966.

Butler, B. D., Ayton, G., Jepps, O. G. and D. J. Evans (1998). *J. Chem. Phys.,* **109**, 6519-6522.

Chialvo, A. A., Simonson, J. M., Kusalik, P. G. and P. T. Cummings (2000). *In preparation,* .

Heinze, H., Borrmann, P., Stamerjohanns, H. and E. R. Hilf (1997). *Z. Phys. D,* **40**, 190-193.

Hirschfelder, J. O. (1960). *J. Chem. Phys.,* **33**, 1462-1466.

Jorgensen, W. L. (1981). *J. Ame. Chem. Soc.,* **103**, 335-340.

Reed, T. M. and K. E. Gubbins (1991). *Applied Statistical Mechanics.* Butterworth, Stoneham.

Rugh, H. H. (1998). *J. Phys. A: Math. Gen.,* **31**, 7761-7770.

Tóth, G. and A. Baranyai (1999). *Molecular Phys.,* **97**, 339-346.

π-HYDROGEN BONDING OF INDOLE AND WATER

Ken R. F. Somers, Luc G. Vanquickenborne, and Arnout Ceulemans
Katholieke Universiteit Leuven
3000 Leuven, Belgium

Abstract

The results of a quantum chemical study on the π-hydrogen bonding between water and indole are compared with a molecular dynamics study. The quantum chemical calculations were based on density functional theory (DFT) and the Becke3LYP functional, with a 6-31+G* basis set. Molecular dynamics is performed using the standard Amber 5.0 force field. Both methods predict the same hydrogen and oxygen distances with respect to the indole plane. Different results are obtained when comparing the preferential, DFT locations, with the most frequented molecular dynamics sites. The DFT results select the five-membered ring as the more probable place for π-hydrogen bonding with water, while the six-membered ring is more populated in the molecular dynamics results. Both simulations prefer opposite sites on the five-membered ring, the molecular dynamics simulation preferring the nitrogen site. We conclude that molecular simulation can be used as an indicator for possible π-hydrogen bonds with indole, but does not offer a reliable prediction of the positions of the bounded waters.

Keywords

Indole, Tryptophan, π-hydrogen bonding, Quantum chemical, Molecular dynamics, Amber.

Introduction

Tryptophan is one of the most widely used spectroscopic probes in protein chemistry. The sensitivity of both its ultraviolet spectrum and fluorescence lifetime to the local environment makes it a useful source of information. Both protein structure and solvent accessibility can be assessed, explaining its frequent use by biochemists. Theoretical work on tryptophan is difficult due to its large conformational freedom. The internal interactions between the aromatic chromophore and the peptide fragment change the spectroscopic properties. To circumvent this problem the present study will be limited to the aromatic chromophore only.

The indole molecule, C_8NH_7, corresponds to the aromatic chromophore of tryptophan, it is responsible for tryptophans specific spectroscopic properties. Both experimentalists and theoreticians show a wide interest in its photophysics and electronic spectroscopy (Carney, 1998; Muiño, 1994; Simonson, 1997) Most theoretical studies did not attempt to incorporate interactions with other molecules. Recent experimental and theoretical work by Carney (1999) on indole and water in the gas phase confirms the existence of a π-hydrogen bond between the five-membered ring of 1-methyl-indole and water. A similar π-hydrogen bond between indole and water was calculated. Carney also claims the theoretical existence of a less preferable π-hydrogen bond between the six-membered ring of indole and water. The interaction between indole and two water molecules creates a third type of π-hydrogen bond, with the π-bonded water molecule standing above the plane, aside of the indole molecule.

Using these results as a starting point a thorough search of the indole-(water)$_2$ conformation space, using the DFT method, was executed, with special focus on the π-hydrogen bonded waters. In addition we investigated the behavior of indole in a box of water using molecular dynamics. It was our purpose to verify the existence behavior of π-hydrogen bonds in the molecular dynamics package Amber 5.0.

Computational Methods

Quantum Chemical

Density functional theory calculations of the structures, dipole moments, binding energies and harmonic vibrational frequencies were carried out in Gaussian 98 (Frisch, 1998) using the Becke3LYP functional with a 6-31+G* basis set.

Optimizations of the indole water complexes were made for a wide variety of starting geometries, based on former results and completed by a scanning procedure. Vibrational frequencies were calculated to confirm the location of true minima and to distinguish between the structures.

Molecular Dynamics

Molecular dynamics simulations were performed with the program package Amber 5.0.(Case, 1997). A waterbox (42.7 Å, 41.3 Å, 37.2 Å), with periodic boundary conditions, containing 1510 water molecules was created. For all water molecules, the TIP3P water molecule was used, with an oxygen charge of -0.834 e.

Indole was formed starting from tryptophan as provided by Amber 5.0. We choose not to modify the parameters of the newly created indole, so to keep a maximal resemblance between the indole and the tryptophan properties, especially concerning the π-hydrogen bonding. The added hydrogen was of the amber type "H". It was necessary to define one bond, H-C* (367.0 kcal/mol ; 1.080 Å) and two angles, H-C*-CW (35.0 kcal/mol ; 120.00°) and H-C*-CB (35.0 kcal/mol ; 120.00°).

The extra charge introduced by replacing the peptide part by a hydrogen is averaged out over the entire molecule, as to keep the specific tryptophan charge pattern.

The structures were equilibrated using the sander module of Amber 5.0, while constraining bond lengths with the SHAKE algoritme. The equilibrations were run at a constant temperature of 300K, using a pressure of 1 bar. A cutoff radius of 8.0 Å was used, with a pairlist update every 20 steps. Equilibration was executed using 200,000 steps of 5 fs. The next run took 200,000 steps of 2 fs during which every 100 steps samples were taken.

Results and Analysis

Counting of atoms starts at the nitrogen, moves over the five-membered ring, to continue on the next non-bridged atom of the six-membered ring and to end at the two bridged atoms. Only the nitrogen and the carbon atoms will be considered as they contain the information about the π-electron cloud. All indole molecules are placed in the XY-plane, with atom 8 in the origin. The nitrogen, atom 1, defines the positive Y-direction and atom 9 lays on the X-axis, defining the positive X-direction.

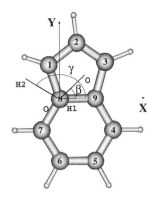

Figure 1.   Position of indole in the XYZ-space.

The water molecules are treated liked fixed triangles, only leaving six degrees of freedom to the indole water system. All waters can therefore be described by six parameters corresponding to the position of oxygen in the XYZ-space and three angles between the water molecule and the indole molecule. The two hydrogen atoms of the water distinguished as "H1" for the hydrogen most closely to the indole-plane and "H2" for the other. The angle between H1-O and the indole plane is called "α" and the angle between the X-axis and the projection of H1-O is called "β". Leaving "γ" to be defined as the angle between the projection of O-H2 in the indole-plane and the X-axis.

Quantum Chemical

An extensive search of the conformation space resulted in 13 stable minima (24 when including mirror images), three indole-(water)$_1$ complexes and 10 indole-(water)$_2$ complexes. Of those complexes only the water molecules that formed direct π-hydrogen bonds were selected, leaving 17 indole-water complexes (34 when including mirror images).

The two remaining indole-(water)$_1$ complexes are formed by π-hydrogen bonds between water and the five- or six-membered ring and have binding energies of 2.5 kcal/mol and 2.3 kcal/mol respectively (Somers, 2000). These results are in agreement with Carney (1999), who found binding energies of 2.60 kcal/mol for the former and 2.45 kcal/mol for the latter. Slightly higher values, 2.9 kcal/mol, were obtained by Feller (1999) for the binding energies between benzene and water.

The most stable indole-(water)$_2$ complexes have the water dimer acting as a bridge between indole's N-H and the π-electron cloud (indole N-H···O$_1$-H···O$_2$-H···π-indole), thus placing the waters outside the borders defined by the carbon skelton.

Projection of the 17 water molecules makes it possible to distinguish three groups of water complexes: the waters above the indole carbon skeleton, bound to the five- or six-membered ring and the waters outside the borders defined by the carbon skeleton.

Fig. 2 and Fig.3 will have carbon and nitrogen positions of indole indicated by black dots.

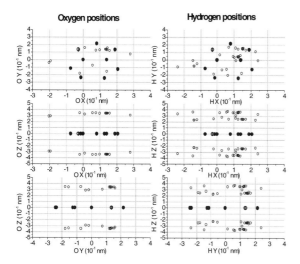

Figure 2.   Projections of quantum chemical results for both oxygens and hydrogens of water.

The position of those 17 waters is summarized in Table 1.

Table 1.   Positions of Water Molecules.

O X(Å)	O Y(Å)	O Z(Å)	α (°)	β (°)	γ (°)
2.806	0.749	3.063	50.3	-31.9	-114.4
1.570	1.201	3.425	80.6	-137.8	-129.4
1.438	1.540	3.406	83.7	-168.2	-92.8
1.422	1.315	3.423	79.6	-131.1	-99.3
1.405	-1.558	3.395	76.8	102.4	-147.2
1.389	1.398	3.413	80.7	-151.7	-96.7
1.343	1.383	3.399	78.9	-140.9	-95.8
1.335	1.366	3.413	79.1	-149.3	-96.8
1.324	1.367	3.414	78.9	-150.3	-96.8
0.982	1.293	3.382	71.9	-156.2	-71.1
0.764	1.273	3.477	61.2	-138.3	-74.2
0.562	-0.723	3.469	46.4	66.7	48.5
0.105	1.607	3.246	51.8	-171.6	-86.4
-0.198	-1.790	3.465	83.3	88.0	35.3
-1.934	-0.187	2.937	36.2	153.7	-119.9
-2.017	-0.385	2.903	39.5	144.2	28.8
-2.020	-0.396	2.897	39.4	144.1	29.5

Molecular Dynamics

Our specific interest in π-hydrogen bonds made it necessary for analysis to restrict the space to a box just surrounding indole. One more restriction could be applied: only water molecules with a hydrogen closer to the indole plane than the oxygen were retained. The box was divided into subboxes, the total number of oxygens in each of these boxes was determined as to determine the water density in each of these boxes. These values were used to create Fig. 3, using a proportional scaling between the density and the radii of the circles.

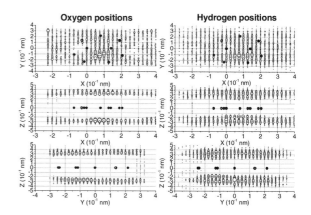

Figure 3.   Projections of density of molecular dynamics results for both oxygens and hydrogens of water.

Parameters were determined for all the remaining water molecules and plotted as histograms, Fig. 4.

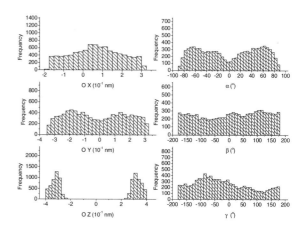

Figure 4.   Histogram of molecular dynamics parameters.

Combination of Fig. 3 and Fig. 4 leads to the definition of three different zones of interaction. The first zone lies above the six-membered ring and is strongly preferred, the second zone lies above the five-membered ring and is less preferred and the third zone lies in the pockets between five- and six-membered ring of the indole molecule and is deprived of water molecules.

The preferred α-values are situated around -65° and +65°. The γ-values show a preference for angles near -75°. Closer inspection of the β- and γ-values showed that there exists a distinct correlation between both of them: $\beta = \gamma \pm 80°$. These values clearly define the preferential position of the water molecules.

Discussion

It must be noted that the distribution of waters in the Z direction reaches a close match when comparing both techniques. The positive interval described by the quantum chemical results is limited by 2.9 Å and 3.5 Å.

The same results are obtained for the molecular dynamics simulation with a lower limit of 2.75 Å and a steep decrease in frequency of appearance when reaching 4.00 Å and a maximum frequency at 3.25 Å. We may therefore conclude that a molecular dynamics simulation will provide useful information concerning the distance between the water molecules and the indole-plane.

The water molecules in the molecular dynamics simulation did prefer the six-membered ring. The preference for the six-membered ring is probably an artifact of the parameterization of tryptophan, the peptide fragment will partly prohibit the formation of the π-hydrogen bond with the five-membered ring.

As far as the ring positions are concerned molecular dynamics and quantum chemical results are in conflict. Quantum mechanical calculations predict that, at room temperature, the most preferable water positions are outside the carbon skeleton, forming a hydrogen bond with the π-electron cloud and sharing the oxygen with a water that forms hydrogen bound with the hydrogen bounded to the nitrogen. Less preferable are water positions with a tendency to form hydrogen bond with the five-membered ring. The least stable hydrogen bonds are complexes with water binding to the six-membered ring.

A third zone of interest is located in the pockets between the five- and six-membered rings. We believe that the relative absence of oxygen atoms in this zone is due a steric effect. The water molecules situated on the five- and six-membered ring will both push away the waters in the bulk, thus creating a water depleted zone on the intersection of the five- and six-membered rings.

The strong correlation between β and γ in the molecular dynamics run, can be found in five of the 16 quantum chemically calculated structures, with three of them also having α, β and γ equal to the values determined by molecular dynamics. The other two structures have either α or γ being similar to the molecular dynamics angles. Those five structures contain representatives from each of the three groups found using DFT. The only water indole interaction not represented is the interaction of a water outside the carbon skeleton with the six-membered ring. We may therefore conclude that two techniques are in good agreement concerning the intermolecular orientation of the water molecules and indole, where it not for the second correlation function between β and γ, which is not represented in the DFT results.

Conclusions

Our results lead us to conclude that Amber5.0 can be used to predict the presence of water molecules around the indole moiety of a protein. Neither the exact water location nor the preferential water location can be predicted by Amber5.0. Thus making quantum chemical calculations still necessary to predict the appropriate position of the water molecule towards the indole plane.

Tryptophan is an amino acid often found in structure stabilizing sites. The interaction between tryptophan, more specific the indole part of tryptophan, and the surroundings does often play an important role in the stability of the protein. It is therefore necessary to know where tryptophan will be localized and how it interacts with the surroundings. Our results show that the π-hydrogen bonding sites, which are preferred, are localized around the five-membered ring. A result which is not simulated correctly by the Amber5.0 package. Therefore care has to be taken when using Amber5.0 to predict the fine structure of proteins around tryptophan.

Acknowledgments

This investigation has been supported by a grant from the Institute for the Promotion of Innovation by Science and Technology in Flanders (IWT).

References

Carney, J. R., Hagemeister, F. C., and Zwier, T. S. (1998). The hydrogen-bonding topologies of indole-(water)$_n$ clusters from resonant ion-dip infrared spectroscopy. *J. Chem. Phys.*, **108**, 3379-3382.

Carney, J. R., and Zwier, T. S. (1999). Infrared and ultraviolet spectroscopy of water-containing clusters of indole, 1-methylindole, and 3-methylindole. *J. Phys. Chem.*, **103**, 9943-9957.

Case, D. A., et al. (1997). *AMBER 5*. University of California, San Francisco.

Feller, D. (1999). Strength of the benzene-water hydrogen bond. *J. Phys. Chem.*, **103**, 7558-7561.

Frisch, M. J., *et al.* (1998). *Gaussian 98*, Revision A.7. Gaussian Inc., Pittsburgh PA.

Muiño, P. L., and Callis, P. R. (1994). Hybrid simulations of solvation effects on electronic spectra: indoles in water. *J. Chem. Phys.*, **100**, 4093-4109.

Simonson, T., Wong, C. F., and Brünger, A. T. (1997). Classical and quantum simulations of tryptophan in solution. *J. Phys Chem.*, **101**, 1935-1945.

Somers, K. R. F., Kryachko, E. S., and Ceulemans, A. (2000). Theoretical study of indole in the ground and excited states. *J. Phys. Chem. A* (to be submitted).

AB INITIO MOLECULAR DYNAMICS SIMULATIONS OF AN OH RADICAL / OH ANION WITH A SOLVATED FORMALDEHYDE MOLECULE

Hideaki Takahashi, Takumi Hori, Tadafumi Wakabayashi, and Tomoshige Nitta
Division of Chemical Engineering
Osaka University
Toyonaka Osaka 560-8531, Japan

Abstract

In this presentation, we employ the ab initio molecular dynamics(AIMD) method to simulate the reaction of a formaldehyde molecule with OH radical or OH⁻ ion in the supercritical water solvent. We have developed a novel code for the AIMD simulation employing the real space grid method. The aim of the calculation is to examine which reaction path is dominant between the radical and ionic processes in the supercritical water. The AIMD simulation for the system of the OH radical and the formaldehyde molecule, in which no water solvent is involved, has shown that the hydrogen abstraction reaction takes place without potential energy barriers within the level of local density approximation(LDA) and generalized gradient approximation(GGA). The calculated heat of reaction is about 33 kcal/mol, which is in good agreement with the experimental value of 33.2 kcal/mol. In the case of the reaction of the OH anion and the formaldehyde molecule, the formation of the molecular complex, (OH-HCHO)⁻ has been verified, which will lead to the second stage of the Cannizzaro reaction via H⁻ ion transfer. It has also been proven that the molecular complex is generated with no energy barriers. The effect of the solvent water molecules on the reaction process of the OH radical or OH anion with the formaldehyde molecule will be examined by the first principles simulation coupled with the molecular mechanical method, in which solvent water molecules are treated classically.

Keywords

Ab initio molecular dynamics, Density functional theory, Supercritical water, Formaldehyde.

Introduction

The ab initio electronic structure calculation is becoming a fundamental method in recent years in the studies of chemistry and physics supported by the rapid growth of computational ability. The density functional theory (DFT) approach coupled with the pseudopotential and planewave basis is one of the most successful methods of the ab initio calculations for the periodic systems. In recent years, a number of works that employ real space grids have been proposed to expand electronic wavefunctions by Chelikowsky(1994). The real space grid approach has several advantages compared with the planewave basis approach. First of all, it can treat non-periodic system such as clusters or charged systems as

well as semi-periodic systems. Within the plane wave basis approach, it is not trivial to get rid of the periodicity. The second is that one can drastically reduce the number of fast Fourier transformations(FFT) that are required for each occupied molecular orbitals because only total electron density must be transformed by FFT and it becomes more amenable to parallel implementations.

Despite the many advantages of the real space grid method, its applications to the investigations of physical or chemical problems can be found scarcely. In this work, we apply the real space grid method to the simulations of chemical reactions by the ab initio molecular dynamics(

AIMD) method pioneered by Car and Parrinello(1985). The AIMD approach enables us to simulate the chemical reactions that involve the bond formations or breakings during the course of the molecular dynamics because the forces acted on the nuclei are explicitly determined from the electronic density of the ground state.

The major purposes of the present real space ab initio molecular dynamics (RS-AIMD) simulation are to examine the reaction of OH radical and OH anion with an organic molecule to get insight into their reactivity under the supercritical conditions of water and further to test the applicability of the real space grid method to the AIMD simulations. Supercritical water(SCW) is well known as a strong reaction field and there has been so many literatures that deal with the organic chemical reactions in water near or above its critical point. It is reported that HCOOH is produced from formaldehyde in hot water without a catalyst. In this paper, we consider following radical and ionic reaction processes of formaldehyde, which both lead to the same product HCOOH.

radical process

$$CHO^{\bullet} + OH^{\bullet} \rightarrow HCOOH .$$

ionic process

$$HCHO + OH^{-} \rightarrow [HCHO \cdots OH]^{-}$$

$$HCHO + [HCHO \cdots OH]^{-} \rightarrow$$
$$HCHO^{-} + HCOOH .$$

The first process is composed of the electrophilic attacks of OH radical. The ionic process is known as a kind of Cannizzaro reaction, the first step of which is the nucleophilic addition of OH ion.

In this work we focus on the first step of each reaction to clarify the difference in the dynamics, energetics and reactivity between radical and ionic process and to get more insight into the origin of the high reactivity of the SCW.

Methods

The electronic ground state of the system is computed by the Kohn-Sham density functional scheme. The Kohn-Sham equation for the orbital φ_i with eigenvalue ε_i of the system can be written as

$$\left[-\frac{1}{2}\nabla^2 + \int \frac{\rho(\mathbf{r})}{|\mathbf{r}-\mathbf{r}'|}d\mathbf{r}' + v_{ps}(\mathbf{r}) + v_{xc}(\mathbf{r}) \right]\psi_i$$
$$= \varepsilon_i \psi_i$$

where the first term inside the parenthesis is the kinetic energy operator, the second is the Hartree potential, the third is the pseudopotentials of the atoms, and the fourth is the exchange and the correlation functional, respectively. In our calculations we use real space grids to

expand wavefunctions and employ fourth-order finite difference method to represent the kinetic energy operator. The Hartree potential is computed in the Fourier space. The non-local atomic pseudopotentials are separated into a local potential and the Kleinman-Bylander non-local form in a real space. As a special treatment of the non-local pseudopotentials, we employ the double grid technique that enables us to realize the rapid behavior of the pseudopotentials near the atomic core regions. The most important feature of the present double grid technique is that it requires only modest increase of computational cost. This is because the wavefunctions on the dense grid points are simply computed by analytical interpolations of the wavefunctions on the coarse grid points as a benefit of the smooth nature of the pseudowavefunctions. The dense grid spacing is taken as one-fifth of the coarse grid spacing h in this calculation. The h is set at 0.314 au, which is equivalent to cut off energy of 50 au in the planewave calculations. The electron exchange and correlation functional is evaluated based on the local density approximation(LDA) and the equation used in

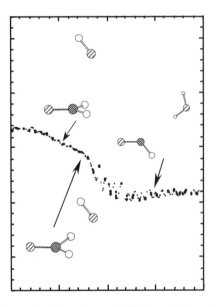

Figure 1. Reaction of HCHO + OH radical.

the present calculation is that developed by Perdew and Zunger. In order to examine the influence of the gradient corrections of the electron density on the potential energy, the BLYP functional is also employed. In the course of the SCF procedure the wavefunctions are updated by the steepest descent(SD) algorithm. After every SCF step, Gram-Schmidt procedure is performed so that the orbitals $\{\varphi_i\}$ obey the orthogonalization constraints,

The MD simulations are performed by velocity-Verlet algorithm with the time step of 20 au (0.484 fs). The temperature of the system is associated with the nuclear kinetic energy and the constant temperature simulation is achieved by rescaling the velocity after every 50 MD steps. The random values on the Gaussian distribution are assigned to the initial velocities of the nuclei and they are scaled so that T becomes 300 K. Calculations are done using a cubic periodic simulation cell of the size L = 20.1 au.

Results and Discussions

HCHO and OH Radical

We first performed the simulation of the radical reaction in order to investigate the reaction dynamics. The time profile of the potential energy obtained in this RS-AIMD simulation are drawn in figure 1, in which molecular configurations after 40, 80, and 150 fs are also depicted. After 40 fs, the OH radical is substantially tilted because the negative charge on the C=O bond of HCHO attract the positively charged hydrogen of OH radical. At about t=80 fs, one of the hydrogen of HCHO is approaching the OH radical as shown by the broken line and one of the C-H bond of HCHO is stretching. Hydrogen abstraction reaction from HCHO to OH radical takes place after about 90 fs. After 150 fs the strained H_2O molecule formed in the reaction is relaxed to its stable geometry. The potential energy of the product is estimated by averaging the energies of the last 50 MD steps, from which the reaction heat of this process is found to be 35.0 kcal/mol. The reaction energy determined by experiment is 33.2 kcal/mol, which is slightly smaller than that obtained by this work. To examine the effect of the electron exchange and correlation functional on the heat of reaction, additional calculation was performed employing the BLYP functional that takes into consideration of the density gradient corrections. The heat of reaction computed by the BLYP functional is 33.0 kcal/mol, which is much closer to the experimental value than that of LDA. The reaction dynamics at the BLYP level does not qualitatively differ from that at the LDA results.

HCHO and OH Anion

In the same way as the radical process we have studied the dynamics of the nucleophilic addition of the OH anion to formaldehyde starting from the same molecular configuration and velocities. The time profile of the potential energy and the snapshots of molecular dynamics after 30, 70, and 150 fs are shown in figure 2. After about 30 fs, the OH anion is tilted in the same way as in the case of OH radical due to the electrostatic attractive force between the negative charge on the carbonyl group and the positive charge on the hydrogen

of OH anion. During the next 40 fs, the HCHO molecule begins to distort from planar to tetrahedral conformation, where the OH anion is approaching the carbon atom of HCHO. The OH bond moves so that the direction of the OH bond coincides with the CO bond direction in order to gain the electrostatic stabilization energy. After 150 fs, we obtain the relaxed tetrahedral complex formed by addition of OH anion to HCHO. The heat of reaction for the ionic process computed in the same way as the radical process is 46.6 kcal/mol within the LDA, while that obtained by the BLYP functional is 29.1 kcal/mol. It is surprising that the inclusion of the gradient corrections greatly diminishes the heat of reaction for the ionic process.

For the investigation of the solvation effects of the water molecules, we have to take substantial number of water molecules into consideration. Recently we have developed a coupled approach, where the reactants are treated quantum mechanically while the solvent molecules are represented by classical molecules with point charges.

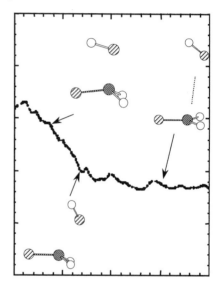

Figure 2. Reaction of HCHO + OH anion.

References

Chelikowsky, J. R., N. Troullier, and Y. Saad (1994). Finite-difference-pseudopotential method: electronic structure calculations without a basis. *Phys. Rev. Lett.*, **72**, 1240-1243.

Car, R., and M. Parrinello (1985). Unified approach for molecular dynamics and density-functional theory. *Phys. Rev. Lett.*, **55**, 2471-2474.

BINDING FREE ENERGY CALCULATIONS FOR BENZAMIDINE-TRYPSIN COMPLEXES

Yiannis Kaznessis[1,2], Lakshmi Narasimhan[2], and Mark Snow[2]

[1]Department of Chemical Engineering
University of Michigan
Ann Arbor, MI 48109
[2]Parke-Davis Pharmaceutical Research
Ann Arbor, MI 48105

Abstract

We employ the linear interaction energy (LIE) approximation to examine the inhibitory activity of a series of substituted benzamidines against the serine protease trypsin. The fundamental premise of the LIE approximation is that the binding free energy of ligands to proteins can be calculated as a linear combination of ligand environment interaction differences between the bound and unbound states. Monte Carlo (MC) sampling is employed to calculate the ensemble averaged electrostatic and van der Waals interactions between the benzamidines and their respective environments in the two physical states. These descriptors are then used to calculate the free energy of binding. The results demonstrate the scientific merit of the approximation. Perhaps more importantly, the results allow the assessment of the combination of LIE and MC simulations as an additional tool in computer-assisted drug design.

Keywords

Binding affinity, Linear interaction energy approximation, Computer-assisted drug design.

Introduction

The most prominent pharmacological response model consists of the specific binding of drugs to target receptors. Based on this model, the first stages of the rational design of new therapeutic agents can make extensive use of computational modeling methods. These methods allow the study of phenomena occurring at molecular time and length scales and can shed light onto the nature of ligand (drug)- receptor interactions.

Aqvist et al. (1994) proposed a semi-empirical computational approach for calculating relative and absolute binding free energies. The method, termed the linear interaction energy (LIE) approximation, postulates that the binding affinity can be calculated as a linear combination of ligand environment interaction differences between the bound and unbound states:

$$\Delta G_{bind} = \alpha \left(\left\langle V_{l-s}^{vdW} \right\rangle_{bound} - \left\langle V_{l-s}^{vdW} \right\rangle_{free} \right) + \beta \left(\left\langle V_{l-s}^{elec} \right\rangle_{bound} - \left\langle V_{l-s}^{elec} \right\rangle_{free} \right) \quad (1)$$

where $\left\langle V_{l-s}^{vdW} \right\rangle_{free}$ and $\left\langle V_{l-s}^{elec} \right\rangle_{free}$ are the van der Waals and electrostatic interaction energies between the ligand and the solvent, and $\left\langle V_{l-s}^{vdW} \right\rangle_{bound}$ and $\left\langle V_{l-s}^{elec} \right\rangle_{bound}$ are the van der Waals and electrostatic interaction energies between the ligand and the solvent-protein environment. A theoretical approach based on assuming a linear response behavior of the solvent with respect to electrostatic forces suggests a value of $\beta=0.5$ (Aqvist, 1994). α is an adjustable empirical parameter. It has been determined, through the study of different protein-ligand complexes, that there is no universal value for the coefficients (Aqvist, 1994, 1995, Wang, 1999, 1999b, Smith, 1998). Moreover, Jones-Hertzog et al. (1997) and Smith et al. (1998) suggested that the cost of creating a cavity in the solvent, for the inhibitor to be solvated, should be accounted for by adding a third term in Eq. 1. This leads to Eq. 2.

$$\Delta G_{bind} = \alpha \left(\left\langle V_{1-s}^{vdW} \right\rangle_{bound} - \left\langle V_{1-s}^{vdW} \right\rangle_{free} \right) + \beta \left(\left\langle V_{1-s}^{elec} \right\rangle_{bound} - \left\langle V_{1-s}^{elec} \right\rangle_{free} \right) + \gamma \left(\left\langle SASA \right\rangle_{bound} - \left\langle SASA \right\rangle_{free} \right) \quad (2)$$

where SASA is the solvent-accessible surface area. Detailed accounts on the LIE approximation and the derivation of Eq. 1 and Eq. 2 have been given elsewhere (Aqvist, 1995, Smith, 1998, Wang, 1999).

The brackets in Eq. 2 indicate ensemble averages. These interaction energy ensemble averages can be calculated from the trajectories of molecular dynamics (MD) simulations or from the configurations produced by a Monte Carlo (MC) sampling scheme. In principle, two sets of simulations are required for each ligand, one for each physical state (bound and free ligand).

In this paper we present the results of Monte Carlo simulations of trypsin complexes with a number of benzamidine-based inhibitors. Trypsin is a member of the serine proteases, a family of digestive enzymes that catalyze the hydrolysis of peptidic bonds (Walsh, 1979). The primary goal of our study is the evaluation of the methodology and its incorporation in the panoply of structure-based drug design tools. The systems were chosen for the reliability of experimentally calculated binding affinities, and the availability of non-proprietary, high quality crystal structures of the enzyme-inhibitor complexes.

Simulations Methodology

The program Monte Carlo Simulations for Proteins and Peptides (MCPRO) (Jorgensen, 2000) was employed for the simulations, in conjunction with the OPLS all-atom (OPLS-AA) force field (Jorgensen, 1996). The methodology has been described in detail elsewhere (Smith, 1998). Here we present the outline of the simulations protocol. In Fig. 1, the 10 benzamidine–based inhibitors used in this study are presented.

For each of the inhibitors, two simulations are performed at 25 °C and 1 atm. Initially, single inhibitor molecules are simulated in aqueous solution. The molecule is solvated in a 22 Å water droplet (no periodic boundary conditions are used), centered at the center of mass of the inhibitor. Schematically, this is shown in Fig. 2a). In the second simulation the protein-inhibitor complex is hydrated with a 22 Å sphere of water, centered again at the center of mass of the inhibitor, as depicted in Fig. 2b). For the simulation of each physical state, some details are as follows.

Free Inhibitors

Since OPLS-AA parameters are not available for all of the functional groups, PM3 quantum mechanical calculations are performed using the CM1P procedure (Storer, 1995). All of the inhibitors carry a charge of +1.00. The inhibitors are considered to be fully flexible,

whereas the TIP4P water (Jorgensen, 1983) molecules undergo only rigid-body translations. A Metropolis Monte Carlo scheme is employed to generate configurations. 4×10^6 equilibration configurations are followed by sampling 10×10^6 configurations, where averages of the inhibitor/solvent interaction terms are accumulated. The parameters for the attempted moves were such that led to an acceptance ratio of approximately 40%.

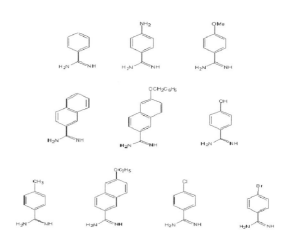

Figure 1. Benzamidine-based trypsin inhibitors. From top-to-bottom, left-to right: benzamidine, p-aminobenzamidine, p-methoxybenzamidine, p-naphthamidine, 6-benzyloxy2-naphthamidine, 4-hydroxybenzamidine, 4-methylbenzamidine, 6-ethoxy,2-naphthamidine, p-chlorobenzamidine, 4-bromobenzamidine.

Trypsin Inhibitor Complex

The crystal structure of the trypsin-benzamidine complex was obtained from the Protein Data Bank (Bode, 1975) (PDB entry 3ptb). Based on the atomic coordinates of the benzamidine, the rest of the substituted benzamidines were positioned inside the active pocket of the enzyme by superimposing common atom positions. The protein-inhibitor complex was then solvated, maintaining the crystallographic waters. The backbone of the protein was kept frozen during the simulations. The internal coordinates of side chains that were within 15 Å of the inhibitor were sampled. The simulations consisted of 5×10^6 equilibration steps followed by 5×10^7 production run configurations. The acceptance ratio of attempted moves was around 25%.

For a meaningful comparison with the unbound inhibitor simulations, the environment of the benzamidines should be electrically neutral. At physiological pH, trypsin has a global charge of +8 (considering the histidines to be neutral). In principle, one could add counterions to neutralize the system appropriately. However, the size of the system does not allow for equilibration of the ion positions in an acceptable amount of time. Hence, a simple protocol was chosen to neutralize the protein: all the residues were kept neutral, except for the ones that participate in salt bridges with each other and the two glutamic acids that ligated with the calcium ion present in the crystal structure. Detailed accounts on the legitimacy of such approaches have been given elsewhere (Wang, 1999). We further discuss the implications of such an approach in the next section.

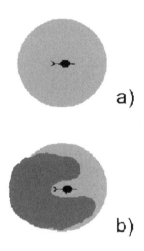

Figure 2. Schematic representation of the two simulated physical states: a) free inhibitor in a sphere of solvent, and b) solvated protein-inhibitor complex

Results-Discussion

The average electrostatic and van der Waals interaction between the inhibitors and their respective environments in the bound and free states were determined from the simulations. In addition, the SASA was determined for all the inhibitors. The results are summarized in Table 1, in the form of differences between the two states. In addition, the experimentally measured (Mares-Guia, 1970, 1977, Wagner, 1977) binding free energies for all the inhibitors are presented.

Employing a least squares algorithm and using the data in Table 1, we can calculate the coefficients of Eq. 1. We find $\alpha = -0.185$ and $\beta = 0.177$ with a root mean square (rms) error of 0.512 kcal/mol. In Fig. 3, the calculated free energies of binding are plotted against the experimental ones.

The value of β is not close to the linear response theoretical suggestion of 0.5. Moreover, one can argue, on physical grounds, that a negative value for any of the coefficients is unreasonable. We attempt to improve the values of the coefficients by using Eq. 2 and including the SASA data in the fitting. We obtain $\alpha = -0.129$, $\beta = 0.089$ and $\gamma = 0.011$, with a rms error of 0.414 kcal/mol.

Although there is considerable improvement in the prediction of the binding affinities, the coefficient for the van der Waals interactions is still negative. It is interesting to note that Smith et al. (1998) were also faced with negative values for α. They used the same methodology and force field as the ones used for the present study to calculate binding affinities for TIBO inhibitors of HIV-1 reverse transcriptase. Since the employment of other force fields and methodologies lead to positive coefficients, even without the use of a third term, one can conclude that the force field does not properly account for the interactions between the inhibitors and their environments. In addition, one has to keep in mind that a different protocol for neutralizing the system results to different values for the coefficients. We have employed three different protocols for neutralizing the protein and found that the actual coefficient values do vary considerably. Nevertheless, the trends between the physical descriptors and the binding affinity remain largely unaffected.

Table 1. Electrostatic Energy (Δele, kcal/mol), van der Waals Energy (ΔvdW, kcal/mol), and Solvent-Accessible Surface Area ($\Delta SASA$, Å^2) Differences between Bound and Unbound States for Trypsin Inhibitors. ΔG_{exp} are the Experimentally Measured Binding Affinities (kcal/mol)

Inhibitor	Δele	ΔvdW	$\Delta SASA$	ΔG_{exp}
aminobenzamidine	-47.49	-12.09	-311.86	-6.70
bromobenzamidine	-49.17	-12.37	-305.73	-6.22
benzamidine	-47.42	-10.32	-280.74	-6.30
benzyloxynaphthamidine	-49.86	-7.12	-396.71	-7.74
chlorobenzamidine	-48.50	-12.71	-304.04	-5.62
ethoxynaphthamidine	-48.98	-9.84	-325.60	-7.23
hydroxybenzamidine	-45.44	-10.55	-294.76	-6.40
methoxybenzamidine	-49.39	-16.75	-307.05	-5.55
methylbenzamidine	-44.05	-12.87	-323.18	-6.03
naphthamidine	-53.95	-11.46	-327.90	-6.60

In addition, useful correlations between the nature of the inhibitors, their environment interactions and their potency can be extracted from the results. The simulations provide a clear picture of the interactions between the benzamidines and the enzymatic site residues, suggesting ways of rationally designing potent inhibitors.

Currently, the main hurdle for the employment of the LIE approximation in computer-assisted drug design in scales of industrial interest stems from the time consuming simulation set up. Special care has to be taken for properly setting up each of the simulations, so that meaningful results and comparisons are available. Moreover, each simulation requires many hours of CPU rendering the methodology difficult to implement for large-scale predictions and screening. Specifically, simulations of the unbound state required 0.36 CPU hours per 10^6 steps on a LINUX based PentiumIII (500MHz), whereas for the bound state 2.8 hours of CPU time were required. Nevertheless, automation of the setup process is possible, and in conjunction with high-performance computational resources the methodology can potentially be amenable to high-throughput computing

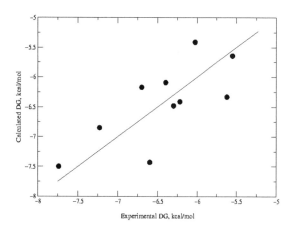

Figure 3.   Calculated free energies of binding versus experimental ones.

Conclusions

The linear interaction energy approximation was examined for 10 trypsin inhibitors using Monte Carlo simulations. Electrostatic and van der Waals interaction energies were calculated as well as solvent-accessible surface areas for the ligands. The numerical fitting processes resulted to reasonable predictions of binding affinities.

Nevertheless, the values of the linear response coefficients do not coincide with simple theoretical predictions, rendering the assignment of physical meaning problematic. In view of the limitations of the employed force field (no polarizability, optimized for a limited set of physical conditions, etc.), the results suggest that LIE be solely considered as an empirical method. Subsequently, the employment of additional molecular descriptors, such as those used in quantitative structure-analysis relationships (number of hydrogen bond acceptors/donors, number of aromatic groups, etc.), in the equations of binding free energies should be considered. Such an approach might improve the predictability of the method and shed light to the molecular mechanisms that contribute to the free energy of binding.

Furthermore, one could envision the LIE approximation in a hierarchy of computer-based tools for screening drugs, once a protein has been identified as a potential therapeutic target. High-volume docking

calculations could suggest a relatively small number of ligands with high affinity. Using a subset of those compounds as training set and performing LIE calculations, in conjunction with binding affinity experiments, might allow for very accurate predictions of binding affinities and more efficient screening of therapeutic agents.

Acknowledgments

YNK is grateful to Parke-Davis Pharmaceutical Research for funding of his postdoctoral fellowship at the University of Michigan.

References

Aqvist, J., C. Medina, and J. E. Samuelsson (1994). A new method for predicting binding-affinity in computer-aided drug design. *Protein Eng.*, **7**, 385-391.

Aqvist, J., and S. L. Mowbray. C. (1995). Sugar recognition by a glucose-galactose receptor – evaluation of binding energetics from molecular dynamics simulations. *J. Biol. Chem.*, **270**, 9978-9981.

Bode, W., and P. Schwager (1975). The refined crystal structure of bovine beta-trypsin at 1.8 A resolution. II. Crystallographic refinement, calcium binding site, benzamidine binding site and active site at pH 7.0. *J. Mol. Biol.*, **98**, 693-717.

Jones-Hertzog, D. K., and W. L. Jorgensen (1997). Binding affinities for sulfonamide inhibitors with human thrombin using Monte Carlo simulations with a linear response method. *J. Med. Chem.*, **40**, 1539-1549.

Jorgensen, W. L., J. Chandrasekhar, J. D. Madura, R. W. Impey, and M. L. Klein (1983). Comparison of simple potential functions for simulating liquid water. *J. Chem. Phys.*, **79**, 926-935.

Jorgensen, W. L., D. S. Maxwell, and J. Tirado-Rives (1996). Development and testing of the OPLS all-atom force field on conformational energetics and properties of organic liquids. *J. Am. Chem. Soc*, **118**, 11225-11236.

Jorgensen, W. L. (2000). *MCPRO*, version 1.65. Yale University, New Haven, CT.

Mares-Guia, M., and A. F. S. Figueiredo (1970). Thermodynamics of the hydrophobic interaction in the active center of trypsin. Investigation with amidines and guanidines. *Biochemistry*, **9**, 3223-3227.

Mares-Guia, M., D. L. Nelson, and E. Rogana (1977). Electronic effects in the interaction of para-substituted benzamidines with trypsin: the involvement of the p-electronic density at the central atom of the substituent in binding. *J. Am. Chem. Soc.*, **99**, 2331-2336.

Smith, R. H. Jr, W. L. Jorgensen, J. Tirado-Rives, M. L. Lamb, P. A. Janssen, C. J. Michejda, and M. B. Kroeger Smith (1998). Prediction of binding affinities for TIBO inhibitors of HIV-1 reverse transcriptase using Monte Carlo simulations in a linear response method. *J. Med. Chem.*, **41**, 5272-86.

Storer J. W., D. J. Giesen, C. J. Cramer, and D. G. Truhlar (1995). Class IV charge models: a new semiempirical approach in quantum chemistry. *J. Comput. Aided Mol. Des.*, **9**, 87-110.

Wagner, G., I. Lischke, F. Markwardt, P. Richter, J. Stuerzebecher, and P. Walsmann (1977). Synthetic inhibitors of serum proteinases. Part 15: Antitrypsin, antiplasmin, and antithrombin activity of 2-amidino-6-alkoxy- and aralkoxynaphthalenes. *Pharmazie*, **32**, 761-763

Walsh, C. (1979). *Enzymatic Reaction Mechanisms*. W.H. Freeman and Company, San Francisco.

Wang, J., R. Dixon, and P. A. Kollman (1999). Ranking ligand binding affinities with avidin: a molecular dynamics-based interaction energy study. *Proteins: Struct. Funct. Genet.*, **34**, 69-81.

Wang, W., J. Wang, and P. A. Kollman (1999b). What determines the van der Waals coefficient b in the LIE (Linear Interaction Energy) method to estimate binding free energies using molecular dynamics simulations. *Proteins: Struct. Funct. Genet.*, **34**, 395-402.

MOLECULAR SIMULATION FOR ADSORPTION OF HALOCARBONS IN HIGH-SILICA ZEOLITE

Kazuyuki Chihara*, Ryuichi Kamiyama, Hirotaka Mangyo and Masaki Omote
Department of Industrial Chemistry
Meiji University
Higashimita, Tama-ku, Kawasaki, 214-8571, Japan

Caroline F. Mellot
Institut Lavoisier
Universite de Versailles
78035 Versailles Cedex, France

Anthony K. Cheetham
Material Research Laboratory
University of California
Santa Barbara, CA 93106-5130

Abstract

Equilibrium and isosteric heat of adsorption for the system of chloroform and USY-type zeolite were studied. The USY-type zeolites (PQ Co., SiO_2/Al_2O_3=70) was used both as a pure crystalline powder and as granulated particles with binder. Chloroform was reagent grade .The adsorption equilibria were measured using a gravimetric method and were expressed as isotherms. A chromatographic method (i.e. pulse response of chloroform through the USY column with helium carrier) was used to get the initial slope of the isotherms. In the simulation, the GCMC method was used to calculate amounts adsorbed for various conditions. Forcefield parameters were confidently applied. And modified structure models are effective for simulation.

Keywords

Molecular simulation, Grand Canonical Monte Carlo method, High silica zeolite, Adsorption equilibrium.

Introduction

Molecular simulation has now become powerful tool for the study of adsorbed molecules in zeolites, and the Grand Canonical Monte Carlo (GCMC) method is especially useful for predicting adsorption equilibria. However, information on forcefield parameters and charges are often inadequate, even in systems where the structure is well known.

From the environmental point of view, the adsorption of chlorinated hydrocarbons by the use of zeolites may have some potential utility in ground water or soil remediation and other areas. Mellot et al. (1998a) recently reported new forcefield parameters and charges for chlorinated hydrocarbons in the faujasite zeolite: NaX, NaY and siliceous Y. These yield heats of adsorption that are in good agreement with calorimetric data (Mellot et al. (1998b)) .

In this study, their forcefield was used to simulate adsorption isotherms and isosteric heats of adsorption for chloroform in USY-type zeolite. The results are compared with gravimetric and chromatographic experiments.

Experiment

Gravimetric Method

Fig. 1 shows experimental apparatus for gravimetric analysis. The zeolite sample (about 1 [g]) was placed in a quartz basket (K) . Then the adsorbate in flask(B) was fed to adsorption tube (N) . Whole the apparatus was in a constant temperature air bath.   The temperature range was   303[K] □323[K] .   The amount adsorbed was determined from the change in the length of quartz spring(G) . The pressure of the atmosphere in the tube was measured by pressure sensor(O) .

In this way, adsorption isotherms were obtained (Fig. 2 and Fig. 7) .

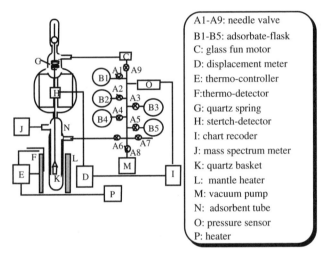

Figure 1.   Experimental apparatus to measure adsorption equiliblium.

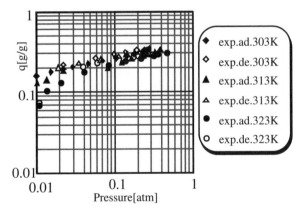

Figure 2.   Adsorption isotherms for PQ-USY-chloroform system.

Chromatographic Method (Chihara et al . 1978)

In order to study the isotherm at lower pressure range, we used a pulse response method with a gas chromatograph.

The gas chromatograph (GC 9-A;Shimadzu Co., Ltd.) was used with helium carrier. Siliceous zeolite pellets (with binder) were crushed and screened to obtain particle size between 4.95 x 10^{-4} to 8.33 x 10^{-4}[m] (an average particle radius of 6.64 x 10^{-4}[m]) and packed to the column (length, 30 [cm]) . The experimental column pressure was kept 200[kPa], the experimental column temperature was kept at 393, 423 and 453[K] , respectively. Pulse responses of vaporized chloroform with helium were detected by TCD. Response data were stored and processed by a personal computer.

From retention time of pulse response, adsorption equilibrium constant at zero coverage was obtained and plotted in Fig. 3 as van't Hoff plot to get heat of adsorption (39.71 [kJ/mol]) and adsorption equilibrium constant extrapolated to 303 [K] . This extrapolated value was plotted in Fig. 7 as initial slope.

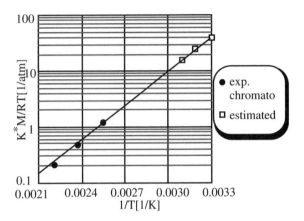

Figure 3.   van't Hoff plot for PQ-USY-chloroform system.

Simulation

Cerius2, version 3.5 (MSI Inc.,) was used throughout the simulations. Forcefield parameters obtained by Mellot et al. 1998a are listed in Table 1. The Grand Canonical Monte Calro method (under constant chemical potential(μ) , volume (V) , temperature (T)) was used to get the equilibrium amount adsorbed.

Models

Three models were considered here. Pure siliceous faujasaite (Y-type) (Fig. 4) was in the database of Cerius2. Dummy atoms were put in sodalite cages to avoid impossible occupation of adsorbates. PQ-USY was dealuminated in preparation. So it definitely contains silanol nest. PQ-USY has the silica-alumina ratio of SiO_2/Al_2O_3=70. So there remains aluminum.Then we put 7 silanol nest as in Fig. 5. And 4 Acid site as in Fig.6. Charges for Si and O were set to be +2.4[e.u] and -1.2[e.u] , respectively, as usual. H and O in silanol nest were assumed to have the charges as +0.1[e.u] and -0.7[e.u] , respectively. Al, O and H in replaced Al were assumed to have the charges as +1.4[e.u] , -0.59[e.u] and 0.39[e.u] . Si charges were adjusted after these setting to neutral·by averaging.

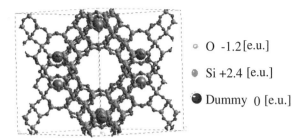

Figure 4. *Pure Siliceous Y Type model.*

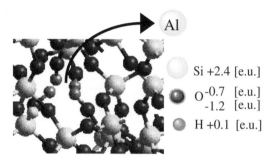

Figure 5. *Silanol nest model.*

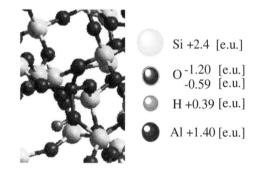

Figure 6. *Acid site model.*

Results and Discussion

In Fig. 7, experimental adsorption isotherm for chloroform in PQ-USY at 303 [K] is shown. Initial slopes for 303[K] are also shown. This whole range of adsorption isotherm at 303[K] could be compared with simulation.

At higher pressure (0.1 [atm]), all the simulation are coincident and almost correspond to gravimetric data .

At lower pressure, a simulation for the acid site model was good agreement with chromatographic date.

Experimental heat of adsorption obtain by chromatography at zero coverage is compared with simulated heat of adsorption for 3 models in Table 2. Here simulation for the acid site model was closer to the experimental value than the pure siliceous model and the silanol nest model.

In Fig. 8, snap shot of MC adsorption simulation for chloroform shows that each one of chlorine in chloroform faces to acid site. This fact seems to be the cause of

difference in initial slope. Using molecular mechanics calculation up to eight chloroform or tetrachloroethylene in zeolite cage, orientation of chlorine in each molecule to acid site were checked and shown in Fig. 9 and 10, which denote distance between each atom and acid site. In case of polar chloroform molecule, chlorines are near to acid site compared with hydrogen. On the other hand, chlorines and hydrogens in tetrachloroethylene show no difference in distance to acid site.

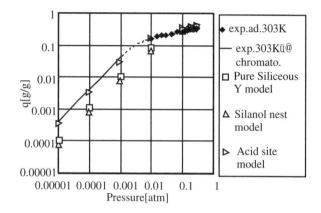

Figure 7. *Comparison between exp. and simulations for adsorption isotherms of PQ-USY-chloroform system.*

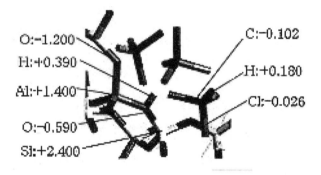

Figure 8. *Condition of adsorption by chloroform.*

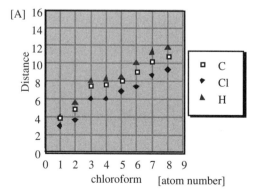

Figure 9. *Distance between each atom and acid site.*

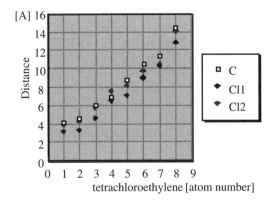

Figure 10. Distance between each atom and acid site.

Table 1. van der Waals Parameter.

	ε/k[K]	R_Min[ü.]
Si - Si	0.000	0.000
O - O	0.000	0.000
O - C	87.058	3.253
O - Cl	162.000	3.430
O - H	90.535	2.698
C - C	25.860	3.753
C - Cl	55.650	3.787
C - H	26.730	3.358
Cl - Cl	119.800	3.822
Cl - H	57.530	3.392
H - H	27.630	2.963
X - X	0.000	0.000
X - H	0.010	6.000
X- Cl	0.010	6.000
X- C	0.010	6.000

Table 2. Heat of Adsorption [kJ/mol].

Exp. Chromato.	39.71
Pure Siliceous Y model	34.21
Silanol nest model	33.82
Acid site model	40.17

Conclusion

Equilibrium and isosteric heat of adsorption for the system of chloroform -USY-type zeolite were studied.

The adsorption equilibria were measured using a gravimetric method and were expressed as isotherms. A chromatographic method was used to get the initial slope of the isotherms. In the simulation, GCMC method was used to calculate amounts adsorbed for various conditions. To get better coincidence between experimental data and simulation, we tried simulation of silanol nest model and acid site model, which was the modification from pure siliceous Y zeolite. Then, it was found to be necessary to account for aluminum rather than silanol nest. Further, MM calculation confirmed acid site effect.

As a conclusion, FF parameters were confidently applied. And modified structure models are effective for simulation.

Nomenclature

K^* = adsorption equilibrium constant, cc/g
M = molecular Weight, g/mol
q = amount adsorbed, g/g
R = gas constant, J/K mol
R_Min = Bond Length Equilibrium , $\Box$
T = temperature, K
ε = potential energy minimum, eV
k = Boltzmann constant, J/K

References

C. F. Mellot, A. M. Davidson, J. Eckert and A. K. Cheetham (1998a). Adsorption of chloroform in NaY zeolite: A computational and vibrational spectroscopy study. *J. Phys. Chem. B*, **102**, 2530-2535.

C. F. Mellot, A. K. Cheetham, S. Harms, S. Savitz, R. J. Gorte and L. A. Myers (1998b). Calorimetric and computational studies of chlorocarbon adsorption in zeolites. *J. Am. Chem. Soc.*, **120**, 5788.

Chihara, K., Suzuki, M.and Kawazoe, K (1978). Adsorption rate on molecular sieving carbon by chromatography. *AIChE J.*, **24**, 237-246.

SIMULATIONS OF CAPILLARY
CONDENSATION IN POROUS GLASSES

Lev D. Gelb
Florida State University
Tallahassee, FL 32306-4390

Keith E. Gubbins
North Carolina State University
Raleigh, NC 27695-7905

Abstract

Molecular simulation studies of the adsorption and desorption of gases in models of controlled-pore glasses are presented. Both the preparation of our model glasses and their structural properties are discussed. The connnections between materials characterization methods, capillary phase equilibria, and structural models are examined. Computer simulation studies of this type offer the possibility of significantly advancing our understanding of nano-scale phenomena and nano-structured materials in ways which are not presently possible through experimental studies.

Keywords

Molecular simulation, Capillary condensation, Porosity, Materials, Phase equilibria, Adsorption, Vycor.

Introduction

For some time, molecular simulation studies have contributed to our understanding of the effects of confinement on phase equilibria and liquid properties, as reviewed in Evans, 1990, and Gelb, *et al,* 1999. More recently, simulation studies have begun to contribute to our understanding of the structure and properties of porous media themselves, and how they can be described and characterized. Crystalline porous media, such as zeolites, can usually be fully described either through single-crystal x-ray diffraction experiments or by the "solution" of their structure from power-diffraction data. In the case of amorphous porous media, although some materials can be directly characterized using imaging methods such STM, for those materials with pores too small or delicate to be imaged the primary characterization method is the analysis of gas adsorption and desorption isotherm data.

Characterization of porous materials by gas adsorption is non-trivial because the relationship between the materials' structures and their adsorptive properties is complex. The use of adsorption data for characterization invariably involves the use of an adsorption model (or theory) coupled with some assumptions about the structure of the material. The value of any given method, then, depends both on the quality of the underlying theory, and the applicability of the structural model (Gregg and Sing, 1982).

Most characterization methods rely on explicit models. For instance, methods for obtaining *pore size distributions* generally assume that the material can be described as a collection of pores of simple geometry (say, cylindrical) varying in a single parameter (diameter), and then fit the experimental isotherm data to a distribution of such pores, by using an adsorption model to generate the adsorption isotherm for every pore size.

In the case of mesoporous media, which have pore sizes large enough that many gases exhibit *capillary condensation* and isotherm hysteresis at sufficiently low temperatures (indeed, this is the historical source of the definition of the "mesoporous" range of pore size) the adsorption model must accurately describe capillary phenomena as well as multilayer adsorption. Thus, the BJH method for pore size distributions models the adsorption in each (cylindrical) pore as a multilayer adsorption isotherm taken from a non-porous reference material plus the Kelvin equation prediction for the pressure of capillary condensation (Barrett, Joyner and Halenda, 1951) More modern methods rely on density functional theories of the liquid state to predict the isotherm in each model pore, which is somewhat more accurate (Lastoskie, Gubbins and Quirke, 1993).

In cases where the material structure is complex and perhaps not well described by a collection of pores of simple geometry, the accuracy of the adsorption theory is not the limiting factor in improving materials characterization. Indeed, since modern molecular simulation methods could be used to generate isotherms that are essentially exact for the model used (at high computational cost), attention should now be focused on improving our structural models and thus our descriptions of these materials. In addition, the availability of more physically meaningful models will improve our ability to relate a materials' characterization to its useful properties.

(Note that the preceding discussion is also wholly applicable to other characterization methods such as thermoporometry and mercury-intrusion porosimetry.)

Computer simulation is an excellent technique for addressing these issues, because only in a computer model can one have "complete" information with which to investigate the accuracy and effectiveness of different structural models and characterization methods[†]. That is, one can *simulate* a complex, amorphous material structure, and then attempt to *characterize* the simulated structure by using *simulated* adsorption isotherms, diffraction, or other data. Since the model structure is completely known, the effects of different approximations or models can be systematically analyzed. In addition, this approach brings a better understanding of the precise meanings of macroscopic (and classical) concepts such as surface area and porosity on a microscopic scale.

Controlled-pore glass and vycor glass

The preparation of controlled-pore glasses (CPGs) is based on the near-critical phase separation of a binary liquid mixture via *spinodal decomposition*, which produces complex networked structures. The original preparations of CPGs were done by Haller (Haller, 1965), who partially phase separated a mixture of SiO_2, Na_2O and B_2O_3 and etched out the boron-rich phase, leaving a nearly pure silica matrix with a porosity between 50% and 75% and an average pore size controllable between 10 nm and 400 nm. Vycor glasses are made in a similar manner, with

the main differences being in the washing and processing of the etched glass (Elmer, 1991). CPGs and vycor have narrow pore size distribution and are a popular substrate for experimental studies of confined phase equilibria (Gelb, *et al*, 1999).

Computer Models of Porous Glasses

We generate models of controlled-pore glasses using molecular dynamics simulations which imitate the experimental preparation recipe (Gelb and Gubbins, 1998). In order to reduce computational cost, the borosilicate mixture is replaced with a symmetric binary Lennard-Jones fluid mixture which exhibits similar phase separation patterns. Using large-scale parallelized simulations, we can obtain models with networked pore structures with surface areas and pore sizes quantitatively comparable with those of experimental materials.

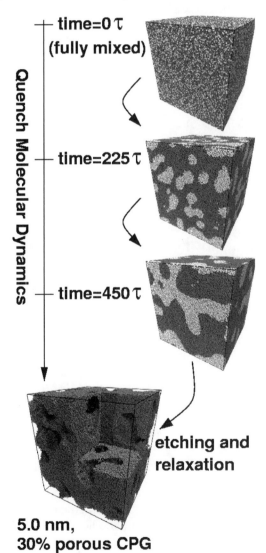

Figure 1. Schematic of procedure for modeling controlled-pore glasses.

This process is shown schematically in Figure 1. In order to obtain plausible CPG models using this approach,

---

[†] In experimental studies of crystalline materials, this level of information can be achieved through x-ray diffraction studies. However, the crystal structure itself becomes the characterization, and there is no need to resort to descriptions based on pore sizes or surface areas. In addition, crystalline materials have only very simple network topologies so that their use as "reference systems" in evaluation of measures of pore connectivity, etc., is of limited value.

relatively large simulation cells must be used for two reasons. The first is that the cell must be sufficiently large to contain a statistically meaningful "chunk" of pore network, e.g., that its average properties (porosity, surface area, pore size, connectivity) do not change if we repeat the preparation procedure from a different initial condition. The second size constraint is that the cell must be large enough that its periodic boundaries do not influence the physics of phase separation (and thus cause artifacts in pore geometry).

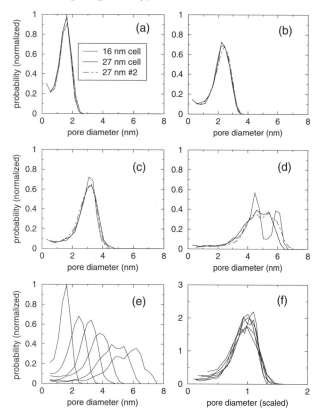

Figure 2. (a-d) Geometric pore size distributions of 30% porous models generated in three different size simulation cells; in (a) and (b) the simulation cell size does not affect the PSD, while in (c) small differences are visible between the small cell and the large ones, and in (d) large differences are visible. Graph (e) shows a series of six PSDs for models from the same quench run, of different sizes, and graph (f) shows the same data scaled by average pore size; these materials all have similar morphologies by this measure.

In order to investigate this issue we prepared three series of models; one using a cell of 16 nm periodic size, and two in cells of 27 nm periodic size, starting from different initial conditions. Geometrically-derived pore size distributions (see Gelb and Gubbins, 1999) for these models are shown in Figure 2. For pores of smaller than 4 nm average size all three systems have similar PSDs, with better agreement between the two large cells, while for the largest pore sizes, shown in graph (d), the shape of the PSD for the small system is markedly different (probably

due to boundary condition issues) and the two large-cell PSDs are visibly somewhat different. Graphs (e) and (f) indicate that the morphology of the pore structure does not qualitatively change even up to 6.5 nm average pore size in the large cells, although statistical quality drops off quickly as the pore size is increased at constant cell size.

A useful rule of thumb, at least for these systems, appears to be that the simulation cell must be at least four times larger than the desired pore size in order to not induce artifacts, and at least five times larger to insure a smooth and reproducible pore size distribution.

Simulations of Adsorption Isotherms

Grand canonical Monte Carlo simulations can be used to obtain adsorption and desorption isotherms in model porous materials. Because the simulation cells must be quite large in order to contain a statistically reasonable section of pore network, these simulations can be time-consuming. For the 27 nm cells already described, at pore filling (high pressures) the model pore network contains between 100,000 and 200,000 adsorbed molecules, depending on the porosity and the size of the adsorbate. In order to accomplish these simulations in reasonable time we have implemented a domain-decomposed parallel version of the GCMC algorithm (Heffelfinger and Lewitt, 1996). We have found that at low pressures, and for relatively small changes in pressure between successive points, simulations of approximately 10^8 MC moves are sufficient. Near to the capillary rise, the pressure must only be changed in very small increments, and significantly longer equilibration runs are often necessary.

We have simulated xenon adsorption in these materials using a Lennard-Jones potential; this choice of system was made in order to compare with the extensive experimental studies of the xenon/vycor system made by Nuttall (Nuttall, 1974). In addition, we have prepared idealized single-pore models with the same pore-wall properties by simply excavating cylindrical channels out of a similarly quenched single-component liquid; since most previous simulation (and theoretical) studies of capillary condensation effects have relied on idealized pore models, these simulations provide a useful connection to previous work. Lastly, these ideal-pore systems may provide a useful connection between studies of capillary condensation phenomena in controlled-pore glasses and in the recently developed MCM-41 and other templated materials (for an overview, see Barton, *et al*, 1999), which have pores of cylindrical or other regular geometry and amorphous silica wall structure.

Preliminary simulation results for materials with pores of 5 nm size are shown in Figure 3. These are both adsorption and desorption data from simulated materials and experimental ones, at three different temperatures. Note that no attempt has been made to fit the simulated systems to the experimental one, other than the choice of a similar pore size and porosity.

Lennard-Jones potential model used, which likely does not describe the surface tension of xenon very well.

Discussion

Access to modern high-performance computers, coupled with recent developments in simulation methodology, allow the modeling of porous media and capillary phenomena on a scale large enough to quantitatively reproduce many experimental results. We are using simulation methods to probe the structure of controlled-pore glasses and to better understand gas adsorption thermodynamics in these materials. While many refinements have yet to be made in our models, methodologies, surfaces, and gas and gas-solid potentials, recent results are encouraging and demonstrate both that simulations on this scale are feasible and that they can contribute substantially to our fundamental understanding of confined fluids.

Acknowledgments

We thank the National Science Foundation (grant no. CTS-9896195) for their support of this work and for the Metacenter grant (no. MCA93S011P) and NRAC allocation (no. MCA93S011) which made these calculations possible.

References

Barrett, E. P., Joyner, L. G. and P. P. Halenda (1951). The determination of pore volume and area distributions in porous substances. I. Computations from nitrogen isotherms. *J. Am. Chem. Soc.*, **73**, 373-380.

Barton, T. J., *et al* (1999). Tailored porous materials. *Chem. Mater.*, **11**, 2633-2656.

Elmer, T. H. (1991). Porous and Reconstructed Glasses. In S. J. Schneider, Jr. (Ed.), *ASM Engineered Materials Handbook*, vol. 4. ASM, Materials Park, OH., 427-432.

Evans, R. (1990). Fluids adsorbed in narrow pores: phase equilibria and structure. *J. Phys. Condens. Matter*, **2**, 8989-9007.

Gelb, L. D. and K. E. Gubbins (1998). Characterization of porous glasses: Simulation models, adsorption isotherms, and the Brunauer-Emmett-Teller analysis method. *Langmuir*, **14**, 2097-2111.

Gelb, L. D. and K. E. Gubbins (1999). Pore size distributions in porous glasses: A computer simulation study. *Langmuir*, **15**, 305-308.

Gelb, L. D., Gubbins, K. E., Radhakrishnan, R., and M. Sliwinska-Bartkowiak (1999). Phase separation in confined systems. *Rep. Prog. Phys.*, **62**, 1573-1659.

Gregg, S. J. and K. S. W. Sing (1982). *Adsorption, Surface Area and Porosity*, 2nd ed. Academic Press, Inc., London.

Haller, W. (1965). Chromatography on glass of controlled pore size. *Nature*, **206**, 693-696.

Heffelfinger, G. S. and M. E. Lewitt (1996). A comparison between two massively parallel algorithms for Monte Carlo computer simulation: An investigation in the grand canonical ensemble. *J. Comp. Chem.*, **17**, 250-265.

Lastoskie, C., Gubbins, K. E., and N. Quirke (1993). Pore size heterogeneity and the carbon slit pore: A density functional theory model. *Langmuir*, **9**, 2693-2702.

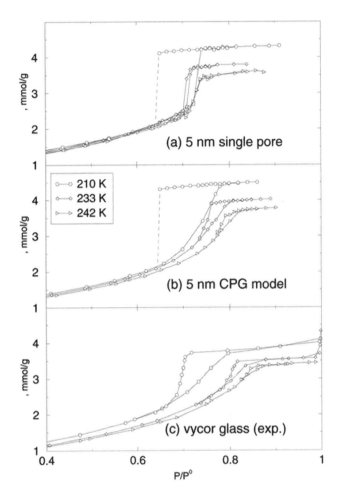

Figure 3. Xenon adsorption and desorption isotherms in (a) a single 5 nm diameter silica pore, (b) 30% porous, 5 nm model CPG, and (c) experimental data for xenon adsorption in vycor glass, from (Nuttall, 1974). Dashed lines are approximate locations of desorption drops. The horizontal scale and temperature key are the same in each plot.

There are a number of significant points to be made regarding these plots. The first is that the CPG model does a significantly better job of reproducing the *shape* of the hysteresis loops of the experimental system than does the single pore model. Indeed, the single pore model does not even show a hysteresis loop at the highest temperature, although it does still exhibit a capillary rise. The second significant point is that the simulated isotherms exhibit very pronounced hysteresis and extremely sharp desorption at the lowest temperature; this is likely due to "artifical" metastability due to the short length of a GCMC simulation run. At higher temperatures this does not seem to be an issue, probably because the free energy landscape is flatter and the sampling correspondingly better. Thirdly, the capillary rises in the simulated materials occur at somewhat lower pressures than those in the experimental material; we expect that this is due to deficiencies in the

GROUP-CONTRIBUTION THEORY FOR
COADSORPTION OF GASES AND
VAPORS ON SOLID SURFACES

M. Douglas LeVan and Giuseppe Pigorini
Department of Chemical Engineering
Vanderbilt University
Nashville, TN 37235

Abstract

This paper reports on the development of a group contribution theory for adsorption of n-alkanes of various lengths and some of their binary mixtures on BPL activated carbon. The basis for the approach is contained in Russell and LeVan (1996). The theory is derived from Guggenheim's lattice theory of solutions. Molecules are broken into "elements," and interactions between elements of various molecules and between elements and a heterogeneous surface are considered. Pure component isotherms for methane, ethane, n-butane, and n-hexane are treated first to determine parameters for the elements and their interaction energies with the surface. Then, using only the parameters determined from the pure component isotherms, experimental results are analyzed for mixtures of methane and n-hexane and ethane and n-hexane.

Keywords

Adsorption, Adsorption equilibrium, Group-contribution theory.

Introduction

The development of a group-contribution theory for adsorp-tion equilibrium would make possible the prediction of adsorption equilibrium for pure components and mixtures on which data are not available. This paper considers such a theory for simple molecules.

The basis for the approach is contained in Russell and LeVan (1996). There, experimental data for multicomponent mixtures of gases and vapors adsorbed on a nonuniform surface are analyzed using a model of localized adsorption for a patchwise heterogeneous adsorbent. The theory is derived from Guggenheim's lattice theory of solutions (Guggenheim, 1952). Molecules are broken into functional "elements." Interactions among adsorbed molecules on each patch are treated in the Bragg-Williams approximation. The analysis accounts for adsorbent heterogeneity, molecular size differences of the mixture components, and the polyfunctionality of the adsorbate molecules, all of which contribute to nonideal behavior of adsorbed solutions. The final equations of the theory appear as a combination of the multisite Langmuir equation, adsorbate-adsorbate interactions taken on an element-by-element basis between molecules adsorbed on neighboring sites, adsorbate-adsorbent interactions deter-mined by pair-wise group contributions of elements of unequal sizes and unequal energies, and energy distributions for a heterogeneous surface.

Systems were chosen in this paper to emphasize the importance of molecular size differences on adsorption equilibria. Previous studies (Kaminsky and Monson, 1994; Dunne and Myers, 1994; Sircar, 1995) have shown that such differences lead to negative deviations from Raoult's law; that is, activity coefficients are less than unity referenced to the ideal adsorbed solution theory (Myers and Prausnitz, 1963).

Several lattice models for adsorption equilibria of unequal sized molecules have been developed (Marczewski et al., 1986; Jaroniec and Madey, 1988; Rudzinski and Everett, 1992; O'Brien, 1993; Al-Muhtaseb and Ritter, 1999). Of particular relevance to the present study is the work of Nitta and coworkers (Nitta et al., 1984ab; Nitta and Yamaguchi, 1992, 1993) on the multisite Langmuir isotherm and on accounting for interactions among adsorbed molecules. (See also Golden and Sircar, 1994, Sircar, 1994, 1995; Sircar and Rao, 1999).

In this paper, we consider a group-contribution theory for coadsorption of light and moderately heavy n-alkanes. Parameters are determined from pure component isotherms of methane, ethane, n-butane, and n-hexane. Using these parameters, predictions for coadsorption of methane and n-hexane and ethane and n-hexane are compared with experimental measurements.

Theory

A two-dimensional hexagonal lattice with adsorbed ethane and n-hexane is depicted in Figure 1. In this paper, molecules are broken up into three types of elements: CH4 (for methane), CH3, and CH2. We consider interactions of these elements with the surface and with each other for nearest neighbor contacts.

The adaptation of Guggenheim's lattice theory of solutions by Russell and LeVan involved changing the dimensionality from three to two, allowing for unfilled sites (in configurational properties), and having the elements interact with the sites. Thermodynamic properties were deduced from a canonical ensemble partition function in the Bragg-Williams approximation. The adsorption isotherm is

$$p_j = \frac{1}{K_j} \frac{\theta_j}{1 - \sum_i r_i \theta_i} \left[\frac{1 - \sum_i (r_i - s_i)\theta_i}{1 - \sum_i r_i \theta_i} \right]^{r_j - 1}$$

$$\times \exp\left\{ \frac{z}{2kT} \sum_\beta \sum_\alpha w_{\alpha\beta} \left[\frac{\sum_i \theta_i s_i s_j (t_{i\alpha} t_{j\beta} + t_{i\beta} t_{j\alpha})}{1 - \sum_i (r_i - s_i)\theta_i} \right. \right. \quad (1)$$

$$\left. \left. + \frac{(r_j - s_j)(\sum_i \theta_i s_i t_{i\alpha})(\sum_i \theta_i s_i t_{i\beta})}{[1 - \sum_i (r_i - s_i)\theta_i]^2} \right] \right\}$$

where

$$K_j = K_{j0} \exp(\varepsilon_j / kT) \quad (2)$$

$$\varepsilon_j = r_j \sum_\alpha t'_{j\alpha} \varepsilon_\alpha \quad (3)$$

$$\theta_j \equiv n_j / m \quad (4)$$

Symbols are defined in the nomenclature.

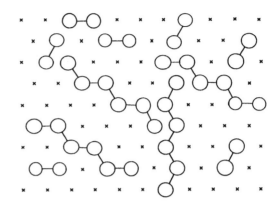

Figure 1. Ethane and n-hexane coadsorbed on hexagonal lattice.

For our purpose here, we have used a special case of eq. 1. In terms of Flory's approximation ($r_i - s_i = 0$), eq. 1 becomes (Russell and LeVan, 1996)

$$p_j = \frac{1}{K_j} \frac{\theta_j}{(1 - \sum_i r_i \theta_i)^{r_j}}$$

$$\times \exp\left[\frac{z}{2kT} \sum_\beta \sum_\alpha \sum_i w_{\alpha\beta} \theta_i r_i r_j (t_{i\alpha} t_{j\beta} + t_{i\beta} t_{j\alpha}) \right] \quad (5)$$

For molecules composed of the same element, eq. 5 reduces to the equation derived by Nitta et al. (1984a). If $w_{\alpha\beta} = 0$, the equation becomes the multisite Langmuir equation.

For the heterogeneous surface, potential energies for the element-site interactions are given by the beta distribution. Furthermore, in this paper, because of the similarity of the elements, we assume that a single dimensionless distribution applies. Thus, we have

$$n_j^{tot} = \int_0^1 n_j(\eta)\, h(\eta)\, d\eta \quad (6)$$

$$\eta \equiv \varepsilon_j / \varepsilon_{j,\max} \quad (7)$$

$$h(\eta) = \frac{\eta^{a-1}(1-\eta)^{b-1}}{B(a,b)} \quad (8)$$

where $B(a,b)$ is the beta function.

Experiments

Experiments were performed to measure pure component isotherms for methane, ethane, n-butane, and n-hexane and mixtures of methane and n-hexane, and ethane and n-hexane adsorbed on BPL activated carbon at 25°C. This carbon has a surface area of approximately 1200 m²/g and a mean pore width of about 1 nm. In mixture experiments, the loading of the heavy component was held constant while that for the light component was varied. Details are provided by Pigorini (2000).

Results and Discussion

Parameters to be determined in solving eqs 2-8 are K_{j0}, m, r_α, $\varepsilon_{\alpha,max}/kT$, $w_{\alpha\beta}/kT$, a and b. r_α for CH2 was set to unity to normalize the lattice, which was taken to be hexagonal ($z = 6$). Eq. 6 was evaluated using a 16 point Gaussian integration. Eq. 5 was solved for each n_j using the Powell hybrid method implemented in hybrd1 of minpack. The parameters were obtained from the pure component data by minimizing a least squares objective function based on differences between predicted and measured values of $\ln n_j^{tot}$ using the Levenberg-Marquardt algorithm in lmdif1 of minpack.

Pure Components

Pure component isotherms predicted at 25°C are shown in Figure 2. The description of the data is very good. All parameters used to describe the binary systems were obtained from this analysis of pure component data. Figure 3 shows the adsorption energy distribution described by eq. 8.

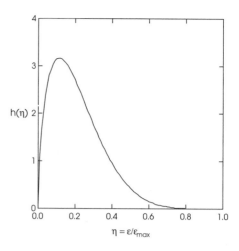

Figure 3. Distribution function for adsorption energies on the heterogeneous surface. Parameters: a = 1.67, b = 6.24

Mixtures

An adsorbed heavy component can ordinarily be expected to diminish the adsorption of a light component, relative to its adsorption as a pure component. Figures 4 and 5 show methane and ethane, respectively, coadsorbed with n-hexane. The plots are in the form of methane and ethane isotherms at constant loadings of n-hexane. The loadings of these light compounds are significantly suppressed by the coadsorbed n-hexane. Comparing Figures 4 and 5 with Figure 2, hexane present at a loading of about 1.0 mol/kg gives loadings of methane and ethane that are about 30% and 60%, respectively, of their loadings as pure components at the same pressure. Predictions of the group contribution theory are reasonably good.

The hexane partial pressure is sensitive to the loading of the light component. Also, these systems show negative deviations from Raoult's law (Pigorini, 2000).

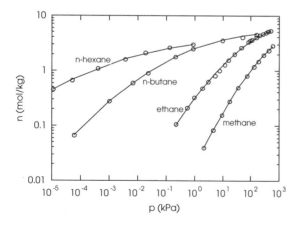

Figure 2. Pure component isotherms. Experimental data shown as open circles. Solid lines are group contribution theory. Parameters: m=28.8 mol/kg, r_CH2 = 1.0, r_CH3 = 1.4, r_CH4 = 2.5, $w_{\alpha\beta}/kT$ = -0.018.

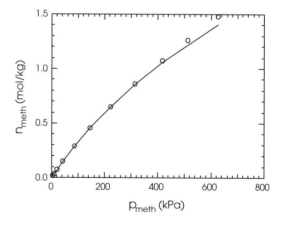

Figure 4. Methane coadsorbed with n-hexane at n_{hex} = 1.08 mol/kg. Open circles: experimental data. Solid curve: group contribution theory.

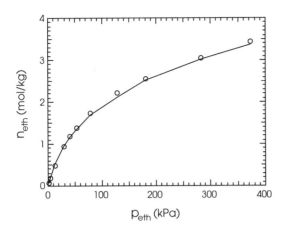

Figure 5. Ethane coadsorbed with n-hexane at $n_{hex} = 0.99$ mol/kg. Open circles: experimental data. Solid curve: group contribution theory.

Conclusions

The group contribution theory of Russell and LeVan (1996) has been used here to describe pure component isotherms of methane, ethane, n-butane, and n-hexane and mixtures of methane and n-hexane and ethane and n-hexane. Parameters used with binary mixtures were obtained solely from analysis of the pure components. The theory breaks molecules into elements and accounts for element-element and element-surface interactions. The theory describes the adsorption equilibrium with reasonably good accuracy.

Acknowledgment

The authors are grateful to the U.S. Army ECBC for financial support.

Nomenclature

i, j = component index
m = lattice site density, mol/g
n = adsorbed concentration, mol/g
r = number of lattice sites covered
s = contacts per molecule per z
$t_{j\alpha}$ = fraction of molecule j contacts from its α elements
$t'_{j\alpha}$ = fraction of elements in molecule j of type α
$w_{\alpha\beta}$ = potential energy of interaction, J
α, β = element index
ε = adsorption potential energy, J

References

Al-Mutaseb, S. A. and J. A. Ritter (1999). New model that describes adsorption of laterally interacting gas mixtures on random heterogeneous surfaces. *Langmuir*, **15**, 7732-7744.

Dunne, J. and A. L. Myers (1994). Adsorption of gas mixtures in micropores: the effect of difference in size of adsorbate molecules. *Chem. Eng. Sci.*, **49**, 2941-2951.

Golden, T. C. and S. Sircar (1994). Gas adsorption on silicalite. *J. Colloid Interface Sci.*, **162**, 182-188.

Guggenheim, E. A. (1952). *Mixtures*. Oxford University Press, London.

Jaroniec, M. and R. Madey (1988). *Physical Adsorption on Heterogeneous Solids*. Elsevier, New York.

Kaminsky, R. D. and P. A. Monson (1994). Modeling the influence of heterogeneous adsorbent microstructure upon adsorption equilibria for binary mixtures. *Langmuir*, **10**, 530-537.

Marczewski, A. W., A. Derylo-Marczewska and M. Jaroniec (1986). Energetic heterogeniety and molecular size effects in physical adsorption on solid surfaces. *J. Colloid Interface Sci.*, **109**, 310-324.

Myers, A. L. and J. M. Prausnitz (1965). Thermodynamics of mixed-gas adsorption. *AIChE J.*, **11**, 121-127.

Nitta, T., T. Shigetomi, M. Kuro-Oka and T. Katayama (1984a). An adsorption isotherm of multi-site occupancy model for homogeneous surface. *J. Chem. Eng. Japan*, **17**, 39-45.

Nitta, T., M. Kuro-Oka and T. Katayama (1984b). An adsorption isotherm for multi-site occupancy model for heterogeneous surface. *J. Chem. Eng. Japan*, **17**, 45-52.

Nitta, T. and A. Yamaguchi (1992). A hybrid isotherm equation for mobile molecules adsorbed on heterogeneous surface or random topography. *J. Chem. Eng. Japan*, **25**, 420-426.

Nitta, T. and A. Yamaguchi (1993). Effects of surface topography on adsorption isotherms of mobile molecules: comparison of patchwise and random surfaces. *Langmuir*, **9**, 2618-2623.

O'Brien, J. A. (1993). Adsorption of laterally interacting mixtures on heterogeneous solid surfaces: a model study. In M. Suzuki (Ed.), *Fundamentals of Adsorption*. Kodansha, Tokyo, pp. 483-490.

Pigorini, G. (2000). Ph.D. dissertation, University of Virginia.

Rudzinski, W. and D. H. Everett (1992). *Adsorption of Gases on Heterogeneous Surfaces*. Academic Press, San Diego.

Russell, B. P. and M. D. LeVan (1996). Group-contribution theory for adsorption of gas mixtures on solid surfaces. *Chem. Eng. Sci.*, **51**, 4025-4038.

Sircar, S. (1995). Influence of adsorbate size and adsorbent heterogeneity on IAST. *AIChE J.*, **41**, 1135-1145.

Sircar, S. and M. B. Rao, (1999). Effect of adsorbate size on adsorption of gas mixtures on homogeneous adsorbents. *AIChE J.*, **45**, 2657-2661.

MONTE CARLO STUDY OF CONFINEMENT
EFFECTS ON ZEOLITE CRACKING MECHANISMS

Michael D. Macedonia and Edward J. Maginn
University of Notre Dame
Notre Dame, IN 46556

Abstract

Results of a Monte Carlo study that examines the impact of confinement within zeolite pores on the competitive cracking mechanism of *n*-hexane and 3-methylpentane are presented. A methodology is developed in which free energies of reactants and transition states are computed through use of configurational-bias Monte Carlo integration of a fully-flexible all-atom model for the adsorbed species. Computed free energies are used to predict relative reaction rates, which are directly related to the experimentally determined constraint index (CI). For small pore zeolites, the results indicate that a monomolecular cracking mechanism is responsible for selectivity. For larger pores, a bimolecular mechanism becomes important. The experimentally observed reduction in the CI for ZSM-5 with temperature is captured by the simulations and explained in terms of entropic factors that govern the stability of the transition state.

Keywords

Monte Carlo, Catalytic cracking, Zeolites, Transition state, Constraint index.

Introduction

Zeolites are crystalline nanoporous aluminosilicates that are used widely as catalysts. Because the pores of zeolites are commensurate with molecular dimensions, catalytic selectivity and reactivity are strongly dependent on pore size and topography. Zeolites can control product formation via three different mechanisms: 1) *reactant shape selectivity*, where only certain molecules in a reactant mixture can access the active sites inside the zeolite pores due to size and/or shape exclusion; 2) *transition state selectivity*, where only products having transition states small enough to form in the pores are formed; and 3) *product shape selectivity*, where only molecules small enough to leave the pores of the zeolite end up in the product mixture. Any one or a combination of these mechanisms can lead to the selectivity observed in an experiment.

One of the most important zeolite catalyzed reactions is hydrocarbon cracking. There are two classical mechanisms that have been proposed for this reaction (Pines, 1981). The first, which we will refer to as *bimolecular* cracking, involves a hydrogen transfer between a carbocation and the feed paraffin, followed by a rapid rearrangement and beta-scission to form an olefinic product and another carbocation. The second cracking mechanism, referred to as *monomolecular*, proceeds through direct protonation of the feed molecule by the zeolitic acid site. It is observed that the bimolecular mechanism predominates at lower temperatures, higher feed partial pressures and higher conversions. The monomolecular mechanism has been found to predominate at higher temperatures, lower feed partial pressures and lower conversions. The structure of the zeolite also plays a major role in determining which mechanism is operative. Haag and Dessau (1984) noted that the bimolecular mechanism is more important in large pore zeolites, while the monomolecular mechanism is more important in small to medium pore zeolites. Presumably this is due to the lack of space available for forming the bulky bimolecular transition states in the smaller zeolite pores.

To identify which effects dominate in a given zeolite, catalytic test reactions such as the competitive cracking of *n*-hexane / 3-methylpentane mixtures has been used to investigate reaction mechanisms in different zeolites (Frillette, Haag and Lago, 1981). Selectivity is quantified in terms of the disappearance of reactants via a "constraint index" (CI). From the CI, one can determine which of the selectivity mechanisms are dominant and thus gain information on the likely pore size of the material. Many other test reactions have been developed (Weitkamp and Ernst, 1994), but it is still difficult to determine from these experiments exactly which mechanism is operative.

Molecular modeling studies have also been carried out to elucidate the cracking mechanisms in zeolites. Electronic structure calculations have been employed most often (Kazansky, Senchenya, Frash and van Santen, 1994), but limitations on the size of cluster models have prohibited these calculations from giving much insight into the impact *pore confinement* has on the reaction mechanism. Since the CI experiments demonstrate the importance of confinement effects on cracking selectivity, methods must be developed that are capable of treating these effects. In this paper, we present a technique in which relatively simple Monte Carlo integration methods can be used in conjunction with a classical forcefield to gain insight into the role confinement plays on the cracking mechanism of *n*-hexane and 3-methylpentane in twelve different zeolite structures.

Theory

The CI experiment is conducted by passing an equimolar stream of *n*-hexane ("A") and 3-methylpentane ("B") over a catalyst. By measuring the conversion and assuming first order kinetics, the CI is given by

$$CI \equiv \frac{r_A}{r_B} = \frac{k_A[A]}{k_B[B]} = \frac{k_A}{k_B} \qquad (1)$$

where r_A and r_B are the rates of reaction of *n*-hexane and 3-methylpentane, respectively. k_A and k_B are *apparent* first order rate constants based on the gas phase concentrations [A] and [B], which are equal in this case. The CI is thus the ratio of these apparent rate constants.

The experimentally measured CI (and thus ratio of apparent rate constants) varies between different zeolites. This can be due to either intrinsic differences in the activities of the zeolites for the two reactions, or the complex interplay between diffusional limitations and confinement effects imposed by the zeolite lattice on the reactants and transition states. The latter is generally thought to be the most important. Indeed, the CI reaction is used to assess the pore size of unknown zeolitic materials. A CI of less than 1 indicates large pores, values from 1-10 indicate medium pores, while CI values greater

than 10 indicate small pores. It is known that in the absence of confinement effects (i.e. in liquid acids), branched hydrocarbons crack faster than linear molecules. It is thus understandable why the CI is less than 1 for large pore zeolites where confinement effects are negligible. As the pore size decreases, the bulkier branched hydrocarbon is hindered from reacting while the smaller linear molecule is not so hindered. The CI rises as a result.

To gauge the relative importance of confinement on the rates of cracking for these two species, the reaction processes may be broken down into five steps: 1) adsorption of reactants; 2) diffusion of reactants to active sites; 3) formation of transition states; 4) diffusion of products; 5) desorption of products. The formation of the transition state is assumed to be the rate-limiting step in the process. The rate of reaction for species A is assumed to be governed by the following rate expression

$$r_A^{\ i} = k_A^{\ i}[A*][X^+] \qquad (2)$$

where $k_A^{\ i}$ is the intrinsic rate constant for the reaction, [A*] is the adsorbed phase concentration of species A at the active site, and [X+] is the concentration of charged species (carbocation for the bimolecular case, acidic proton for the monomolecular case). A similar expression can be written for species B. Note that the rate expression in Eqn. (2) differs from that in Eqn. (1). It is a second order rate expression involving the actual concentration of the adsorbed species and carbocation, along with an intrinsic rate constant.

The constraint index, which is simply the ratio of the rates of reaction for species A and B, is given by

$$CI = \frac{k_A^{\ i}\ [A*]}{k_B^{\ i}\ [B*]} \qquad (3)$$

where the concentration of carbocations cancels. Eqn. (3) will apply to either monomolecular or bimolecular reactions. We further assume that the intrinsic rate constants in Eqn. (3) can be broken down into *electronic* and a *confinement* terms. That is

$$k_A^{\ i} = k_A^{\ e} k_A^{\ c} \qquad (4)$$

with a similar expression for species B. We anticipate that $k_B^{\ e} > k_A^{\ e}$, since we know branched species crack faster than linear species in the absence of confinement. From previous quantum calculations (Sauer, 1994), we expect that the electronic terms will not vary much between different zeolites *for a given reactant*. The confinement terms, however, should be strongly dependent on the zeolite topology. It is these terms we wish to determine in the present study.

Within the classical transition state theory formalism, the confinement rate constants may be computed by evaluating the configurational integrals of the reactants and transition states (Macedonia and Maginn, 2000). This can be expressed as

$$k_A^{\ i} \propto \frac{Z_{A+}^{\ \mp}}{Z_A Z_+} \qquad (5)$$

where $Z_{A+}^{\ \mp}$ is the configurational integral of the transition state, Z_A is the configurational integral of the adsorbed reactant, and Z_+ is the carbocation configurational integral.

Configurational integrals may be evaluated for each adsorbed species using configurational-bias Monte Carlo integration techniques (Macedonia and Maginn, 1999). After a bit of manipulation (Macedonia and Maginn, 2000), one can show that the CI reduces to

$$CI = S \left(\frac{k_A^{\ e}}{k_B^{\ e}} \right) \left(\frac{Z_{A+}^{\ \mp} Z_B^{\ g}}{Z_{B+}^{\ \mp} Z_A^{\ g}} \right) \qquad (6)$$

where S is a constant specific to the reaction and the superscript g refers to the gas phase. Eqn. (6) clearly shows that the CI is made up of electronic and adsorption terms. The electronic terms in the first set of parenthesis may not be determined via simple classical simulations; to compute these rate constants one must resort to quantum mechanical calculations. In addition, the constant S is difficult to evaluate for complex species. This makes direct comparison of the CI computed from Eqn. (6) with experiment difficult. To circumvent this problem, the computed and experimental CI values for each zeolite will be normalized by the CI for a reference zeolite that is known to exhibit cracking by one of the two mechanisms. If the electronic terms are roughly constant between different zeolites for a given mechanism they will cancel, yielding a normalized CI for zeolite i given by

$$CI'_{ref}(i) = \left(\frac{Z_{A+}^{\ \mp}(i) Z_B^{\ g}(i)}{Z_{B+}^{\ \mp}(i) Z_A^{\ g}(i)} \right) \left(\frac{Z_{B+}^{\ \mp}(ref) Z_A^{\ g}(ref)}{Z_{A+}^{\ \mp}(ref) Z_B^{\ g}(ref)} \right) \qquad (7)$$

where $CI'_{ref}(i)$ is the ratio of the CI for zeolite i to that of the reference zeolite. Eqn. (7) shows that this normalization eliminates the electronic terms, to yield a normalized CI that depends only upon configurational integrals for the adsorbed and gas phase species in the two zeolites. These integrals are computed in the present work using Monte Carlo integration. By normalizing the experimentally measured CIs in the same way, a comparison can be made between the computed and measured quantities. For the monomolecular mechanism, we chose the MTT zeolite as the reference structure, since the calculations clearly show a hindrance for the formation of the bimolecular transition state. MFI, the structure type of the zeolite ZSM-5, was chosen as the reference zeolite for the bimolecular mechanism. Haag, Lago and Weisz (1982) have presented convincing evidence that the cracking mechanism for these reactants is bimolecular in ZSM-5.

Results and Discussion

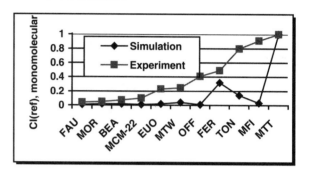

Figure 1. Comparison of calculated and measured monomolecular CI.

To gauge the confinement effects on the CI reaction, twelve different zeolite structures representing several types of cages and pores were investigated. Fig. 1 shows a comparison between computed and measured CIs referenced against MTT for a monomolecular mechanism. Agreement between computed and measured monomolecular CIs in Fig. 1 indicates that selectivity is due to adsorption effects. Lack of agreement does not indicate a failure in the model; rather this signifies that adsorption selectivity is not the main factor governing the observed CI. The result indicates that FER and MTT crack the reactants by a monomolecular mechanism, with selectivity arising mainly from preferential adsorption of reactants. The selectivity in the other zeolite structures appears to be due to other factors. Results for ERI are not shown in Fig. 1, since the CI for this zeolite is very high.

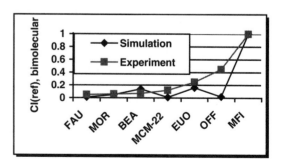

Figure 2. Comparision of calculated and measured bimolecular CI.

Fig. 2 shows a comparison of computed and measured CI values for the bimolecular mechanism referenced against MFI. Agreement between computed and measured CI indicates that bimolecular transition state selectivity governs the rate of the reaction. Computed values for MOR, BEA, EUO, (and the reference MFI) all show good agreement with experiment, indicating that these zeolites crack the reactants according to a bimolecular mechanism. Confinement effects on the transition state drive the selectivity for these zeolites. Note that not all zeolites are listed in Fig. 2 because some zeolites could not support the formation of the bimolecular transition state.

Even when computed and measured reference CI values disagree, the calculations provide useful insight. For example, transport effects were identified as the source of selectivity for ERI and OFF. In both cases, the bimolecular transition state could form, yet the analysis of conformations clearly indicated diffusional limitations for the reactants passing through narrow pore apertures.

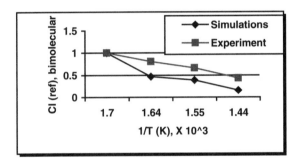

Figure 3. Temperature dependence of bimolecular CI.

The temperature dependence of the CI can also be studied with this method. For MFI, the experimental CI is observed to decrease with increasing temperature. This was attributed by Haag and Dessau (1984) to a change in the reaction mechanism from bimolecular to monomolecular. Fig. 3 shows the temperature dependence of the computed reference CI for the binary reaction compared with experimental measurements made by Frillette, Haag and Lago (1981). It agreement with experiment, the computed CI decreases with temperature *without changing from a bimolecular mechanism*. The computed monomolecular CI (results not shown) does not show nearly as dramatic a decrease with temperature. Thus, it is not necessary to invoke a change in mechanism to explain the reduction in the CI for MFI with temperature. The current work shows that confinement effects decrease as temperature increases, which leads to a reduction in selectivity (and hence CI). This is due to an entropic effect, in which the energetic penalty associated with the confinement of the transition state becomes less important at high temperatures. This illustrates the

importance of using methods that are based on free energies rather than just potential energies when examining problems of this type.

Conclusions

A Monte Carlo method has been developed which can help gauge the impact adsorption and restricted transition state selectivity have on *n*-hexane and 3-methylpentane cracking reactions. The method is based on classical transition state theory and requires that configurational integrals of reactants and transition states be computed. The results indicate that selectivity in MTT and FER is due to adsorption selectivity of the reactants. Selectivity in MFI, MOR, BEA and EUO arises from restricted transition state selectivity associated with the formation of a reactant-carbocation complex. The selectivity mechanisms operative in other zeolites could not be unambiguously determined, but evidence of transport limitations for ERI and OFF was observed. The temperature dependence of the CI was rationalized in terms of a competition between energetic and entropic factors which govern the formation of the transition states.

References

Frillette, V. J., W. O. Haag, and R. M. Lago (1956). Catalysis by crystalline aluminosilicates: Characterization of intermediate pore-size zeolites by the 'constraint index'. *J. Catal.*, **67**, 218.

Haag, W. O., and R. M. Dessau (1984). Duality of mechanism for acid-catalyzed paraffin cracking. *Proceedings of the 8th International Congress on Catalysis (II)*. Verlag Chemie, Weinham, 305.

Haag, W. O., R. M. Lago, and P. B. Weisz (1982). Transport and reactivity of hydrocarbon molecules in a shape-selective zeolite. *Faraday Disc. Chem. Soc.*, **72**, 317.

Kazansky, V. B., I. N. Senchenya, M. Frash, and R. A. van Santen (1994). A quantum-chemical study of adsorbed nonclassical carbonium ions as active intermediates in catalytic transformations of paraffins 1. Protolytic cracking of ethane on high silica zeolites. *Catal. Lett.*, **27**, 345.

Macedonia, M. D., and E. J. Maginn (1999). A biased grand canonical Monte Carlo method for simulating adsorption using all-atom and branched united atom models. *Molec. Phys.*, **96**, 1375.

Macedonia, M. D., and E. J. Maginn (2000). Gauging the impact of molecular confinement on hydrocracking selectivity in zeolites through Monte Carlo integration. *AIChE J.* (submitted).

Pines, H. (1981). *The Chemistry of Hydrocarbon Conversions*. Academic Press, New York, 83.

Sauer, J. (1994). Structure and reactivity of zeolite catalysts: Atomistic modeling using ab initio techniques. *Zeolites and Related Microporous Materials: State of the Art 1994*. Elsevier, Amsterdam, 2039.

STRUCTURE AND ULTRAVIOLET/VISIBLE SPECTRA OF COBALT-EXCHANGED FERRIERITE: A COMPUTATIONAL STUDY

Scott A. McMillan, Linda J. Broadbelt, and Randall Q. Snurr
Department of Chemical Engineering
Northwestern University
Evanston, IL 60208

Abstract

An obstacle for quantum chemical reaction studies over many transition-metal exchanged zeolites has been the unknown extraframework siting of the transition metal. Three extraframework sites for divalent cobalt ions in dehydrated ferrierite have been recently proposed based on diffuse reflectance ultraviolet/visible (UV/Vis) spectroscopy (Kauck | et al., 1999). The structure of one of these sites, α-Co^{2+}-ferrierite, has been studied with quantum chemical calculations on three sizes of zeolite cluster models. The calculated UV/Vis spectrum is very sensitive to the structure of the α-Co^{2+} site and has been found to agree reasonably well with experiment for ground-state geometries calculated with a sufficiently high level of theory regardless of the cluster model size.

Keywords

Cobalt-ferrierite, Transition-metal siting, Zeolites, Ultraviolet/visible spectra, Density functional theory, Hartree-Fock, Configuration interaction.

Introduction

Zeolites and modified zeolites are very effective catalysts for a number of chemical processes (Armor, 1998; Davis, 1998). In particular, cobalt-exchanged zeolites are uniquely active for nitrogen oxide reduction with methane (Li and Armor, 1992) and ethane ammoxidation to acetonitrile (Li and Armor, 1997). One obstacle towards understanding these reactions at the atomic level is the lack of information on cobalt extraframework siting within zeolites. Without prior knowledge of the transition-metal extraframework sites, quantum chemical computational studies of these materials are handicapped.

Recently Wichterlová and co-workers have proposed three extraframework sites for divalent cobalt ions in dehydrated mordenite (Dědeček and Wichterlová, 1999), ferrierite (Kaucký et al., 1999), and ZSM-5 (Dědeček et al., 2000) by diffuse reflectance UV/Vis spectroscopy. The extraframework sites in mordenite were suggested based on the known preference of non-transition metal atoms for specific extraframework sites. The extraframework sites in ferrierite and ZSM-5 were identified by analogy to the extraframework sites in mordenite due to similarities in the spectra.

The three cobalt extraframework sites, designated α, β, and γ, each correspond to different zeolite environments and coordinations. In ferrierite, α-Co^{2+} is coordinated to four oxygens in a rectangular plane located at the intersection of the eight- and ten-member ring channels. β-Co^{2+} is coordinated to six oxygens in a deformed six-member ring that forms the pore wall dividing two eight-member channels. γ-Co^{2+} is located in a boat-shaped site with an unknown coordination. From a reaction engineering viewpoint, the α and β sites are more interesting since they are more accessible to reactants than γ and are the preferred cobalt sites for a wide range of cobalt loadings. We present here results for one of the sites of interest, the α site.

Calculation

Three different sizes of α-Co^{2+}-ferrerite clusters were investigated, small, medium, and large (Fig. 1-3, respectively). Each cluster contained two aluminum atoms at the central bridging positions. The scheme used to label the atoms is consistent for all three cluster models. The small and medium clusters were allowed to relax completely, while the terminal SiH_x groups were fixed to their ferrierite crystallographic positions for the large cluster. The terminating hydrogen atoms were placed in the direction of what would normally be the next oxygen atom at a distance of 1.488 Å. The terminating hydrogen atoms are not shown for clarity. All three cluster sizes were constrained to C_{2v} symmetry. The energy difference between selected symmetry constrained and unconstrained clusters was negligible; accordingly symmetry was used for computational efficiency.

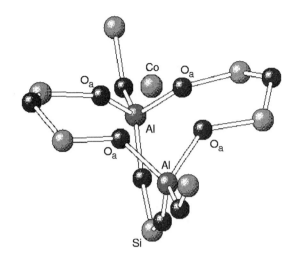

Figure 2.   Medium α-Co^{2+}-ferrerite model.

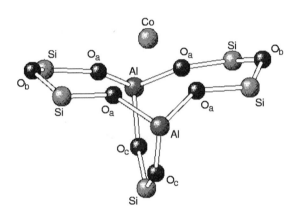

Figure 1.   Small α-Co^{2+}-ferrerite model.

The ground-state structure of each zeolite cluster model was calculated by a variety of methods. Density functional theory (DFT) calculations utilized the Becke exchange and Perdew correlation functionals (BP86) and a double zeta quality numerical basis set with polarization (DNP) and were performed with DMol (Delley, 1990) from Molecular Simulations, Inc. Hartree-Fock (HF) and singly and doubly excited configuration interaction (CISD) calculations used Gaussian basis sets.

UV/Vis spectra corresponding to each small and medium size ground state cluster geometry were computed with CI-Singles (CIS) with a STO-3G basis set. Using larger basis sets results in numerous 'false' excitations. The use of the STO-3G basis set emphasizes cobalt transitions over artificial transitions arising from the artificial termination of the zeolite cluster models. Research is in progress that selectively applies additional basis functions to cobalt and the immediate oxygen atoms. The HF and CI calculations were performed with Gaussian 98 (Frisch et al., 1998).

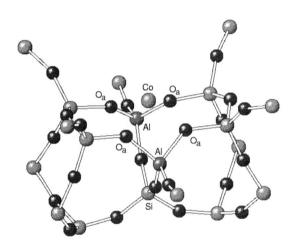

Figure 3.   Large α-Co^{2+}-ferrerite model.

Results and Discussion

Structure

The optimized structures of α-Co^{2+}-ferrerite for each cluster size are reported in Table 1. The geometry of the α-Co^{2+}-ferrerite cluster depends strongly on the basis set. The larger basis set calculations consistently predict O_a-O_a distances across the ring larger than the STO-3G results for the small and medium clusters. As the ring widens, the O_a-Co-O_a angles also increase. This result is not due to the neglect of exchange-correlation in HF theory since CISD also underpredicts this distance. Including exchange-correlation via CISD has little impact on the structure compared to HF with the same basis set. The structure of the large cluster model is less dependent on the basis set choice, most likely due to the constraints placed on the geometrical relaxation.

The basis set also strongly influences the Mulliken charge population and spin density (Table 2). The STO-3G calculations predict a negative charge and a localization of spin-up electron density on cobalt for the small size clusters. Again, incorporation of exchange-correlation does little to change this result. For all of the other calculations the unpaired electron is predicted to reside almost solely on cobalt. For all three cluster sizes, the HF Mulliken populations differ significantly for STO-3G and 6-31G.

Ultraviolet/Visible Spectra

The calculations of UV/Vis spectra provide the transition energy and oscillator strength. The oscillator strength is proportional to the intensity of the UV/Vis transition (Orchin and Jaffé, 1971). The experimental UV/Vis spectra of solids are rarely a set of sharp peaks; instead each transition peak has a width. The width of a transition peak cannot be determined via calculation so for each computed transition a Gaussian function with a maximum located at the transition wavenumber and area equal to the oscillator strength was created. An uniform half-width of 100 cm^{-1} was chosen so that individual transitions could be resolved and the strength of each transition could be easily compared by the relative peak heights. The sum of the individual Gaussian functions is the total spectrum.

The characteristic UV/Vis transition wavenumber for α-Co^{2+}-ferrierite from experiment is 15,000 cm^{-1} (Kauck et al., 1999). The calculated spectra are shown in Fig. 4. The calculated spectra depend strongly on the ground-state geometry. The optimized cluster models with larger

Table 1. α-Co^{2+}-ferrierite Geometrical Parameters.

Cluster	Calculation	Interatomic Distances (Å)					Interatomic Angles (degrees)		
		Co-O_a	O_a-O_a^1	O_a-O_a^2	O_a-O_a^3	Co-Al	O_a-Co-O_a^1	O_a-Co-O_a^2	O_a-Co-O_a^3
small	BP86/DNP	2.024	3.075	2.528	3.981	2.766	98.9	77.3	159.3
small	HF/STO-3G	2.214	2.501	2.539	3.564	2.954	68.8	69.9	107.1
small	HF/6-31G	2.114	3.145	2.583	4.069	2.847	96.1	75.3	148.4
small	CISD/STO-3G	2.212	2.504	2.531	3.560	2.970	68.9	69.8	107.2
medium	BP86/DNP	1.978	3.066	2.497	3.954	2.596	101.6	78.3	176.9
medium	HF/STO-3G	1.912	2.645	2.309	3.511	2.777	87.5	74.3	133.3
medium	HF/6-31G	2.113	3.117	2.580	4.046	2.845	95.1	75.2	146.5
medium	CISD/STO-3G	2.006	2.674	2.375	3.577	2.859	83.6	72.6	126.2
large	HF/STO-3G	2.009	2.764	2.456	3.697	2.786	86.9	75.4	133.9
large	HF/6-31G	2.012	2.775	2.473	3.717	2.795	87.2	75.8	134.9

1. Across the ring
2. With shared aluminum
3. Diagonally across the ring

Table 2. α-Co^{2+}-ferrierite Mulliken Population Analysis.

Cluster	Calculation	Mulliken Charge (au)					Mulliken Spin Density (au)				
		Co	Al	O_a	O_b	O_c	Co	Al	O_a	O_b	O_c
small	BP86/DNP	0.69	0.78	-0.71	-0.64	-0.66	1.01	0.00	-0.01	0.00	0.01
small	HF/STO-3G	-0.03	1.12	-0.63	-0.63	-0.54	2.28	0.04	-0.07	0.00	-0.49
small	HF/6-31G	1.11	1.82	-1.32	-1.22	-1.30	1.01	0.00	0.00	0.00	0.00
small	CISD/STO-3G	-0.02	1.10	-0.62	-0.60	-0.52	2.28	0.04	-0.07	0.00	-0.50
medium	BP86/DNP	0.66	1.02	-0.70	-0.64	-0.66	0.98	0.00	-0.01	0.00	0.00
medium	HF/STO-3G	0.46	1.22	-0.66	-0.63	-0.74	0.95	0.00	0.01	0.00	0.00
medium	HF/6-31G	1.09	2.23	-1.32	-1.22	-1.30	1.01	0.00	0.00	0.00	0.00
Medium	CISD/STO-3G	0.50	1.20	-0.64	-0.60	-0.73	0.89	0.00	0.03	0.00	0.00
large	HF/STO-3G	0.45	1.12	-0.65	-0.71	-0.78	1.12	0.00	-0.04	0.00	0.00
large	HF/6-31G	1.27	2.17	-1.36	-1.18	-1.34	1.07	0.00	0.00	0.00	0.00

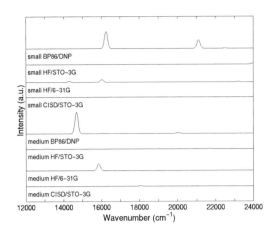

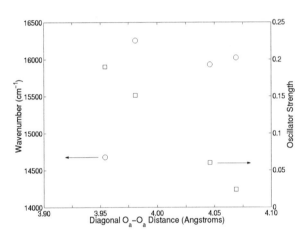

Figure 4. CIS/STO-3G ultraviolet/visible spectra for geometries obtained at various levels of theory.

Figure 5. Sensitivity of ultraviolet/visible spectra to the cluster model geometry.

O_a-O_a diagonal distances (greater than 3.9 Å) have transitions with strong oscillator strengths near the experimental transition. The optimized zeolite cluster models with smaller O_a-O_a distances (less than 3.6 Å) do not have a transition near 15,000 cm^{-1} with a strong oscillator strength. The diagonal O_a-O_a interatomic distance in the large cluster models is between these extremes. Due to the size of the large cluster models, calculation of the UV/Vis spectra is still in progress.

The correlation between the diagonal O_a-O_a distance and the UV/Vis spectra is shown in Fig. 5. Similar correlations can also be obtained for the other geometric parameters relating to the coordination of cobalt and the rectangular plane of oxygen atoms. Extrapolating these results for diagonal O_a-O_a distances characteristic of the large cluster models (around 3.7 Å) is speculative, but potentially enlightening. The oscillator strength is linearly increasing with decreasing interatomic distance, indicating that if a transition does occur for the large

cluster models, it will likely be strong. Predicting the wavenumber of the transition is not as straightforward since it has a more complex dependence on interatomic distance.

These results indicate that the UV/Vis spectrum is very sensitive to the geometry of the rectangular plane of oxygen atoms coordinated to cobalt. The general agreement between the calculated and experimental spectra also lends further support for the proposed siting of α-Co^{2+}-ferrierite. These zeolite cluster models will be used in future quantum chemical studies of catalytic reactions.

References

Armor, J. N. (1998). Metal-exchanged zeolites as catalysts. *Micropor. Mesopor. Mat.*, **22**, 451-456.

Davis, M. E. (1998). Zeolite-based catalysts for chemicals synthesis. *Micropor. Mesopor. Mat.*, **21**, 173-182.

Dědeček, J. and B. Wichterlová (1999). Co^{2+} ion siting in pentasil-containing zeolites. I. Co^{2+} ion sites and their occupation in mordenite. A vis-NIR diffuse reflectance spectroscopy study. *J. Phys. Chem. B*, **103**, 1462-1476.

Dědeček, J., D. Kaucký, and B. Wichterlová (2000). Co^{2+} ion siting in pentasil-containing zeolites, Part 3. Co^{2+} ion sites and their occupation in ZSM-5: A VIS diffuse reflectance spectroscopy study. *Micropor. Mesopor. Mat.*, **35-36**, 483-494.

Delley, B. (1990). An all-electron numerical method for solving the local density functional for polyatomic molecules. *J. Chem. Phys.*, **92**, 508-517.

Frisch, M. J., G. W. Trucks, H. B. Schlegel, G. E. Scuseria, M. A. Robb, J. R. Cheeseman, V. G. Zakrzewski, J. A. Montgomery, Jr., R. E. Stratmann, J. C. Burant, S. Dapprich, J. M. Millam, A. D. Daniels, K. N. Kudin, M. C. Strain, O. Farkas, J. Tomasi, V. Barone, M. Cossi, R. Cammi, B. Mennucci, C. Pomelli, C. Adamo, S. Clifford, J. Ochterski, G. A. Petersson, P. Y. Ayala, Q. Cui, K. Morokuma, D. K. Malick, A. D. Rabuck, K. Raghavachari, J. B. Foresman, J. Cioslowski, J. V. Ortiz, A. G. Baboul, B. B. Stefanov, G. Liu, A. Liashenko, P. Piskorz, I. Komaromi, R. Gomperts, R. L. Martin, D. J. Fox, T. Keith, M. A. Al-Laham, C. Y. Peng, A. Nanayakkara, C. Gonzalez, M. Challacombe, P. M. W. Gill, B. Johnson, W. Chen, M. W. Wong, J. L. Andres, C. Gonzalez, M. Head-Gordon, E. S. Replogle, and J. A. Pople (1998). *Gaussian 98, Revision A.7.* Gaussian Inc., Pittsburgh.

Kaucký, D., J. Dědeček, and B. Wichterlová (1999). Co^{2+} ion siting in pentasil-containing zeolites. II. Co^{2+} ion sites and their occupation in ferrierite. A VIS diffuse reflectance spectroscopy study. *Micropor. Mesopor. Mat.*, **31**, 75-87.

Li, Y. and J. N. Armor (1992). Catalytic reduction of nitrogen oxides with methane in the presence of excess oxygen. *Appl. Catal. B*, **1**, L31-40.

Li, Y. and J. N. Armor (1997). Ammoxidation of ethane to acetonitrile over Co-Beta zeolites. *J. Chem. Soc. Chem. Comm.*, **20**, 2013-2014.

Orchin, M. and H. H. Jaffé (1971). *Symmetry, Orbitals, and Spectra*. John Wiley & Sons, New York.

MOLECULAR MODELING OF
MULTICOMPONENT DIFFUSION IN ZEOLITES

Randall Q. Snurr, Amit Gupta, and Martin J. Sanborn
Department of Chemical Engineering
Northwestern University
Evanston, IL 60208 USA

Abstract

Molecular dynamics (MD) simulations have been used to calculate self-diffusivities for binary systems in zeolites. In addition, non-equilibrium properties, in particular the phenomenological coefficients and the Fickian "transport" diffusivities, have been obtained from the MD simulations through appropriate correlation functions. For some systems, MD cannot access the relevant time scales for diffusion, and more appropriate simulation techniques are under development.

Keywords

Molecular dynamics, Zeolites, Diffusion, Phenomenological coefficients.

Introduction

Diffusion plays a vital role in many applications of zeolites in adsorption separations and catalysis. Molecular simulations have been quite successful in predicting self-diffusivities of molecules in zeolites and elucidating the nature of molecular motion (Theodorou et al., 1996). However, most simulations to date have focussed on single-component systems, while applications always involve multicomponent systems. In this work, we use molecular modeling to investigate binary diffusion in several important zeolites.

In previous work, we have used atomistic molecular dynamics (MD) simulations to investigate binary mixtures of CF_4 with C_1 through C_{10} n-alkanes in siliceous faujasite zeolite (Sanborn and Snurr, in press). Faujasite is a large-pore zeolite widely used in fluidized catalytic cracking and other industrial processes. It is composed of supercages of 12 Å diameter, tetrahedrally connected by windows of 7.5 Å diameter. Simulations of the pure-component alkane systems in faujasite (Clark et al., 1999) yield heats of adsorption, self-diffusivities, and activation energies that agree well with experiment, particularly pulsed field gradient NMR self-diffusivities (Kärger and Ruthven, 1992). Analysis of the chain conformations and

motional frequencies indicates that the bulk liquid and adsorbed phases are quite similar, most likely due to the open nature of the faujasite structure. The preferred siting of molecules within the zeolite is observed to shift as chain length is increased. For binary mixtures with CF_4, we find that scaling of self-diffusivities with mixture composition is not pronounced for a given system and may be explained in terms of available volume and single-component molecular mobilities. Evidence of clustering of similar species is observed in systems containing the longer alkane species.

Our previous work also included preliminary results for the mutual transport diffusion coefficients for methane-CF_4 mixtures in faujasite (Sanborn and Snurr, in press). In this paper, we present further investigations of the transport diffusivities. In addition, MD simulations of methane co-adsorbed with C_6 species n-hexane, cyclohexane, and benzene in silicalite zeolites are presented. The C_6 species "block" diffusion of methane through the narrow channel-type pores of silicalite, and the MD simulations provide insight into the effect that the nature of the blocking species has on methane diffusion.

Theory

The self-diffusivity can be easily calculated from MD simulations by tracking the molecules' mean-squared displacements and using the Einstein equation (Theodorou et al., 1996). This can be done for both single-component and multicomponent systems and is usually reported for systems at equilibrium. For multicomponent systems, it is also of interest to know the mutual "transport" diffusivities for systems having concentration gradients (non-equilibrium). The flux $\mathbf{J}_i$ of component i in the usual Fickian formulation is

$$\mathbf{J}_i = -D_{ii}\nabla c_i - D_{ij}\nabla c_j \qquad (1)$$

where D_{ii} is the main-term diffusivity, D_{ij} is the cross-term diffusivity, and c_i is the concentration of species i in the zeolite. This can also be expressed in terms of the phenomenological coefficients and gradients of chemical potential:

$$\mathbf{J}_i = -L_{ii}\nabla\mu_i - L_{ij}\nabla\mu_j \qquad (2)$$

where L is the phenomenological coefficient and μ_i is the chemical potential of species i. The phenomenological coefficients obey the Onsager relation

$$L_{ij} = L_{ji} \qquad (3)$$

and can be calculated using linear response theory (Theodorou et al., 1996). In general D_{ij} is not equal to D_{ji}. In the Green-Kubo form, the equation for L_{ij} is

$$L_{ij} = \frac{1}{3VkT}\int_0^\infty \langle \mathbf{j}_i(0) \cdot \mathbf{j}_j(t)\rangle dt \qquad (4)$$

where k is Boltzmann's constant, T is the system temperature, and $\mathbf{j}_i$ is the microscopic current of component i

$$\mathbf{j}_i = \sum_{l=1}^{N_i} \mathbf{v}_{li} \qquad (5)$$

In Eqn. (5) $\mathbf{v}_{li}$ is the center-of-mass velocity vector of molecule l of species i and N_i is the total number of molecules of species i within an element of volume V of the system. The L coefficients and Eqn. (2) can be used directly in macroscopic engineering calculations of mass transport, or the diffusion coefficients D can be calculated from the L's as described elsewhere (Sanborn and Snurr, in press).

Simulation Details

For mixtures of methane and CF_4 in siliceous faujasite, the simulation parameters and methods were identical to those reported previously (Sanborn and Snurr, in press) except for the following. To obtain better statistics, a total of 25 MD runs (vs. 14 earlier) were performed for each set of conditions, and results were averaged together. The time step was 5 fs. Each system was equilibrated for 1.5 ns, followed by a production run of 1 ns. It should be noted that while the data at total loadings of 6 molecules/supercage (s.c.). would require very high pressures to reproduce in the laboratory, the data is included here to obtain a clearer picture of the loading dependence of the transport properties.

MD simulations were also performed for methane in silicalite with various C_6 species. Methane was modeled as a single Lennard-Jones site with potential parameters from Goodbody et al. (1991). The potentials for n-hexane and benzene in silicalite were also taken from the literature (June et al., 1992; Snurr et al., 1993). Following Rowley and Ely (1992) we used the same force field for cyclohexane as n-hexane but without the non-bonded intramolecular Lennard-Jones term. To study the effects of co-adsorbate flexibility on methane diffusivity, we also modeled a planar, rigid "pseudo-cyclohexane." Intermolecular binary interaction parameters were obtained from the Lorentz-Bertholet mixing rules.

The silicalite simulations were done in the microcanonical ensemble using Gear's six value, five parameter predictor-corrector method. The bond lengths were constrained using the method of Edberg, Evans, and Morriss (1986) based on Gauss's principle of least constraint. We used the algorithm of Ciccotti et al. (1982) for dealing with the singular constraint matrices that arise for benzene and pseudo-cyclohexane. Their algorithm, however, was recast for use with the Edberg, Evans, Morriss method instead of SHAKE (Gupta and Snurr, in preparation). Production runs were 0.5 – 2 ns with an integration time step of 1.5 – 2 fs.

Results and Discussion

Figures 1, 2, and 3 show the phenomenological coefficients plotted versus system composition for methane-CF_4 mixtures in faujasite at 300 K. Throughout the discussion methane will be referred to as component i and CF_4 as component j. The figures show the L coefficients at overall loadings of 2, 4, and 6 molecules/s.c., respectively. In all cases, it is clear that the cross-term coefficients are equal, as required by the Onsager relation. Both main term coefficients appear to be functions of composition; L_{ii} increases with methane mole fraction, while L_{ij} decreases (except at 6 molecules/s.c. as discussed below). The cross terms, on the other hand, are relatively constant with composition and are smaller in magnitude than the main term coefficients. It is also interesting to note that the values of the cross terms are relatively constant over the range of loadings.

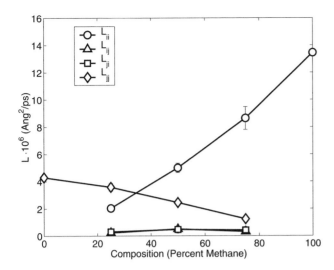

Figure 1. *L coefficients for methane (i) and CF₄ (j) mixtures in faujasite at 300 K and a total loading of 2 molecules per supercage.*

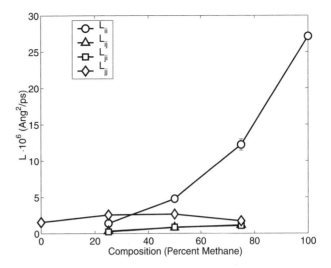

Figure 3. *L coefficients for methane (i) and CF₄ (j) mixtures in faujasite at 300 K and a total loading of 6 molecules per supercage.*

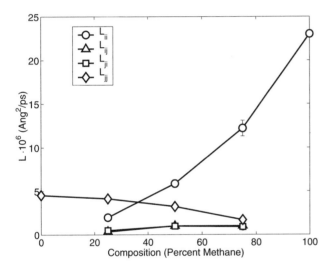

Figure 2. *L coefficients for methane (i) and CF₄ (j) mixtures in faujasite at 300 K and a total loading of 4 molecules per supercage.*

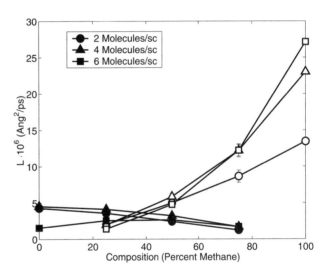

Figure 4. *Main term L coefficients for loadings of 2, 4, and 6 molecules per supercage. The open symbols are L_{ii} coefficients, and the filled symbols are L_{jj} coefficients.*

In Figure 4 L_{ii} and L_{jj} at all loadings are plotted together versus composition. While the L_{ii} values for 2 molecules/s.c. are always less than those for 4 molecules/s.c., the values for 6 molecules/s.c. do not follow this monotonic tendency with total loading. Similarly, the L_{jj} values for 4 molecules/s.c. are larger than those for 2 molecules/s.c., but the values at 6 molecules/s.c. violate the trend with loading. Interestingly, the L_{jj} values at 6 molecules/s.c. exhibit a small maximum as a function of composition. The tendency at the lower loadings is for L_{jj} to increase with increasing CF₄ mole fraction in the system. However, at 6 molecules/s.c., the zeolite is quite full of molecules, and replacing methane with larger CF₄ molecules reduces

the amount of volume available for movement to the point that molecular mobility is reduced, causing the drop in L_{jj} at the compositions with the most CF₄. Currently, we are using the calculated L coefficients to model macroscopic fluxes through a faujasite membrane in order to assess the importance of the cross terms for practical applications. Given the coefficients as functions of composition and loading, one can solve the macroscopic equations for systems involving co-diffusion, counter-diffusion, etc.

Figure 5 shows the self-diffusivity of methane in silicalite at 300 K as a function of the loading of co-adsorbed C₆ molecules. Methane loading was held constant at 3 molecules per unit cell, while loading of the co-adsorbed molecules was varied from 0 to 4 molecules per unit cell. The methane diffusivity decreases as the loading of the larger, slower-moving co-adsorbed molecules is increased. Since n-hexane is the most mobile – and the most flexible – we expect it to have the least blocking effect on methane diffusion, and indeed the decrease in methane diffusivity is much more gradual with n-hexane compared to the other co-adsorbates.

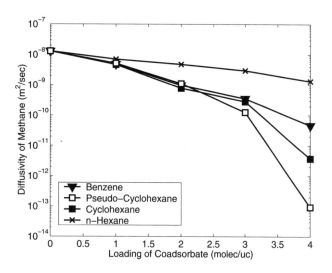

Figure 5.   Self-diffusivity of methane in silicalite with varying loadings of co-adsorbed C_6 species at 300 K.

The preferred adsorption sites of the co-adsorbed molecules also play a role in determining the decrease in methane diffusivity. n-Hexane preferentially adsorbs in the silicalite channels (Gupta et al., 2000), where it can affect the methane traffic solely through that channel. The other co-adsorbates, on the other hand, adsorb strongly in the channel intersections, where they can block traffic from the four channels that meet at that intersection, thus providing greater pore blockage. Also striking is the large decrease in the diffusivity in going from 3 to 4 molecules per unit cell. Since MD is not very efficient at tracking diffusivities lower than about 10^{-10} m^2/s, the diffusivities of methane at 4 molecules per unit cell are at best rough estimates of the methane mobility.

Note that there are 4 intersections per unit cell, so for cyclohexane, pseudo-cyclohexane, and benzene, a loading of 4 molecules per unit cell means that essentially all intersections are blocked. Methane motion is only possible if an empty intersection is found (e.g. because the C_6 molecule has hopped to an adjacent channel site temporarily) or by "passing" the blocking molecule. Current work in our group focusses on simulating these passing events using free-energy perturbation techniques.

References

Ciccotti, G., M. Ferrario, and J. P. Ryckaert (1982). Molecular dynamics of rigid systems in Cartesian coordinates: a general formulation. *Mol. Phys.*, **47**, 1253-1264.

Clark, L.A., G.T. Ye, A. Gupta, L.L. Hall, and R.Q. Snurr (1999). Diffusion mechanisms of normal alkanes in faujasite zeolites. *J. Chem. Phys.*, **111**, 1209-1222.

Edberg, R., D.J. Evans, and G. P. Morriss (1986). Constrained molecular dynamics: simulations of liquid alkanes with a new algorithm. *J. Chem. Phys.*, **84**, 6933-6939.

Goodbody, S.J., K. Watanabe, D. MacGowan, J.P.R.B. Walton, and N. Quirke (1991). Molecular simulation of methane and butane in silicalite. *J. Chem. Soc. Faraday Trans.*, **87**, 1951-1958.

Gupta, A., L.A. Clark, and R.Q. Snurr (2000). Grand canonical Monte Carlo simulations of nonrigid molecules: siting and segregation in silicalite zeolite. *Langmuir*, **16**, 3910-3919.

Gupta, A. and R.Q. Snurr, in preparation.

June, R.L., A.T. Bell, and D.N. Theodorou (1992). Molecular dynamics studies of butane and hexane in silicalite. *J. Phys. Chem.*, **96**, 1051-1060.

Kärger, J. and D.M. Ruthven (1992). *Diffusion in Zeolites and other Microporous Solids.* Wiley-Interscience, New York.

Rowley, R.L. and J.F. Ely (1992). Non-equilibrium molecular dynamics simulations of structured molecules II. Isomeric effects on the viscosity of models for n-hexane, cyclohexane, and benzene. *Mol. Phys.*, **75**, 713-730.

Sanborn, M.J. and R.Q. Snurr (in press). Diffusion of binary mixtures of CF$_4$ and n-alkanes in faujasite. *Sepn. Purif. Technol.*

Snurr, R.Q., A.T. Bell, and D.N. Theodorou (1993). Prediction of adsorption of aromatic hydrocarbons in silicalite from grand canonical Monte Carlo simulations with biased insertions. *J. Phys. Chem.*, **97**, 13742-13752.

Theodorou, D.N., R.Q. Snurr, and A.T. Bell (1996). Molecular dynamics and diffusion in microporous materials. In G. Alberti and T. Bein (Eds.), *Comprehensive Supramolecular Chemistry*, Vol. 7. Pergamon, Oxford. 507-548.

QUANTUM CALCULATIONS AND CAR-PARRINELLO SIMULATIONS OF PROCESSES IN THE ZEOLITE CHABAZITE

Claudiu A. Giurumescu and Bernhardt L. Trout
Department of Chemical Engineering
Massachusetts Institute of Technology
Cambridge, MA 02139

Abstract

This extended abstract describes results of Car-Parrinello simulations of methanol coupling in the zeolite chabazite. We have found that direct coupling of two methanol molecules without a methoxy intermediate is plausible. The mechanism that resulted from this study involves stable intermediates of methane and protonated formaldehyde. The stabilization presumably is brought about by the zeolite lattice.

Keywords

FOMMS 2000, Car-Parrinello, Density functional theory, Quantum simulations, Quantum calculations, Zeolite, Chabazite, Methanol coupling.

Introduction

There has recently been considerable interest in coupling reactions of methanol in zeolites, such as ZSM-5 and chabazite, and zeotypes, such as SAPO-34, both from industry and academics (Stocker, 1999). From the industrial standpoint, such interest has been spurred from the possibilities of synthesizing olefins using methanol to olefins (MTO) processes and of developing more environmentally friendly processes for synthesizing gasoline using methanol to gasoline (MTG) processes. Methanol adsorption on zeolitic acid sites has also been studied theoretically and computationally as a probe to characterize these sites. In addition, the formation of surface methoxonium, dimethyl ether, and of C-C bonds have been studied via first-principles computational methods in order to gain an enhanced understanding of the catalytic properties of zeolitic acid sites and to elucidate the reaction mechanisms of those reactions (Blaszkowski and van Santen, 1995; Haase and Sauer, 1995; Blaszkowski and van Santen, 1996; Shah et al., 1996; Shah et al., 1996; Blaszkowski and van Santen, 1997; Blaszkowski and van Santen, 1997; Haase et al., 1997; Shah et al., 1997; Shah et al., 1997; Shah et al.,

1997; van Santen, 1997; Stich et al., 1998; Gale et al., 1999; Stich et al., 1999).

Most of these studies have been performed using small, isolated, gas-phase clusters to model the zeolitic acid sites. Recently, researchers have used periodic models to represent the zeolite as an infinite crystalline system. Moreover, use of periodic models combined with recent advances in computational power have allowed the incorporation of dynamics and simulations at non-zero temperatures in the study of the interaction of methanol with zeolitic acid sites. Such studies have resolved important questions that were not able to be answered by performing static calculations.

Formation of the first C-C bond is thought to be the rate-limiting step for the MTO and MTG processes. Because of the complexity of the reaction processes involving methanol in zeolites and of the limitations of applying experimental methods to these reactions, the reaction or reactions involving the formation of the first C-C bond have never been isolated experimentally nor has a mechanism been definitively agreed upon. In fact, there have been more than 20 proposed mechanisms for the formation of the first C-C bond (Stocker, 1999). In order to study these dynamic effects and to gain insight into the methanol coupling process, we have performed a series of Car-Parrinello simulations (Car and Parrinello, 1985) to sample a reaction pathway for the formation of the first C-C bond.

Methodology

In order to determine structural, thermodynamic and kinetic properties of characteristic zeolites and processes in zeolites with as much accuracy as possible, we have chosen to use density functional theory (Hohenberg and Kohn, 1964; Kohn and Sham, 1965; Parr and Yang, 1989) to calculate electronic structure and energetics. We have used a plane-wave basis set code with periodic boundary conditions (Hutter et al., 1990-1996) in order to model the zeolite as an infinite crystalline system. The solid acid catalyst that we have chosen to study is the zeolite chabazite. Not only is chabazite active for the MTO and MTG processes, it has been well characterized (Yuen et al., 1994; Smith et al., 1997). Moreover, it has the decisive computational advantage of having 36 atoms per unit cell, making it tractable to treat periodically.

We have chosen the Car-Parrinello method (Car and Parrinello, 1985) for our non-zero temperature simulations. This method combines a quantum mechanical treatment of the electrons via density functional theory and a classical treatment of the nuclei, which are moved via a molecular dynamics simulation. All simulations were performed at 400 °C, a typical industrially used temperature for MTO and MTG processes.

In order to observe the formation of the first C-C bond with as few *a priori* restrictions as possible, we chose as general a reaction coordinate as possible and performed non-zero temperature simulations to sample across that reaction coordinate. In particular, we chose two methanol molecules in our zeolite systems and the C-C distance to be our reaction coordinate. Starting with the C-C distance, optimized at 0 K for two methanol molecules on the acid site, we decreased this distance and constrained its values. From these simulations, the free energy of reaction as a function of the value of the reaction coordinate can be computed (Carter et al., 1989; Sprik and Ciccotti, 1998).

Results and Discussion

The free energy as a function of reaction coordinate can be seen in Figure 1. By the time, the constraint is r_{CC}= 1.8 Å, the sign of the average force on the constraint has changed, indicating that the transition state (saddle point along the reaction coordinate) has been surpassed. Thus, the transition state occurs between r_{CC}= 1.8 Å and r_{CC}= 2.0 Å.

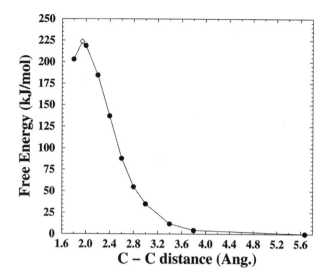

Figure 1. Free energy as a function of value of constraint.

In order to determine the free energy barrier height and the C-C bond length at the transition state, we linearly interpolated the value of the average force on the constraint between 2.0 and 1.8 , to find out where that force is 0. The result was a C-C^{TS} bond length of 1.94 , and a free energy barrier height of 223.5 kJ/mol. This point is presented as an open circle on Figure 1. In order to separated the energetic from the entropic terms in the free energy, we selected an equilibrated frame of the trajectory at r_{CC}= 2.0 Å, reduced the constrained bond length to 1.94 , and performed a geometry optimization. This yielded for Reaction 1:

$$\Delta U^{TS} (0 \text{ K}) = 154.4 \text{ kJ} / \text{mol}$$

$$\Delta S^{TS} (673 \text{ K}) \approx -0.1025 \text{ kJ/(mol} \cdot \text{K)}.$$

It is interesting to note that the change in entropy from the reactants to the transition state is negative, even though the transition state consists of two free species solvated by the lattice, but not bound near the Al site, whereas the reactant state consists of two methanol molecules bound to the acid site.

Additional information can be found in Giurumescu and Trout (2000).

Two published studies presenting results on formation of C-C bonds from methanol at acid sites in zeolites are by Blaszkowski and van Santen (1997) and by Tajima et al. (1998). Blaszkowski and van Santen performed a variety of transition state searches on very small zeolite clusters, varying intermediates. The pathway that they found for methanol coupling with the lowest barrier height involved the formation of a methoxy intermediate with H_2O participating by reducing the barrier height. The barrier that they found was 251 kJ/mol. Tajima et al. (1998) performed similar calculations with a somewhat larger cluster model, and found that a pathway involving free CH_4 and an CH_2O species bound to the acid site had the lowest barrier height. The height of this barrier is 184.8 kJ/mol. A summary of these results is presented in Table 1.

Table 1. Comparison of Barrier Height of Transition States Computed via Various Methods.

Mechanism and Study	Barrier Height (kJ/mol)
Direct reaction, this study, found via reaction temperature constrained dynamics	154.5
Direct reaction, this study, found via 0 K synchronous transit method	221
Surface methoxy intermediate (Blaszkowski and van Santen, 1997), transition state search at 0 K	251
Free methane, bound formaldehyde intermediate (Tajima et al., 1998), transition state search at 0 K	184

The barrier that we found using reaction temperature constrained dynamics is 30 kJ/mol lower than that found by Tajimi et al. (1998) and 96 kJ/mol lower than that found by Blaszkowski and van Santen (1997). We believe that our method gives us the ability to find lower energy saddle points. However, this could be due to spurious dipole-dipole interactions among periodic images. In our simulations, in contrast to the assumptions made by the other two researchers, we do not observe the formation of a surface methoxy species. These have been observed by NMR to be present at reaction conditions. We conclude, however, that a surface methoxy is not a necessary intermediate in the methanol coupling reaction. Similarly, Blaszkowski and van Santen (1997) and Shah et al. (1997) found that surface methoxy intermediates do not play a role in the condensation of methanol to form dimethyl ether.

Conclusions

Our constrained Car-Parrinello dynamics simulations of the coupling of methanol in the zeolite chabazite yielded a mechanism based on a methane and protonated formaldehyde intermediate. The energy barrier height was calculated to be 154.5 kJ/mol, and the free energy barrier height at 400 °C was calculated to be 223.5 kJ/mol.

Acknowledgements

We would like to thank the National Center for Supercomputing Applications for computing time and the Sloan Foundation and the MIT UROP office for partial support.

References

Blaszkowski, S. R. and R. A. van Santen (1995). Density-functional theory calculations of the activation of methanol by a Bronsted zeolitic proton. *J. Phys. Chem.* **99**, 11728-11738.

Blaszkowski, S. R. and R. A. van Santen (1996). The mechanism of dimethyl ether formation from methanol catalyzed by zeolitic protons. *J. Am. Chem. Soc.* **118**, 5152-5153.

Blaszkowski, S. R. and R. A. van Santen (1997). Methanol adsorption and activation by zeolitic protons. in *Prog. in Zeolite and Microporous Mat., Parts A-C*, **105**, 1707-1714.

Blaszkowski, S. R. and R. A. van Santen (1997). Theoretical study of C-C bond formation in the methanol-to-gasoline process. *J. Am. Chem. Soc.*, **119**, 5020-5027.

Car, R. and M. Parrinello (1985). Unified approach for molecular dynamics and density-functional theory. **55**, 2471.

Carter, E. A., Ciccotti, G., Hynes, J. T., and R. Kapral (1989). Constrained reaction coordinate dynamics for the simulation of rare events. *Chem. Phys. Lett.*, **156**, 472.

Gale, J. D., Shah, R., Payne, M. C., Stich, I., and K. Terakura (1999). Methanol in microporous materials from first principles. *Catal. Today,* **50**, 525-532.

Giurumescu, C. and B. L. Trout (2000). Methanol Coupling in the zeolite chabazite studied via first-principles molecular dynamics. in preparation.

Haase, F. and J. Sauer (1995). Interaction of methanol with Bronsted acid sites of zeolite catalysts - an *ab-initio* study. *J. Am. Chem. Soc.*, **117**, 3780-3789.

Haase, F., Sauer, J., and J. Hutter (1997). *Ab-initio* molecular dynamics simulation of methanol adsorbed in chabazite. *Chem. Phys. Lett.* **266**, 397-402.

Hohenberg, P. and W. Kohn (1964). Inhomogeneous electron gas. **136**, B864-B871.

Hutter, J., Ballone, P., Bernasconi, P., Focher, P., Fois, E., Goedecker, S., Parrinello, M., and M. Tuckerman (1990-1996). *CPMD*, version 3.0. MPI fur Festkorperforschung and IBM Zurich Research Laboratory.

Kohn, W. and L. J. Sham (1965). Self-Consistent equations including including exchange and correlation effects. **140**, A1133.

Parr, R. G. and W. Yang (1989). *Density Functional Theory of Atoms and Molecules*. Oxford University Press, New York.

Shah, R., Gale, J. D., and M. C. Payne (1996). Methanol adsorption in zeolites - A first-principles study. *J. Phys. Chem.*, **100**, 11688-11697.

Shah, R., Gale, J. D., and M. C. Payne (1997). The active sites of microporous solid acid catalysts. *Phase Transit.*, **61**, 67.

Shah, R., Gale, J. D., and M. C. Payne (1997). *In situ* study of reactive intermediates of methanol in zeolites from first principles calculations. *J. Phys. Chem. B,* **101**, 4787-4797.

Shah, R., Payne, M. C., and J. D. Gale (1997). Acid-base catalysis in zeolites from first principles. *Int. J. Quantum Chem.,* **61**, 393-398.

Shah, R., Payne, M. C., Lee, M. H., and J. D. Gale (1996). Understanding the catalytic behavior of zeolites: A first- principles study of the adsorption of methanol. *Science,* **271**, 1395-1397.

Smith, L. J., Davidson, A., and A. K. Cheetham (1997). A neutron diffraction and infrared spectroscopy study of the acid form of the aluminosilicate zeolite, chabazite (H-SSZ-13). *Catal. Lett.,* **49**, 143-146.

Sprik, M. and G. Ciccotti (1998). Free energy from constrained molecular dynamics. *J. Chem. Phys.,* **109**, 7737-7744.

Stich, I., Gale, J. D., Terakura, K., and M. C. Payne (1998). Dynamical observation of the catalytic activation of methanol in zeolites. *Chem. Phys. Lett.,* **283**, 402-408.

Stich, I., Gale, J. D., Terakura, K., and M. C. Payne (1999). Role of the zeolitic environment in catalytic activation of methanol. *J. Am. Chem. Soc.,* **121**, 3292-3302.

Stocker, M. (1999). Methanol-to-hydrocarbons: catalytic materials and their behavior. *Microporous Mesoporous Mat.,* **29**, 3-48.

Tajima, N., Tsuneda, T., Toyama, F., and K. Hirao (1998). A new mechanism for the first carbon-carbon bond formation in the MTG process: A theoretical study. *J. Am. Chem. Soc.,* **120**, 8222-8229.

van Santen, R. A. (1997). Quantum-chemistry of zeolite acidity. *Catal. Today,* **38**, 377-390.

Yuen, L.-T., Zones, S. I., Harris, T. V., Gallegos, E. J., and A. Auroux (1994). Product selectivity in methanol to hydrocarbon conversion for isostructural compositions of AFI and CHA molecular sieves. **2**, 105-117.

MOLECULAR MODEL OF γ-ALUMINA: NITROGEN ADSORPTION AND PORE SIZE DISTRIBUTION

S. Figueroa-Gerstenmaier and L. F. Vega
Departament d'Enginyeria Química, ETSEQ
Universitat Rovira i Virgili
43006 Tarragona, Spain

F. J. Blas
Departamento de Física Aplicada e Ingeniería Eléctrica
EPS, Universidad de Huelva
Huelva, Spain

K. E. Gubbins
Department of Chemical Engineering
North Carolina State University
Raleigh, NC 27695-7905

Abstract

We present adsorption isotherms for nitrogen in γ-alumina as obtained from density functional theory (DFT) and Grand Canonical Monte Carlo simulations. DFT is also used to estimate a pore size distribution function of these amorphous materials. Nitrogen molecules are described as single Lennard-Jones interaction units, while γ-alumina is modeled as a collection of non-interacting individual cylindrical structured pores of Lennard-Jones spheres with different diameters. Three different adsorption behaviors are observed, depending on the pore size range, corresponding to capillary condensation, $0 \rightarrow 1$ layering transition, and continuous pore filling. It is seen that each one of these size ranges contributes to different parts of the global adsorption isotherm, depending on the pressure considered. The pore size distribution of the adsorbent material is obtained by fitting the theoretical adsorption integral to the experimental adsorption isotherm of nitrogen on γ-alumina at 77.35K. We also report the same function as obtained by the Barret-Joyner-Halenda method for comparison.

Keywords

Pore size distribution, Nitrogen adsorption, γ-Alumina, Density functional theory, Monte Carlo simulations.

Introduction

γ-Alumina, one of the transition aluminas, is a common material used for industrial applications. The fact that it is abundant in nature, has a relative low cost and can be used as an adsorbent for separation processes, as a catalyst and for the fabrication of membranes make it a very attractive material. However, there is no systematic study available to characterize it from a molecular perspective.

Our interest in the material it is based on its application for olefin/paraffin separations. Recently, Yang and Kikkinides (1995) have published promising results where they show that γ-alumina activated with CuCl presents greater capacity and separation power than previously investigated materials. We have proposed a simple molecular model to describe the adsorption of pure ethane and ethylene on bare γ-Al$_2$O$_3$ and CuCl/γ-Al$_2$O$_3$ (Blas, Vega and Gubbins, 1998). This model gives good agreement with the experimental data of Yang and Kikkinides (1995) except at extreme conditions: for the case of bare alumina, the adsorption of ethane is underestimated at low pressures, and for activated alumina the adsorption of ethylene is underestimated at the higher pressures studied. The discrepancies between the model and the experiments are mainly attributed to the fact that this material is polydisperse, while the model considers a unique pore with averaged diameter.

The objective of the present paper is to propose a simple molecular model for γ-alumina, which explicitly includes the pore size dispersity. It is known that the behavior of fluids adsorbed on substrates strongly depends on the number, size and geometry of the pores, as well as on pore connectivity and chemical heterogeneity of the surface. Although such information is essential for optimal use of these adsorbents in specific applications, finding this data becomes difficult and some simplifications are needed.

A standard method to experimentally characterize adsorbent materials is nitrogen adsorption at 77.35K and normal pressure. A model is needed to interpret the experimental adsorption data and to obtain the pore size distribution (PSD) and the surface area. There are several models available in the literature. Some of them, such as the Kelvin equation and modifications of it, are based on classical thermodynamics. Although straightforward to use, the PSD determined by these methods underestimates the sizes of the pores present in the material. On the other hand, the molecular modeling approach has been used to interpret adsorption behavior on model materials with remarkable success. In particular, molecular simulation and density functional theory (DFT) have showen to be very powerful tools to attack this kind of problem (Gubbins, 1997 and references therein), and for systematic studies of the relevant variables in the adsorption processes.

Here we use DFT to model the adsorption behavior of nitrogen in γ-alumina at temperature T=77.35K and normal pressure and to propose a PSD function for this material. The accuracy of the theory in predicting the adsorption behavior is tested against Grand Canonical Monte Carlo (GCMC) simulations for the same molecular model.

Molecular Model

The model assumes that the internal structure of the porous material is made up of a set of individual pores with homogeneous chemical surface, identical geometry, but different pore sizes. The experimental adsorption isotherm is expressed in terms of a sum over the individual adsorption isotherms for each pore diameter (H):

$$\Gamma_T(P) = \int_{H_{min}}^{H_{max}} \rho_T(P,H) f(H) \, dH \qquad (1)$$

where, $\Gamma_T(P)$ is the experimental adsorption isotherm of nitrogen on the adsorbent measured at temperature T; H_{min} and H_{max} are the widths of the smallest and largest pores considered in the analysis; $\rho_T(P,H)$ is the mean density of nitrogen in a pore of width H at pressure P and temperature T; and $f(H)$ is the PSD of the material considered.

To determine the PSD function, $f(H)$, a large amount of pressures and pore sizes must be considered. There are different ways to obtain the mean density of N$_2$ inside the pore, depending on P and H. In this work a non-local DFT approach is used, based on the functional proposed by Kierlik and Rosinberg (1991). Once $\rho_T(P,H)$ is available for a suitable range of P and H, $f(H)$ can be obtained by solving the integral equation, Eqn. (1).

Nitrogen molecules are modeled as single Lennard-Jones (LJ) interaction units, with parameters obtained by fitting the DFT equation of state to the bulk nitrogen saturated liquid density of 0.02887mol cm^{-3}, and saturation pressure of 1atm at the normal boiling point of 77.35K. The potential parameters obtained in this way are σ=0.3544nm and ε/k_B=96.17K.

The molecular model of the material is taken after Peri (1965). γ-Alumina is modeled as structured cylinders made up of oxygen atoms taken as spherical LJ sites. The effective parameters of these oxygen atoms take into account the presence of aluminum atoms. The LJ spheres are arranged in 16 concentric cylindrical layers separated by a distance Δ=0.2125nm. The face of each layer is formed as a square lattice of side d=0.30nm with the oxygen atoms placed at the sites. The Δ and d values are chosen to reproduce the true density of the alumina reported by Cascarini de Torre et al. (1995), of 52 O^{2-} ions/nm^3. Each oxygen ion is displaced 0.15nm in the angular direction between consecutive layers. The solid-solid interaction molecular parameters, σ_{ss}=0.303nm and ε_{ss}/k_B=108.46K, are taken from Cascarini de Torre et al. (1995). The Lennard Jones potential also gives the interaction between a single unit of an adsorbate molecule and an oxygen ion of the surface, following the Lorentz-Berthelot combining rules.

Non-local Density Functional Theory

The non-local version of the DFT used in this work is due to Kierlik and Rosinberg, which has been successfully used to model the adsorption of fluids on carbons (Evans, 1992). The details of the equations and the procedure can be found elsewhere (Kierlik and Rosinberg, 1991). In our case, the potential exerted by the wall is previously calculated with a sufficient number of LJ spheres so that the interactions between the adsorbed molecules and the outer oxygen ions of the pore are negligible. This potential is then averaged over the axial and angular direction in such a way that only the dependence in radial direction is conserved. The individual isotherms of nitrogen on γ-Al$_2$O$_3$ have been obtained for a range of pore sizes from H=0.5 to 12.8nm with an increment of 0.5nm, at T=77.35K.

Grand Canonical Monte Carlo Simulations

To test the accuracy of the DFT, we have performed GCMC simulations to obtain the adsorption isotherm for some selected diameters. We have generated adsorption and desorption isotherms for the following diameter sizes: H=0.7, 0.8, 1.0, 1.15, 1.35, 3.725, 3.84, 4.0 and 6.05nm. All simulations were equilibrated with between 5 and 12 million moves, depending on the system considered, *i. e.*, P and H values, followed by 20-30 million production moves. The relationship between the activity and the bulk pressure of the fluid was established from GCMC simulations performed in the bulk phase.

Results

Nitrogen Adsorption Isotherms

Although we have performed a systematic study of the adsorption behavior at several pore sizes, we present here only the more relevant results. The adsorption isotherm of a cylindrical pore of diameter H=4.0nm shows a hysteresis loop and is presented in Fig. 1. This pore size is in the mesopore range. Vertical lines on the figure show the location of capillary condensation (right), capillary evaporation (left), and equilibrium (middle) transitions in the pore.

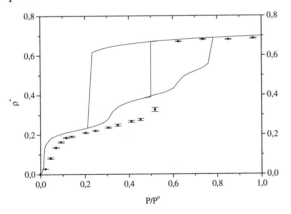

Figure 1. N$_2$ adsorption isotherm at 77.35K on γ-Al$_2$O$_3$; H=4.0nm. Symbols: GCMC results, solid line: DFT results.

The DFT and GCMC isotherms agree well with each other, although the simulation results exhibit a lower adsorption. The same agreement is observed for smaller pores, up to 1nm, and results are not shown here. The

adsorption behavior in a narrower pore, H=1nm, is presented in Fig. 2. DFT results show a $0 \rightarrow 1$ layering transition while GCMC exhibits continuous pore filling. This is the pore size limit where the DFT and GCMC results begin to disagree. The disagreement becomes more pronounced when the pore size is decreased.

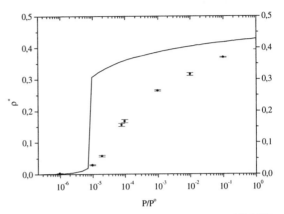

Figure 2. N$_2$ adsorption isotherm at 77.35K on γ-Al$_2$O$_3$; H=1.0nm. Symbols as in Fig. 1.

Pore Size Distribution Function

To determine the PSD function of γ-alumina it is necessary to solve the adsorption integral equation, Eqn. (1). We have chosen a functional form for the function $f(H)$, as a multimodal logarithm normal distribution, and then we fitted the parameters defining $f(H)$ to the experimental adsorption isotherm by a least-squares procedure. To solve the integral equation we used 50 diameters and 40 experimental points. Following this procedure, we obtained the optimized parameter values of $f(H)$ with three modes. Figure 3 shows the adsorption isotherm of nitrogen on γ-alumina at T=77.35K. As can be seen, the model provides a good fit to the full pressure

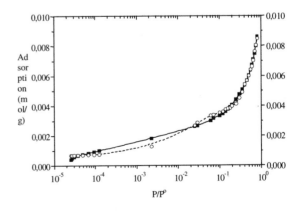

Figure 3. N$_2$ adsorption isotherm on γ-Al$_2$O$_3$ at 77.35K. Symbols: experimental data (Yang, 1998), solid line: DFT results.

range of the experimental adsorption isotherm, especially at high pressures.

In Fig. 4 the PSD function obtained by DFT is shown (solid line). The same function has been calculated using a classical method, the BJH method (Gregg and Sing, 1982) and results are shown here for comparison (dashed line). DFT results present three peaks distributed in a wide range of pore sizes: one centered on $H=0.8$nm (ultramicropore region), the second one with a maximum at $H=1.35$nm (micropore region) and the third one centered at $H=3.84$nm. This latter peak is associated with a wide distribution of pore widths, and is the mode that mainly contributes to the adsorption isotherm. In the PSD obtained by the BJH method we found only two peaks and the whole distribution is displaced to lower pore sizes.

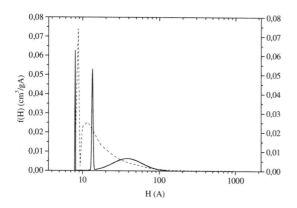

Figure 4.   PSD function of γ-Al2O3 obtained using the DFT analysis (solid line) and from BJH method (dashed line).

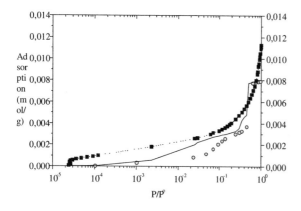

Figure 5.   N₂ adsorption isotherm on γ-Al₂O₃ at 77.35K. Squares: experimental data (Yang, 1998), circles: GCMC results and solid line: DFT, using both only 3 values of H.

Since the molecular modeling approach gives one the possibility of studying the influence of microscopic properties on the macroscopic behavior, we have analyzed the contribution to the global adsorption isotherm of N₂ on γ-alumina of different pore sizes. The integral

adsorption (Eqn. 1) is linear with respect to $f(H)$, which means that the whole adsorption behavior can be considered as a sum of three adsorption isotherms, with pore diameters corresponding to the peaks of each mode of the distribution, biased with their statistical weight. The GCMC and DFT results obtained with this simplification are shown in Fig. 5. It is seen that this crude approximation is able to qualitatively describe the adsorption of N₂ on γ-alumina, although it underestimates it at low pressures. The capillary condensation observed at high pressures is expected, since a single pore size is used to account for the whole mesopores region.

Conclusions

A molecular model of γ-alumina has been presented, and used to describe the adsorption behavior of nitrogen, as well as to obtain a PSD function of this material.

Different adsorption behaviors have been observed, depending on the pore size. The good agreement between theoretical and simulation results deteriorates as the pore size is decreased, especially in the micropore region. A trimodal PSD function has been obtained by DFT. The contribution of each peak to the total isotherm shows that the experimental isotherm is mainly dominated by the mesopore region.

Once the material has been characterized, the PSD can be used to model the adsorption behavior of several fluids on this material. It can also be used to model other phenomena where γ-alumina may be involved.

References

Blas, F. J., Vega, L. F. and K. E. Gubbins (1998). Modeling new adsorbents for ethylene/ethane separations by adsorption via π-complexation. *Fluid Phase Equil.*, **150-151**, 117-124.

Cascarini de Torre, L. E., Flores, E. S., Llanos, J. L. and E. Bottani (1995). Gas-solid potentials for N₂, O₂, and CO₂ adsorbed on graphite, amorphous carbons, Al₂O₃, and TiO₂. *Langmuir*, **11**, 4742-4747.

Evans, R. (1992). Density functional in the theory of nonuniform fluids. In Henderson, D. (Ed.), *Fundamentals of Inhomogeneous Fluids*. Marcel Dekker, Inc., New York, 85-175.

Gregg, S. J. and K. S. W. Sing (1982). *Adsorption, Surface Area and Porosity*. Academic Presss, London, 138-142.

Gubbins, K. E. (1997). Theory and simulation of adsorption in micropores. In Fraissard, J. and Connor, W. C. (Eds.), *Physical Adsorption: Experiment, Theory and Applications*. Kluwer, Dordrecht, 65.

Kierlik, E. and M. L. Rosinberg (1991). Density-functional theory for inhomogeneous fluids: Adsorption of binary mixtures. *Phys. Rev. A*, **44**(8), 5025-5037.

Peri, J. B. (1965). Infrared study of adsorption of ammonia on dry γ-alumina. *J. Phys. Chem.*, **69**, 231-239.

Yang, R. T. (1998). *Private communication*.

Yang, R. T. and E. S. Kikkinides (1995). New sorbents for olefin/paraffin separations by adsorption via π-complexations. *AIChE J.*, **41**(3), 509-517.

AUTHOR INDEX

SUBJECT INDEX